Mima Mounds:
The Case for Polygenesis and Bioturbation

edited by

Jennifer L. Horwath Burnham
Department of Geography
Augustana College
Rock Island, Illinois 61201-2296
USA

Donald L. Johnson
Department of Geography
University of Illinois
Urbana, Illinois 61801
USA

THE
GEOLOGICAL
SOCIETY
OF AMERICA®

Special Paper 490

3300 Penrose Place, P.O. Box 9140 ▪ Boulder, Colorado 80301-9140, USA

2012

Published by The Geological Society of America, Inc.
3300 Penrose Place, P.O. Box 9140, Boulder, Colorado 80301-9140, USA
www.geosociety.org

Printed in U.S.A.

GSA Books Science Editors: Kent Condie and F. Edwin Harvey

Library of Congress Cataloging-in-Publication Data

Mima mounds : the case for polygenesis and bioturbation / edited by Jennifer L. Horwath
   Burnham and Donald L. Johnson.
       p. cm. — (Special paper ; 490)
    Papers mostly from Geological Society of America Annual Meetings and field trips held in
   Houston Texas, October 4–9, 2008.
    Includes bibliographical references.
    ISBN 978-0-8137-2490-4 (pbk.)
       1. Paleontology—West (U.S.) 2. Mounds—West (U.S.) 3. Mounds—Northwest, Canadian.
    4. Mima Mounds (Wash.) 5. Animal burrowing—West (U.S.) I. Horwath Burnham, Jennifer L.,
    1975–. II. Johnson, Donald L. (Donald Lee)
QE720.2.U6M56 2012
551.3′9—dc23   2012024959

Cover, top: Aerial view of Mima Mounds Natural Area Preserve, looking north over Bordeaux Road, ~18 km southwest of Olympia, near Littlerock, Washington. One of many mounded tracts in the area, it is viewed by some as the "type locality" for Mima-prairie-pimple mounds. Mounds here are round to ovate, average 13 m diameter, and have average heights of ~1.6 m but vary from several cm to 2.4 m, with densities that approach 34 mounds per hectare. The prairie slopes 1% south toward viewer. The soil is the Spanaway gravelly sandy loam (Typic Melanoxerands), which developed during the last ~16 ka under fire-impacted prairie vegetation. Like nearby mounded prairies, the soil developed in glacial outwash gravels and sands that received episodic dustings of loess and volcanic ash during post-glacial time. Most North American moundfields are less spectacular, display lower mound densities, and commonly have smaller mounds. (Photo courtesy of Chris J. Taylor, *Seattle Times*, used with permission.) **Bottom:** The mounds—prominent expressions of the locally thickened biomantle of the Spanaway soil series—are composed of dark soot- and organic-stained silts and sands with walnut-size and smaller pebbles scattered throughout. The concavo-convex biomantle extends as pocket gopher–size krotovina into subjacent stratified outwash gravels. Although not evident in this photo, a layer of mainly cobble-size clasts forms the base of many mounds, and continues as a pavement between the mounds. The mound-intermound stonelayer-pavement is a common feature of many moundfields formed in gravelly soil, and together with the pebbly nature of the mounds is as predicted by the expanded Dalquest-Scheffer-Cox biogenic model of Mima mound formation (cf. Dedication). Directly or indirectly, this volume's contributors employ, within a general polygenetic framework, versions of this biogenic model in explaining Mima mounds in North America. (Photo courtesy of Diana N. Johnson.)

10 9 8 7 6 5 4 3 2 1

Contents

Dedication

This volume honors Walter W. Dalquest and Victor B. Scheffer (Fig. 1) for their studies of the mounded prairies of the Olympia–Puget Sound area of Washington State (cf. cover photos). Their collective work provided the fields of ecology, geomorphology, and pedology with key concepts on the biogenic role of burrowing animals in forming soil and producing the mounded landscapes of North America (Dalquest and Scheffer, 1942, 1944; Scheffer, 1947, 1948, 1954, 1956a, 1956b, 1958, 1960, 1969, 1981, 1983, 1984; Scheffer and Kruckeberg, 1966; Cox and Scheffer, 1991). Certain others, most notably W. Armstrong Price (1949, 1950), R.J. Arkley (1948), and Arkley and H.C. Brown (1954), also contributed and clarified important details of the Dalquest-Scheffer model. Later, George W. Cox, contributor to this volume (Chapter 3), by himself and with various colleagues consistently tested, clarified, and refined basic tenets of the Dalquest-Scheffer explanatory approach, and in so doing kept their ideas at the forefront. To honor Cox's sustained efforts it seems appropriate to refer to the biogenic approach as the Dalquest-Scheffer-Cox—or DSC—model. This Special Paper continues the clarifying tradition of these scholars, not only in clarifying the role of animals in mound-making but their essential role in soil-making and landscape evolution generally. Science moves forward on the strength of novel, useful, and clarifying ideas. Dalquest and Scheffer, among many others cited in this volume, had many of them.

A synopsis of the expanded DSC model, explained more fully in the Introduction (Johnson and Horwath Burnham, this volume), states that any outward burrowing by soil animals from their nesting centers includes a centripetal component, where some soil is back-transferred to the center. The process, fast-forwarded, creates the mounded landscapes observed today.

Figure 1. Left: Walter Dalquest, summer 1984, age 67. (Photo courtesy Norman Horner, Midwestern State University, Wichita Falls, Texas.) Right: Victor Scheffer, summer 1993, age 86. (Photo courtesy Diana N. Johnson.)

REFERENCES CITED

Arkley, R.J., 1948, The Mima mounds: The Scientific Monthly, v. 66, no. 2, p. 175–176.

Arkley, R.J., and Brown, H.C., 1954, The origin of Mima mound (hogwallow) microrelief in the far western states: Proceedings—Soil Science Society of America, v. 18, no. 2, p. 195–199, doi:10.2136/sssaj1954.03615995001800020021x.

Cox, G.W., and Scheffer, V.B., 1991, Pocket gophers and Mima terrain in North America: Natural Areas Journal, v. 11, no. 4, p. 193–198.

Dalquest, W.W., and Scheffer, V.B., 1942, The origin of the mounds of western Washington: The Journal of Geology, v. 50, no. 1, p. 68–84, doi:10.1086/625026.

Dalquest, W.W., and Scheffer, V.B., 1944, Distribution and variation in pocket gophers, *Thomomys talpoides*, in the state of Washington: American Naturalist, v. 78, p. 308–333, doi:10.1086/281203.

Johnson, D.L., and Horwath Burnham, J.L., 2012, this volume, Introduction: Overview of concepts, definitions, and principles of soil mound studies, *in* Horwath Burnham, J.L., and Johnson, D.L., eds., Mima Mounds: The Case for Polygenesis and Bioturbation: Geological Society of America Special Paper 490, doi:10.1130/2012.2490(00).

Price, W.A., 1949, Pocket gophers as architects of mima (pimple) mounds of the western United States: The Texas Journal of Science, v. 1, no. 1, p. 1–17.

Price, W.A., 1950, Origin of pimple mounds, by E.L. Krinitzsky: American Journal of Science, v. 248, no. 5, p. 355–360, doi:10.2475/ajs.248.5.355.

Scheffer, V.B., 1947, The mystery of the mima mounds: The Scientific Monthly, v. 65, no. 5, p. 283–294.

Scheffer, V.B., 1948, Mima mounds: A reply: The Journal of Geology, v. 56, no. 3, p. 231–234, doi:10.1086/625506.

Scheffer, V.B., 1954, Son exclusivos del oeste de Norteamerica los micromonticulos de tipa Mima?: Investigaciones Zoológicas Chilenas, v. 2, p. 89–94.

Scheffer, V.B., 1956a, Weather or gopher: Natural History, v. 65, no. 5, p. 278.

Scheffer, V.B., 1956b, 1st das mikrorelief der Mima-hugel auf das westliche Nordameriko beschrankt?: Saugetierkundliche Mitteilungen Sonderdruck, v. 4, no. 1, p. 17–21.

Scheffer, V.B., 1958, Do fossorial rodents originate mima-type microrelief?: American Midland Naturalist, v. 59, p. 505–510, doi:10.2307/2422495.

Scheffer, V.B., 1960, Spatial patterns in the world of nature: Pacific Discovery, v. 13, p. 18–20.

Scheffer, V.B., 1969, Mima mounds: Their mysterious origin: Pacific Search, v. 3, no. 5, p. 3.

Scheffer, V.B., 1981, Mima prairie's mystery mounds: Backpacker, v. 4–5, p. 96.

Scheffer, V.B., 1983, Tapestries in nature: Pacific Discovery, v. 36, no. 2, p. 16–24.

Scheffer, V.B., 1984, A case of prairie pimples: Pacific Discovery, v. 37, p. 4–8.

Scheffer, V.B., and Kruckeberg, A., 1966, The mima mounds: Bioscience, v. 16, p. 800–801.

The Geological Society of America
Special Paper 490
2012

Introduction: Overview of concepts, definitions, and principles of soil mound studies

Donald L. Johnson
Department of Geography, University of Illinois, Urbana, Illinois 61801, USA, and
Geoscience Consultants, Champaign, Illinois 61820, USA

Jennifer L. Horwath Burnham
Department of Geography, Augustana College, Rock Island, Illinois 61201-2296, USA

Where controversy exists, science never rests.—Kruckeberg (1991, p. 296)

VOLUME BACKGROUND AND PURPOSE

This volume grew out of a symposium titled: "The origin of Mima mounds and similar micro-relief features: Multidisciplinary perspectives," and an associated similarly themed field trip, both held at the Geological Society of America Annual Meetings, Houston, Texas, 4–9 October 2008. Five of the eight papers in that symposium were expanded for inclusion in this volume; with one nonsymposium paper added later (Johnson and Johnson, Chapter 6). The volume was invited and encouraged by the editors of GSA Special Papers to be one of their series. In regard to soil mounds, the symposium was timely for several reasons.

The first was, among the innumerable theories on how mounds form, evidence has gradually accumulated which confirms that burrowing animals are involved. Involvements would seem to include (1) animals initiating the mounds themselves, where they begin as "activity centers" for basic living purposes (denning, reproduction, food storage, safety, etc.), which are actively bioturbated; (2) where landscape microhighs created by some physical or biological process, or both (e.g., coppicing), become occupied by animals for "activity centers," then augmented through bioturbation to create "hybrid mounds"; or (3) occupy soil-filled joints and fissures in otherwise thin soil or eroded bedrock areas.[1] In any of these three conditions we are concerned with the question: Would activity centers evolve into mounds, and perhaps persist wherever the centers confer living-survival-

reproductive advantages to the animals that inhabit them? Since we examine mounds *after* they form, how can we tell, beyond theorizing, what the initial conditions were that led to mound formation? A purpose of the symposium, and this volume, was to revisit and examine these and other soil mound issues and questions, especially the role of life in landscape evolution.

A second reason is the recent availabilities of useful analytical tools, such as LIDAR (light detection and ranging) and Google Earth technologies, allow new and different light to be shed on soil mound matters. In fact, they are revolutionizing studies of mounded landscapes.

Third, bioturbation- and biomantle-related ideas and formulations on pedogenesis have appeared that are spawning different genetic understandings on how soils form and landscapes evolve (Humphreys and Mitchell, 1988; Johnson et al., 2002, 2003, 2005a, 2005b, 1999; Paton et al., 1995; Schaetzl and Anderson, 2005; Horwath and Johnson, 2006, 2008; Johnson and Lin, 2008; Johnson and Johnson, 2010; Wilkinson et al., 2009; Fey, 2010). These views and models are leading to new questions being asked by Quaternary geologists, geomorphologists, pedologists, and archaeologists, not only about soil mounds but also about the fundamental principles we draw upon to explain soils and landscapes, and to accurately assess archaeological sites. It should be noted that the term "site formation processes," as used by archaeologists and geoarchaeologists, equates to "pedogenesis," as used here, and generally by pedologists and geologists (cf. Finney, this volume, Chapter 5). Contributors to this volume are asking fundamental questions about the role of animals in creating *subaerial* landscapes that ichnologists and marine geologists—whose fields gave us the term "bioturbation" and associated

[1]Good examples of fracture trace Mima-type mounds are Upper Table Rock near Medford, Oregon (42° 28′ 29″ N, 122° 54′ 05″ W) and North Table Mountain near Oroville, California (39° 34′ 57″ N, 121° 33′ 42″ W); cf. also Shlemon et al. (1973).

Johnson, D.L., and Horwath Burnham, J.L., 2012, Introduction: Overview of concepts, definitions, and principles of soil mound studies, *in* Horwath Burnham, J.L., and Johnson, D.L., eds., Mima Mounds: The Case for Polygenesis and Bioturbation: Geological Society of America Special Paper 490, p. 1–19, doi:10.1130/2012.2490(00). For permission to copy, contact editing@geosociety.org. © 2012 The Geological Society of America. All rights reserved.

concepts and principles—once asked about the role of animals in *subaqueous* terrains (Ekdale et al., 1984). Such biodynamic principles are now central to the explanatory-operating-process paradigms of these fields (cf. Bromley, 1996).

The fourth and last reason, in light of these fundamental queries, new technologies, and conceptual advances, and because they seem likely to illuminate prairie-Mima-pimple mound issues that have produced nearly 200 years of lively and spirited genetic discourse, it seemed timely and useful to bring them to bear on the historically sticky issues surrounding the nature and origin of these mounds.

For earth science students, teachers, and other readers who may have somehow escaped familiarity with the subject, or its depth and contentious nature, perusals of end-volume appendices (A–F) coupled with the timeline quotes below, should convey that the above statements "lively and spirited genetic discourse" and "historically sticky issues" are rather mild understatements:

Few subjects have of late years more engaged the minds of scientific antiquaries than the mounds in the valley of the Mississippi . . . there have already been too many wild speculations respecting them throughout this vast region. (Taylor, 1843)

There is a class of mounds west of the Mississippi delta and extending from the Gulf to Arkansas and above, and westward, to the Colorado in Texas, that are to me, after thirty years' familiarity with them, entirely inexplicable. . . . In utter desperation I cease to trouble myself about their origin, and call them "inexplicable mounds." (Forshey, 1851–1852, cited in Foster, 1973, and Veatch, 1906)

These mounds are so uniform in appearance [near Maysville, Arkansas] that they convey the idea of an artificial origin. (Owen, 1858)

The origin of these "peculiar structures" [in Iowa] is a mystery to most people; they believing them to be "Indian mounds" or even "ancient muskrat houses." (Webster, 1897)

They are not confined to any deposit or to any hypsometric level [in Louisiana]. Entirely absent in one locality they are quite abundant in another. (Harris and Veatch, 1899)

The probable origin of these mounds [Mississippi Valley] has been a source of contention since the time they were brought to the attention of the scientific world. (Rice and Griswold, 1904)

It is altogether probable that the mounds which have been noted in various parts of the country are not exactly similar and have not had a common origin. . . . (Campbell, 1906)

The small flat mounds beginning in the Iron Mountain neighborhood in Missouri, and extending southward into Texas and Louisiana, are inexplicable in our present state of knowledge. (Fowke, 1910)

The small, low, flattened mounds of the lower Mississippi Valley are a problem for archaeologists. (Fowke, 1922)

Probably no landform of similar size [in Oregon, Washington, regionally] has occupied such a conspicuous place in geological controversy. (Waters and Flagler, 1929)

[T]he enigmatic origin of these mounds [Mima Prairie] constitutes a continuous embarrassment and a challenge to geological science. (Newcomb, 1952)

The struggle of ideas concerning pimpled plains leans either to physical processes or to biological activity and is tempered by an observer's experience and prejudice. (Malde, 1964)

The literature on their origin is vast and confusing [in Texas-Louisiana Gulf Coast] . . . and the debate seems endless. (Aronow, 1978)

They are a mystery that has been discussed for over 150 years and . . . have generated a greater variety of hypotheses than any other geologic feature. (Higgins, 1990)

[I]f there is a truly perplexing, enigmatic aspect of lower Mississippi Valley Quaternary geomorphology that has defied concerted efforts at explanation and for which there is no consensus [it is] the origin of [pimple] mounds. . . . (Saucier, 1994)

VOLUME JUSTIFICATION

The editors, in sum, firmly believe that any scientific subject or theme, like Mima mounds, with a contentious explanatory history that has covered nearly two centuries—and considered by many to be still contentious—deserves another look.

TERMS AND DEFINITIONS

Mima, Pimple, Prairie, and Natural Mounds

The *Glossary of Geology: Fifth Edition* (Neuendorf et al., 2005), the reference "bible" for earth scientists, has entries for "Mima mound" and "pimple mound" but not "prairie mound" nor "natural mound." "Mima mound" and "pimple mound" once carried regional connotations, "Mima" linked to the Pacific Northwestern states, and "pimple" to the Mississippi Valley–Gulf Coast region. Both terms, along with "prairie mounds" and "natural mounds," are now viewed as synonyms without regional linkages. Entries in the *Glossary* change and evolve as earth sciences evolve. *Glossary* entries for both Mima and pimple mound are presented here, though slightly modified and updated to match our current understandings.[2] Regional connotations of "Mima" and "pimple" remain because they provide a historic context.

Mima mound (Mi'-ma [my'-ma]): A term originally and historically used in the NW U.S., but now also elsewhere, for one of hundreds of thousands, possibly millions of low, roughly circular or elliptical domes, sometimes with flat tops, or low shield-like mounds composed of loose, unstratified, often gravelly silt or loamy soil material, formed on a wide array of soil and landform types, geologic substrates, and ecological environments, from sea level to alpine tree line; basal diameters vary from 1 m to > 30 m, and heights from about 10 cm to more than 2 m. Named after Mima Prairie in western Washington state. Cf: *pimple mound*. Sp: *monticulo de Mima*.

Pimple mound: A term historically used along the Gulf Coast of Texas and SW Louisiana, the lower and upper Mississippi Valley, and more recently in the prairie provinces of Canada for one of hundreds of thousands, and possibly millions of low, roughly circular or elliptical domes or shield-like mounds, often with flat tops, composed of unstratified sandy loam soil coarser than, and distinct from, the surrounding less coarse, often more clayey soil; basal diameters range from 1 m to more than 30 m, and heights from about 10 cm to more than 2 m. Cf: *Mima mound*. Syn: *pimple*.

---

[2]Editors will formally recommend to *Glossary* editor-compilers that current entries be replaced with versions presented here.

Two key descriptors in these definitions are "unstratified" and "coarser." Both are clarified and explained by the role of animals in soil and landscape evolution in sections covered below. The expression "Mima mound," coined by J Harlen Bretz (1913), derives from a small village at the south end of Mima Prairie, which no longer exists. The term was given wide attention by the landmark mound studies of Dalquest and Scheffer (1942, 1944), and by Scheffer's subsequent contributions, and various responses by others to them. Other names have been coined or used for similar microrelief elsewhere (Appendix C), with "prairie mounds" and "pimple mounds" being the more common, and "natural mounds" less so.

Origin of the expression "pimple mound" is uncertain, but dates back at least to Hopkins (1870) who used the term for the "pimpled prairies" of southwestern Louisiana. (It may have been a local term specific to that area.) McMillan and Day (2010), citing previous work (Cross, 1964, and others), applied the term to organism-modified dunes on Virginia's barrier islands, features that might be kindred to "hybrid mounds" (see below). The expression "prairie mound" appears to predate all these terms, and is the more general expression. The terms Mima, natural, pimple, prairie, and soil mounds are used interchangeably in this volume (all are used in chapter titles). Regardless of terminology, all mounds are simply variations on a biodynamic and polygenetic process theme. As defined, all animal-produced mounds in North America which fall within the size dimensions indicated are Mima-pimple mounds.

Clarification of "Mima-Type" and "Mima-Like" Mounds

A clarifying discussion of the difference between "Mima-type" and "Mima-like" mounds is detailed in Chapters 4 and 6 (Burnham et al., this volume; Johnson and Johnson, this volume). Both kinds of Mima mounds are viewed as bioturbated *activity centers* of burrowing animals that represent end members of a microtopographic spectrum of mounds, with all soil mounds falling somewhere on the spectrum. Other things equal, Mima-type mounds are relatively long lasting and "fixed" because they confer some survival-reproductive advantage to the animals that inhabit them. Mima-like mounds are relatively ephemeral and "unfixed" because such advantages are either absent or neutral.

We suppose that "survival-reproductive advantages" for any burrowing animal species may not always be obvious to human investigators and observers. Moreover, the boundary between Mima-type and Mima-like mounds on the microtopographic spectrum is unspecified. Both issues may be unsettling to those who seek limits and boundaries for complex natural phenomena, and brings to mind the Aristotelian postulate: "It is the mark of an instructed mind to rest satisfied with the degree of precision which the nature of the subject permits and not to seek an exactness where only an approximation of the truth is possible." (Cf. *GSA Today,* June 1998, p. 16.) The postulate, in some ways, also relates to the concept of "hybrid Mima mounds."

Hybrid Mima Mounds[3]

The term "hybrid mound" refers to a preexisting microhigh formed by any physical (freeze-thaw, dune) or biological (tree-uprooting) process, or combination (coppices), which becomes inhabited by soil animals (pocket gophers, ground squirrels, badgers and other predators, moles, ants, termites, other insects, etc.) and modified in form and shape to become "Mima mound-like." After some period of habitation the final biogenetic-polygenetic outcome can be a coarser textured mound than surrounding soil. Examples of preexisting microhighs on which Mima mounds commonly form are: low ridges on old meander scrolls; shrink-swell (gilgai) microhighs; small first- or second-order floodplain high-spots and/or their stream levees; low rill divides on slopes; bumps on a terrace; small dunes; knolls; boles of uprooted trees; coppice accumulations; or any small rise. If the rise or microhigh confers a survival-reproduction-dwelling advantage over the surrounding area to any soil animal that occupies it—commonly against wetness—it becomes occupied, destratified, and accordingly rounded-up and modified via the animals' life activities. The rise is now a Mima-type "hybrid mound." Many, perhaps most, hybrid mounds are Mima-type, although reasonably distinguishing them from Mima-like hybrids evokes Aristotelian uncertainties.

The hybrid mound concept embeds notions of polygenesis—multiple processes of formation—and notions of equifinality where different processes produce similar landforms, in this case forms that might resemble Mima mounds but by definition are not (lava blisters, small pingos, microhighs on partially thawed permafrost, small stratified dunes-lunettes-coppices, etc.). The concept is restated with different language in the Introduction to Chapter 4 (Burnham et al., this volume). The editors believe that an *absence* of a formal hybrid mound concept is at least partly, if not largely, responsible for the historic proliferation of mound hypotheses and theories, and for the general confusion about mounds and unsettled views on their genesis.

Hybrid Mima mounds can be either geologically old (Pleistocene) or historically formed. North American examples of historic hybrid Mima mounds are many and notable. One is the extensive and now largely rodent-insect-predator destratified and bioturbated small coppice mounds crossed by Interstate-10 and U.S. Highway 54 in the El Paso-Las Cruces regions of west Texas and New Mexico (Gile, 1966; Hall et al., 2010; Johnson, 1997; personal observations). Another example is the hybrid, partly coppiced, "grassy sand mounds" mapped as Padre and Madre soils on pocket gopher–inhabited North Padre Island, Texas (Brezina, 2007; personal observations), particularly near the National Seashore entrance kiosk. Notable, and genetically telling, is that while Mima (hybrid) mounds and pocket gophers are plentiful on North Padre Island, both are absent on dune-covered South Padre Island.

---

[3]This term was coined and the concept formulated for inclusion in the field guide for the field trip held in conjunction with the 2008 Houston GSA Symposium that led to this volume (Johnson and Johnson, 2008).

Other examples are the Mima-like hybrid mounds in Fernley Sink, Nevada, crossed by Interstate-80. They are rodent-bioturbationally destratified, lake-basin–derived clay- and carbonate-enriched coppice dunes. Similarly bioturbated, lake-basin eolian-produced hybrid mounds exist on the east side of Laguna del Perro crossed by Highway 60 near Silio, New Mexico (cf. Allen and Anderson, 2000; Anderson et al., 2002). Such mounds also occupy portions of the floor of Pleistocene Lake Bonneville (Great Salt Lake), as along much of the western fringe of Logan, Utah, and near its airport (Campbell, 1906; personal observations; Janis Boettinger, 2011, personal commun.). Interstate-15 intersects a similar moundfield formed on low terraces of the Bear River several kilometers north of Brigham City, Utah (personal observations).

Hogwallows (Hog Wallows) and Vernal Pools

The term "hogwallows," used to describe Mima-type moundfields in California and Texas, is misleading. "Wallows" refers to the depressions between mounds, or to vernal pools, not the mounds themselves. Buckman Hogwallows Preserve in California, shown in Figure 1 and discussed below, is an exam-

ple. Further, the term "hogwallows" originally carried an image of clayey shrink-swell grumusol-Vertisol–type landscapes, as in the Mississippi valley and Gulf Coast, for seasonally cracking soils that often displayed gilgai microrelief. Contrary to what some pedologists suppose, sandy Mima-type mounds or "sand spots" were once fairly common on Gulf Coast gilgai microhighs, and in several places still are, and had likewise formed in some California Vertisols that lacked gilgai (Retzer, 1946). A careful reading of the literature, plus personal field observations in these regions, confirm this fact (e.g., Carter and Patrick, 1919; Foster and Moran, 1935; Smith and Marshall, 1938; Watson and Cosby, 1924; Retzer, 1946; McGuff, 1973; Fields et al., 1986; Heinrich, 1986; Ensor et al., 1990). How sandy, largely biogenically produced Mima-type mounds could form on otherwise clay-rich gilgaied or nongilgaied soils is indicated both in the 1975 Soil Taxonomy (Soil Survey Staff, 1975, p. 21) and in Van Duyne and Byers (1915, p. 1078). The process involves a combination of animal bioturbations at activity centers and episodic rainwash (fine particle elutriations) over time. Such processes produce point-centered locally thickened and coarse textured biomantles relative to surrounding soil (cf. Johnson et al., 2003).

Figure 1. Three seasonally different views of Buckman Hogwallows Preserve, a small 4.1 ha (10 acre) low slope (~1%) Mima-type moundfield in hardpan soils that escaped the plow at the eastern edge of San Joaquin Valley, near Visalia, California (36° 21′ 26.58″ N, 119° 05′ 02.33″ W). The tiny preserve is now entirely surrounded by citrus groves (in order to plant trees hardpans were not uncommonly broken with dynamite; cf. Amundson, 1998). The notably domed Mima-type mounds here formed on relict alluvial fans on footslopes of the Sierra Nevada Mountains. Photo A was taken April 2006 after a wet winter. Photo B was taken January 2009 following a burn, several showers, and beginning of regrowth. Myriad surface heaps of the Botta pocket gopher (*Thomomys bottae*) attest to the effective landscape "rounding," mound making, and soil moving role that this seldom seen animal performs on this and many other moundfields. Photo C was taken January 2012 at the end of the dry season. View of photos is north, each taken at slightly different zoom levels from the same 1.8 m high mound. This moundfield, managed by the Tulare County Historical Society, provides a window into how this part of California looked before the plow and dynamite arrived (cf. Reed and Amundson, this volume, Chapter 1). The photos show that one's impressions of mounds and moundfields are strongly colored by the season observed. Photos courtesy D.N. Johnson.

Most vernal pools in Pacific Coast states occur where slopes are low and drainage restricted, often by the presence of Mima-type mounds. Larger vernal ponds and lakes are usually owed to more general geologic processes, such as warping, dune formation, or uneven igneous (basalt) extrusions (cf. Alexander and Schlising, 2000; Holland, 2000b). Such processes, however, are altogether different from, and independent of, the biogenic processes that create Mima mounds. Vernal pools constitute rainy season "hogwallows" of Mima-type mounds. The mound-vernal pool system is usually underlain by an aquiclude, such as hardpan, as is the case with many California moundfields.

An example is Buckman Hogwallow Preserve (Fig. 1), populated presently with one major mound maker, the Botta pocket gopher (*Thomomys bottae*). Here pools exist only because mounds and hardpan impede surface drainage (personal observations). Vernal pools at Buckman, like those in California generally, are typically shallow and seasonally disappear under high evaporation rates in late spring and early summer (Jain and Moyle, 1984; Holland, 2000a). As slopes steepen, as along ravine slopes in the Merced moundfields described in Chapter 1 (Reed and Amundson, this volume), pools become rills, with mounds invariably arrayed paternoster-like on rill divides.

Pixley Vernal Pools, another hardpan moundfield some 44 km south-southwest of Buckman Hogwallows, and four times larger (16.4 versus 4.1 ha), is another low slope Mima-mounded preserve, also in Tulare County (35° 59′ 03.99″ N, 119° 12′ 45.80″ W). In addition to the Botta pocket gopher, Pixley soils include two additional major bioturbators and mound makers, the California ground squirrel (*Spermophilus beecheyi*) and the American badger (*Taxidea taxus*). Other predators, coyotes, weasels, etc. are also likely present. Wild pigs likewise bioturbationally impact this moundfield (personal observations). As a consequence of this collective suite of different bioturbators, instead of smooth dome-shaped gopher-mediated mounds that characterize Buckman Preserve, Pixley mounds are irregular in outline, have hummocky and bumpy surfaces, and are heavily pock-marked by squirrels and badgers.

Environmental emphases and justification for preservation of both these Mima-mounded prairies is invariably on the pools and their often endemic, scientifically valuable, and commonly aesthetically showy and appealing flowers. Ironically, little attention is paid to the animals—primarily pocket gophers—that created the mounds, and by doing so created the pools. The long-term back-transfers of soil to nesting centers by these various animals, especially pocket gophers, during burrow-foraging activities (Dalquest-Scheffer-Cox [DSC] model, see Dedication and sections below) is what creates the mounds—and hence the depressions that hold the pools.

Polygenesis and Mound Complexity

Because all soils and landforms are polygenetic process-wise, so likewise are Mima mounds (cf. hybrid mounds, above). Mima mounds are of *polygenetic* and *complex* origin, but where bioturbation is the common process denominator. Polygenesis as used here[4] refers, literally, to the myriad and innumerable abiotic-biotic conditions, factors, and processes that impact soils and landforms at the many and varied elevations and diverse ecological environments in which Mima mounds are known to occur. Another perspective on polygenesis is that multiple organisms (animals, plants, fungi, protoctists, microbes), in various combinations, inhabit *every* soil and soil mound that exists. These life-forms all move, wriggle, and bioturbate, and in the process impart key biochemical signatures that trigger a cascade of transformations to soils and soil mounds.

Mounds are *complex* because the relative effect of each process and/or condition waxes and wanes with time, and some may be minimal, overprinted, or absent when studied. Because polygenesis and complexity must vary between mounds and moundfields, local and regional contrasts invariably fuel explanatory controversy (cf. Isaacson and Johnson, 1996).

LIDAR AND GOOGLE EARTH TECHNOLOGIES

LIDAR

LIDAR (light detection and ranging) is an active remote sensing technology that uses a pulsating laser sensor to scan the Earth's surface. The reflected pulses, up to 100,000 per second, are detected by instruments that record their location in three dimensions, which can then be used to create topographic resolution of the land surface at decimeter scales (Heidemann, 2012). The technique is excellent for accurately assessing Mima mound sizes, numbers, densities per unit area, estimating mound heights, and other measures. The authors of Chapter 1 (Reed and Amundson, this volume), for example, used LIDAR-produced mound data with other information to very conservatively calculate that 4.4 trillion metric tons of soil were moved by one species of burrowing animal, the Botta pocket gopher (*T. bottae*) in building mounds in an area equivalent to ~10% of California. As the authors note, and if their estimates are close, a small, seldom seen and nondescript burrowing rodent that is passionately detested by home-owners, greens-keepers, and gardeners alike, may be responsible for moving 350 times the annual sediment discharge of all the world's rivers!

While opening new opportunities for gaining high-resolution data on bioturbated and translocated soil and mound formation, LIDAR studies are also annually expanding our view of just how many Mima mounds might actually still exist. In fact, the numbers of mounds and moundfields being discovered with the technique by some volume contributors are attaining surprising levels in spite of several centuries of human landscape modification.

In addition to the LIDAR study in Chapter 1 (Reed and Amundson, this volume), another example of its usefulness in soil mound studies comes from Vandenberg Air Force Base in Santa Barbara County, California. Two volume contributors

---

[4]Polygenesis in this volume varies significantly from how Price (1950, p. 359) used it, which was in place of the much more appropriate current term "equifinality," not then in the lexicon of earth sciences.

(Johnson and Johnson) have, for several decades, been intermittently monitoring a number of Mima-mounded tracts on Vandenberg selected in the 1980s for long-term observations. The base has a diversity of burrowing animals, which in order of most importance are the Botta pocket gopher (*T. bottae*) and California ground squirrel (*Spermophilous beecheyi*). Site selections were based on multiyear airphoto coverage and base-wide mound observations (Johnson, 1988, 1990; Johnson et al., 1991). Yet, when LIDAR became available, the number of known mounds increased exponentially. Indeed, owing to vegetation masking and subtle relief, mounds not seen in the field nor on airphotos were made instantly obvious on LIDAR, as exemplified by Figure 2. The figure shows mounded and nonmounded areas with subtle intergrades.

Irregular bumps at Sites A on Figure 2 are slightly coppiced but unmounded sage vegetation. The largest and most distinct mounds occupy low, seasonally wet swales and depressions that occasionally flood. At such sites soil material is available in sufficient quantity for making large mounds, as at Sites 4 and 7. Fines washed or blown into depressions are back-transferred to mounds by gophers during forage burrowing (DSC model). Comparatively smaller mounds occur: (1) where water accumulates only infrequently or (2) where shallow soils over hardpan slope into gullies and reentrants. Where soil biomantles are thick (>1 m) and water does not accumulate, Mima-type mounds do not form. Mounds in the depression at Site 3 range in height from ~20 cm to ~1 m, and in diameter from ~1 to 10 m, with exposed, virtually soil-free hardpan between. Arrows at Site 4 (inset) identify low (~10–20 cm), broad (~10 m) nascent mounds now forming at sinused perimeters. These evolve from erosionally thinned and centripetally bioturbated perimeter activity centers, and were unobserved before LIDAR. Depression floors receive fines from the surrounding biomantle during storms, which provides new soil material for biogenic growth of mounds; tiny bumps on depression floor are gopher heaps produced during forage runs. Seasonal sediment influxes fuel growth of large mounds as at Site 7. Many mounds are predator-impacted, especially by badgers, whose holes on mounds are quickly filled by gopher bioturbations. Nearby Sites 1 and 2, just off this image along Tangair Road, consist of large, sinus-edged, mound-dotted depressions in which gopher-inhabited mounds become islands during wet winters, as do mounds in depression Sites 3, 4, and 6.

Google Earth

Google Earth, like LIDAR, is also a relatively new research tool, and likewise has revolutionized Earth surface and soil studies, particularly soil mound studies. It draws on and applies satellite imagery, airphotos, topographic and shaded relief maps, and various other resources. Users can rapidly zoom laterally and vertically to any point on Earth, and then quickly determine cursor-point elevations (m, km), locations (latitude-longitude), and scaled distance measures (m, km). Spatial-lateral scanning allows rapid examination of large areas over short periods of time. For mound studies, the Historical Imagery application is particularly useful because a moundfield that is not apparent on one, two, or three airphotos of the same area covering different years may yet be visible on a fourth, fifth, or eighth.

LIDAR and Google Earth Combined

The volume contributors who were monitoring Vandenberg soil mounds (Johnson and Johnson) also initiated a systematic field examination of many documented (published) moundfields in former grassland or open forest tracts of Arkansas, Louisiana (west of the Mississippi), and eastern Texas (e.g., Aronow, 1988; Seifert et al., 2009; Archeuleta, 1980; Holland et al., 1952; Bragg, 2003; Cain, 1974; Owen, 1860; Frye and Leonard, 1963). Field examinations were based on locations and maps of published studies, and on airphoto analysis. While somewhat successful, the work proved to be disproportionately time-consuming, expensive, and complicated due to increased forest cover and human development, and to private property access issues.

Recent LIDAR imagery for some of these earlier examined areas, however, display an astonishingly far greater number, detail, and resolution of mounds and moundfields than had been reported, or expected, especially in light of historic human modifications. Moreover, owing to increases in forest cover, few mounds and moundfields were detected in such areas on Google Earth images. In fact, LIDAR demonstrates that in some areas almost every landscape segment is studded over with mounds, sometimes in surprising densities and detail. This is particularly true for some coastal areas, as at Hackberry Island and the Houston Ridge in Cameron-Calcasieu Parishes (cf. Aronow, 1988; Heinrich, 2007). It is also true for sublevels of the Deweyville and other terraces of the Sabine and Red Rivers (Alford and Holmes, 1985; Frye and Leonard, 1963), and on old terraces along the major rivers of this broad region (e.g., Red, Trinity, Neches, Sabine, Calcasieu, Mermentau, Arkansas, Saline, Little, Tensas, Ouachita, White, and Mississippi). Mounds are absent, as predicted, on floodplains of all these big rivers, at least those that overbank most years (most burrowing animals do not survive deep annual floods).

Figure 3 shows a comparison of LIDAR coverage matched with the best airphoto coverage in the Historical Imagery application of Google Earth. The images clearly demonstrate that LIDAR is unmatched in its effectiveness for identifying mounds in forested areas.

SOME BIOGENIC PRINCIPLES OF SOIL MOUND FORMATION

Expanded DSC Model of Soil Mound Formation

The Dalquest-Scheffer-Cox (DSC) model, which embeds biogenesis as its central concept, is that soil animals—those that spend part of their lives burrowing and living in soil or

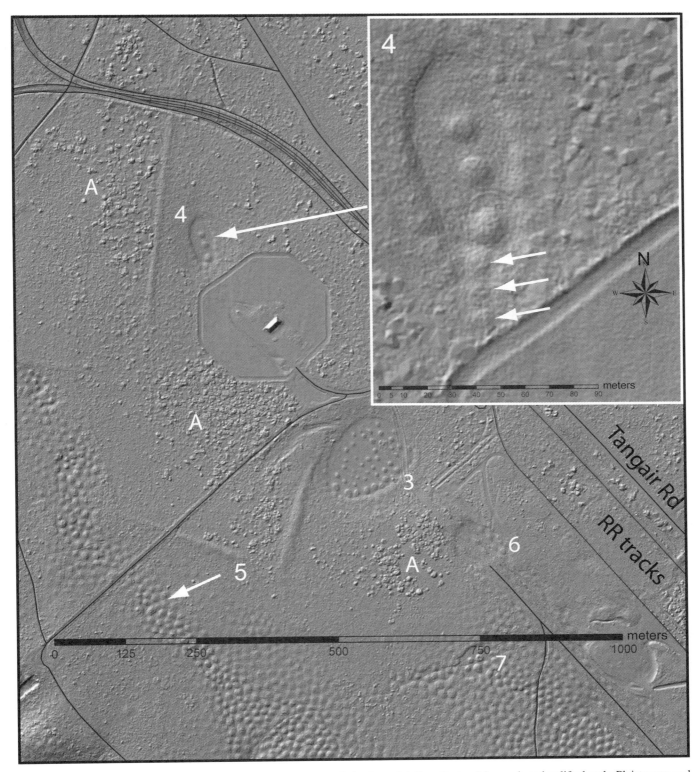

Figure 2. LIDAR imagery reveals a multitude of Mima-type mounds in sharp detail on Burton Mesa, a broad uplifted early Pleistocene and duripan-armored marine terrace, Vandenberg Air Force Base, California. LIDAR imagery helped expand the number of known mounded tracts on the base by several orders of magnitude. Burrowing animals, primarily the Botta pocket gopher (*T. bottae*), are predominantly responsible for producing the mounds, aided by local soil, slope, and runoff conditions and processes. As predicted by the expanded DSC model, mounds form where the biomantle is thin (<1 m) above a dense substrate, in this case silica-cemented hardpan. See text for explanation of numbered and lettered sites. LIDAR courtesy of James Carucci and Environmental Group, Vandenberg Air Force Base, Vandenberg, California.

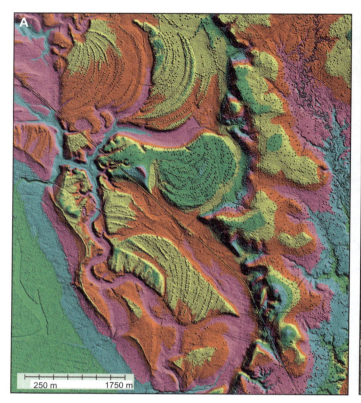

Figure 3. LIDAR image (A) and Google Earth airphoto (B) of the same semiforested border region of Ashley County, Arkansas, and Morehouse Parish, Louisiana. Image center-point is 33° 01′ 42″ N, 92° 00′ 53″ W, ~28 m elevation. The region is the lower confluence area of Saline and Ouachita Rivers. The Ouachita floodplain, on left side of images, ~20 m elevation, lacks mounds because of frequent "big river" overbank flooding. Mima-type (pimple, hybrid) "fixed" mounds occupy the higher and drier spots of very low-relief meander scrolls on two low ter-races, at 25 and 28 m, of late Pleistocene Deweyville Allogroup. Mounds are arrayed in arcuate paternoster patterns, like festooned "beads on a necklace." But, on much older and higher ground nearby to the east, they are seemingly randomly distributed in a nearest-neighbor–like pattern. Mounds also dot higher spots of the small stream basin on the right side of images. These high spots afford greater survival opportuni-ties against wetness for animals that live on them, which leads to mound formation. LIDAR courtesy Paul V. Heinrich, Louisiana Geological Survey, Baton Rouge, Louisiana. *Note:* Each color contour increases 1.524 m (5 ft) elevation; e.g., green = 20 m, blue = 21.524 m, purple = 23.048 m, etc.

sediment, which actually includes most animals—invariably establish activity centers for living, specifically for nesting-reproduction, burrowing, denning, food storage, overwintering, estivation, hibernation, and or safety. Any outward burrowing from these centers must include a lateral component where some amount of soil is centripetally back-transferred to the center (see Centripetal "Law" of Soil Movement to Animal Mounds, next section).

The *expanded* DSC model includes the proviso that all other soil-geomorphic processes, such as water-wind erosions, eolian infall, vegetation growth, mass wasting, shrink-swell, freeze-thaw, earthquakes, and other processes that can impact any landscape may, proportionately, also impact mounds, while they form, after they have formed, and while they are wasting. The collective processes, fast-forwarded, create the complex of soil mounds and mounded landscapes we observe today. DSC processes dominate over all others in low slope, level, or nearly level landscape positions.

Centripetal "Law" of Soil Movement to Animal Mounds

When animals burrow in soil or sediment in establishing activity (nesting) centers, which ultimately leads to mound for-mation, burrowing is never always vertically straight downward. There is invariably a lateral or centripetal component in burrow-ing from any nesting center. The nature of mound making would logically involve the slow radial and centripetal-lateral move-ment of some soil to activity centers from below and surrounding areas. The massive and robust termite and ant mounds that dot the tropics and subtropics, and their smaller mid-latitude equivalents, must be partly, if not largely, composed of such soil, mined and biotransferred from surrounding areas. The same must be true of those animals that play roles in producing Mima mounds in North America, most especially pocket gophers, and to some ex-tent ground squirrels and prairie dogs, but also especially town ants (*A. texana*). Hence, areas surrounding some animal mounds must be topographically slightly lower, and thus wetter. Where

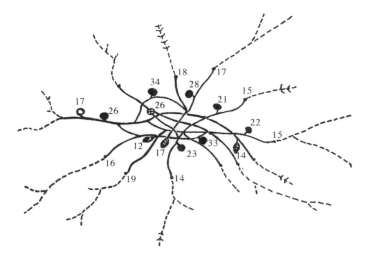

Figure 4. Plan view of pocket gopher (*T. talpoides*) reproductive–food-storage–living center that has outward radiating forage tunnels. (Original diagram unscaled but ~3 m in diameter is a reasonable estimate.) Rounded projections are nests; the one with the cross had three young; those in black were abandoned. Ovate projections, cross hatched, are food-storage chambers. (Defecation chambers not illustrated.) Numbers denote depths (inches) at which features were found (after Criddle, 1930).

Mima mounds are directly associated with, and genetically linked to, vernal pools, this fact becomes obvious. The process in this case is more or less as Price (1949) had described.

Basal Concavities of Some Soil Mounds, and Breaks and/or Depressions in Subsoils of Others

Dalquest and Scheffer (1942) noted that mounds near Olympia, Washington, commonly display concave bases in soil profile (see cover photo). Other workers have noted similar basal concavities (Holliday, 1987), including breaks or holes in hardpans and claypans into subjacent soil parent materials. The concavity, hole, or break—"dimples" as Price called them—often are found below the thickest part of the mound (cf. Nikiforoff, 1941; Retzer, 1946; Price, 1949; Arkley and Brown, 1954). Such features would be expected, if not almost predicted, by mounds predominantly made by burrowing animals. In the case of hardpan holes, it suggests that the animals were present, and bioturbation active, when the hardpans initially began developing and during subsequent formation. It also suggests that activity centers were established during early stages of soil formation, and that bioturbation rates were consistently greater than, or at least as great as, rates of pan formation.

The Principal Mound-Making Animals of North America

The five principle mound-making animals are: pocket gophers, which comprise some 40 species; kangaroo rats (k-rats) with some 21 species; ground squirrels and prairie dogs, with

~40 species; one notable ant, the Town ant (*Atta texana*) of Texas and Louisiana; and the American badger. Here, of necessity, we selectively focus on the two that most link to Mima mound formation: pocket gophers and k-rats. A brief sketch is also provided on the role of badgers, tremendous soil movers and mound-makers in their own right. The focus draws on the literature, and the co-editors' personal observations.

As a prefatory comment, it is instructive to emphasize, strongly, that of the main mound-makers in North America, pocket gophers and k-rats are behaviorally *cryptic*—respectively fossorial and nocturnal—hence rarely if ever observed on, or visually associated with, their mounds. Seldom seen badgers, to a large extent, fall into this category. Uncertainties about who made the mounds invariably arise. Conversely, ground squirrels, prairie dogs, and Town (Atta) ants are *noncryptic*, and thus commonly observed on their mounds, or seen to be immediately associated with them. Hence, observers normally do not question who made such mounds. We submit that these cryptic, noncryptic issues have historically and fundamentally significantly blurred our perceptions of Mima mound genesis in North America. They have, we believe, played a major role in shaping the genetic controversy surrounding the subject.

Pocket Gophers

Information on pocket gopher burrowing is deep and extensive, but two papers that graphically convey burrowing particulars are instructive. Figure 4 is a plan view of a gopher activity center excavated by hand on 14 June 1927 in Manitoba, Canada. The activity center is, of course, the central area of bioturbation. The continuous lines are deep open tunnels; dashed lines are foraging tunnels, some plugged, others extending out unseen into surrounding food-harvesting areas. Every radial foraging event records some soil being back-transferred to the center. The process builds mounds. Because gopher species have short lives, 2–3 years at most, where sites confer living-survival advantages over surrounding areas they become "fixed" and occupied by multiple generations of gophers. Large volume Mima-type mounds can then form.

Figure 5 shows both plan (A) and cross-sectional (B) views of another hand-excavated gopher activity center; with associated forage tunnels (in this case the species, location, and excavation date are unspecified by the author). Here the active nest was tunneled into the subsoil, confirming that this horizon does not limit vertical burrowing. It also suggests that no particular living advantage is likely conferred at this site over any other in this soil. While information is lacking to say for certain, it is possible that the very low, shield-like mound that is forming above the activity center in B will not grow into a much larger mound, and that it is likely a Mima-like mound, and thus will be ephemeral. If it does grow, the site must offer some living-survival advantage, and the biomantle will become locally thickened at the expense of the subsoil, and likely surrounding soil, and thus become a fixed point-centered biomantle (cf. Johnson et al., 2003). (Aristotle's postulate might be pondered here.)

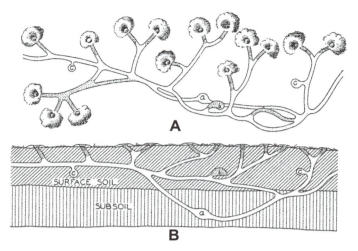

Figure 5. Plan view (A) and cross section (B) of a pocket gopher reproductive–food-storage center with lateral forage tunnels (original diagram unscaled). Surface soil is the biomantle. Symbols: a—active nest chamber; b—abandoned soil-filled nest; c—food-storage chambers (after Crouch, 1933).

Figure 6. Photo of a Mima-like mound "precinct," with multiple ~10-cm-diameter entrance holes, produced by one banner-tailed kangaroo rat (k-rat; *D. spectabilis*) on north side right-of-way of U.S. Highway 60, ~40 km west of Socorro, New Mexico. Many widely distributed k-rat mounds dot the broad, low-slope alluvial fans here that emanate from the Magdalena Mountains. Mounds, which can reach 1+ m high and up to 5+ m in diameter, are built over several years by individual k-rats, normally one animal per mound, except during breeding season. Photo courtesy D.N. Johnson.

Kangaroo Rats

Best (1972) recorded stages in the formation of two "characteristic" banner-tailed kangaroo rat, or k-rat (*D. spectabilis*), mounds in New Mexico, by two adults. One mound was formed in disturbed soil, and the other in undisturbed pastureland. In the disturbed soil, a circular mound 3.1 m in diameter and 36 cm high was developed in 23 months. In the pastureland, an ovate mound 3.7 by 2.2 m in diameter and 41 cm high formed in 30 months. An average of 3–5 entrance holes appeared in the mounds during their constructions. Each mound was constructed and inhabited by one adult k-rat, a male and a female. Figure 6 is an example of a mound built by this species, also in New Mexico.

Best's study shows the rapidity with which this species can create mounds. There are 20 other species of k-rats, and each has its own burrowing/mound-making behaviors and styles. One k-rat, the largest, may possibly challenge the pocket gopher for honors in mound-making propensities and speed, as demonstrated in the Carrizo Plain region of California.

American Badgers

The American badger, *Taxidea taxus*, is a major bioturbator of North American soils and Mima mounds in essentially *every* moundfield that co-editors and several contributors have examined—wherever rodents, a major prey of badgers, live. Ironically, badgers are notably absent, at least presently, in the one region considered by some to be the Mima mound type locality—the Olympia, Washington area (cover photos). As mentioned, the badger is also a major mound-maker in its own right.

Co-editor Johnson was introduced to the tremendous soil volumes, and large cobbles and boulders, that are regularly bioturbated and moved about by this animal while engaged in landscape evolution studies in the Tularosa Basin-Otero Mesa regions

of southern New Mexico and West Texas (Johnson, 1997, 1999). This chief predator of rodents, most notably pocket gophers and ground squirrels, has impacted many, if not most, Saskatchewan mounds discussed by the Chapter 2 authors (Irvine and Dale, this volume). Badgers doubtless have done likewise to most mound-fields discussed in other chapters. Some idea of soil volumes bioturbated and moved by this superburrower while rodent foraging, which could be anywhere in western North America, are summarized for a part of Idaho by Eldridge (2004):

In the western United States, American badgers (*Taxidea taxus*) excavate large volumes of soil and create fan-shaped mounds while foraging for fossorial rodents. Densities of 790 mounds/ha were recorded on the Snake River Plain, west-central Idaho. . . . Mounds and diggings occupied an average of 5–8% of the landscape and the mass of mounded soil averaged 33.8 kg, equivalent to 26 t/ha. The surface cover of plants, cryptogams, and litter increased, and bare ground decreased, as mounds aged. Excavation holes were present at 96% of active and crusted mounds compared with 31% of older recovering mounds. Sites with a greater density of shrubs tended to have a greater density of both badger mounds and ground squirrel diggings. Additionally, increased density of badger mounds was associated with increases in the density of ground squirrel holes and scratchings. These results indicate that badger mounds are a significant landscape structure and that badger activity is likely to have major impacts on soil and ecosystem processes. . . .

Figure 7 shows a badger, with his burrow and tailings, caught in the act of either (1) co-opting a Mima-type mound for making a den (called a "sett"); or (2) preying on the rodent (gopher) occupant of the mound; or (3) perhaps both. The chief badger

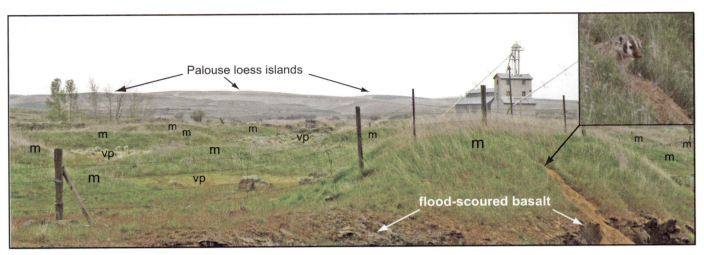

Figure 7. Photo of Mima-type moundfield developed in thin, loess-ash-rich soil above Missoula flood-scoured basalt in Channeled Scablands, Whitman County, Washington. Letter m's identify mounds, and vp's identify intermound swales-vernal pools. The photo of badger (inset) peering from mound on right was taken immediately after the first. Both photos courtesy D.N. Johnson.

prey here is the abundant Northern pocket gopher (*T. talpoides*), a major mound-maker widespread across the Pacific Northwest region (cf. Johnson and Johnson, this volume).

Rapidity with Which Mounds Form and Re-Form

Figure 8 is a photo looking east from the northbound shoulder of Interstate-5 north of Weed, California taken shortly after snowmelt. It displays many principles of the expanded DSC model. Mima-type mounds here, most dotted with abundant fresh and old gopher heaps, are formed in thin soil over hardpan—a duripan, an effective aquiclude developed on glacial outwash. Intermound pools form in very low slope areas where mounds interfere with surface run-off. Mounds are in quasi-dynamic denudation equilibrium, where any erosional removals (negligible)

are offset by biogenic mound formation, or in this case re-formation (Johnson et al., 2011). The farm road over which mounds have re-formed was bulldozed sometime after 1928, the date when bulldozers were invented, and before being used as a footpath. The road fell into disuse when blocked by I-5 construction in 1965, and the mounds have re-formed in the 47 years since. Another instance where mounds re-formed over a road, "an old wagon trail," is near Boulder, Colorado, as noted by Branson et al. (1965, p. 318; cf. also Murray, 1967).

The rapidity (decades) with which Mima mounds can be formed by pocket gophers has been documented by a number of workers (e.g., Koons, 1926, 1948; Brown, *in* Arkley and Brown, 1954, p. 197; Johnson et al., 1999). Personal observations have documented mounds having formed recently, and still forming, along a number of highway right-of-ways. Examples are on both

Figure 8. Shasta Valley, Siskiyou County, California. The ranch road, bulldozed down to the duripan sometime after 1928 and used into the 1960s, fell into disuse when construction of Interstate-5 blocked it in 1965. Mounds have re-formed in the ensuing 47 years (black arrows). Inset shows close-up of duripan developed in glacial outwash, exposed in a diversion ditch. Photo courtesy D.N. Johnson, taken early March 2011.

sides of U.S. Highway 287, near Tie Siding on the Sherman Pene-plain, Albany County, Wyoming, and the east side of California State Highway 139, ~7 km north of Canby on the Modoc Plateau, Modoc County, California (cf. Johnson et al., 2011). And of course, innumerable hybrid mounds are documented as having formed historically (references cited in "Hybrid Mima Mounds" section above).

San Luis Obispo County, California, is home of the Carrizo Plain National Monument, and to hundreds of thousands, if not millions, of Mima mounds, produced primarily by burrowing rodents, perhaps augmented by badgers and other rodent predators. The Carrizo Plain comprises a northwest-southeast structural basin created by the San Andreas Fault, which occupies its eastern side. The Monument consists of 101,000 ha (250,000 acres) of mounded terrain, but mounds occupy a far greater area than the Monument per se. The landscape, and the multitude of burrowing animals that occupy it, are a collective window into what much of the entire western side of the San Joaquin Valley and adjacent Coast Range hills and ridges were like before intensive agriculture, grazing, and oil extractions began (Braun, 1985; Williams, 1992). For those who wonder how rapidly Mima mounds can form by burrowing animals, the Carrizo Plain is the place to learn.

Figure 9 consists of photographs of mounds that formed since the 1980s, prior to which the fields were under cultivation. Mounds here are commonly 7–12 m in diameter, range between 20 and 50 cm in height, and, depending on location, have densities between 45 and 60 ha. Variation from these values, however, can vary considerably depending on location (Braun, 1985). The mounds are formed on broad, low slope alluvial fans that emanate from surrounding mountains. But they also have formed on many other geomorphic surfaces and soils of the surrounding hillslopes, including the adjoining Elkhorn Plain, and slopes of the Temblor Range to the east and south. The mounds, called "precincts" by kangaroo rat specialists, are the self-made activity centers of k-rats, other burrowing rodents (Wallace, 1991), their commensals, and their predators.

The principal mound-makers are kangaroo rats (several species) and pocket gophers, and of these the endangered giant kangaroo rat or GKR (*Dipodomys ingens*) plays a major role. Its main presettlement distribution was over much of the western San Joaquin Valley and adjacent Coast Range hills and slopes, where it is now nearly extirpated. It presently occupies only 2% of its historic range (Prugh and Brashares, 2011). Except during breeding season, each mound is normally occupied by one adult, male or female (cf. Williams and Kilburn, 1991; L.R. Prugh, 2011–2012, personal commun.). Because GKR longevity is 2–4 years, like pocket gophers, many generations must be involved in making and maintaining mounds.

Some Key Biomantle Principles of Soil Bioturbation

Figure 10 summarizes in a tabular and organized fashion what happens, other things equal, when different soil animals bioturbationally operate on soils and substrates that have different textures and particle sizes. Biomantle principles explain why most mounds in the Mississippi Valley and Gulf Coast that are formed in fine materials (Aronow, 1963, 1968, 1976, 1978, 1988) will result in profiles like 3 and 6 (bottom), regardless of bioturbator type. The profiles are biogenically unsorted (i.e., *nonbiostratified*) because the mixture of clast sizes required for sorting is absent.

The principles also explain why gravelly mounds in southwest Missouri (Chapter 4), in Minnesota (Chapter 5), parts of California (Chapters 1 and 6), and in much of the Pacific Northwest (Chapters 4 and 6), which formed in gravelly materials bioturbated by animals such as pocket gophers and moles, result in profiles like 2 and 5. Any stones larger than the animals can move through their burrows settle downward to the base of the biomantle, which is the main zone of bioturbation. Such profiles and mounds invariably display basal stone layers, with walnut-sized and smaller pebbles scattered throughout. These profiles are biogenically sorted, and hence biostratified (cf. Johnson et al., 2002, 2008).

If, on the other hand, mounds form in gravelly soils where the dominant bioturbators are very small, like ants, termites, and or worms, the resulting profiles will resemble 1 or 4. These again are profiles and mounds that have basal stone layers, but lack pebbles or coarse fraction in the upper profile (beyond a few bird gastroliths, which eventually migrate to the stone layer). These profiles likewise are biogenically sorted (biostratified).

What Figure 10 does not display, however, are two other closely related and important biomantle principles. The first is that all bioturbated, biomixed, and biosorted particles (bottom row) will be slightly smaller in size due to *particle comminution*, the grain-to-grain grinding and abrasion that occurs during bioturbation, and in the case of earthworms and other soil organisms, during "ingestion comminution." The second principle, articulated by James Thorp many years ago (Soil Survey Staff, 1975, p. 21) is that whenever loose soil is bioturbationally heaped onto the surface and exposed to rainwash and or snowmelt, the finer particles are washed away—elutriated, which imparts a coarser texture to the surface.

Biomantle processes have operated geologically since burrowing life forms first evolved, likely in the pre-Cambrian. Understanding what they do on landscapes today will help us infer what they did in the past, and more accurately interpret the geologic record.

AN UPDATED MIMA MOUND MAP OF NORTH AMERICA

Mima-like mounds, and smaller bioturbational heaps and spoil that resemble them, were once ubiquitous over North America (light-colored areas of Fig. 11), and still are in uncultivated lands. With exceptions in Illinois and Wisconsin, Mima-type mounds (dark brown color) are unknown east of the Mississippi River. Shaded zones of Figure 11 indicate a transition between the two mound types. Note the abrupt boundary along the central-lower Mississippi River, coincident with eastern faunal limits of two major mound-makers, *Atta texana* and *Geomys bursarius*. Inset map of California Channel Islands shows Mima-like mounds

Figure 9. New Mima mounds that have historically "sprouted" on land cultivated into the 1970s and 1980s, Carrizo Plain National Monument, San Luis Obispo County, California. Photo A: taken from the summit of a small basalt-capped conical hill off Soda Lake Road, the main northwest-southeast road in the monument. View is to the northwest. The giant kangaroo rat, GKR (*Dipodomys ingens*), is primarily implicated in producing and maintaining many or most mounds, called "precincts." But, some mounds are co-inhabited, and thus co-produced and co-maintained, by Botta pocket gophers (*T. bottae*). Photos B and C taken along Soda Lake Road several kilometers north of conical hill. Views are southwest, with Caliente Range in the distance. Mounds are roughly circular shaped and have irregular perimeters, and uneven bumpy surfaces that resemble large grass-covered soil heaps. Photo B displays fresh bioturbated pocket gopher heaps; mound in photo C shows multiple "comet-like" GKR entrances, invariably sprinkled with fecal pellets. Photos courtesy D.N. Johnson.

(tan color) on one single island, Santa Catalina—the only one inhabited by a fossorial rodent (*Spermophilus beecheyi nesioticus*; cf. Johnson and Johnson, 2010). Map is based on the historical record (literature) and on the personal multidecadal observations of volume contributors D.L. and D.N. Johnson.

THE DEEP, EXTENSIVE, AND UNEVEN SOIL MOUND LITERATURE

Aronow (1978, 1988) lamented, understandably—as have others, on the vastness and confusing nature of the literature on Mima-prairie mounds. Price (1950, p. 359) in apparent frustra-

tion said: "Many writers have evidently written as though the mounds were a mere scientific curiosity and have treated them lightly, rushing to give a 'solution' on the basis of hearsay, examination of photographs, or external examination of mounds in the field." Readers who examine this deep and uneven literature might agree.

Rather than comprehensively review this extensive body, which would be painstaking, tedious, and require much space, assembled selections have been included in Appendices C–F. Together with chapter references, the appendices give readers a large selection of the diverse and deep Mima-pimple-prairie-natural-soil mound literature.

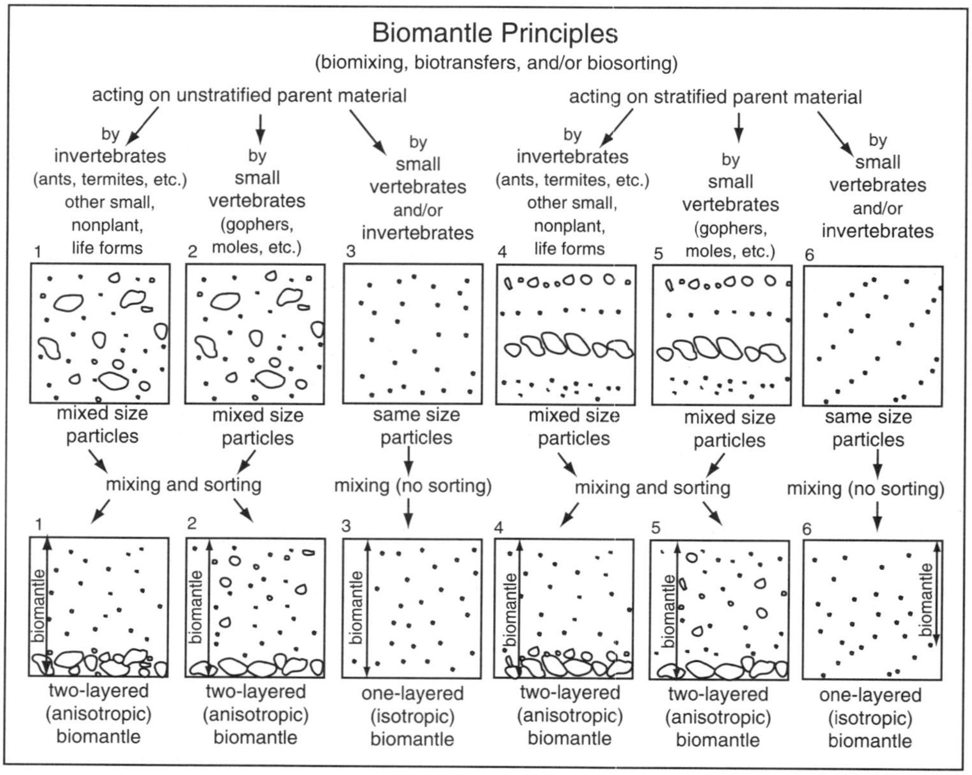

Figure 10. The "key biomantle principles" diagram, which shows the relationships between soil particle sizes and stratified sediments, and types, kinds, and sizes of bioturbators, and predicted textural profiles of soils and soil mounds after bioturbation over some period of time (see text for discussion).

APPENDICES

There are six end-volume appendices. Appendix A consists of a "patterned ground" paper that was mainly about Mima mounds written in the 1960s by geologist Roald H. Fryxell, which, while never published was nevertheless widely circulated and cited by various Pacific Northwest mound researchers (Fryxell, 1964). Appendix B consists of lengthy excerpts of Mima mound papers written nearly a century earlier in the 1870s, by geologists Joseph LeConte and Grove Karl Gilbert (LeConte, 1874, 1875, 1877; Gilbert, 1875). All three distinguished scientists by dint of their persuasive personalities, and their writings, played key and important roles in establishing different strands of genetic thought about the origin of Mima mounds.

Appendix C is an alphabetized selection of the profusion of names and synonyms that have been used to describe Mima mounds. In this regard, every name carries a concept, and thus a message to the reader. It is assumed that the multitude of names and synonyms used to describe mounded landscapes has played a role in creating an aura of confusion and uncertainty that has surrounded Mima mounds. The list provides interested readers access to a wide author-identified sampling of mound literature.

Appendix D is a timeline selection of a wide range of key mound papers, with a synopsis of the author's view or theory on how mounds form, or have formed. The list is expanded and updated from one in Huss (1994). Many of the papers focus on specific areas, but others are more broadly regional or general.

If the author(s) is (are) uncertain as to genesis, uncertainty is indicated. The purpose of the list is to recapitulate the key mound literature in a concise, coherent, and tabular way, but still convey to readers the diversity of genetic views on mound origins which have contributed to the controversy surrounding Mima mounds. As in Appendix C, it also provides readers access to the genetic literature.

Appendix E lists references cited in Appendices C and D. It is not a complete list of mound- related papers, but it is substantial.

Appendix F is an alphabetical list of masters' and doctoral theses produced in North America, at least those of which the editors are aware. There are 48 entries, 34 masters' and 14 doctoral theses. Because the subject is a popular theme for graduate study and likely will remain so, each entry identifies author, date, thesis title, thesis granting institution, each students' major advisor, and a précis of major points and or conclusions drawn. The purpose is to alert readers, especially students and their advisors, to most of the mound-linked and mound-focused theses that exist to date on the subject.

LAST WORDS

The contributors to this volume, plus those whose works and views are highlighted in the appendices—like scientists generally—execute research by means of preferred philosophies, world views, methodologies, research strategies, and explanatory frameworks, the contemplated usefulness of each must vary

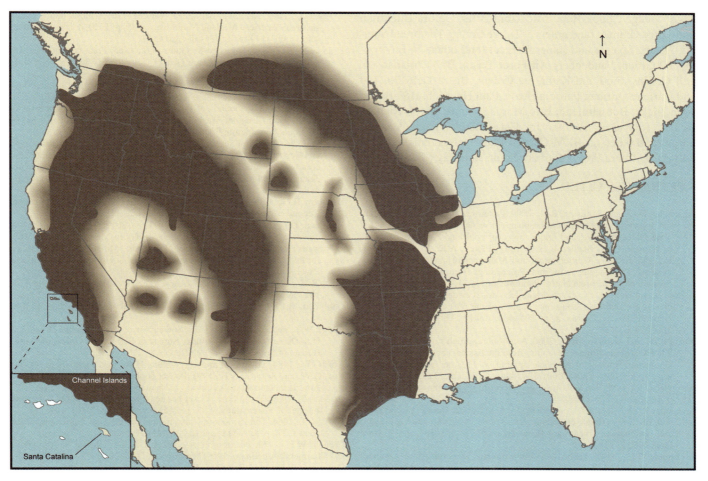

Figure 11. Generalized map of Mima mound distributions in North America. Dark brown denotes areas where *fixed activity centers*, or *Mima-type mounds*, are known to be present and/or commonly found. Light color denotes areas where *unfixed activity centers*, or *Mima-like mounds*, are either present and/or commonly found. Neither mound type is exclusive to its mapped area. (Notably, unlike Santa Catalina, on which mound-makers [ground squirrels] and Mima-like mounds both occur, other California Channel Islands lack mound-makers, and thus—predictively—mounds.) Updated from maps in Washburn (1988), Cox (1984), Price (1949), from personal experience, and from information contained in this volume. (See text for discussion.)

between us all. Such research diversity is partly what moves science forward. Approaches employed and/or formulated by contributors to this volume will, we confidently predict, expand and strengthen our explanatory paradigm in the interconnected fields of archaeology, ecology, geomorphology, soils, and earth surface process studies.

But, will the views, arguments, and evidence assembled in this volume put to rest the long controversy that has swirled around the subject? For some, probably not, but for others it might, and for them, if it does " . . . it will stop a great amount of conjecture and give rest to puzzled brains, for every person who sees such land wonders how it ever got that way" (Whitney, 1921).

ACKNOWLEDGMENTS

We especially acknowledge Diana N. Johnson who contributed in almost every aspect to the success and production of this volume, including thoughtful editing of every chapter. We also acknowledge Charles Fredericks and Rolfe Mandel who, with co-editor Don Johnson, collaborated in conceiving, arranging, and executing the "Origin of Mima mounds . . ." symposium and associated field trip at the 2008 Geological Society of America Annual Meeting in Houston, which ultimately led to this volume. We also especially acknowledge the hard work of thoughtful editing and evaluation of individual chapters by referees Joe (John) Alford, Dick Berg, Tony and Barbara Lewis, R. Bruce McMillan, Jonathan Phillips, Randy Schaetzl, Ken Schlichte, Roy Shlemon, W. Raymond Wood, and Katherine Yansa, each of whom contributed importantly to the final product. Jim Carucci, Jane Domier, Charles Higgins, Paul Heinrich, Marty Kaatz, and Don Luman aided in making available key reference and LIDAR materials. Special thanks also to librarians Lura Joseph, Jo Kibbee, Kathy Danner, their colleagues, and other librarians at the University of Illinois, and Lee Walkling and other librarians in Washington State, Arkansas, California, Louisiana, Oregon, Texas, and Wyoming who assisted in many ways. We also acknowledge Chuck Bancroft, Albin Bills, Delora Buckman and Terry Ommen (and others at Tulare County Historical Society), Larry Spanne and James Carucci (and others in Environmental Group, Vandenberg Air Force Base), Bob Christiansen, Bill Collins, Fred and Katharine Colvin, Bruce Delgado, Ted and Darlene Dittmer, Patricia Healy, Paul Heinrich, B.F. Hicks, John Hicks, Bob Holland, Eric Morgan, James Nardi, Richard Reider, Richad E. Riefner Jr., Cary Stieble, Larry Vredenburgh (and others at BLM-Carrizo Plain in Bakersfield), Debi Salvestrin, Laura Prugh, Barbara Soulé, and Ray (Skip) Soulé.

REFERENCES CITED

Alexander, D.G., and Schlising, R.A., 2000, Patterns in time and space for rare macroinvertebrates and vascular plants in vernal pool ecosystems at the Vina Plains Preserve, and implications for pool landscape management, *in* Witham, C.W., Bauder, E.T., Belk, D., Ferren, W.R., Jr., and Ornduff, R., eds., Ecology, Conservation, and Management of Vernal Pool Ecosystems: Sacramento, California, Proceedings from a 1996 Conference: Sacramento, California Native Plant Society, p. 71–75.

Alford, J., and Holmes, J.C., 1985, Meander scars as evidence of major climate change in southwest Louisiana: Annals of the Association of American Geographers, v. 75, no. 3, p. 395–403, doi:10.1111/j.1467-8306.1985.tb00074.x.

Allen, B.D., and Anderson, R.Y., 2000, A continuous, high resolution record of late Pleistocene climate variability from the Estancia basin, New Mexico: Geological Society of America Bulletin, v. 112, no. 9, p. 1444–1458, doi:10.1130/0016-7606(2000)112<1444:ACHRRO>2.0.CO;2.

Amundson, R., 1998, Do soils need our protection?: Geotimes, v. 43, no. 3, p. 16–20.

Anderson, R.Y., Allen, B.D., and Menking, K.M., 2002, Geomorphic expression of abrupt climate change in southwestern North America at the glacial termination: Quaternary Research, v. 57, p. 371–381, doi:10.1006/qres.2002.2323.

Archeuleta, T.E., 1980, Analysis of mima mounds in northeastern Arkansas [unpublished M.S. thesis]: New Orleans, Louisiana, University of New Orleans, 59 p.

Arkley, R.J., 1948, The Mima mounds: The Scientific Monthly, v. 66, no. 2, p. 175–176.

Arkley, R.J., and Brown, H.C., 1954, The origin of Mima mound (hog-wallow) microrelief in the far western states: Proceedings—Soil Science Society of America, v. 18, no. 2, p. 195–199, doi:10.2136/sssaj1954.03615995001800020021x.

Aronow, S., 1963, Internal characteristics of pimple (prairie) mounds in southeast Texas and southwest Louisiana, *in* Abstracts for 1962: Geological Society of America Special Paper 73, 106 p.

Aronow, S., 1968, Landscape deterioration and the origin of pimple mounds: A paper presented to the Soil Survey Technical Work-Planning Conference, April 1968: College Station, Texas, Texas A&M University.

Aronow, S., 1976, Geology, *in* Mills, J.F., and Wilding, L.P., eds. and compilers, Morphology, classification, and use of selected soils in Harris County, Texas: Department of Soils and Crop Sciences, Departmental Technical Report 76-48: College Station, Texas, Texas A&M University, p. 6–17.

Aronow, S., 1978, Day two stop descriptions, *in* Kaczorowski, R.T., and Aronow, S., eds., The Chenier Plain and Modern Coastal Environments, Southwestern Louisiana and Geomorphology of the Pleistocene Beaumont Trinity River delta plain: Houston, Texas, Field trip guidebook, Houston Geological Society, 87 p.

Aronow, S., 1988, Hackberry salt dome and pimple mounds, *in* Birdseye, R.U., and Aronow, S., eds., Late Quaternary Geology of Southwestern Louisiana and Southeastern Texas: Guidebook for south-central Friends of the Pleistocene Sixth Field Conference, March 25–27: p. 13–25; part 1, 63 p.; part 2, 19 p.

Best, T.L., 1972, Mound development by a pioneer population of the Banner-tailed kangaroo rat, *Dipodomys spectabilis baileyi* Goldman, in eastern New Mexico: American Midland Naturalist, v. 87, no. 1, p. 201–206, doi:10.2307/2423893.

Bragg, D.C., 2003, Natural presettlement features of the Ashley County, Arkansas area: American Midland Naturalist, v. 149, p. 1–20, doi:10.1674/0003-0031(2003)149[0001:NPFOTA]2.0.CO;2.

Branson, F.A., Miller, R.F., and McQueen, I.S., 1965, Plant communities and soil moisture relationships near Denver, Colorado: Ecology, v. 46, p. 311–319, doi:10.2307/1936334.

Braun, S.E., 1985, Home range and activity patterns of the giant kangaroo rat, *Dipodomys ingens*: Journal of Mammalogy, v. 66, no. 1, p. 1–12.

Bretz, JH., 1913, The mounds of the Vashon Outwash, *in* Glaciation of the Puget Sound Region: Washington Geological Society Bulletin No. 8, Olympia, Washington, Frank M. Lamborn, Printer, p. 81–108.

Brezina, D.N., 2007, Soil Survey of Padre Island National Seashore, Texas, Special Report: U.S. Department of Agriculture-Soil Conservation Service in cooperation with U.S. Department of Interior-National Park Service and Texas Agricultural Experiment Station: Washington, D.C., U.S. Government Printing Office, 275 p.

Bromley, R.G., 1996, Trace Fossils, Biology, Taphonomy, and Applications (second edition): London, Chapman and Hall, 351 p.

Burnham, J.H., Johnson, D.L., and Johnson, D.N., 2012, this volume, The biodynamic significance of double stone layers in Mima mounds, *in* Horwath Burnham, J.L., and Johnson, D.L., eds., Mima Mounds: The Case for Polygenesis and Bioturbation: Geological Society of America Special Paper 490, doi:10.1130/2012.2490(04).

Cain, R.H., 1974, Pimple mounds: A new viewpoint: Ecology, v. 55, p. 178–182, doi:10.2307/1934633.

Campbell, M.R., 1906, Natural mounds: The Journal of Geology, v. 14, p. 708–717, doi:10.1086/621357.

Carter, W.T., and Patrick, A.L., 1919, Soil survey of Bryan County, Oklahoma: U.S. Department of Agriculture, Bureau of Soils, Field Operations of the Bureau of Soils for 1914, 16th Report, p. 2165–2212.

Cox, G.W., 1984, Mounds of mystery: Natural History, v. 93, no. 6, p. 36–45.

Cox, G.W., and Scheffer, V.B., 1991, Pocket gophers and Mima terrain in North America: Natural Areas Journal, v. 11, no. 4, p. 193–198.

Criddle, S., 1930, The prairie pocket gopher, *Thomomys talpoides Eufescens*: Journal of Mammalogy, v. 11, p. 265–280, doi:10.2307/1374147.

Cross, C.I., 1964, The Parramore Island mounds of Virginia: Geographical Review, v. 54, p. 502–515, doi:10.2307/212979.

Crouch, W.E., 1933, Pocket gopher control: U.S. Department of Agriculture Farmers' Bulletin no. 1709: Washington, D.C., 20 p.

Dalquest, W.W., and Scheffer, V.B., 1942, The origin of the mounds of western Washington: The Journal of Geology, v. 50, no. 1, p. 68–84, doi:10.1086/625026.

Dalquest, W.W., and Scheffer, V.B., 1944, Distribution and variation in pocket gophers, *Thomomys talpoides*, in the state of Washington: American Naturalist, v. 78, p. 308–333, doi:10.1086/281203.

Ekdale, A.A., Bromley, R.G., and Pemberton, S.G., 1984, Ichnology, trace fossils in sedimentology and stratigraphy: SEPM (Society of Economic Paleontologists and Mineralogists) Short Course no. 15: Tulsa, Oklahoma, 317 p.

Eldridge, D.J., 2004, Mounds of the American Badger (*Taxidea taxus*): Significant features of North American shrub-steppe ecosystems: Journal of Mammalogy, v. 85, no. 6, p. 1060–1067, doi:10.1644/BEH-105.1.

Ensor, H.B., Aronow, S., Freeman, M.D., and Sanchez, J.M., 1990, An archeological survey of the Proposed Greens Bayou Regional Storm Water Detention Facility, Greens Bayou, Harris County, Texas: College Station, Texas, Archaeological Surveys no. 7, Archaeological Research Laboratory, Texas A&M University, 137 p.

Fey, M., 2010, The Soils of South Africa: Cambridge, UK, Cambridge University Press, 287 p.

Fields, R.C., Godwin, M.F., Freeman, M.D., and Lisk, S.V., 1986, Inventory and Assessment of Cultural Resources at Barker Reservoir, Fort Bend and Harris Counties, Texas: Prewitt & Associates, Consulting Archeologists, Inc. (Austin), Reports of Investigations 40, 176 p.

Finney, F.A., 2012, this volume, Natural prairie mounds of the Upper Midwest: Their abundance, distribution, origin, and archaeological implications, *in* Horwath Burnham, J.L., and Johnson, D.L., eds., Mima Mounds: The Case for Polygenesis and Bioturbation: Geological Society of America Special Paper 490, doi:10.1130/2012.2490(05).

Foster, J.W., 1973, Prehistoric Races of the United States (second edition): Chicago, S.C. Griggs and Co., 315 p.

Foster, Z.C., and Moran, W.J., 1935, Soil survey of Galveston County, Texas: U.S. Department of Agriculture Bureau of Chemistry and Soils, in cooperation with Texas Agricultural Experiment Station, Series 1930, no. 31, p. 1–18.

Fowke, G., 1910, Antiquities of Central and South-Eastern Missouri: Bureau of American Ethnology (Smithsonian Institution) Bulletin 37: Washington, D.C., Government Printing Office, 116 p.

Fowke, G., 1922, Archaeological Investigations: Bureau of American Ethnology (Smithsonian Institution) Bulletin 76: Washington, D.C., Government Printing Office, 204 p.

Frye, J.C., and Leonard, A.B., 1963, Pleistocene Geology of Red River Basin in Texas: Report of Investigations, Texas University: Bureau of Economic Geology, Report 49, 48 p.

Fryxell, R., 1964, The late Wisconsin age of mounds on the Columbia Plateau of eastern Washington: Unpublished report: Pullman, Washington, Laboratory of Anthropology, Washington State University, 16 p. (See Appendix A.)

Gilbert, G.K., 1875, Prairie mounds, in Report upon Geographical and Geological Explorations and Surveys West of the 100th Meridian: Washington, D.C., Government Printing Office, pt. 5, v. 3–Geology, p. 539–540. (See Appendix B.)

Gile, L.H., 1966, Coppice dunes and the Rotura soil: Proceedings—Soil Science Society of America, v. 30, p. 657–676, doi:10.2136/sssaj1966.03615995003000050035x.

Hall, S.A., Miller, M.R., and Goble, R.J., 2010, Geochronology of the Bolson sand sheet, New Mexico and Texas, and its archaeological significance: Geological Society of America Bulletin, v. 122, no. 11–12, p. 1950–1967, doi:10.1130/B30173.1.

Harris, G.D., and Veatch, A.C., 1899, A Preliminary Report on the Geology of Louisiana: State Experiment Station, Part 5: Baton Rouge, Louisiana, 354 p.

Heidemann, H.K., 2012, Lidar base specification version 1.0: U.S. Geological Survey Techniques and Methods, Book 11, 63 p.; http://pubs.usgs.gov/tm/11b4/ (accessed 14 September 2012).

Heinrich, P.V., 1986, Geomorphology of the Barker Reservoir area: Appendix D, in Inventory and Assessment of Cultural Resources at Barker Reservoir, Fort Bend and Harris Counties, Texas, by Fields, R.C., Godwin, M.F., Freeman, M.D., and Lisk, S.V.: Prewitt & Associates, Consulting Archeologists (Austin), Reports of Investigations, no. 40, p. 148–176.

Heinrich, P.V., 2007, The Houston Ridge: An ancient shoreline in Calcasieu Parish, Louisiana: News-Insights: Louisiana Geological Survey, v. 17, no. 1, p. 1–3.

Higgins, C.G., 1990, Editorial comment on mounds: Geology, v. 18, p. 284.

Holland, R.F., 2000a, The Flying M. Ranch: Soils, plants, and vernal pools in eastern Merced County: Fremontia, v. 27, no. 4, and 28, no. 1, p. 28–32.

Holland, R.F., 2000b, Great Valley vernal pool distribution, in Witham, C.W., Bauder, E.T., Belk, D., Ferren, W.R., Jr., and Ornduff, R., eds., Ecology, Conservation, and Management of Vernal Pool Ecosystems: Proceedings from a 1996 Conference: Sacramento, California Native Plant Society, p. 71–75.

Holland, W.C., Hough, L.W., and Murray, G.E., 1952, Geology of Beauregard and Allen Parishes: Louisiana Geological Survey: Geological Society of America Bulletin, v. 27, p. 11–68.

Holliday, V.T., 1987, Observations on the stratigraphy and origin of the Cinco Ranch mounds: Appendix 4, in The Cinco Ranch Sites, Barker Reservoir, Fort Bend County, Texas, by H.B. Ensor, Archeological Research Laboratory Report of Investigations 3: College Station, Texas A&M University, p. 275–280.

Hopkins, F.V., 1870, First Annual Report of the Louisiana State Geological Survey, in Annual Report of the Board of Supervisors of the Louisiana State Seminary of Learning and Military Academy, for Year Ending Dec. 31, 1869: New Orleans, Louisiana, A.L. Lee, State Printer, p. 77–109.

Horwath, J.L., and Johnson, D.L., 2006, Mima-type mounds in southwest Missouri: Expressions of point centered and locally thickened biomantles: Geomorphology, v. 77, nos. 3–4, p. 308–319; v. 83, p. 193–194 (errata).

Horwath, J.L., and Johnson, D.L., 2008, The biodynamic significance of double stone-layers at Diamond Grove mima moundfield, southwest Missouri: Geological Society of America Abstracts with Programs, v. 40, no. 6, p. 208.

Humphreys, G.S., and Mitchell, P.B., 1988, Bioturbation: An important pedological and geomorphological process: Abstracts, v. 1, p. 265: Sydney, Australia, 26th Congress, International Geographic Union.

Huss, J.M., 1994, Stratigraphic and sodium analyses of Laramie Basin mima-like mounds, Wyoming [unpublished M.A. thesis]: Laramie, University of Wyoming, 176 p.

Irvine, L.L.-A., and Dale, J.E., 2012, this volume, "Pimple" mound microrelief in southern Saskatchewan, Canada, in Horwath Burnham, J.L., and Johnson, D.L., eds., Mima Mounds: The Case for Polygenesis and Bioturbation: Geological Society of America Special Paper 490, doi: 10.1130/2012.2490(02).

Isaacson, J., and Johnson, D.L., 1996, The polygenetic and complex origins of "mima mounds" in the California coastal area: Perspectives on a scientific controversy: Geological Society of America Abstracts with Programs, v. 28, no. 7, p. A111.

Jain, S., and Moyle, P., eds., 1984, Vernal Pools and Intermittent Streams: Institute of Ecology Publication no. 28: Davis, California, University of California, 280 p.

Johnson, D.L., 1988, Soil Geomorphology, Paleopedology, and Pedogenesis in the Vandenberg-Lompoc Area, Santa Barbara County, California: A field guide for Soils Geomorphology Tour sponsored by the Soil Science Society of America, Annual Meetings of the American Society of Agronomy, Anaheim, California, 27 November–2 December, 79 p.

Johnson, D.L., 1990, Soil geomorphological analysis, pedogenetic interpretation, and paleoenvironmental reconstruction of selected coastal archaeological sites between the Santa Ynez River and Pedernales Point, Vandenberg Air Force Base, California, in Glassow, M., ed., Archaeological Investigations on Vandenberg Air Force Base in Connection with Development of Space Transportation System Facilities: Department of Anthropology, University of California, Santa Barbara, California, San Francisco, California, sponsored by National Park Service, Western Region Interagency Archaeological Services Branch, Appendix 2, p. A2-1 to A2-61.

Johnson, D.L., 1997, Geomorphological, geoecological, geoarchaeological, and surficial mapping study of McGregor Guided Missile Range, Fort Bliss, New Mexico, v. 1–2: U.S. Army Corps of Engineers, Fort Worth District and Fort Bliss Military Reservation Miscellaneous Report of Investigations no. 157: Plano, Texas, Geomarine, Inc.

Johnson, D.L., 1999, Badgers as large clast translocators, petrocalcic rippers, and artifact movers and shakers, in western Nearctica: Geological Society of America Abstracts with Programs, v. 31, no. 5, p. A-24.

Johnson, D.L., and Johnson, D.N., 2008, White paper: Stonelayers, mima mounds and hybrid mounds of Texas and Louisiana—Logical alternative biomantle working hypotheses: Appendix 3, in Johnson, D.L., Mandel, R.D., and Frederick, C.D., eds. and compilers, The Origin of the Sandy Mantle and Mima Mounds of the East Texas Gulf Coastal Plains: Geomorphological, Pedological, and Geoarchaeological Perspectives: Field trip guidebook, Geological Society of America Annual Meeting, Houston, Texas, 4 October 2008.

Johnson, D.L., and Johnson, D.N., 2010, Why are mima mounds and pedogenic stone lines (stonelayers) absent on the California Channel Islands?: Geological Society of America Abstracts with Programs, v. 42, no. 5, p. 365.

Johnson, D.L., and Johnson, D.N., 2012, this volume, The polygenetic origin of prairie mounds in northeastern California, in Horwath Burnham, J.L., and Johnson, D.L., eds., Mima Mounds: The Case for Polygenesis and Bioturbation: Geological Society of America Special Paper 490, doi:10.1130/2012.2490(06).

Johnson, D.L., and Lin, H., 2008, The biomantle-critical zone model: Abstracts, American Geophysical Union (AGU) National Meetings, 11–15 December, San Francisco, California: The Smithsonian/NASA Astrophysics Data System: http://adsabs.harvard.edu/abs/2006AGUFM.H11G.06J (accessed 23 January 2012).

Johnson, D.L., Glassow, M.J., and Graettinger, K.K., 1991, A field guide to the geoarchaeology of the Vandenberg-Lompoc-Point Conception area, Santa Barbara County, California, in Walawender, M.J., and Hanan, B.B., eds., Geological Excursions in Southern California and Mexico: San Diego, California, San Diego State University, p. 255–289.

Johnson, D.L., Johnson, D.N., and West, R.C., 1999, Pocket gopher origins of some midcontinental Mima-type mounds: Regional and interregional genetic implications: Geological Society of America Abstracts with Programs, v. 31, no. 7, p. A232.

Johnson, D.L., Horwath, J., and Johnson, D.N., 2002, In praise of the coarse fraction and bioturbation: Gravelly Mima mounds as two-layered biomantles: Geological Society of America Abstracts with Programs, v. 34, no. 6, p. 369.

Johnson, D.L., Horwath, J., and Johnson, D.N., 2003, Mima and other animal mounds as point-centered biomantles: Geological Society of America Abstracts with Programs, v. 34, no. 7, p. 258.

Johnson, D.L., Domier, J.E.J., and Johnson, D.N., 2005a, Animating the bio-dynamics of soil thickness using process vector analysis: A dynamic denudation approach to soil formation: Geomorphology, v. 67, no. 1–2, p. 23–46, doi:10.1016/j.geomorph.2004.08.014.

Johnson, D.L., Domier, J.E.J., and Johnson, D.N., 2005b, Reflections on the nature of soil and its biomantle: Annals of the Association of American Geographers, v. 95, no. 1, p. 11–31, doi:10.1111/j.1467-8306.2005.00448.x.

Johnson, D.L., Johnson, D.N., Horwath, J.L., Wang, H., Hackley, K.C., and Cahill, R.A., 2008, Predictive biodynamic principles resolve two long-standing topographic-landform-soil issues: Mima mounds and soil stone-layers: Geological Society of America Abstracts with Programs, v. 40, no. 6, p. 209.

Johnson, D.L., Horwath Burnam, J., and Johnson, D.N., 2011, Historic formation and re-formation of mima mounds: Geological Society of America Abstracts with Programs, v. 43, no. 5, p. 509–510.

Koons, F.C., 1926, Origin of the sand mounds of the pimpled plains of Louisiana and Texas [unpublished master's thesis]: Chicago, University of Chicago, 26 p.

Koons, F.C., 1948, The sand mounds of Louisiana and Texas: The Scientific Monthly, v. 66, no. 4, p. 297–300.

Kruckeberg, A.R., 1991, The Natural History of Puget Sound: Seattle, University of Washington Press, 468.

LeConte, J., 1874, On the great lava-flood of the Northwest, and on the structure and age of the Cascade Mountains: American Journal of Science (series 3), v. 7, p. 167–180, 259–267. (See Appendix B.)

LeConte, J., 1875, On the great lava-flood of the Northwest, and on the structure and age of the Cascade Mountains: Proceedings of the California Academy of Sciences, v. 5, p. 215–220.

LeConte, J., 1877, Hog wallows or prairie mounds?: Nature, v. 15, no. 390, p. 530–531, doi:10.1038/015530d0.

Malde, H.E., 1964, Patterned ground in the western Snake River Plain, Idaho, and its possible cold climate origin: Geological Society of America Bulletin, v. 75, p. 191–208, doi:10.1130/0016-7606(1964)75[191:PGITWS]2.0.CO;2.

McGuff, P.R., 1973, Association of archeological material with pimple mounds: Appendix III, *in* McGuff, P.R., and Cox, W.N., A Survey of the Archeological and Historical Resources of Areas to be Affected by the Clear Creek Flood Control Project, Texas: Research Report no. 28, Texas Archeological Survey: Austin, University of Texas, p. 62–72.

McMillan, B.A., and Day, F.P., 2010, Micro-environment and plant assemblage structure on Virginia's barrier island "pimple" dunes: Northeastern Naturalist, v. 17, no. 3, p. 473–492, doi:10.1656/045.017.0308.

Murray, D.F., 1967, Gravel mounds at Rocky Flats, Colorado: The Mountain Geologist, v. 4, p. 99–107.

Neuendorf, K.K.E., Mehl, J.P., Jr., and Jackson, J.A., eds., 2005, Glossary of Geology (fifth edition): Alexandria, Virginia, American Geological Institute, p. 412. Online version: http://www.agiweb.org/pubs/glossary/index.html (accessed 22 January 2012).

Newcomb, R.C., 1952, On the origin of the Mima mounds, Thurston County region, Washington: The Journal of Geology, v. 60, no. 5, p. 461–472, doi:10.1086/625998.

Nikiforoff, C.C., 1941, Hardpan and Microrelief in Certain Soil Complexes of California: U.S. Department of Agriculture Technical Bulletin no. 745, 45 p.

Owen, D.D., 1858, First Report of a Geological Reconnaissance of the Northern Counties of Arkansas (Made During Years 1857 and 1858): Little Rock, Arkansas, Johnson & Yerkes, State Printers, 141 p.

Owen, D.D., 1860, Second Report of a Geological Reconnaissance of the Middle and Southern Counties of Arkansas Made During the Years 1859 and 1860: Philadelphia, C. Sherman & Son, Printers, 433 p.

Paton, T.R., Humphreys, G.S., and Mitchell, P.B., 1995, Soils: A New Global View: New Haven, Connecticut, Yale University Press, 213 p.

Price, W.A., 1949, Pocket gophers as architects of mima (pimple) mounds of the western United States: The Texas Journal of Science, v. 1, no. 1, p. 1–17.

Price, W.A., 1950, Origin of pimple mounds, by E.L. Krinitsky: American Journal of Science, v. 248, no. 5, p. 355–360, doi:10.2475/ajs.248.5.355.

Prugh, L.R., and Brashares, J., 2011, Partitioning the effects of an ecosystem engineer: Kangaroo rats control community structure via multiple pathways: Journal of Animal Ecology, v. 81, p. 667–678, doi:10.1111/j.1365-2656.2011.01930.x.

Reed, S., and Amundson, R., 2012, this volume, Using LIDAR to model Mima mound evolution and regional energy balances in the Great Central Valley, California, *in* Horwath Burnham, J.L., and Johnson, D.L., eds., Mima

Mounds: The Case for Polygenesis and Bioturbation: Geological Society of America Special Paper 490, doi:10.1130/2012.2490(01).

Retzer, J.L., 1946, Morphology and origin of some California mounds: Proceedings—Soil Science Society of America, v. 10, p. 360–367, doi:10.2136/sssaj1946.03615995001000C00062x.

Rice, T.D., and Griswold, L., 1904, Soil survey of Acadia Parish, Louisiana: U.S. Department of Agriculture Field Operations of Bureau of Soils for 1903, 5th Report, p. 461–485.

Saucier, R.T., 1994, Geomorphology and Quaternary Geologic History of the Lower Mississippi Valley: Volume 1: Vicksburg, Mississippi, Waterways Experimental Station, Corps of Engineers, 364 p.

Schaetzl, R.J., and Anderson, S., 2005, Soils—Genesis and Geomorphology: Cambridge, UK, Cambridge University Press, 817 p.

Scheffer, V.B., 1947, The mystery of the mima mounds: The Scientific Monthly, v. 65, no. 5, p. 283–294.

Scheffer, V.B., 1948, Mima mounds: A reply: The Journal of Geology, v. 56, no. 3, p. 231–234, doi:10.1086/625506.

Scheffer, V.B., 1954, Son exclusivos del oeste de Norteamerica los micro-monticulos de tipa Mima?: Investigaciones Zoológicas Chilenas, v. 2, p. 89–94.

Scheffer, V.B., 1956a, Weather or gopher: Natural History, v. 65, no. 5, p. 278.

Scheffer, V.B., 1956b, 1st das mikrorelief der Mima-hugel auf das westliche Nordameriko beschrankt?: Saugetierkundliche Mitteilungen Sonderdruck, v. 4, no. 1, p. 17–21.

Scheffer, V.B., 1958, Do fossorial rodents originate mima-type microrelief?: American Midland Naturalist, v. 59, p. 505–510, doi:10.2307/2422495.

Scheffer, V.B., 1960, Spatial patterns in the world of nature: Pacific Discovery, v. 13, p. 18–20.

Scheffer, V.B., 1969, Mima mounds: Their mysterious origin: Pacific Search, v. 3, no. 5, p. 3.

Scheffer, V.B., 1981, Mima prairie's mystery mounds: Backpacker, v. 4–5, p. 96.

Scheffer, V.B., 1983, Tapestries in nature: Pacific Discovery, v. 36, no. 2, p. 16–24.

Scheffer, V.B., 1984, A case of prairie pimples: Pacific Discovery, v. 37, p. 4–8.

Scheffer, V.B., and Kruckeberg, A., 1966, The mima mounds: Bioscience, v. 16, p. 800–801.

Seifert, C.L., Cox, R.T., Foreman, S.L., Foti, T.L., Waskiewicz, T.A., and McColgan, A.T., 2009, Relict nebkhas (pimple mounds) record prolonged late Holocene drought in the forested region of south-central United States: Quaternary Research, v. 71, p. 329–339, doi:10.1016/j.yqres.2009.01.006.

Shlemon, R.J., Begg, E.L., and Huntington, G.L., 1973, Fracture traces: Pacific Discovery, v. 26, p. 31–32.

Smith, H.M., and Marshall, R.M., 1938, Soil Survey of Bee County, Texas: U.S. Department of Agriculture Bureau of Chemistry and Soils, in cooperation with Texas Agricultural Experiment Station, Series 1932, no. 30, 34 p.

Soil Survey Staff, 1975, Soil Taxonomy: A Basic System of Soil Classification for Making and Interpreting Soil Surveys: U.S. Department of Agriculture Handbook 436, Washington, D.C., 754 p.

Taylor, S., 1843, Description of ancient remains, animal mounds, and embankments, principally in the counties of Grant, Iowa, and Richmond, in Wisconsin Territory: American Journal of Science, v. 44, p. 21–40.

Van Duyne, C., and Byers, W.C., 1915, Soil survey of Harrison County, Texas: U.S. Department of Agriculture Bureau of Soils, Field Operations of Bureau of Soils for 1912, 14th Report, p. 1055–1097.

Veatch, A.C., 1906, On the human origin of the small mounds of the lower Mississippi Valley and Texas: Science, v. 23, no. 575, p. 34–36, doi:10.1126/science.23.575.34-a.

Wallace, R.E., 1991, Ground-Squirrel Mounds and Related Patterned Ground along the San Andreas Fault in Central California: U.S. Geological Survey Open-File Report 91-0149, 25 p.

Washburn, A.L., 1988, Mima Mounds, an Evaluation of Proposed Origins with Special Reference to the Puget Lowlands: Report of Investigations, State of Washington Department of Natural Resources, Division of Geology and Earth Resources Report no. 29, Olympia, Washington, 53 p.

Waters, A.C., and Flagler, C.W., 1929, Origin of small mounds on the Columbia River Plateau: American Journal of Science (fifth series), v. 18, no. 105, p. 209–224.

Watson, E.B., and Cosby, S.W., 1924, Soil Survey of the Big Valley [Lassen-Modoc Cos.], California: U.S. Department of Agriculture, Bureau of Soils, Field Operations of the Bureau of Soils for Year 1920, 22nd Report, p. 1005–1031.

Webster, C.L., 1897, History of Floyd County, Iowa: Intelligencer Print (newspaper), Charles City, Iowa, Part 1, p. 14–27.

Whitney, D.J., 1921, Water made hog wallows: California Cultivator, v. 56, p. 40.

Wilkinson, M.T., Richards, P.J., and Humphreys, G.S., 2009, Breaking ground: Pedological, geological, and ecological implications of soil bioturbation: Earth-Science Reviews, v. 97, p. 257–272, doi:10.1016/j.earscirev .2009.09.005.

Williams, D.F., 1992, Geographic distribution and population status of the giant kangaroo rat, *Dipodomys ingens* (Rodentia, Heteromyidae), *in* Williams, D.F., Byme, S., and Rado, T.A., eds., Endangered and Sensitive Species of the San Joaquin Valley, California: Their Biology, Management, and Conservation: Sacramento, California, The California Energy Commission, 388 p.

Williams, D.F., and Kilburn, K.S., 1991, Dipodomys ingens: Mammalian Species, v. 377, p. 1–7, doi:10.2307/3504176.

MANUSCRIPT ACCEPTED BY THE SOCIETY 5 MARCH 2012

The Geological Society of America
Special Paper 490
2012

Using LIDAR to model Mima mound evolution and regional energy balances in the Great Central Valley, California

Sarah Reed*
Ronald Amundson*

Department of Environmental Science, Policy, and Management, University of California, Berkeley, California 94720-3114, USA

ABSTRACT

Mima mounds, often associated with vernal pools, have historically been shrouded in genetic uncertainty. Nevertheless, emerging from the array of explanations proposed, a biological mechanism for mound formation has steadily gained strength. We use innovations in remote sensing and geomorphic modeling to develop a new approach to evaluate the microtopography. Using a digital elevation model created from LIDAR (light detection and ranging) data, morphometric values—average mound diameters, heights, slopes, and curvatures—were calculated across an 18 km^2 sector of a mound-pool region that covers an ancient river terrace near Merced, California. The terrain information was applied to a sediment transport model to estimate mound erosion and swale deposition rates. The mean net erosion rate was 38 cm kyr^{-1}, using a diffusion coefficient of 50 cm^2 yr^{-1}. At steady state, erosion must be balanced by a restorative upslope transport, and this estimate of erosion is comparable to observed rates of sediment mounding via pocket gopher burrowing (61 cm kyr^{-1}). These data suggest that bioturbation may play a dominant role in maintaining Mima mound terrain. LIDAR measurements were also used to develop a model that approximates the energy required for the formation of Mima mounds (shearing, pushing, and uplifting soil) and their maintenance (counteractions to erosion). This energy estimate was compared to estimates of energy available to gopher populations in the region. Our results indicate that gophers have ample energy to build typical Mima mounds in as little as 100 years, thus strongly supporting a biotic mechanism of Mima mound development and maintenance.

INTRODUCTION

The Pleistocene and Pliocene landscapes of the Great Valley of California support an assemblage of peculiar and compelling topographic features (Fig. 1). The gently sloping alluvial fans and stream terraces in the region are, in many locations, continuously mantled with a multitude of low mounds and intervening depressions. Where slopes are low, the depressions fill with water during winter and spring months, and become vernal pools, which

support an endemic and endangered suite of plant and animal species (Holland, 1978; Zedler, 1987). These microtopographical features once covered several million acres in California, but have been largely eradicated through conversion to agriculture and urban land uses (Holland, 1978; Trust and Holland, 2009). The remaining tracts, particularly large expanses of rangeland near Merced, California, are protected by the Endangered Species Act (16 U.S. Code 1531 et seq. [2000]) and, as a result, are at the center of an intensive regulatory and conservation effort (e.g., Witham et al., 1998; U.S. Fish and Wildlife Service, 2004; Schlising and Alexander, 2007). Yet, despite their biologic and

---

*E-mails: s_reed@berkeley.edu, earthy@berkeley.edu.

Reed, S., and Amundson, R., 2012, Using LIDAR to model Mima mound evolution and regional energy balances in the Great Central Valley, California, *in* Horwath Burnham, J.L., and Johnson, D.L., eds., Mima Mounds: The Case for Polygenesis and Bioturbation: Geological Society of America Special Paper 490, p. 21–41, doi:10.1130/2012.2490(01). For permission to copy, contact editing@geosociety.org. © 2012 The Geological Society of America. All rights reserved.

Figure 1. Photographs showing the Merced-area Mima mound land-scape. (A) A wide view of the mounds on the ca. 3 m.y. old Laguna formation. B details the linear alignment on hillslopes. In C, cobbles fill the intermound regions and have been shown to form a continuous layer extending underneath the mounds. (D) The seasonal wetlands, called vernal pools, which are commonly found in the intermound swales in the winter and spring. Cattle ranching is a common land use practice, especially on the older terraces. Here, the cattle are grazing on the 1–3 m.y. old North Merced Gravels formation. The Sierra Nevada mountain range is visible in the background.

historical importance, questions remain about the details of why and how these features formed, how they are maintained, and the role of burrowing animals in the mounds' evolution.

Arkley and Brown (1954) studied the mounds on the alluvial terraces near Merced and concluded that a hypothesis proposed by Dalquest and Scheffer (1942), which attributes the microtopography to the work of fossorial (burrowing) rodents, mainly pocket gophers (Rodentia: Geomyidae, hereafter referred to as "gophers"), is "undoubtedly correct." A key element of the Dalquest-Scheffer model is that the existence of either restrictive hardpan layers (caused by long-term changes in vadose zone hydrology) or coarse gravel layers compels burrowing animals, being unable to dig downward, to escape wet and/or shallow soil and minimize predation by burrowing laterally. Lateral burrowing preferentially moves soil such that more soil is transported toward the animal's nest than away from it, and it is in this way that biologically created mound and swale topography is proposed to occur. Despite Arkley and Brown's conviction, questions remain about the origin of the Merced and similar microtopography, questions that lie at the heart of several historic genetic uncertainties. The Mima mound type-locality is found in Washington State, but many other mound regions similar in form and distribution are found in the grasslands and open forests of North America and elsewhere around the world. We note that while it has not been shown conclusively that the different mound regions are related, in this study we discuss the Merced-area mounds in the context of other mound regions because of their potential genetic associations. For simplicity, we refer to all similar microtopography, like "pimple," "prairie," and similarly named mounds, as Mima mounds, while noting that differences of origin and environment may exist between different mounded terrains.

The origin of Mima mounds has long been at the center of a lively debate (e.g., Washburn, 1988; Berg, 1990), with many, diverse explanations proposed for the mounds since their discovery (Cox, 1984a). Of these suggested mechanisms for formation, five general groupings of viable hypotheses exist: erosional, depositional, shrink-swell, seismic, and biotic. Washburn (1988) conducted an analysis of the most feasible hypotheses and deduced that an erosion-vegetation anchoring hypothesis was most consistent with the evidence for the majority of mound sites in the United States. It is clear from Washburn's writings that he was generally not supportive of the idea that small burrowing animals, like gophers, constructed the mounds. He pointed out that much of the evidence in support of gopher-built mounds suggests that the animals may have simply used preexisting topography to their advantage. Further, he outlined the arguments of several authors who proposed that typical gopher burrowing patterns would accelerate mound erosion rather than lead to the pimpled topography (Hubbs, 1957; Nikiforoff, 1941; Paeth, 1967). Despite these arguments, there is growing support for a biological interpretation. For example, Cox (1984a, 1984b, 1990), Cox and Allen (1987), and other workers (Horwath and Johnson, 2006; Reed and Amundson, 2007) have conducted studies that support the proposal for burrowing animal-built mounds.

The possibility that small burrowing animals, in response to slowly changing soil conditions that restrict burrowing and promote periodic soil wetness, have formed these unique features is particularly relevant given the increasing awareness of the importance of life on geomorphic and soil processes (e.g., Butler, 1995; Johnson et al., 2005; Meysman et al., 2006; Dietrich and Perron, 2006). Additionally, new geomorphic tools and models make it possible to quantitatively test hypotheses of Mima mound origin in novel ways. In this context, we revisit the issue of Mima mound landscapes, with a focus on the Merced-area mounds, using recent advances in remote sensing technology and geomorphic transport theory. This work extends our previous plot observations (Reed and Amundson, 2007) to the landscape scale using LIDAR (light detection and ranging) topographic data of Mima mounds and their intervening vernal pools (herein we use "pools" synonymously with "swale," regardless of water status, to refer to the intermound depressions). The high-resolution LIDAR data allow us to precisely quantify the morphology of the mounds and to better investigate the processes responsible for their formation and maintenance. Furthermore, the analysis is inspired by recent studies (e.g., Yoo et al., 2005; Phillips, 2009) that have included quantitative considerations of biological energy within geomorphic models. Our work thus includes conservation of energy analyses to assess whether populations of burrowing mammals have produced these features.

METHODS

Study Site

An early record of California Mima mound topography was written by Hilgard (1884), who noted that certain landscapes of the Central Valley

are dotted with the singular rounded hillocks, popularly known as 'hog-wallows,' from 10 to 30 feet in diameter and from 1 to 2 feet high, which are evidently the result of erosion, but precisely under what conditions it is difficult to explain. These hillocks are usually most abundant near the foothills. … They occur on all kinds of soil, and even on the rolling foothill lands themselves. … In some cases they are so thickly set, abrupt, and resistant as to render the land valueless for ordinary cultivation.

Eastern Merced County harbors one of the largest remaining tracts of the landscape that Hilgard described, located in the watershed of the Merced River (Fig. 2). The study area, bordered on the east by the Sierra Nevada mountain range and on the west by the Central Valley, consists of a series of dissected river terraces and alluvial fans, formed from glacial outwash from the Sierra Nevada (Marchand, 1976). The terraces comprise a chronosequence of soils, and range in age from a few hundred years (the active floodplain of the Merced River) to the several million-year-old China Hat member of the Laguna formation (Marchand, 1976). The fluvial sediments consist primarily of granitic alluvium, and the soils are classified as Entisols on the lower, younger terraces and as Alfisols on the higher, older ter-

races (Arkley, 1962). On the older surfaces, claypans or duripans (associated with restrictive *Bt* and *Bqm* soil horizons, respectively) have formed (Arkley, 1962; Harden, 1982).

The climate of the region is Mediterranean, and it receives a mean 310 mm of precipitation annually, 90% of which falls between November and April. During the summer, the mean daytime temperature is 37 °C, while the winter average is 7 °C (Arkley, 1962; National Oceanic and Atmospheric Administration, 2009 [Station 045535]). The vegetation is dominated by European annual grasses and forbs (e.g., *Bromus* spp.; *Hordeum* spp.), which largely overtook the native perennial bunchgrass community (e.g., *Poa secunda* spp. *secunda*; *Aristida oligantha*) (Heady, 1988). In addition, due to the seasonal influx of rainfall and the development of restrictive soil layers, ephemeral wetlands support a variety of terrestrial and aquatic plant and animal species (e.g., *Lepidurus packardi*; *Neostapfia colusana*) (Laabs et al., 2002; Dittes and Guardino, 2002). Burrowing rodents are abundant, particularly the California ground squirrel (*Spermophilus beecheyi*) and the Botta pocket gopher (*Thomomys bottae*) (Laabs and Allaback, 2002) (Fig. 3). Cattle ranching and agriculture are the most common land use practices in the region (Robins and Vollmar, 2002) (Fig. 1D).

Field Observations

Baisden et al. (2002) dug a trench through a Mima mound on the Laguna Formation (latitude 37.55° N; longitude 120.44° W; elevation = 220 m) using a backhoe. The trench was left open after the research concluded. For this study, the outer 6–12 inches of material were removed from the trench face in order to re-expose and examine the profile of a typical mound. In addition, the adjacent intermound region was excavated, using shovels, to the depth of the apparent water-restrictive layer. Preliminary descriptions of the soil profile horizons were made based on visual observations and manual texturing following procedures recommended by Schoenenberger et al. (2002).

In 2007, Reed and Amundson measured the aboveground evidence of gopher bioturbation (number and average mass of their tailings heaps—characteristically, small semicircular piles of loose soil which cover burrowing openings [Grinnell and Storer, 1924]) of a 3507 m² region on the Laguna formation (latitude 37.40° N; longitude 120.39° W; elevation 520 m). For this study, the aboveground extrusions were measured one additional time, and the average rate of aboveground soil movement (cm/1000 years—assuming that tailings pile production occurred in a 6 month period, based on last-recorded rainfall) was calculated by combining our 2007 and later (this study) data. (We report on the results of our additional gopher survey in the Modeling section.)

LIDAR Characterization of Mounded Terrain

The advent of commercially available LIDAR technology, which uses lasers to measure distance from a laser source to a target, has provided unprecedented resolution and accuracy in

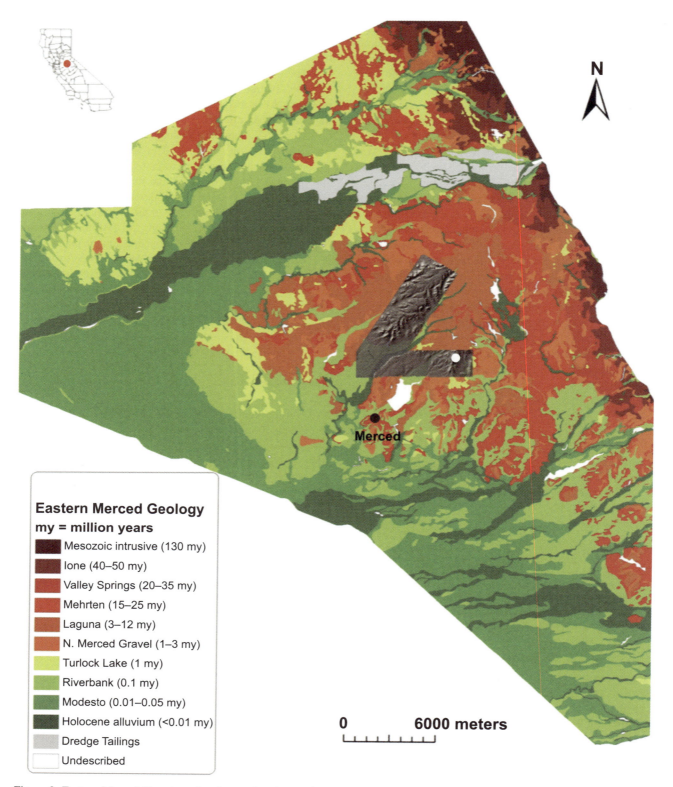

Figure 2. Eastern Merced County regional map showing geologic surfaces as mapped by Marchand (1976) and the outline of the LIDAR survey area (totaling 30 km²). Terrace age generally increases from west to east, with younger terraces in green and older in red. The LIDAR survey area is centered in the Yosemite Lake USGS (United States Geological Survey) Quadrangle, at latitude ~37.42° N and longitude 120.42° W. The gopher survey is denoted by a white circle and is located in the lower right-hand corner of the LIDAR survey.

Figure 3. (A) Typical tailings piles from pocket gopher (*Thomomys bottae*) tunneling in the region. An average gopher tailings pile is ~15 cm in diameter. (B) A tailings heap created by ground squirrels (*Spermophilus beecheyi*), which normally leave their burrows open (as opposed to the closed openings of gophers). The long axis of the ground squirrel tailings pile shown is 75 cm.

determining surface profiles (Slatton et al., 2007). This technology is especially useful for remote sensing of Mima mounds, which cannot be adequately resolved with other remote sensing technology such as the National Elevation Data set (resolution = 3–30 m).

We used high-resolution topographic data, generated by a 30 km^2 airborne LIDAR survey, to analyze the mounded terrain. In September 2006, the National Center for Airborne Laser Mapping (NCALM) conducted a laser altimetry survey of a region of mounds ~10 km northeast of the city of Merced, California (Fig. 2). The data (resolution = 1 m) were collected with an Optech 1233 airborne laser terrain mapper (ALTM) mounted in a Cessna 337. The scan angle was a constant ±20° and the scan frequency was 28 Hz for the duration of the flight. Two GPS (global positioning system) ground stations were used as reference stations for the flight. No filtering was performed on the data set given the absence of large vegetation or other obstructions. The data was gridded by kriging using Surfer software v.8 and 1 m grid spacing.

The data set was classified by geological formation using a digitized version of Marchand's (1976) 1:24,000 geologic maps of the region. We focused our study on the oldest alluvial terrace, the Laguna formation, on which the most developed mounds reside. The study region encompasses a range of elevations in land surface (~60–250 m above sea level), due to regional slope and postdepositional incision, even within this one geologic forma-

tion. This variation hinders morphometric analysis of the mounds (i.e., the large-scale topographic changes can obscure the microtopography). To resolve this issue, a 20×20 m moving window filter was used to smooth out the low-frequency signals and accentuate the mounded features. This made it possible to analyze the mounds across the survey region without obfuscation by the underlying topography. All subsequent terrain data were collected from the filtered data set.

Attempts to automatically identify mounds (e.g., using object-oriented image analysis [Hay and Castilla, 2006]), yielded relatively low-accuracy (<65%) results; therefore, all mound parameters reported for this study were measured on mounds and swales that were manually identified. For mound size and shape measurements, 80 points were randomly generated (using ArcGIS) on the Laguna formation (relatively level terrain only— local slope <15%), and the mound and pool nearest to the point were digitally delineated and measured using Surfer software. Average mound and pool slope and curvature (negative Laplacian) values were calculated, based on values at 10 locations evenly distributed on each of the 80 sample mounds and pools.

If gopher transport of soil formed the mounds, the intermound and mound volumes should be roughly equal (assuming the original surface was relatively level) because gophers necessarily moved sediment from intermound areas in order to build the mounds. To test this notion, mound dispersion (variation in intermound distance) was measured. Fifty-two points were randomly located (on slopes <15%), and one large mound and one small mound adjacent to a given point were identified. Then the distances from each mound to all directly adjacent mounds were measured (mound apex to mound apex) and intermound distances were compared with mound diameters. Mound-pool volumetric relationships were assessed by randomly locating 20 mounds and using Surfer to calculate the volume of each mound and the pool nearest to it. In several cases, the pools were not distinctly closed. In those instances, the volume of the depression reaching to the midpoint of all adjacent mounds was measured. Finally, to document the differences in mound morphology with underlying slope, elongation (ratio of semimajor [a] and semiminor [b] axes) was measured for 10 mounds on each of five regional slope classes.

Modeling Mima Mound Development

According to the argument for rodent-built mounds, a restrictive stratum must exist in the soil. This layer hinders vertical transport and prompts fossorial rodents to construct the mounds in an effort to secure protection from predators in the thin soils and/or to avoid a perched water table or low permeability conditions. In many cases, the obstructive layer is a consequence of long-term soil formation processes. These layers include argillic horizons and duripans, which are present in the study area.

Harden (1982) found that the soils of the Merced River chronosequence show consistent changes in soil development with landscape age, particularly an increase in clay in the B horizons. Fine-textured subsoils generally have lower permeability, leading to soil water retention and, potentially, to saturation or perched water tables. Gophers spend nearly the entirety of their lives in intimate contact with the soil (Ingles, 1951) and must be keenly sensitive to differences in permeability. Gophers have high rates of gas exchange because of their thermoregulation requirements (McNab, 1966). As soil textures become finer, air exchange decreases, and gophers would likely need to make behavioral or physiological adaptations to compensate. For instance, McNab (1966) observed that in areas cohabitated by large and small gophers, the small gophers were relegated to the shallower soils due to the greater gas exchange requirements of the larger gophers, even though smaller gophers would choose deeper soils in areas without competition for habitat. In addition, McNab (1966) concluded that increased burrowing by gophers during wet periods (Miller, 1948; Ingles, 1949; Hansen and Morris, 1968; Cox and Hunt, 1992) could be a response to decreased air exchange in burrows, due to decreased permeability. Furthermore, in a simulated gopher environment, Romañach et al. (2007) found that gopher survival decreased significantly as soil clay content increased. Thus, soil permeability is of particular importance to gophers because the gas required by fossorial rodents must be derived via soil diffusion.

How, then, do the gophers on the Merced River chronosequence adapt to the limitations imposed by decreased permea-

bility? Gophers are plentiful on even the oldest and most developed of the alluvial terraces (Laabs and Allaback, 2002). We require methods to test whether preferential soil movement is their response, and if so, we need to develop a detailed understanding of how gophers may work to redirect soil in order to create mounds. To approach a solution, we begin with a review of previous work, which describes and models the processes by which rodents may build the mounds.

Dalquest and Scheffer's (1942) original description of their gopher-based hypothesis was largely qualitative. They observed that gophers inhabiting mounded regions on Scott's Prairie of Western Washington (near the type locality Mima Prairie, which is not currently inhabited by gophers) regularly burrow out "exploratory" tunnels that reach roughly 10 feet out from the mound center. They assumed the mound centers are located at points that may have originally been better drained or had thicker soil profiles. Dalquest and Scheffer measured five of these active tunnels; more than half of the soil emptied from the passageways did not get transported to the surface and was thus assumed to be incorporated into the mound. According to the authors, "this shuttling of earth, long continued, accounts for the sunken contour of the inter-mound area and for the greatest part of the volume of the Mima mound." Five years later, Scheffer (1948) conceded that the model they presented did not wholly account for the building of the mounds:

In fancy, it is easy to picture the start of a Mima Mound. It is less easy to account for its growth. For reasons that may never be known, the gophers carried more dirt toward the nest than away from it. Perhaps some biologist will suggest an experiment whereby the growth of a Mima-type mound can be studied from start to finish. At present, we do not know whether the mounds on the Puget Sound and other prairies are still growing, whether they are in equilibrium with the forces tending to reduce them, or whether they are shrinking.

A group of researchers responded to this challenge by closely monitoring gopher activity in mounds near San Diego. In 1987, Cox and Allen tracked, over the course of one year, the directional movement of small iron pellets emplaced in soil plugs that were implanted into gopher burrows. They found that the net movement of the markers was largely moundward (average 41.3 cm) and slightly upward (average 4.9 cm). In a subsequent study, Cox (1990) measured the quantity of soil mined (and deposited both aboveground and belowground) in the mounded terrain and found that deposition of surface heaps per unit area was greatest at the center of the mounds, and was reduced toward the mound edge and intermound areas. Cox (1990) combined the results of these two studies to develop a model which predicts the net deposition or removal of soil at given points along the mound-intermound gradient, and for mounds with differing heights.

Horwath and Johnson (2006) proposed a qualitative model of Mima mound evolution based on observations and analyses of gravelly Mima mounds in seasonally wet Diamond Grove Prairie Preserve, Newton County of southwestern Missouri. The mounded prairie is underlain by a dense claypan, and a substantial stonelayer that, like the Merced moundfield, is exposed in seasonally wet and stony intermound swales. The authors drew on particle size analyses, included gravel fractions, and trench-exposed horizonation assessments to hypothesize how a combination of biotic, pedogenic, aeolian depositional, and erosional (surface washing) forces led to the current mound profiles, which include bimodal clay-rich horizons and two stonelayers (see Horwath Burnham et al., this volume, Chapter 4). The authors concluded that gophers, which had recently inhabited the moundfield—evidenced by gopher-sized soil krotovina (refilled burrows) and the presence of gophers in nearby counties—were the primary force in creating the polygenetic mound-intermound terrain.

These observations and models provide conceptual representations of biologically built Mima mounds, and they are supported by considerable evidence. However, Cox's empirical model only directly applies to the mounds in the San Diego region, and also does not address mound erosion. Furthermore, burrowing is a costly activity that makes up a large fraction of rodents' energy budgets (Vleck, 1979; Du Toit et al., 1985), and these models do not demonstrate that mound creation or maintenance by gophers is energetically feasible. Thus, we use energy as the metric to test the possibility that gophers build and sustain Mima mounds. The model outlined below, which combines geomorphic and biologic models and observations, is a simple representation of mound evolution, but is a step toward an analytic model that could be applied to other regions where mounded topography exists.

Diffusive Sediment Transport

In the last half century, hillslope geomorphology, building on G.K. Gilbert's (1909) conceptual model of soil-mantled hillslopes, has moved toward a quantitative, process-based theory of landscape evolution (e.g., Culling, 1960, 1963; Dietrich et al., 1995; Roering et al., 1999; Yoo et al., 2005). Based on these developments, we consider Mima mounds to be miniature, convex, soil-mantled hillslopes. However, unlike most soil-mantled hillslopes, the mounds in the Merced study area are not underlain by bedrock, but by several soil layers, which we refer to as "parent material" for purposes of description. Transport in the upper layer is assumed to be dominated by slope-dependent creep, driven by biotic disturbance (mixing and perturbation by rodents). Biogenic sediment transport is commonly modeled using a linear, diffusive transport law (Equation 1), in which sediment flux is proportional to hillslope gradient. The model was originally introduced by Davis (1892) and Gilbert (1909) and was formalized by Culling (1960, 1963), using an analogy with Fick's law of diffusion (in which flux is proportional to a concentration gradient):

$$q_s = -K\nabla z, \qquad (1)$$

where the sediment flux q_s [(length)^2 (time)$^{-1}$] is proportional to the hillslope gradient, ∇z, and K is equivalent to a diffusion coefficient with dimensions (length)^2 (time)$^{-1}$ and embodies fac-

tors such as climate, soil type, and vegetation. While the vegetation in the study ecosystem is important from both a conservation and a geomorphic perspective, we simplistically incorporate the entire effect of vegetation on erosional processes into the diffusion coefficient. We hope to develop an approach that more directly quantifies the role of vegetation on mound development and maintenance (and that includes feedbacks between gopher behavior and vegetation status) in future iterations of this modeling work. Equation 1 was applied to the Mima mounds in order to model soil movement on them. Based on studies conducted in California grasslands where, similar to the Merced region, bioturbation is the main transporter of sediment (Dietrich et al., 1995; Heimsath et al., 1997; McKean et al., 1993; Reneau, 1988), K values in the Merced region are assumed to range between 25 and 75 cm^2 yr^{-1}.

There has been considerable recent work showing that transport on soil mantled slopes may be more accurately modeled with nonlinear functions that involve some dependency on soil thickness or disturbance magnitude (Roering et al., 1999, 2001; Gabet, 2000; Roering, 2004). Thus, soil creep on mounds is likely not fully described by a simple linear model. However, the traditional creep model is used here to conduct a first-order assessment of transport with the LIDAR data. In future work, we plan to explore additional soil transport models and address nonlinear mechanisms.

In order to estimate the erosion rates of convex Mima mounds and the subsequent deposition into the concave vernal pool regions, we use the mass continuity equation (Equation 2) of soil thickness:

$$\rho_s \frac{\partial h}{\partial t} = -\rho_{pm}\frac{\partial e}{\partial t} - \rho_s \nabla \cdot q_s, \qquad (2)$$

where ρ_s and ρ_{pm} are the bulk densities of the mobile soil layer and the "parent material," respectively, h is the soil thickness (measured from the soil surface to the top of the parent material), e is the elevation of the soil-parent material interface, and q_s is the sediment transport vector. Here, $\rho_{pm}\,\partial e/\partial t$ represents the production of soil from the parent material (assumed here to be primarily due to disruption and transport of sediment by rodents) and $\rho_s\nabla \cdot q_s$ represents soil removal. If the mounds are persistent, and thus at steady-state (constant mobile layer thickness and hillslope shape, $\partial h/\partial t = 0$), then the expression (Equation 3) for landscape lowering ($E = \partial z/\partial t$):

$$E = -\left(\rho_s/\rho_{pm}\right)K\nabla^2 z, \qquad (3)$$

shows that erosion can be calculated using curvature values ($\nabla^2 z$). The ratio of the bulk density of soil in the mobile layer (the upper layer, above the parent material) and the bulk density of the parent material (ρ_s/ρ_{pm}) was assumed to be equal to 0.75, based on work by Reichman and Seabloom (2002), who stated that the density of material in gopher tailings piles is generally 10%–40% lower than the underlying consolidated soil.

Recently, Reed and Amundson (2007) used topographic data collected via a differential GPS to calculate curvature values on a small (<1 ha) area of mounds and pools, in order to use Equation 3 to determine how rapidly mounds would erode, in the absence of a preserving force. It was found that, without an active agent maintaining this topography, the landscape would be largely flattened by gravity-driven erosion in ~10^3 years (assuming minimal aeolian deposition). This initial study, conducted on a very small area, with soils and terrain characteristic of the greater study area (see Fig. 2), serves as a basis for this work.

Mima Mound Energy Balance

The mass balance approach to hillslope evolution described above, in which land surface changes are represented by mass fluxes in the form of erosion and deposition, does not directly consider the effect of biota on resultant topography. Thus, we combine the mass balance model with studies of gopher behavior and physiology in order to determine if there is sufficient energy available to the burrowing animals to produce and maintain the mounds.

Energy Requirements for Mound Building and Maintenance

We assume that the mounds are built when gophers forage centripetally (radially) outward from what were perhaps originally better-drained portions of the landscape. Such locations would likely function as nest areas, and thus become main activity centers, not only for gophers but other organisms as well. We further assume that the very act of burrowing outward from the center should result in slightly more soil being moved back to their starting point, than their distal end point. Repeating this process over multiple generations produces Mima mounds.

While the model below does not address exactly how this preferential transport occurs on a particle-by-particle basis, it does represent aggregate transport. It was constructed so as to represent a maximum of the energy required for mound building, providing a test of the feasibility of gopher built mounds on an energetic basis. To begin, the energy required to build and preserve a mound is defined:

$$E_{mound} = E_{lift} + E_{burrow} + E_{maintain}. \qquad (4)$$

The component energy terms are converted into energy density values $[EL^{-2}T^{-1}]$ and used to express biologic and geomorphic energy storage and expenditures. The definition and estimation of these terms are discussed below.

Lift energy (E_{lift}). We begin by assuming that the original terrace surfaces were approximately level and unmounded, representing the local base level. The change in potential energy required to lift sediment from below base level (in what ultimately becomes a swale) to a position above the base level (to form the mound) is determined by the following equation:

$$\Delta PE = mg\Delta h = \rho V \Delta h, \qquad (5)$$

where m is the soil mass, ρ is soil bulk density, V is soil volume, g is the gravitational constant, and Δh is the change in height of the sediment. Restated, the mound sediment is assumed to be derived from the intermound depressions, and the model allows one to calculate the change in potential energy for uplifting sediment. The mounds in the Merced region are slightly ellipsoid in shape (according to the LIDAR survey); thus by integrating over the volume of a half ellipsoid, the potential energy change involved in lifting a mound's volume of sediment against the force of gravity is

$$E_{lift} = \tfrac{1}{2}\rho g \iiint z\, dx\, dy\, dz, \qquad (6)$$

where $z = c\sqrt{\left(1 - x^2/a^2 - y^2/b^2\right)}$ (the formula for an ellipse), a is the semimajor axis of the mound, b is the semiminor axis, and c is mound height. Bulk density values ($\rho_{avg} = 1.7$ g cm^{-3}) are from Harden (1987).

Solving this integral from base level to the top of a mound yields

$$E_{lift} = \tfrac{1}{4}\pi\rho gabc^2. \qquad (7)$$

This value was simply doubled to estimate the total change in potential energy due to the lift from below base level to above base level. The actual E_{lift} would be less because when the soil volume is inverted from concave up (below base level) to concave down (above base level), the potential energy change diminishes with distance from mound center. Thus, the calculated E_{lift} represents a maximum because we are calculating the energy required to raise a concave down volume of soil from below to above base level. The energy expressed in Equation 7 (after doubling) represents the lift energy for one mound. To measure the total change in potential energy caused by soil redistribution across the survey area (J m^{-2}), Equation 7 was multiplied by the average mound concentration (mounds per area, as estimated from the LIDAR data) on the Laguna formation. In order to calculate the energy density of mound lifting work (J m^{-2} yr^{-1}), it is necessary to estimate the time required for the mound-building process. A range of time scales (10–10^4 years) is applied to the energy calculations in order to assess the likelihood of different mound-building scenarios.

Burrowing energy (E_{burrow}). Several studies have shown that shearing and pushing soil are the most energetically costly activities associated with biotic soil transport (Vleck, 1979; Du Toit et al., 1985; Seabloom et al., 2000). Vleck (1979) developed a model to estimate the energy needed to shear the soil from the face of an individual burrow (cross-sectional area approximately equal to gopher body diameter) and to push loose soil along the burrow length. He parameterized the model by measuring gopher oxygen consumption during burrowing for different distances and in different soils. In order to estimate mound-building costs, we modified Vleck's model and assumed a gopher creates multiple burrows within the swale soil, such that the sum cross section of all burrows is, over time, equal to the cross section of a mound or swale. Then, the gopher pushes that soil from the basin area to the mound area. Cox (1990) estimated, based on his observations of the direction and magnitude of gopher-driven soil movement, that the mean soil displacement distance during mound formation is 4.4 m. In order to place upper bounds on the amount of soil sheared and the distance it is subsequently pushed, the average mound diameter is used to estimate the shear and push distance parameters. This value is meant to represent an upper-limit average on the distance of soil movement, including upward movement (through surface-access tunnels) and movement from pools on all sides of a given mound. The energy required for a gopher to fully burrow out an area equal to the size of a mound cross section and transport it (ultimately building a mound) is

$$E_{burrow} = \pi \rho acd \left(K_s d + \tfrac{1}{2} K_p d \right), \qquad (8)$$

where a is the semimajor axis of the mound, c is mound height, d is the mound diameter (assumed to be equal to swale diameter), and ρ is soil bulk density. K_s represents the energy cost (J m^{-2} yr^{-1}) of shearing 1 g of loose soil, while K_p represents the energy cost of pushing 1 g of loose soil 1 m in distance. The energy required to move soil for a range of time frames (10–10^4 yr), and for both sandy loam and clay soils, was calculated. While this model clearly does not offer a detailed particle-based representation of precisely *how* gophers may move soil from basin areas to form mounds (e.g., how burrow architecture changes with time), it does represent an upper bound on the energy required to loosen and push an equivalent amount of soil from source (pool) to resultant form (mound).

Maintenance energy (E$_{maintain}$). Apart from the nuanced dynamics of how gophers might form mounds, modeling (Reed and Amundson, 2007) indicates that, assuming zero or minimal aeolian infall, the mounds would not persist long-term against erosion without a conserving force. We use the erosion rates calculated with Equations 1 and 2 to estimate the energy required to maintain mound form against erosive forces. Given an erosion rate (m yr^{-1}), the rate of release of potential energy (J m^{-2} yr^{-1}) is given by calculating the product of the erosion rate, soil bulk density, and gravitational constant (following Phillips, 2009). The geomorphic model we described neglects the change in mound form (curvature values) that would result as erosion occurs over time and thus likely overestimates the maintenance energy required.

Energy Available for Mound Building and Maintenance (E$_{gopher}$)

Using the approach outlined above, the energy required for the formation of mounds from a level surface can be estimated. The question we hope to address is, "Given the energy budget for mound formation, are biologically built mounds a possibility?" To determine whether gophers are capable of doing this work, we drew from the literature to estimate total ecosystem biological energy stores and the proportion of that energy which is partitioned to gophers. Net primary productivity (NPP) is the amount of biomass generated from photosynthesis, minus the energy

used for cellular respiration. NPP measurements from Callaway et al. (1991) (in a central California grassland ~170 km southwest of the study site) were used to estimate the total stored biological energy available in the study region. Previous studies, which have compared biotic and geomorphic processes, have used NPP as the primary biologic comparison metric. These have shown that NPP stores are many orders of magnitude greater than the energy required for most geomorphic processes (Devlin, 2003; Yoo et al., 2005; Phillips, 2009), supporting the notion of biologic influence on landscape evolution.

However, because the energy densities of NPP are generally so much greater than that of geomorphic work, it is necessary to further constrain the bioavailable energy in order to determine if gophers, in particular, have enough energy in reserve to build (and maintain) mounds. Thus, a budget of the energy available to gophers for sediment transport processes was calculated, using previous studies of gopher physiology and Merced-area field measurements, and compared to the energy required for mound building. Gettinger (1984) measured energy flux through gopher populations being fed a natural diet in the laboratory. He partitioned assimilated energy (energy consumed less energy excreted) into "existence" energy (resting metabolism and nonthermal regulatory maintenance) and "activity" energy. He further estimated that ~20% of the activity energy is used for locomotion (nonburrowing movement, including through burrows and, rarely, aboveground). The remaining activity energy (activity energy less locomotion) is the energy available for all burrowing activities, from which the energy for mound building would be derived.

In order to convert the burrowing–available energy value into an energy density, an estimate of the population density of gophers in the Merced region is required. Richens (1965) demonstrated that periodic measurements of gopher tailings heaps are highly correlated with gopher population density in a Utah grassland. Although gopher populations will likely fluctuate in both space and time, the Merced-area gopher density can be estimated by dividing measurements of the aboveground gopher transport (tailings heaps m^{-2} yr^{-1}) by an average estimate of tailings pile production per gopher (110 tailings heaps yr^{-1}) (Miller, 1948; Bandoli, 1981). The burrowing–available energy estimated from Gettinger (1984) is then divided by the gopher concentration estimate to calculate the density of energy (J m^{-2} yr^{-1}) available to gophers for soil movement (E_{gopher}) in the Merced region. The energy expenditure (E_{mound}) and storage values (E_{gopher}) were then compared, on various time scales and within different substrates, to explore whether gophers could be responsible for both building and/or maintaining the mounded terrain.

RESULTS AND DISCUSSION

…when one contemplates the fact that there are about 15 mounds per acre, or nearly 10,000 mounds per square mile, the amount of work expended in this one small area is seen to be tremendous. To attribute this colossal earth-moving feat to small rodents certainly appears ridiculous at first glance. —R.J. Arkley (1948)

Reed and Amundson

Field Observations

Figure 4 shows the profile of the excavated mound on the Laguna formation. The upper 25 cm form the *A* horizon and are characterized by a relatively darker color value and more homogeneous matrix than the lower horizons. The *A* horizon also appeared to contain a higher concentration of small gravels and relative absence of large gravels and cobbles compared to the horizon directly below it. These characteristics suggest significant bioturbation and sorting by burrowing organisms, likely gophers. Although the trenched mound did not appear to be currently occupied by gophers, aboveground gopher tailings nearby indicate that they are active in the region (Fig. 3). In addition, gopher-sized krotovina were observed at several locations in the soil profile. Our observations were consistent with Page et al.'s (1977) analysis of a Mima mound cross section on the China Hat Laguna formation.

A *Bt* horizon, located beneath the *A* horizon (25–88 cm; ~30% clay), has columnar structure and an irregular lower boundary. Such features generally require 10^4–10^5 years to develop (if Birkeland's [1999] estimates hold in this region). The *Bt* layer does not appear to be continuous into the intermound regions. Based on these observations, several scenarios could be proposed. For instance, the mounds may have existed in roughly their current form since the late Pleistocene. On the other hand, the mound stratigraphy could also reflect a time when the formation rates of the columnar structure of the *Bt* horizon exceeded bioturbation rates, or could reflect the attenuation of biotic activity with depth. In addition, a *Bt* layer could have been already in place across the landscape when mound-building began, and the gophers may have extracted the *Bt* horizon sediment from the intermound regions and redeposited it in the mounded areas.

Figure 4. Mima mound cross section (latitude 37.46° N; longitude 120.37° W) with pick, 65 cm in length, for scale. Mound apex is near left-hand side of photo and photographer is facing approximately southeast. Krotovina (not visible here) were observed in the *A* and *B* horizons, down to 70 cm. Mounded region in the background shows tailings pile from trenching. The cobble layer was shown to be continuous from the base of this mound to an adjacent swale. (Photo courtesy of A. Heimsath.)

Further constraining the age controls on mound formation and comparing them to records of gopher activity and inhabitation could help untangle the factors of formation. Davis (1999) conducted a pollen analysis of sediments in Tulare Lake, a currently dry lakebed ~140 km south of the field region. To our knowledge the results of the analysis are the best proxy to provide a chronology of vegetation and climate for this study region.

The study found that Great Basin-like vegetation (giant sequoia and greasewood) dominated the low elevation interior from roughly 28,000 to 8500 yr B.P., with vegetation similar to present-day vegetation (grassland) occurring around 7000 yr B.P. Gophers likely began to inhabit the region around the time of this vegetation shift. Potter Creek Cave and Samwel Cave, both located in north-central California in Shasta County, are the source of most gopher fossils collected in the northern half of the state (Museum of Vertebrate Zoology at Berkeley, 2009). *Thomomys* emplacement is dated to the late Pleistocene, during the last glacial maximum (Feranec et al., 2007). These observations support gopher presence on a time scale long enough to build the mounds according to the 2007 model and this study's stratigraphic observations.

The layer (88–120 cm) underlying the *Bt* horizon is comprised of large cobbles. This layer is the same elevation as the cobble layer in the adjacent basin and represents a nearly continuous stone line, which also has been observed by other studies in the region (e.g., Arkley, 1962). Several authors have pointed out the importance of particle size analyses for testing hypotheses of mound origin (Dalquest and Scheffer, 1942; Washburn, 1988; Cox, 1984a, 1984b; Horwath and Johnson, 2006). A biotic mechanism of mound creation should result in a two-layered biomantle: a concentration of small clasts, those that can readily be transported by animals, in this case gophers, in the mounds, and a layer of cobbles and large clasts below the mounds (Horwath and Johnson, 2006). Gophers have been observed to burrow around and under unmanageable clasts (Grinnell and Storer, 1924; Murray, 1967) which, when this action continues over time, can undermine the large particles and cause them to settle into a lower stone layer. At the same time, gophers preferentially move smaller sediments into their zone of burrowing (the typical size range of particles moved by gophers is 0.6–2.5 cm [Hansen and Morris, 1968] and the maximum particle size transported by gophers is 5–7 cm [Dalquest and Scheffer, 1942; Murray, 1967; Johnson, 1989]). Cox (1984b) found a higher concentration (1.8 times) of small stones (long axis <5 cm) in the mounds near San Diego as compared with the intermound regions. Finally, the observed particle size distribution is inconsistent with several alternative proposals for mound origin. If the mounds were formed by wind deposition, they would be composed of sand-sized particles and smaller and would not contain gravel. Differential erosion would indeed concentrate gravels and cobbles in the intermound zone; however, it would not cause that concentration to extend continuously underneath the mounds.

Below the cobble layer lays a dense, hard, clay-rich layer which impedes water percolation and burrowing by gophers.

The presence of a hardpan at the base of both the mound and the pool at roughly the same level suggests it may be continuous throughout the region. In 1941, Nikiforoff made observations of hundreds of mounds in grasslands ~175 km south of the study region. Both he and Arkley and Brown (1954) noted that the water-restrictive layers did not parallel the microrelief surface. Further, Nikiforoff noted that many mounds had at their base a "window" in the hardpan surface, which ranged in size from 0.3 to 3.7 m across. Generally, the soil underlying the mound at the level of the hardpan was more friable than that in the depressions. He also observed that gophers were "able to bore even through the firmest hardpan." (Nikiforoff was an opponent of a biotic mechanism of mound formation.) We interpret the sum of these observations to indicate that the windows were either preexisting and expanded by gophers or originally created by them in an effort to establish a nest. A significant advantage was then conferred to subsequent generations to reuse such nests, creating point centers from which generations of gophers foraged, eventually creating the mounds.

In conclusion, the clear presence of a continuous cobble layer at depth, a concentration of small clasts in the A horizon, krotovina throughout the profile, and a dense, clay-rich layer at the base of the mounds all support the possibility that the mounded terrain results from directed (lateral) soil movement by gophers in response to low soil permeability and hardpan restrictions to burrowing.

Mound Morphology

The LIDAR survey area is largely comprised (60%) of the Laguna formation, and the values reported below are for mounds on that geologic formation only. A sample of the digital elevation model (DEM) derived from the survey is shown in Figure 5, and morphometric results are summarized in Table 1. The average Mima mound height on the Laguna formation is 0.44 m, and the average diameter is 7.3 m. The mound density is roughly 4800 mounds km^{-2}, which is greater than Arkley and Brown's (1954) estimate of 3900 mounds km^{-2} for the greater Merced region (which includes other, younger terraces). Generally, the mounds in this portion of California's Great Valley are shorter,

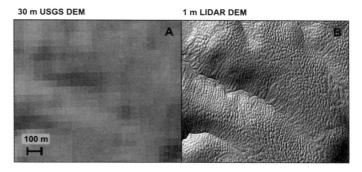

Figure 5. (A) Shaded-relief imagery from a 30 m digital elevation model derived from USGS survey data. (B) Shaded-relief imagery derived from a NCALM (National Center for Airborne Laser Mapping) LIDAR survey of the same region. The comparison highlights the utility of high-resolution survey tools to adequately analyze small-scale topographic features such as Mima mounds. Note the dense mound distribution, the apparent merging of mounds in some cases, and the linearity of the formations on hillslopes.

smaller, and more densely spaced than other Mima-like mounds in North America (Fig. 6). A typical mound profile on level terrain (Fig. 7) approximates a convexoplanar shape, with convex mound tops and more planar sideslopes. The average mound-slope was 0.08 and curvature was –0.10 m^{-1}, with standard deviations of 0.04 and 0.07 m^{-1}, respectively (Fig. 8).

Slopes of the underlying terrain range from 0% to 55%, with approximately two-fifths of the survey region having slopes of 3%–10%. Mounds on level terrain (slope <5%) are slightly elliptical in shape (average ratio of semimajor to semiminor axis length = 1.33), whereas mounds on sloping terrain are particularly elongated downslope (Fig. 9). Mounds were observed on hillslopes with slopes as great as 40%. Many of these hillslope mounds were aligned linearly in near-parallel rows downslope (Figs. 1B and 5). In some cases, the mounds near channels and streambeds seemed to be oriented along the lines of drainage, although this was not always the case. Arkley and Brown (1954) and others (see Washburn, 1988) have noted that mounds sometimes appear to be aligned with regional drainage patterns. It is possible that the observed patterns are indicative of a response of burrowing rodents to preexisting patterns in the terrain—for

TABLE 1. DATA SUMMARY HIGHLIGHTING THE PHYSICAL CHARACTERISTICS OF MIMA MOUND TOPOGRAPHY, NEAR MERCED, CALIFORNIA

Mound concentration (mounds km^{-2})	4800
Estimated total number of mounds in survey area	86,400
Average mound diameter (m)	7.3 ± 1.9
Average mound elongation (semimajor/semiminor axes)	1.33
Average mound height (m)	0.44 ± 0.09
Average mound volume (m^3)	13 ± 5
Survey-wide mound volume (m^3)	1.1 × 10^6
Average mound surface area (m^2)	45 ± 15
Survey-wide mound surface area (m^2)	3.9 × 10^6
Average mound slope	0.08 ± 0.04
Average mound curvature (negative Laplacian) (m^{-1})	–0.10 ± 0.07

Note: The data are compiled from mounds sampled from the 18 km^2 Laguna portion of the LIDAR survey.

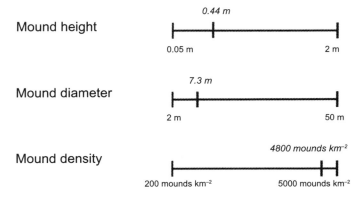

Figure 6. Diagram comparing Merced-area mounds with Mima-like mounds across North America. Data from the Merced-area mounds are the average of measurements made using LIDAR data for the Laguna formation only. Data for the mound continuum are derived from various sources including Cox (1984a), Dalquest and Scheffer (1942), and Washburn (1988).

instance, the animals may take advantage of slight elevational differences associated with drainage patterns, or may adapt to restrictive boundaries, such as streams or rock outcrops (Scheffer, 1958; Washburn, 1988).

Mounds are typically uniformly spaced at an average intermound distance of 12 m. This distance is similar to the observation by Zinnel and Tester (1994), based on a two-year study in Minnesota grasslands, that gophers maintain a distance of ~10 m between neighboring individuals. The sum of average combined mound and swale plan areas in the Merced region is 156 m². Bandoli (1987) measured an average gopher home range of 150 m². The close correspondence between mound spacing and gopher spacing, while circumstantial, is consistent with a biologic explanation of Mima mound creation, in that each mound may represent the territorial sphere of gopher burrowing and foraging activity. Several studies (Dalquest and Scheffer, 1942; Hansen, 1962; Murray, 1967) have observed that each Mima mound is usually occupied by a single gopher. However, the LIDAR data show some mounds which are not distinct units, but rather appear to be the merger of two smaller mounds (see Fig. 5). This may be explained by observations of Cox and Hunt (1990), who investigated 18 mounds in the San Diego, California, area and found that the number of gophers per mound (ranging from 1 to 6) was strongly correlated with mound size. Cox and Hunt hypothesized that within larger mounds, there is enough space for occupation by two animals, which tend to locate themselves on the mound peripheries. Differences in mound morphology, then, may reflect differences in gopher behavior and interactions. Where intermound spacing did vary, it was correlated with mound diameter (Fig. 10). In addition, mound and pool volumes were highly correlated (Fig. 11). These findings are also compatible with the model of gopher-built mounds, in that the model predicts that the rodents use soil from the intermound areas to build the mounds.

Biogeomorphic Model of Mound Evolution

Considering that the 18 km² Laguna LIDAR region contains ~86,400 mounds, whatever forces are responsible for creating this topographic phenomenon are certainly consequential. Based on mound distribution maps (Cox, 1984a; Washburn, 1988), ~10% of California (42,000 km²) was historically (conservatively) covered by Mima mounds. LIDAR results indicate that mounds in the Laguna survey area are comprised of nearly two million metric tons of sediment. Extrapolating this figure to the estimated Mima mound coverage in California and assuming the same concentration as the Laguna formation mounds (likely a maximum), the mounds in the state represent roughly 4.4 *trillion* metric tons of soil. In contrast, a single gopher (*Thomomys bottae*) weighs an average of 147 g (Vleck, 1979) and measures roughly 15–27 cm in length (Daly and Patton, 1986). To imagine such small organisms creating the mounded topography by moving an amount of sediment equal to more than 350 times the annual sediment discharge of the world's rivers (12.2 billion metric tons, Milliman and Meade, 1983) can be perplexing. However, the results of the modeling work show that mound creation and maintenance by gophers is certainly feasible under reasonable environmental scenarios.

Mass Balance Results

Using the average mound curvature value (0.10 m⁻¹), Equation 3, and the previously mentioned range of K values, erosion rates off of the Mima mounds range from 19 to 56 cm kyr⁻¹. A sample map of mound erosion and pool deposition is shown in Figure 12, using $K = 50$ cm² yr⁻¹. Curvature values and erosion rates calculated from the landscape-wide LIDAR terrain data are higher than those of the 2007 study, indicating that, in the absence of a conserving force, the mounds would be leveled even faster than previously estimated. Over three seasons, we measured an average rate of aboveground transport by gophers of 61 cm kyr⁻¹. Thus, on a mass basis, these region-wide erosion rates are still within the range of erosion that can be feasibly compensated for by gopher soil transport, assuming that a portion of that transport is directed moundward and upward. The aboveground sediment transport measurements also likely underestimate the excavation work done by gophers, as several studies (e.g., Andersen, 1978) have shown that up to 69% of excavated sediment gets redeposited belowground and is therefore not represented on the surface.

Energy Analysis Results

Just as the mass balance suggests gophers could build and maintain mounds, the energy analysis also supports the hypothesis. The relationship between the biological energy allocated to gopher burrowing activities and the energy estimated for mound-building costs is summarized in Figure 13 for the case of a mound composed of sandy loam sediment that is built in 10³ years. Figure 14 summarizes the energy budgets for a selection of mound-building scenarios. With respect to mound-building

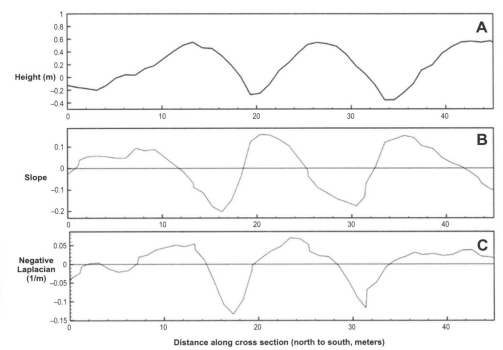

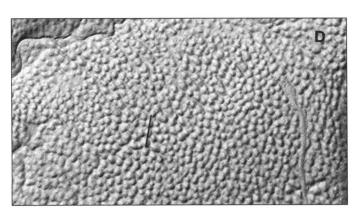

Figure 7. (A) A cross section of typical mound and swale topography on the Laguna formation (location highlighted with a black line in the shaded relief image 7D, where north is at the top of the image). Insets B and C show the changes in slope and curvature, respectively, along the cross section.

costs, the energy required for lifting soil against gravity to build mounds is the smallest component of the mound-building budget and ranges from 95 to 0.095 J m^{-2} yr^{-1} for 10–10^4 year construction scenarios, respectively. The shear and push costs for mound building (sandy loam) range from 4.4 × 10^5 to 440 J m^{-2} yr^{-1} for the same range in time periods. For a mound composed of clayey sediment, the range is 2.8 × 10^6–2800 J m^{-2} yr^{-1}. Using the average erosion rate reported above, the cost of counteracting diffusive erosion of existing mounds is 3.0–9.1 J m^{-2} yr^{-1}, for a range of diffusivity values (25–75 cm^2 kyr^{-1}). Thus the total cost of lifting, shearing, pushing, and maintaining soil in the shape of a mound ranges from 4.4 × 10^5–440 J m^{-2} yr^{-1} for sandy loam and 2.8 × 10^6–2800 J m^{-2} yr^{-1} for clay.

The energy available to gophers for burrowing activities was estimated using gopher sediment transport measurements as a proxy for gopher population density. As discussed above, the three-season average for aboveground sediment movement was 61 cm kyr^{-1} and the associated gopher population was calculated to be 52 gophers ha^{-1}. From this, the remaining activity energy available for sediment transport processes, E_{gopher}, was estimated to be 5.4 × 10^4 J m^{-2} yr^{-1}, which is 1.6% of NPP for the area.

For a sandy loam mound built in 1000 years, the costs of building are 8% and 0.1% of burrowing–available energy and NPP, respectively (Fig. 13). For a clay mound built in the same time period, the costs are 52% and 0.8% of burrowing–available energy and NPP. Based on this analysis then, there is *in*sufficient energy for gophers to build mounds within 10-year (both clay and sandy loam) (not shown in Fig. 14) and 100-year clay mound scenarios (Fig. 14). However, for thousand-year time scales and longer, there clearly does exist adequate biological energy to produce and preserve these landscape features.

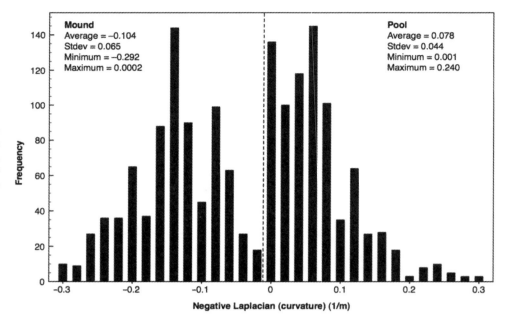

Figure 8. Histogram showing the distribution of curvature (negative Laplacian) values calculated at 1600 points on both mounds and pools. Mound values are more widely distributed and have an overall higher mean than pool values.

Considerations for Model Improvement

These results disagree with the only two known accounts of direct mound building in the United States. In 1949, Price reported that "a mature Mima mound has been built in 5 years." Similarly, Ross et al. (1968) report accounts of black silt loam mounds reappearing 4–5 years after being plowed. Based on this time frame and the reported size of the mound discussed in Price (1949) (19 m diameter and 0.6 m height [sizes reported in Ross et al. are comparable]), the energy required for construction by a single gopher is 2.23×10^6 J m^{-2} yr^{-1}. This is nearly 4000% of energy that a gopher can expend for burrowing, as calculated as

described above, and is impossible under our scenarios unless more than one gopher contributed to mound building (which is possible given Cox and Hunt's [1990] observations of multiple gophers occupying larger mounds), unless gophers in these regions are more efficient burrowers, and/or unless the calculations overestimate the costs or underestimate available energy. Clearly, direct accounts of mound building need more rigorous documentation.

Cox and Allen (1987) focused on recording direct mound building by gophers and showed that soil movement by gophers is net upward and moundward on times scales of 1–2 years. This

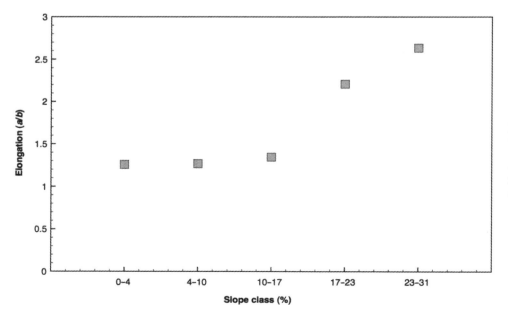

Figure 9. Graph showing mound elongation (*a/b*) increase with slope. On each slope class, 10 mounds were randomly chosen and semimajor (*a*) and semiminor (*b*) axes were measured and averaged.

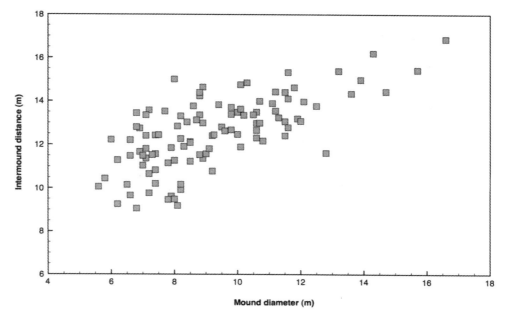

Figure 10. Comparison of mound diameter (semimajor axis) and intermound distance (average span between given mound apex and all directly adjacent mound apexes) for 104 mounds randomly located within the terrain data.

type of approach is crucial to documenting whether mounds are built or maintained by rodents and a similar study is being planned in the Merced-area mounds. Such measurements will also help improve models of mound evolution by better quantifying gopher sediment transport parameters.

In addition to the inconsistency between the model results and direct reports of mound building, other issues associated with the model should be addressed. First, estimated gopher densities for the Merced region (52 gopher ha^{-1}) are considerably lower than has been reported in other regions (e.g., Seabloom et al., 2000, N = 119; Smallwood and Morrison, 1999, N = 53 ± 49). However, the low population density is reasonable considering the eastern Merced County grasslands receive less rainfall than most other U.S. grassland regions and are less productive. If the gopher population density is underestimated here, then it is possible that E_{gopher} is even higher than reported, more strongly supporting the conclusion that rodent-built mounds are energetically possible.

Next, the analysis assumes that gophers are the region's primary subterranean organism. However, other fossorial rodents are found in some regions of Mima mounds and their presence in an ecosystem undoubtedly impacts sediment transport processes and community ecology. Ground squirrels (*Spermophilus beecheyi*), for example, are prevalent in the Merced region (Laabs and Allaback, 2002) and while their burrowing behavior is different from gophers in that their burrows are generally larger and left open after excavation (Fig. 3), they also must affect land surface processes. Other organisms also likely impact the dynamics of mound erosion and upbuilding. American badgers (*Taxidea taxus*) prey on gophers and ground squirrels in the area and also excavate burrows for their own protection (Orloff, 2002). Several studies indicate that the San Joaquin kit

fox (*Vulpes macrotis mutica*) may enlarge ground squirrel and badger burrows for their own use, especially in regions where hardpans exist (Jensen, 1972; Morrell, 1971; Orloff et al., 1986). The digging activities of these other animals would likely affect rates of mound erosion/construction, although the extent of the effects remains to be quantified.

The interactions between gophers and other organisms likely have an impact on population dynamics and mound evolution. For example, Gallie and Drickamer (2008) found that gopher density was significantly lower in the presence of Gunnison's prairie dog, while, conversely, the prairie dog population more

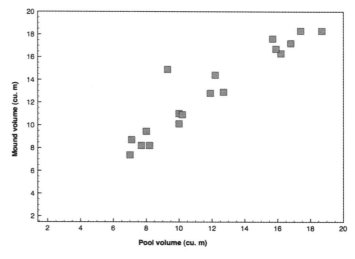

Figure 11. Relationship between mound volume and swale volume as calculated from the LIDAR dimensions of 20 randomly chosen mound-pool pairs within the survey area.

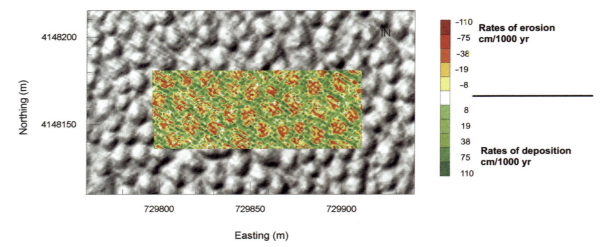

Figure 12. Map showing results of modeled erosion off of mounds (in red tones) and deposition into swales (in green tones). Erosion patterns are overlaid on a shaded relief image of the local terrain. Region-wide erosion values (average = 38 cm kyr^{-1}) were significantly higher than those previously calculated (Reed and Amundson, 2007, average = 15 cm kyr^{-1}). Values shown were calculated using a diffusion coefficient equal to 50 cm^2 yr^{-1} and a soil-to-parent material bulk density ratio of 0.75.

than doubled when they inhabited the same region as gophers. While no prairie dogs exist in the Merced region, similar types of interspecies interactions may occur in the mound-pool ecosystems. Differences in mound size, shape, and density in different locations may reflect local interactions between gophers and other organisms.

Furthermore, nearly 150,000 acres within and surrounding the study region are used for cattle grazing (Robins and Vollmar, 2002), and the effects of this practice are likely to affect sediment transport processes and gopher behavior and ecology. For instance, we have observed that some mounds near our soil excavation (at least 20%) exhibit mid-mound deflation, in which the top center of the mound is a concave depression. Nikiforoff (1941) speculated these were caused by the sinking of gopher tunnels at the mound apex, perhaps due to the weight of cattle, although this sinking could also be due to burrowing by other rodents or predators. He concluded that both the burrow collapse and the tailings piles created by gophers "should be expected to cause a gradual destruction and leveling of the mounds rather than their upbuilding." Grazing may also affect gopher density (and mound density) (e.g., through changes in soil density or foraging supply), although the reported impacts of grazing on gopher populations are contradictory. Several studies have shown a decrease in gopher densities in grazed areas (Hansen, 1965; Turner et al., 1973; Hunter, 1991), while others have reported an increase (Buechner, 1942; Richens, 1965) or equal densities in grazed and ungrazed areas (Turner, 1969).

Although the model warrants improvements, several other studies of gophers support the results reported here. Previous workers have shown that although burrowing is expensive, gopher diet generally provides ample calories for soil digging, even in challenging environments (e.g., Gettinger, 1984;

Romañach et al., 2007—simulation). For instance, Seabloom et al. (2000) showed that gophers do not adjust their sediment transport behavior even on slopes as steep as 30% because the costs associated with shearing and pushing the soil are so much greater than lifting soil against gravity. Huntly and Inouye (1988) call attention to the fact that the energy flow-through rates of gopher populations rival those of large grazers, indicating that, indeed, gopher populations have considerably large amounts of energy available to them. However, many of the gopher calculations described above were derived from studies of the mammals in laboratory settings or in other grassland environments. Detailed investigations of gopher behavior and physiology in the Merced region seem an important avenue for more accurately linking animals to geomorphic evolution of landscapes.

Implications for Conservation Biology

Independent of mound formation, bioturbation has a measurable effect on contemporary mound form and evolution. Any effects would also likely be experienced in the adjacent vernal pool depressions, refugia for more than 10 federally listed endangered species and nearly 100 endemic terrestrial and aquatic species (e.g., Witham et al., 1998). Bauder (2005) showed that small changes (cm-scale) in mound and pool topography (which could occur through changes in gopher sediment transport) can have significant impacts on pool hydrology and on the resident plants and animals, which are finely calibrated to specific hydrological conditions.

Gopher presence can have other effects on life within vernal pools. It has been shown (Loredo et al., 1996, and references within) that tiger salamanders, endangered vernal pool amphibians, use ground squirrel and potentially gopher burrows as habitat

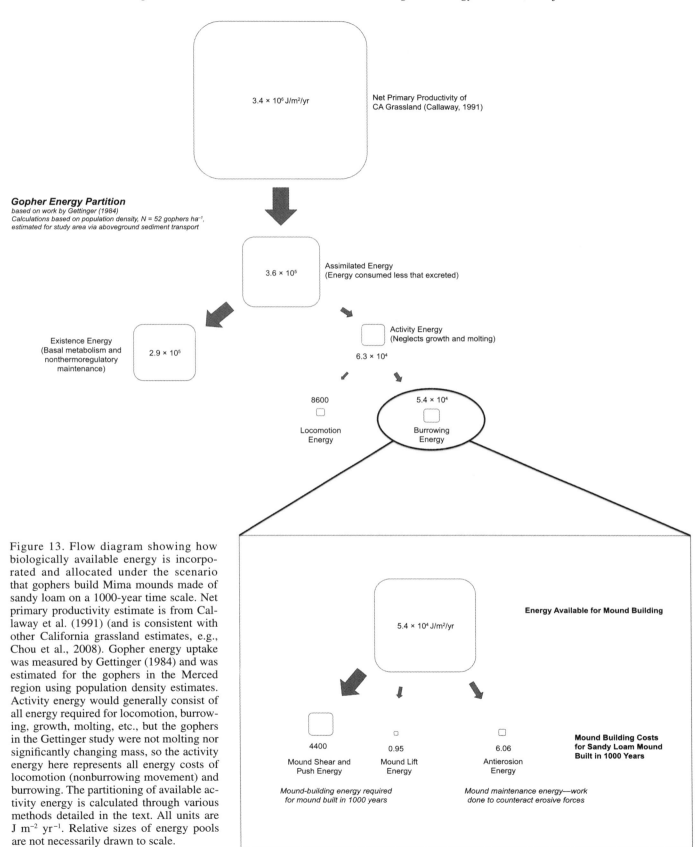

Figure 13. Flow diagram showing how biologically available energy is incorporated and allocated under the scenario that gophers build Mima mounds made of sandy loam on a 1000-year time scale. Net primary productivity estimate is from Callaway et al. (1991) (and is consistent with other California grassland estimates, e.g., Chou et al., 2008). Gopher energy uptake was measured by Gettinger (1984) and was estimated for the gophers in the Merced region using population density estimates. Activity energy would generally consist of all energy required for locomotion, burrowing, growth, molting, etc., but the gophers in the Gettinger study were not molting nor significantly changing mass, so the activity energy here represents all energy costs of locomotion (nonburrowing movement) and burrowing. The partitioning of available activity energy is calculated through various methods detailed in the text. All units are J m^{-2} yr^{-1}. Relative sizes of energy pools are not necessarily drawn to scale.

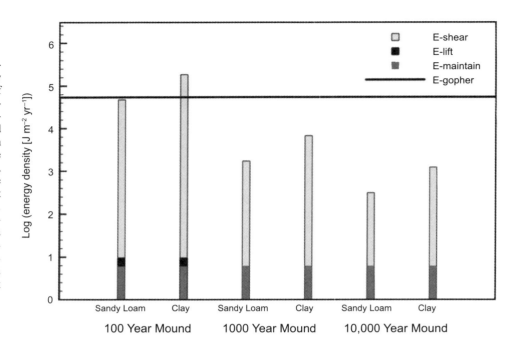

Figure 14. The energy costs of various scenarios of mound building by gophers (expressed as the logarithm of energy density [J m^{-2} yr^{-1}]) for different soil types and time scales. The lightest gray represents the energy required for gophers to shear and push soil from the intermound region into the shape of a mound (based on model by Vleck, 1979). The black quantity indicates the energy required to lift the soil against the force of gravity to build mounds. Finally, the dark gray indicates the energy required to maintain the mounds against diffusive erosion. The black horizontal line represents the estimated energy available to gophers from biomass intake to do the mound-building work (based on Gettinger, 1984).

during their nonbreeding season. Furthermore, gophers may have an effect on maintaining vernal pool plant life via their dietary habits. Hunt (1992) showed that gophers eat vernal pool plants sparingly, mainly in late spring, and instead focus their foraging on grass and forb shoots. This feeding behavior could help keep European grasses and forbs from encroaching on vernal pools, similar to the effect that is observed to occur via livestock grazing (Marty, 2005; Robins and Vollmar, 2002).

The trapping, gassing, and poisoning of gophers has been practiced by farmers and ranchers in California's Central Valley for more than a century because gophers are considered pests that eat and bury agricultural and feed plants, disrupt irrigation flow, and disturb the soil surface, causing problems for farm machinery (Rhoads, 1898; Miller, 1950; Smallwood et al., 2001; Laabs and Allaback, 2002). The practice of gopher eradication may be reflected in observed gopher densities, and potentially in mound and pool form and overall ecosystem function (e.g., Smallwood and Morrison, 1999).

The state of California and various nonprofit agencies have waged a sizeable effort to protect remaining vernal pool ecosystems (UC-Merced and The Nature Conservancy, 2002). One common mitigation tactic is to construct artificial pools to replace natural pools lost through development (Sutter and Francisco, 1998). While this approach is important to the pools' preservation, there is, as yet, no fundamental scientific understanding of how these landscapes have developed, or what mechanism maintains the topography that is critical to the species that live in the pools. Given that gophers may be the key species that both forms and preserves the terrain, these mitigation efforts may lack a full understanding of mound-pool function, which could hamper long-term conservation success.

CONCLUSIONS

Mima mounds, which have captivated researchers for more than a century, along with their adjacent vernal pools, are vibrant templates in which to study linkages between landscape and life. LIDAR data of mound and swale surfaces provided an exceptional opportunity both to quantify mound form and distribution on regional scales, as well as to build and test models to better understand landform evolution, and the rates and energy involved. The LIDAR data and our models have strengthened the hypothesis that the unusual Mima mounded landscapes of central California were both produced, and are being maintained, primarily by pocket gophers. While a quantitative theory detailing how gophers move sediment to build mounds is yet underdeveloped, this work demonstrates that, from an energetic perspective, gopher-created mounds are clearly possible. The hypothesis can be further tested and the models can be improved by implementing studies of sediment tracers, thereby providing more detail to the rates at which the landscape is biotically transformed.

ACKNOWLEDGMENTS

The authors thank Jeff Chance, Chris Robinson, Ralph Fagundes, and the University of California at Merced for generous access to their property. We thank two anonymous reviewers for their careful evaluation and insightful comments, which have greatly improved the content and quality of this manuscript. The Department of Energy's Global Change Education Program, the National Science Foundation, and the University of California at Berkeley (UCB)'s Department of Environmental Science, Policy, and Management provided graduate fellowship support. The LIDAR data was made possible via a graduate student seed award from the National Center for Airborne Laser Mapping (NCALM). UCB's Geospatial Information Facility and the Department of Earth and Planetary Science's NCALM Distribution Center provided valuable technical guidance.

REFERENCES CITED

Andersen, D.C., 1978, Observations on reproduction, growth, and behavior of the northern pocket gopher (*Thomomys talpoides*): Journal of Mammalogy, v. 59, p. 418–422, doi:10.2307/1379927.

Arkley, R.J., 1948, The Mima mounds—Reply: The Scientific Monthly, v. 66, p. 175–176.

Arkley, R.J., 1962, Soil Survey of the Merced area, California, California Agricultural Experiment Station: United States Department of Agriculture Soil Conservation Service, v. 1950, no. 7, p. 131.

Arkley, R.J., and Brown, H.C., 1954, The origin of Mima mound (hogwallow) microrelief in the far western states: Soil Science Society of America Journal, v. 18, p. 195–199, doi:10.2136/sssaj1954.03615995001800020021x.

Baisden, W.T., Amundson, R., Brenner, D.L., Cook, A.C., Kendall, C., and Harden, J.W., 2002, A multiisotope C and N modeling analysis of soil organic matter turnover and transport as a function of soil depth in a California annual grassland soil chronosequence: Global Biogeochemical Cycles, v. 16, p. 1135–1160, doi:10.1029/2001GB001823.

Bandoli, J.H., 1981, Factors influencing seasonal burrowing activity in the pocket gopher, *Thomomys bottae*: Journal of Mammalogy, v. 62, p. 293–303, doi:10.2307/1380706.

Bandoli, J.H., 1987, Activity and plural occupancy of burrows in Botta's pocket gopher *Thomomys Bottae*: American Midland Naturalist, v. 118, p. 10–14, doi:10.2307/2425623.

Bauder, E.T., 2005, The effects of an unpredictable precipitation regime on vernal pool hydrology: Freshwater Biology, v. 50, p. 2129–2135, doi:10.1111/j.1365-2427.2005.01471.x.

Berg, A.W., 1990, Formation of Mima mounds—A seismic hypothesis: Geology, v. 18, p. 281–284, doi:10.1130/0091-7613(1990)018<0281:FOMMAS>2.3.CO;2.

Birkeland, P.W., 1999, Soils and Geomorphology: New York, Oxford University Press, 448 p.

Buechner, H.K., 1942, Interrelationships between the pocket gopher and land use: Journal of Mammalogy, v. 23, p. 346–348.

Butler, D.R., 1995, Zoogeomorphology: Animals as Geomorphic Agents: Cambridge, UK, Cambridge University Press.

Callaway, R.M., Nadkarni, N.M., and Mahall, B.E., 1991, Facilitation and interference of *Quercus douglasii* on understory productivity in central California: Ecology, v. 72, p. 1484–1499, doi:10.2307/1941122.

Chou, W.W., Silver, W.L., Jackson, R.D., Thompson, A.W., and Allen-Diaz, B., 2008, The sensitivity of annual grassland carbon cycling to the quantity and timing of rainfall: Global Change Biology, v. 14, p. 1382–1394, doi:10.1111/j.1365-2486.2008.01572.x.

Cox, G.W., 1984a, The distribution and origin of Mima mound grasslands in San-Diego County, California: Ecology, v. 65, p. 1397–1405, doi:10.2307/1939120.

Cox, G.W., 1984b, Mounds of mystery: Natural History, v. 6, p. 36–45.

Cox, G.W., 1990, Soil mining by pocket gophers along topographic gradients in a Mima moundfield: Ecology, v. 71, p. 837–843, doi:10.2307/1937355.

Cox, G.W., and Allen, D.W., 1987, Soil translocation by pocket gophers in a Mima moundfield: Oecologia, v. 72, p. 207–210, doi:10.1007/BF00379269.

Cox, G.W., and Hunt, J., 1990, Form of Mima mounds in relation to occupancy by pocket gophers: Journal of Mammalogy, v. 71, p. 90–94, doi:10.2307/1381323.

Cox, G.W., and Hunt, J., 1992, Relation of seasonal activity patterns of Valley pocket gophers to temperature, rainfall, and food availability: Journal of Mammalogy, v. 73, p. 123–134, doi:10.2307/1381873.

Culling, W., 1960, Analytical theory of erosion: The Journal of Geology, v. 68, p. 336–344, doi:10.1086/626663.

Culling, W., 1963, Soil creep and the development of hillside slopes: The Journal of Geology, v. 71, p. 127–161, doi:10.1086/626891.

Dalquest, W.W., and Scheffer, V.B., 1942, The origin of the Mima mounds of western Washington: The Journal of Geology, v. 50, p. 68–84, doi:10.1086/625026.

Daly, J.C., and Patton, J.L., 1986, Growth, reproduction, and sexual dimorphism in *Thomomys Bottae* pocket gophers: Journal of Mammalogy, v. 67, p. 256–265, doi:10.2307/1380878.

Davis, O.K., 1999, Pollen analysis of Tulare Lake, California: Great Basin-like vegetation in Central California during the full-glacial and early Holocene: Review of Palaeobotany and Palynology, v. 107, p. 249–257, doi:10.1016/S0034-6667(99)00020-2.

Davis, W., 1892, The convex profile of bad-land divides: Science, v. 508, p. 245, doi:10.1126/science.ns-20.508.245.

Devlin, J.F., 2003, Rationalizing geomorphology with an energy balance: Journal of Geoscience Education, v. 51, p. 398–409.

Dietrich, W.E., and Perron, J.T., 2006, The search for a topographic signature of life: Nature, v. 439, p. 411–418, doi:10.1038/nature04452.

Dietrich, W.E., Reiss, R., Hsu, M.L., and Montgomery, D.R., 1995, A process-based model for colluvial soil depth and shallow landsliding using digital elevation data: Hydrological Processes, v. 9, p. 383–400, doi:10.1002/hyp.3360090311.

Dittes, J.C., and Guardino, J.L., 2002, Rare plants (Chapter 3), *in* Vollmar, J.E., ed., Wildlife and Rare Plant Ecology of Eastern Merced County's Vernal Pool Grasslands: Merced, University of California, p. 58–131.

DuToit, J.T., Jarvis, J.U.M., and Louw, G.N., 1985, Nutrition and burrowing energetics of the Cape mole-rat *Georychus capensis*: Oecologia, v. 66, p. 81–87, doi:10.1007/BF00378556.

Feranec, R.S., Hadly, E.A., Blois, J.L., Barnosky, A.D., and Paytan, A., 2007, Radiocarbon dates from the Pleistocene fossil deposits of Samwel Cave, Shasta County, California, USA: Radiocarbon, v. 49, p. 117–121.

Gabet, E.J., 2000, Gopher bioturbation: Field evidence for non-linear hillslope diffusion: Earth Surface Processes and Landforms, v. 25, p. 1419–1428, doi:10.1002/1096-9837(200012)25:13<1419::AID-ESP148>3.0.CO;2-1.

Gallie, J.A., and Drickamer, L.C., 2008, Ecological interactions between two ecosystem engineers: Gunnison's prairie dog and Botta's pocket gopher: The Southwestern Naturalist, v. 53, p. 51–60, doi:10.1894/0038-4909(2008)53[51:EIBTEE]2.0.CO;2.

Gettinger, R.D., 1984, Energy and water metabolism of free-ranging pocket gophers, *Thomomys bottae*: Ecology, v. 65, p. 740–751, doi:10.2307/1938046.

Gilbert, G.K., 1909, The convexity of hilltops: The Journal of Geology, v. 17, p. 344–350, doi:10.1086/621620.

Grinnell, J., and Storer, T.I., 1924, Animal Life in the Yosemite: An Account of the Mammals, Birds, Reptiles, and Amphibians in a Cross-Section of the Sierra Nevada: Berkeley, University of California Press, 752 p.

Hansen, R.M., 1962, Movements and survival of *Thomomys talpoides* in a Mima-mound habitat: Ecology, v. 43, p. 151–154, doi:10.2307/1932058.

Hansen, R.M., 1965, Pocket gopher density in an enclosure of native habitat: Journal of Mammalogy, v. 46, p. 508–509, doi:10.2307/1377654.

Hansen, R.M., and Morris, M.J., 1968, Movement of rocks by northern pocket gophers: Journal of Mammalogy, v. 49, p. 391–399, doi:10.2307/1378197.

Harden, J.W., 1982, A quantitative index of soil development from field descriptions: Examples from a chronosequence in central California: Geoderma, v. 28, p. 1–28, doi:10.1016/0016-7061(82)90037-4.

Harden, J.W., 1987, Soils developed in granitic alluvium near Merced, California: Soil Chronosequences in the Western United States: U.S. Geological Survey Bulletin 1590-A, p. A1–A65.

Hay, G.J., and Castilla, G., 2006, Object-based image analysis: Strengths, weaknesses, opportunities and threats (SWOT): International Archives of Photogrammetry: Remote Sensing and Spatial Information Sciences, v. 36, p. 1–4.

Heady, H., 1988, Valley grassland, *in* Barbour, M., and Major, J., eds., Terrestrial Vegetation of California: California Native Plant Society Special Publication, Sacramento, California, v. 9, p. 419–514.

Heimsath, A.M., Dietrich, W.E., Nishiizumi, K., and Finkel, R.C., 1997, The soil production function and landscape equilibrium: Nature, v. 388, p. 358–361, doi:10.1038/41056.

Hilgard, E.W., 1884, Report on the physical and agricultural features of the State of California: U.S. Census Office Reports: Tenth Census, v. 6, p. 649–796.

Holland, R.F., 1978, The Geographic and Edaphic Distribution of Vernal Pools in the Great Central Valley, California: California Native Plant Society Special Publication, v. 4, Sacramento, California, p. 1–12.

Horwath, J.L., and Johnson, D.L., 2006, Mima-type mounds in southwest Missouri: Expressions of point-centered and locally thickened biomantles: Geomorphology, v. 77, p. 308–319, doi:10.1016/j.geomorph.2006.01.009.

Horwath Burnham, J.L., Johnson, D.L., and Johnson, D.N., 2012, this volume, The biodynamic significance of double stone layers in Mima mounds, *in* Horwath Burnham, J.L., and Johnson, D.L., eds., Mima Mounds: The Case for Polygenesis and Bioturbation: Geological Society of America Special Paper 490, doi:10.1130/2012.2490(04).

Hubbs, C.L., 1957, Recent climatic history in California and adjacent areas, *in* Craig, H., ed., Proceedings, Conference on Recent Research in Climatology, La Jolla, California: La Jolla, University of California, Committee on Research in Water Resources, p. 25–26.

Hunt, J., 1992, Feeding ecology of Valley pocket gophers (*Thomomys-Bottae-Sanctidiegi*) on a California coastal grassland: American Midland Naturalist, v. 127, p. 41–51, doi:10.2307/2426320.

Hunter, J.E., 1991, Grazing and pocket gopher abundance in a California annual grassland: The Southwestern Naturalist, v. 36, p. 117–118, doi:10.2307/3672126.

Huntly, N., and Inouye, R., 1988, Pocket gophers in ecosystems: Patterns and mechanisms: Bioscience, v. 38, p. 786–793, doi:10.2307/1310788.

Ingles, L.G., 1949, Ground water and snow as factors affecting the seasonal distribution of pocket gophers, *Thomomys monticola*: Journal of Mammalogy, v. 30, p. 343–350, doi:10.2307/1375210.

Ingles, L.G., 1951, Outline for an ecological life history of pocket gophers and other fossorial mammals: Ecology, v. 32, p. 537–544, doi:10.2307/1931730.

Jensen, C.C., 1972, San Joaquin kit fox distribution: Sacramento, California, U.S. Fish and Wildlife Service, p. 1–22.

Johnson, D.L., 1989, Subsurface stone lines, stone zones, artifact-manuport layers, and biomantles produced by bioturbation via pocket gophers (*Thomomys Bottae*): American Antiquity, v. 54, p. 370–389, doi:10.2307/281712.

Johnson, D.L., Domier, J.E.J., and Johnson, D.N., 2005, Reflections on the nature of soil and its biomantle: Annals of the Association of American Geographers, v. 95, p. 11–31, doi:10.1111/j.1467-8306.2005.00448.x.

Laabs, D.M., and Allaback, M.L., 2002, Small mammals (Chapter 8), *in* Vollmar, J.E., ed., Wildlife and Rare Plant Ecology of Eastern Merced County's Vernal Pool Grasslands: Merced, University of California, p. 315–336.

Laabs, D.M., Allaback, M.L., and Orloff, S.G., 2002, Pond and stream breeding amphibians (Chapter 5), *in* Vollmar, J.E., ed., Wildlife and Rare Plant Ecology of Eastern Merced County's Vernal Pool Grasslands: Merced, University of California, p. 193–229.

Loredo, I., Van Vuren, D., and Morrison, M.L., 1996, Habitat use and migration behavior of the California tiger salamander: Journal of Herpetology, v. 30, p. 282–285, doi:10.2307/1565527.

Marchand, D., 1976, Preliminary geologic map showing Quaternary deposits of the Merced area, eastern San Joaquin Valley, California: U.S. Geological Survey Open-File Report 76-8365, scale 1:24,000.

Marty, J.T., 2005, Effects of cattle grazing on diversity in ephemeral wetlands: Conservation Biology, v. 19, p. 1626–1632, doi:10.1111/j.1523-1739.2005.00198.x.

McKean, J.A., Dietrich, W.E., Finkel, R.C., Southon, J.R., and Caffee, M.W., 1993, Quantification of soil production and downslope creep rates from cosmogenic Be-10 accumulations on a hillslope profile: Geology, v. 21, p. 343–346, doi:10.1130/0091-7613(1993)021<0343:QOSPAD>2.3.CO;2.

McNab, B.K., 1966, The metabolism of fossorial rodents: A study of convergence: Ecology, v. 47, p. 712–733, doi:10.2307/1934259.

Meysman, F.J.R., Middelburg, J.J., and Heip, C.H.R., 2006, Bioturbation: A fresh look at Darwin's last idea: Trends in Ecology & Evolution, v. 21, p. 688–695, doi:10.1016/j.tree.2006.08.002.

Miller, M.A., 1948, Seasonal trends in burrowing of pocket gophers (*Thomomys*): Journal of Mammalogy, v. 29, p. 38–44, doi:10.2307/1375279.

Miller, M.A., 1950, Eradication of pocket gophers: California Agriculture, v. 4, no. 12, p. 8–10.

Milliman, J.D., and Meade, R.H., 1983, World-wide delivery of river sediment to the oceans: The Journal of Geology, v. 91, p. 1–21, doi:10.1086/628741.

Morrell, S.H., 1971, The life history of the San Joaquin kit fox: California Fish and Game, v. 58, p. 162–174.

Murray, D.F., 1967, Gravel mounds at Rocky Flats, Colorado: The Mountain Geologist, v. 4, p. 99–107.

Museum of Vertebrate Zoology at Berkeley (MVZB), 2009, Mammal Collection: http://mvz.berkeley.edu/Mammal_Collection.html (accessed 30 November 2009).

National Oceanic and Atmospheric Administration (NOAA), 2009, National Climatic Data Center: Asheville, North Carolina: http://www.ncdc.noaa.gov/oa/ncdc.html (accessed 30 November 2009).

Nikiforoff, C.C., 1941, Hardpan and micro relief in certain soil complexes of California: U.S. Department of Agriculture Technical Bulletin 745.

Orloff, S.G., 2002, Medium to large mammals (Chapter 9), *in* Vollmar, J.E., ed., Wildlife and Rare Plant Ecology of Eastern Merced County's Vernal Pool Grasslands: Merced, University of California, p. 339–384.

Orloff, S., Hall, F., and Spiegel, L., 1986, Distribution and habitat requirements of the San Joaquin kit fox in the northern extreme of their range: Transactions of the Western Section of the Wildlife Society, v. 22, p. 60–70.

Paeth, R.C., 1967, Depositional origin of Mima mounds [thesis]: Corvallis, Oregon State University, 122 p.

Page, W.D., Swan, F.H., III, and Hanson, K.L., 1977, Prairie mounds (Mima mounds, hog wallows) in the Central Valley, *in* Singer, M.J., ed., Soil Development, Geomorphology, and Cenozoic History of the Northeastern San Joaquin Valley and Adjacent Areas, Guidebook for Joint SCS-GSA Field Trip: Davis, Department of Land, Air and Water Resources, University of California, p. 247–266.

Phillips, J.D., 2009, Biological energy in landscape evolution: American Journal of Science, v. 309, p. 271–289, doi:10.2475/04.2009.01.

Price, W.A., 1949, Pocket gophers as architects of Mima (pimple) mounds of the western United States: The Texas Journal of Science, v. 1, p. 1–17.

Reed, S.E., and Amundson, R.G., 2007, Sediment, gophers, and time: A model for the origin and persistence of Mima mound—Vernal pool topography in the Great Central Valley, *in* Schlising, R.A., and Alexander, D.G., eds., Vernal Pool Landscapes: Chico, California State University, p. 15–27.

Reichman, O.J., and Seabloom, E.W., 2002, The role of pocket gophers as subterranean ecosystem engineers: Trends in Ecology & Evolution, v. 17, p. 44–49, doi:10.1016/S0169-5347(01)02329-1.

Reneau, S.L., 1988, Depositional and erosional history of hollows: Application to landslide location and frequency, long-term erosion rates, and the effects of climatic change [Ph.D. thesis]: Berkeley, University of California, 654 p.

Rhoads, S.N., 1898, "Noxious" or "beneficial"? False premises in economic zoology: American Naturalist, v. 32, p. 571–581, doi:10.1086/276971.

Richens, V.B., 1965, An evaluation of control on the Wasatch pocket gopher: The Journal of Wildlife Management, v. 29, p. 413–425, doi:10.2307/3798038.

Robins, J.D., and Vollmar, J.E., 2002, Livestock grazing and vernal pools (Chapter 11), *in* Vollmar, J.E., ed., Wildlife and Rare Plant Ecology of Eastern Merced County's Vernal Pool Grasslands: Merced, University of California, p. 401–430.

Roering, J.J., 2004, Soil creep and convex-upward velocity profiles: Theoretical and experimental investigation of disturbance-driven sediment transport on hillslopes: Earth Surface Processes and Landforms, v. 29, p. 1597–1612, doi:10.1002/esp.1112.

Roering, J.J., Kirchner, J.W., and Dietrich, W.E., 1999, Evidence for nonlinear, diffusive sediment transport on hillslopes and implications for landscape morphology: Water Resources Research, v. 35, p. 853–870, doi:10.1029/1998WR900090.

Roering, J.J., Kirchner, J.W., Sklar, L.S., and Dietrich, W.E., 2001, Hillslope evolution by nonlinear creep and landsliding: An experimental study: Geology, v. 29, p. 143–146, doi:10.1130/0091-7613(2001)029<0143:HEBNCA>2.0.CO;2.

Romañach, S.S., Seabloom, E.W., and Reichman, O.J., 2007, Costs and benefits of pocket gopher foraging: Linking behavior and physiology: Ecology, v. 88, p. 2047–2057, doi:10.1890/06-1461.1.

Ross, B.A., Tester, J.R., and Breckenridge, W.J., 1968, Ecology of Mima-type mounds in northwestern Minnesota: Ecology, v. 49, p. 172–177, doi:10.2307/1933579.

Scheffer, V.B., 1948, Mima mounds: A reply: The Journal of Geology, v. 56, p. 231–234, doi:10.1086/625506.

Scheffer, V.B., 1958, Do fossorial rodents originate Mima-type microrelief? American Midland Naturalist, v. 59, p. 505–510, doi:10.2307/2422495.

Schlising, R.A., and Alexander, D.G., eds., 2007, Vernal Pool Landscapes: Studies from the Herbarium: Chico, California State University, 213 p.

Schoenenberger, P.J., Wysocki, D.A., Benham, E.C., and Broderson, W.D., 2002, Field book for describing and sampling soils: National Soil Survey Center, Lincoln, Nebraska, Natural Resources Conservation Service.

Seabloom, E.W., Reichman, O.J., and Gabet, E.J., 2000, The effect of hillslope angle on pocket gopher (*Thomomys bottae*) burrow geometry: Oecologia, v. 125, p. 26–34, doi:10.1007/PL00008888.

Slatton, K.C., Carter, W.E., Shrestha, R.L., and Dietrich, W., 2007, Airborne laser swath mapping: Achieving the resolution and accuracy required for geosurficial research: Geophysical Research Letters, v. 34, p. L23S10, doi:10.1029/2007GL031939.

Smallwood, K.S., and Morrison, M.L., 1999, Spatial scaling of pocket gopher (*Geomyidae*) density: The Southwestern Naturalist, v. 44, p. 73–82.

Smallwood, K.S., Geng, S., and Zhang, M., 2001, Comparing pocket gopher (*Thomomys bottae*) density in alfalfa stands to assess management and conservation goals in northern California: Agriculture Ecosystems & Environment, v. 87, p. 93–109, doi:10.1016/S0167-8809(00)00300-5.

Sutter, G., and Francisco, R., 1998, Vernal pool creation in the Sacramento Valley: A review of the issues surrounding its role as a conservation tool: Ecology, Conservation and Management of Vernal Pool Ecosystems: Sacramento, California Native Plant Society, p. 190–194.

Turner, G.T., 1969, Responses of mountain grassland vegetation to gopher control, reduced grazing, and herbicide: Journal of Range Management, v. 22, p. 377–383, doi:10.2307/3895846.

Turner, G.T., Hansen, R.M., Reid, V.H., Tietjen, H.P., and Ward, A.L., 1973, Pocket gophers and Colorado mountain rangeland: Colorado Agricultural Experiment Station Bulletin, v. 5545, Fort Collins, Colorado, Colorado State University.

Trust, P.L., and Holland, R.F., 2009, California's Great Valley Vernal Pool Habitat Status and Loss: Rephotorevised 2005: Prepared for Placer Land Trust, Auburn, California, 14 p.

United States Endangered Species Act of 1973, 2000, 16 United States Code Section 1531 et seq.

U.S. Fish and Wildlife Service (USFWS), 2004, Draft Recovery Plan for Vernal Pool Ecosystems of California and Southern Oregon, v. 69, p. 67,601–67,602.

University of California at Merced and The Nature Conservancy, 2002, 5,030 acres in eastern Merced County preserved: Joint press release (4 September 2002).

Vleck, D., 1979, The energy cost of burrowing by the pocket gopher *Thomomys bottae*: Physiological Zoology, v. 52, p. 122–136.

Washburn, A.L., 1988, Mima Mounds: An Evaluation of Proposed Origins with Special Reference to the Puget Lowlands: Washington State Dept. of Natural Resources, Division of Geology and Earth Resources, Report 29, 53 p.

Witham, C.W., Bauder, E.T., Belk, D., Ferren, W.R., Jr., and Ornduff, R., eds., 1998, Ecology, Conservation, and Management of Vernal Pool Ecosystems, Proceedings from a 1996 Conference: Sacramento, California Native Plant Society, 296 p.

Yoo, K., Amundson, R., Heimsath, A.M., and Dietrich, W.E., 2005, Process-based model linking pocket gopher (*Thomomys bottae*) activity to sediment transport and soil thickness: Geology, v. 33, p. 917–920, doi:10.1130/G21831.1.

Zedler, P.H., 1987, The ecology of southern California vernal pools: A community profile: U.S. Fish and Wildlife Service, Biological Report 85 (7.11), 136 p.

Zinnel, K.C., and Tester, J.R., 1994, Plains pocket gopher social behavior, *in* Proceedings of the Thirteenth North American Prairie Conference: Windsor, Ontario, Canada, p. 95–101.

MANUSCRIPT ACCEPTED BY THE SOCIETY 5 MARCH 2012

The Geological Society of America
Special Paper 490
2012

"Pimple" mound microrelief in southern Saskatchewan, Canada

L. Lee-Ann Irvine*
Royal Saskatchewan Museum, 2340 Albert Street, Regina, SK S4P 2V7, Canada

Janis E. Dale†
Department of Geology, University of Regina, 3737 Wascana Parkway, Regina, SK S4S 0A2, Canada

ABSTRACT

Gravelly soil mounds less than 1 m high and up to 20 m in diameter, generally referred to as "pimple mounds," are found in the prairies and aspen parklands of southern Saskatchewan, Canada. These low-relief mounds are the northernmost occurrence in North America of this geomorphic feature documented to date. When truncated by cultivation, former mounds are evident on air photos as a pattern of small, light-toned patches called "mound scars" that contrast with the surrounding darker soil. In many cases, these photographs are the only existing evidence of former mound topography.

Using air photos and direct field observation for identifying both intact and truncated mounds, the spatial distribution and characteristics of 10 pimple mound sites were examined, with mounds at the Little Manitou site studied in detail. Examination of the morphology and stratigraphy of multiple mounds indicates that bioturbation by burrowing animals has had, and continues to have, a major impact on the size, shape, nature, and origin of present-day mounds. Statistical analysis of 124 intact and 190 truncated mounds at the Little Manitou Lake site indicates that mounds form a more regular than random pattern, with strong biological implications. Saskatchewan mounds were compared to 30 other documented mound sites in North America. While most Saskatchewan mounds have greater relief and denser spacing, mounds from Texas, Wyoming, and Colorado are spatially most similar, also occur in predominantly prairie landscapes in areas with shallow soils, and are heavily bioturbated by burrowing animals.

INTRODUCTION

Background

The pimple mound microrelief discussed and assessed here is distributed across the southern part of the province of Saskatchewan, Canada (Fig. 1). Microrelief was first identified on air photos of the Watrous Spillway and Little Manitou Lake Fan southeast of Saskatoon (Fig. 2) by Mollard (1978). It appeared as a pattern of small, light-toned spots and patches in cultivated fields that contrasted with the surrounding darker soil. These patches, or "mound scars"—remnants of mounds truncated and flattened by cultivation—are so abundant that they have been compared to an "outbreak of prairie pimples" (McDougall, 2000). Mollard (1978, 1996, 1999) later observed the same pattern at several other localities in southern Saskatchewan. Although sometimes indistinct at ground level, especially when obscured by vegetation, we confirmed that the pattern is caused by small, low-relief gravelly soil mounds, some truncated by cultivation, and others

*lirvine@royalsaskmuseum.ca
†Janis.Dale@uregina.ca

Irvine, L.L.-A., and Dale, J.E., 2012, "Pimple" mound microrelief in southern Saskatchewan, Canada, *in* Horwath Burnham, J.L., and Johnson, D.L., eds., Mima Mounds: The Case for Polygenesis and Bioturbation: Geological Society of America Special Paper 490, p. 43–61, doi:10.1130/2012.2490(02).

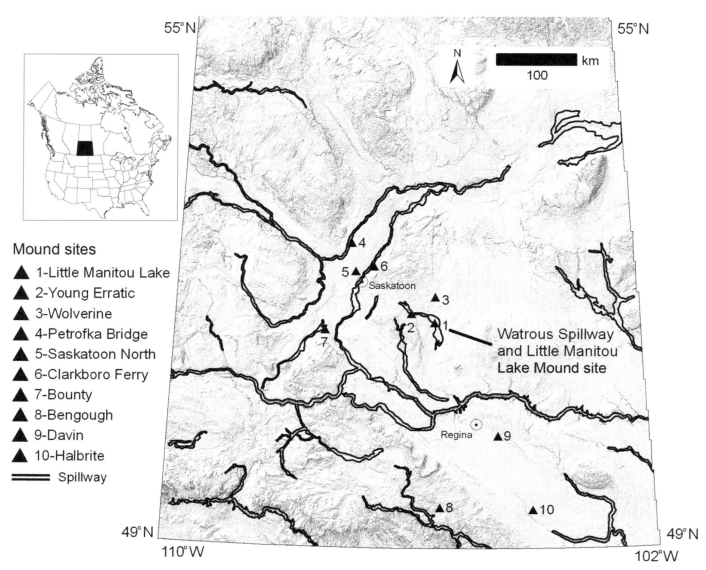

Figure 1. Hill-shaded digital elevation model (DEM) of southern Saskatchewan showing the location of the 10 pimple mound sites and their proximity to glacial spillways. The Watrous spillway contains Little Manitou Lake and Young Erratic mound sites (cf. Table 1). DEM courtesy of the Saskatchewan Heritage Branch, Regina. Modified from Irvine (2005).

partially eroded to reveal their lighter colored, caliche (carbonate) bearing B and C horizons. At some sites, however, intact natural mounds were found in nearby uncultivated areas (Figs. 2 and 3), most displaying clear evidence of small mammal bioturbation, mainly by rodents and their predators.

Purpose

This study documents the regional and spatial distribution, and general geomorphic and soil features of 10 pimple mound sites in southern Saskatchewan prairies and aspen parklands (Fig. 1). Documentation is based on air photos, field observations, and measurements of both intact and truncated mounds. Of these 10 sites, the Little Manitou Lake tract was studied in detail. This research represents the first time mounds have been regionally documented and studied in any detail in Saskatchewan, and thus far represents the northernmost occurrence of this interesting soil-geomorphic feature.

Much microrelief has been modified, truncated, and or lost (destroyed) by cultivation, leaving only mound scars as a blurred and indistinct legacy. Regular cropping will doubtless insure that this legacy becomes less discernible with time, and likely will destroy it completely. Consequently, it is important to document and study intact (uncultivated) and still preserved pristine microrelief, and the mound scars, before both are completely obliterated.

Figure 2. 1995 air photo of the North and South Field study sites in the Watrous Spillway (T31-R23-W2). White arrows point to mound scars that appear as small white spots in cultivated areas. CS 08 144 (1:60,000). Intact (untruncated) mounds can also be seen in uncultivated fields. Modified from Irvine (2005).

Terminology

The term "pimple mound" as used here has been applied to a wide range of natural soil-geomorphic features to which various other names have been given by many authors in an extensive and temporally deep literature (overview in Johnson and Horwath Burnham, this volume, Introduction; cf. also Washburn, 1988). Besides "pimple mounds," two other common names are "prairie mounds" and "Mima mounds."

In Saskatchewan, both intact and truncated mounds have been variously called Venn mounds, prairie mounds, pimple mounds, and Mima mounds (Mollard, 1978, 1996; McDougall, 2000). In this study, the term "pimple mound"—or mound—

is used in reference to intact (untruncated) mounds. The term "mound scar" refers to cultivation-truncated mounds, and or to partially eroded remnant mounds, both of which are visible as light-toned spots or patches on air photos (Irvine, 2005).

STUDY SITES AND METHODS

The location and extent of each of the 10 Saskatchewan mound sites (Fig. 1, Table 1) were determined using a combination of air photos, global positioning system (GPS) coordinates, National Topographic System (NTS) map sheets for the area, and field observations. A mound site is defined as a geographic area where either intact mounds and/or mound scars are present.

Figure 3. Typical intact natural Saskatchewan pimple mound in the Watrous Spillway with a backpack (40 cm long) for scale. (Photo courtesy L.L.-A. Irvine.)

neighbor statistics were then calculated using Microsoft Excel and ArcView with the Nearest Neighbor Analyst Extension.

Measurements and observations were made of the mound height, diameter, and volume, as well as vegetation cover, soil type, and degree of animal bioturbation. The height of each mound was measured from the highest central point to what was considered the outlying mound perimeter in eight directions (S, SE, E, NE, N, NW, W, and SW). A series of shallow, test pits (50 cm sq. × 50 cm deep) were hand dug at a number of mound-intermound locations at the Little Manitou Lake study site (Fig. 4). At one mound field (A-SE17), backhoe trenches were excavated across three representative mounds to describe and document internal mound nature and composition. Soil horizons were identified and samples taken for laboratory analyses to determine dry Munsell soil color, soil texture, and percent calcium carbonate ($CaCO_3$).

RESULTS AND INTERPRETATIONS

Regional Distribution and Physical Setting of Pimple Mound Sites in Southern Saskatchewan

The mound sites identified on air photos by Mollard (1978, 1996, 1999, 2000, personal commun.) formed the initial research areas in southern Saskatchewan (Figs. 1 and 5, Table 1). Based on additional air photo analyses and our field investigations, many new mound fields were identified. Further, our work expanded the geographic extent of Mollard's original sites, plus delineated the boundaries of the sites.

All 10 mound sites in Table 1 lie within the Canadian Prairie Ecozone (Fig. 5), which consists of level to gently rolling grassland plains marked by numerous and dispersed low-elevation "uplands." The climate is characterized by long, cold winters and short, very warm summers, with most precipitation derived from frontal thunderstorms during the summer (Padbury and Acton, 1994; Acton et al., 1998). Precipitation increases from southwest to northeast across the Prairie Ecozone resulting in several Ecoregion subzones (Fig. 5). Eight mound sites lie within the somewhat drier Mixed Grassland Ecoregion, whereas two sites lie within the southern edge of the somewhat cooler and moister Aspen Parkland Ecoregion (Fig. 5). The burrowing animals that live in these areas and inhabit and/or impact the mounds include the northern pocket gopher (*Thomomys talpoides*), Richardson's ground squirrel (*Spermophilus richardsonii*), thirteen-lined ground squirrel (*Spermophilus tridecemlineatus*), and their major predators, the American badger (*Taxidea taxus*), coyote (*Canis latrans*), plus other mustelids and canids that are adapted to the semiarid climate and grassland vegetation of south-western Saskatchewan (Acton et al., 1998).

The present-day topography of Saskatchewan is largely the result of geomorphic processes that occurred during and after the last glaciation, initiated by the Laurentide Ice Sheet. From ca. 18,000 to 8000 ^{14}C yr B.P. ice retreated from the southwest to the northeast corner of Saskatchewan following the regional slope of the landscape (Clayton and Moran, 1982; Saskatchewan Geological Survey, 2003). Ice vacated the study area in south-central

Within each mound site, there may be several distinct groups of mounds, termed "mound fields" (Irvine, 2005). Graticule coordinates of the center of each of the 10 mound sites were recorded after site boundaries were determined. Using a combination of topographic maps, air photos, and Geographic Information Systems (GIS) mapping software (ArcView), mound sites were studied to determine local environmental characteristics of the soils and vegetation, and the relationships of these to the regional and local landforms, primarily glaciogenic.

The Little Manitou Lake mound site (Figs. 1 and 4, Table 1) was selected for detailed study because it is located in the Usborne Prairie Farm Rehabilitation Administration (PFRA) community pasture, and thus had experienced minimal human impact. Regionally, this site contains the largest number of both intact and truncated mounds (mound scars). This well mounded tract is located within the eastern portion of the Watrous Spillway and the Little Manitou Outwash Fan (Fung, 1999), which, in this part of Saskatchewan, formed following the recession of the Late Wisconsinan Laurentide Ice Sheet, ca. 11,500 ^{14}C yr B.P. (Saskatchewan Geological Survey, 2003). Around this time, large volumes of glacial meltwater scoured out what is now Little Manitou Lake and deposited a diverse mix of boulders, gravels, and other coarse-grained materials before ultimately turning southeast into the Last Mountain Channel (Kehew and Teller, 1994).

Three mound fields in uncultivated tracts (A-SE17, B-Angle Road, and C-Driller) and two mound scar fields in nearby cultivated areas (North Field and South Field) were studied within the Little Manitou Lake mound site (Figs. 2 and 4). Following Vitek (1978), a nearest-neighbor approach was used to determine whether mound patterns were more regular than random. The location of each intact mound was recorded at its highest point using a GARMIN hand-held GPS receiver. For the two mound scar fields, the location of each mound was determined using a combination of GIS mapping software and air photos. Mound spacing, field density (mounds/hectare), and nearest-

TABLE 1. PIMPLE MOUND SITES IN SOUTHERN SASKATCHEWAN INDICATING LOCATION, ECOREGION, GEOMORPHIC SETTING, SURFICIAL GEOLOGY, AND DOMINANT SOIL TYPE

Mound site	Latitude, Longitude (center of site)	Ecozone: Ecoregion[§,§§]	Geomorphic setting; surficial geology[***]	Dominant Canadian Soil Type[†] (U.S. Soil Type)[##]: Soil texture[†]
1. Little Manitou Lake[**]	51°38'57" N, 105°17'12" W	Prairie: Moist Mixed Grassland	Glacial lake plain with coarse grained fan[#] within major glacial spillway[***]; glaciofluvial and alluvial deposits	Regosolic (Entisolic): Loam; Dark Brown Chernozemic (Aridic Boroll): Sandy loam
2. Young Erratic[**]	51°45'00" N, 105°42'30" W	Prairie: Moist Mixed Grassland	Glacial lake plain within major glacial spillway[***]; glaciofluvial and alluvial deposits	Regosolic (Entisolic): Loam
3. Wolverine[**]	51°56'45" N, 105°17'00" W	Prairie: Aspen Parkland	Glacial till plain with gravelly outwash deltaic plain[**]; glaciofluvial and alluvial deposits	Black Chernozemic (Udic Boroll): Loam
4. Petrofka Bridge	52°33'30" N, 106°50'00" W	Prairie: Aspen Parkland	Glacial lake plain adjacent to major glacial spillway[***]; glacial till (moraine deposits)	Black Chernozemic (Udic Boroll): Loam
5. North Saskatoon	52°14'00" N, 106°45'00" W	Prairie: Moist Mixed Grassland	Glacial lake and eroded glacial till plain[*]; glacial till (moraine deposits)	Dark Brown Chernozemic (Typic Boroll): Loam
6. Clarkboro Ferry	52°17'30" N, 106°24'30" W	Prairie: Moist Mixed Grassland	Glacial lake and eroded glacial till plain adjacent to major glacial spillway[***]; glacial till (moraine deposits)	Dark Brown Chernozemic (Typic Boroll): Loam
7. Bounty	51°33'00" N, 107°16'00" W	Prairie: Moist Mixed Grassland	Glacial lake plain near major glacial spillway[***]; glacial till (moraine deposits)	Dark Brown Chernozemic (Typic Boroll): Loam, sandy loam
8. Bengough[**]	49°29'00" N, 105°09'30" W	Prairie: Mixed Grassland	Hummocky morainal upland[§§]; glacial till (moraine deposits)	Brown Chernozemic (Aridic Boroll): Clay loam
9. Davin[**]	50°19'30" N, 104°09'00" W	Prairie: Aspen Parkland	Hummocky morainal upland[§§]; glacial till (moraine deposits)	Dark Brown Chernozemic (Typic Boroll): Loam
10. Halbrite[††]	49°27'30" N, 103°32'30" W	Prairie: Moist Mixed Grassland	Glacial till plain near major glacial spillway[***]; glacial till (moraine deposits)	Dark Brown Chernozemic (Typic Boroll): Clay loam

*Acton and Ellis (1978).
†Acton et al. (1990).
§Padbury and Acton (1994).
#Kehew and Teller (1994).
**Mollard (1996).
††Mollard (1999).
§§Acton et al. (1998).
##Canada. Soil Classification Working Group (1998).
***Saskatchewan Geological Society (2002).

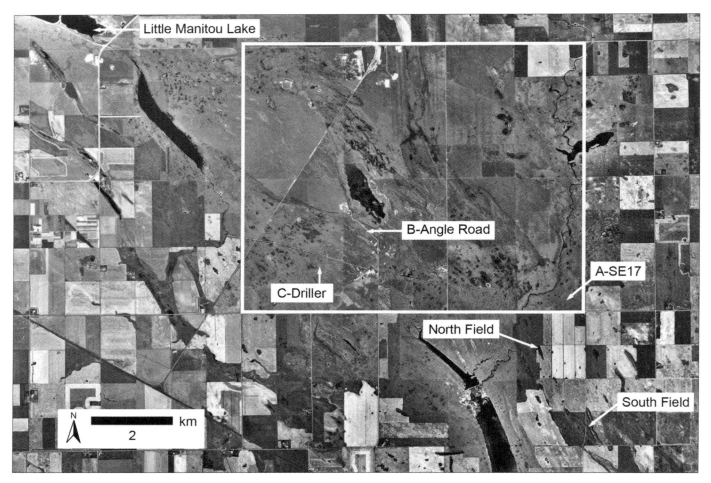

Figure 4. Air photo of the Little Manitou Lake study area showing the A-SE17, B-Angle Road, and C-Driller mound fields and two mound scar fields, North Field and South Field. (Fig. 2 is an enlargement of this area.) The white outline indicates the Usborne Prairie Farm Rehabilitation Administration (PFRA) community pasture. Modified from Irvine (2005).

Saskatchewan between ca. 11,500 and 11,000 ^{14}C yr B.P. Ice recession left a network of lakes, spillways, and meltwater channels that connect a series of glacial lakes and till plains on the Saskatchewan Plain (Saskatchewan Geological Survey, 2003).

Eight of the ten mound sites are within or adjacent to major meltwater channels and gravelly outwash plains associated with glacial spillways (Fig. 1, Table 1). These include spillways associated with Glacial Lake Saskatchewan (Saskatoon North, Petrofka Bridge, Clarkboro Ferry, and Bounty mound sites), Glacial Lake Elstow (Little Manitou Lake, Young Erratic, and Wolverine mound sites), and Glacial Lake Souris (Halbrite mound site). The two remaining mound sites (Bengough and Davin) are associated with hummocky morainal uplands comprised of glacial till deposits and stratified glaciofluvial and glaciolacustrine deposits (Saskatchewan Geological Survey, 2003).

Soils in general reflect prevailing climate and biota, but locally vary in relation to parent material and topography. Under the Canadian soil classification system, mounds occur in areas with Brown, Dark Brown, or Black Chernozemic soils on loam,

sandy loam, and clay loam, which are typical prairie soils (Sites 3–10, Table 1). In the U.S. soil taxonomy system the classification of soils would be, respectively, Aridic Boroll, Typic Boroll, and Udic Boroll (Table 1) (Canada. Soil Classification Working Group, 1998). Regosolic soils, which are comparable to Entisols of the U.S. system, are found in the Little Manitou Lake and Young Erratic sites (Sites 1–2, Table 1). The mounds and mound scars at all 10 sites are composed of gravelly soils formed on either glacial till or glaciofluvial deposits. In general, mound soils have poorly expressed horizons, are strongly bioturbated, and are typically underlain by densely bedded gravels that, at least under the mounds, are commonly caliche (carbonate) coated.

Spatial Distribution of Mounds in the Little Manitou Lake Site

In total, 124 intact mounds in the three uncultivated mound fields (C-Driller, B-Angle, and A-SE17), and 190 mounds scars in the two cultivated fields (North, South) at the Little Mani-

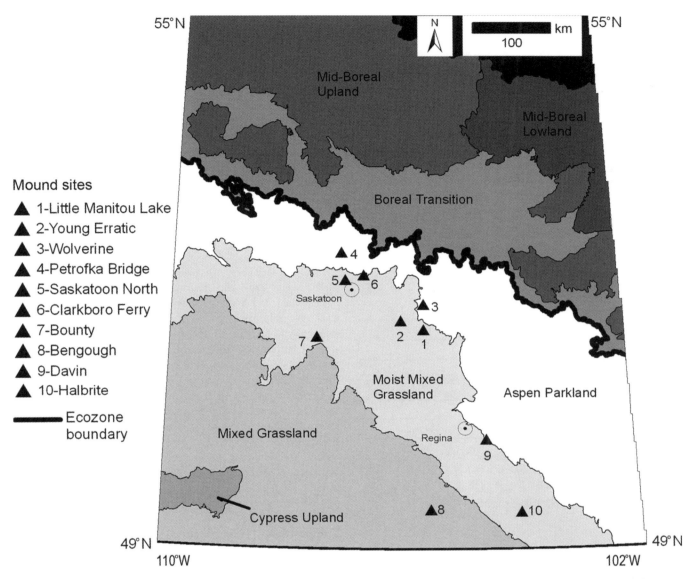

Figure 5. Southern Saskatchewan showing the location of the 10 pimple mound sites and associated ecoregions. The thick black Ecozone boundary line separates the Prairie Ecozone to the south, from the Boreal Plain Ecozone to the north.

tou Lake site were mapped (Figs. 1 and 4, Table 1). The densities of mounds and mound scars range between 2.25 and 16.96 mounds/ha, and are, respectively, spaced between 48 and 22 m apart (Fig. 6, Table 2). The five mound fields show remarkable consistency with respect to densities, except for the B-Angle Road field. This field was much smaller than the other fields (1.3 ha), and yet contained 22 intact mounds, which results in a much higher mound density at 16.96 mounds/ha compared to much lower mound densities of the other fields (2.25–2.75 mounds/ha), and lower mean distance between mounds (22.0 m). In Table 2, the R values calculated for nearest-neighbor statistics in all five mound fields are greater than 1, indicating that the mounds form a more regular than random pattern (cf. Fig. 6). B-Angle Road field has the highest R value at 1.55. This regu-

larity suggests that there is a controlling process that underlies mound formation and location at the Little Manitou Lake site, and that mound densities and spacing are not random occurrences. The consistency of Z-transform values (which measures the significance of the R value based on standard deviation) for intact mounds and mound scars also suggests that these are not random patterns.

Morphology of Intact Mounds in the Little Manitou Lake Site

Intact mounds in the three uncultivated fields A-SE17, B-Angle Road, and C-Driller were measured and assessed to document their general shape, relief (height), volume, and

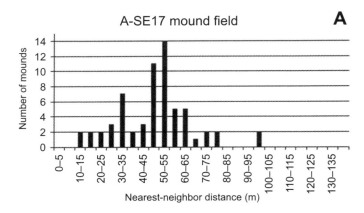

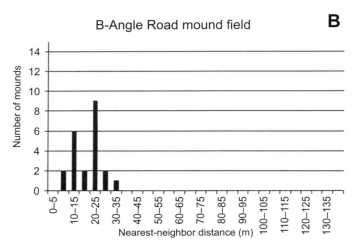

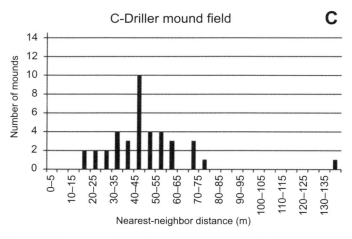

Figure 6. Nearest-neighbor distances for the (A) A-SE17 mound field (mean = 48.03 m), (B) B-Angle Road mound field (mean = 22 m), and (C) C-Driller mound field (mean = 45.88 m). Modified from Irvine (2005).

degree of rodent bioturbation (Table 3). The average mound relief is 0.36 m, the average diameter is 14.7 m, and the average volume 9.8 m³. B-Angle Road has the smallest mounds, with the lowest mean diameter (11.03 m), mean area (108.1 m²), and mean volume (5.0 m³). All three fields have mounds of similar heights, averaging 0.36 m. As indicated by the representative mound in Figure 7, the general mound shape is quasi-circular with no obvious directional orientation, with a distinct central elevation, and somewhat irregular perimeters. The secondary high spot on the contour map indicates the frequently bumpy surfaces of the strongly bioturbated pimple mounds on these three tracts.

The general species composition of mound flora and fauna were recorded and can be used as a measure of bioturbational disturbance. All mounds, including those in aspen parklands, are generally covered with prairie vegetation, including mixed grasses, wildflowers such as low goldenrod (*Solidago missouriensis*), smooth fleabane (*Erigeron glabelus*), and prairie coneflower (*Ratibida columnifera*), and shrubs such as wolf-willow (*Elaeagnus commutata*) and Western snowberry (*Symphoricarpos occidentalis*). Notably, such species are all common to areas of bioturbationally disturbed soil. In this regard, most mound surfaces are uneven, bumpy, and bioturbationally perforated with abundant burrow holes, indicating strong bioturbation by rodents (Table 3) and their predators, mainly badgers. Rodent holes range from a mean of 6 holes/mound for the B-Angle Road to 21, and 24 holes/mound for A-SE17 and C-Driller fields, respectively (Table 3). The presence of numerous pocket gopher, ground squirrel, and badger holes and their respective back dirt (spoil) piles, as stated, give each mound an undulating uneven profile (cf. Fig. 7). Badger activity is especially extensive on the mounds, illustrating the importance of large predator impacts on mounds, and doubtless in the overall mound formation process (Johnson, 1999; cf. Eldridge, 2004).

Composition of Intact Mounds in the Little Manitou Lake Site

Soil Types

Observations determined from the hand-dug test pits and backhoe trenches excavated through representative mounds at the A-SE17 mound field collectively indicate that mound soils are classified under the Canadian soil taxonomic system as Dark Brown Chernozems (Canada. Soil Classification Working Group, 1998) (Table 1). This is the dominant soil type in this area and has developed on till and glaciofluvial parent material. Test pits in mounds at A-Angle Road and C-Driller mound fields indicate that A horizons across the mounds have highly variable thicknesses. Typically A horizons average ~15 cm thick, but thicken to as much as 41 cm toward mound centers, indicating maximal bioturbation at centers.

Gravels

Soils at nearly every mound studied at the Little Manitou Lake site have an abundance of gravels. They are mainly pebble size, as confirmed in the pits and trenches, and abundant in ro-

TABLE 2. SPATIAL DISTRIBUTION AND NEAREST-NEIGHBOR CALCULATIONS FOR MOUNDS IN THE A-SE17, B-ANGLE ROAD, AND C-DRILLER FIELDS, AND FOR MOUND SCARS IN THE NORTH AND SOUTH FIELDS, IN THE LITTLE MANITOU LAKE MOUND SITE

Field name	Number of mounds	Area (ha)	Density (mounds/ha)	Nearest-neighbor analysis			
				Mean distance (m)	R value	Z value	Result
A-SE17 (mounds)	63	27.99	2.25	48.03	1.45	6.81	Tendency toward regularity
B-Angle Road (mounds)	22	1.30	16.96	22.00	1.55	4.91	Tendency toward regularity
C-Driller (mounds)	39	14.16	2.75	45.88	1.51	6.08	Tendency toward regularity
North Field ("mound scars")	82	34.80	2.36	*	1.48	8.38	Tendency toward regularity
South Field ("mound scars")	108	46.80	2.31	*	1.43	8.45	Tendency toward regularity

Note: Modified from Irvine (2005).
*Values not provided with nearest-neighbor extension with ArcView program.

TABLE 3. SUMMARY OF MOUND MORPHOLOGY MEASUREMENTS FOR UNCULTIVATED AND INTACT A-SE17, B-ANGLE ROAD, AND C-DRILLER MOUND FIELDS

Field name	Number of mounds	Mean diameter (m)	Mean area (m²)	Mean height (m)	Mean volume (m³)	Mean number of rodent holes/mound
A-SE17	63 (58*)	15.54	200.3	0.37	11	21
B-Angle Road	22	11.03	108.1	0.35	5.0	6
C-Driller	39	17.54	251.5	0.35	13.4	24
Total/average	124	14.70	186.6	0.36	9.8	17

Note: Modified from Irvine (2005).
*Due to partial truncation and/or excessive disturbance, five mounds were not included in the mean volume calculations.

dent and predator back dirt piles that typify the surface of many mounds. Although coarse gravels (large cobbles, boulders) are abundant in intermound areas and in lower B horizons (stonelayers) of mound soils, fine gravels—mainly pebbles and small cobbles—are observed within the mounds per se. Most mounds, however, are dominated by pebbles <5 cm in diameter. In southwest Missouri pimple mounds, concentrations of small gravels, mainly pebbles (<6.5 cm), were attributed to bioturbational sorting by pocket gophers because only pebbles that size or smaller can fit through the burrows of these small, sausage-size animals (Horwath et al., 2002; Johnson et al., 2002; Horwath and Johnson, 2006). Larger clasts in the Missouri study were assumed to have bioturbationally subsided to form the basal stonelayer that underlies the mounds there. Similarly, the basal coarse gravelly layers—stonelayers—in Saskatchewan mounds likewise suggest bioturbational sorting by the rodent-dominated burrowing animals that inhabit the mounds. Many mound gravels exhibit white coatings (precipitates) of calcium carbonate ($CaCO_3$) that may have formed in lower profile levels prior to bioturbational mixing. Notably, trench observations indicate that such carbonate precipitates are more prevalent in B and C horizons.

Mound Profiles and Bioturbation

Figure 8 depicts a half-mound profile of A-M0028 that is reasonably representative of the three mounds trenched in the A-SE17 mound field. Although the trench was ~20 m in length, the A–B profile displays only the area from the mound center at B, extending 10 m northwest toward the intermound area at A (cf. Fig. 7). Soil profiles display marked disturbance bioturbation, especially near mound centers. The large, ovate in-filled burrows (krotovina) measuring up to 78 cm diameter, are attributed to the broad, dorso-ventrally flattened bodies of burrowing badgers, a major predatory species of the various burrowing rodents that inhabit the mounds.

Owing to such high bioturbation levels, soil horizon boundaries are invariably blurred, and not as distinct or abrupt as displayed in Figure 8. Indeed, toward the centers of the mounds, horizon boundaries range from gradual (10–20 cm) to diffuse (>20 cm). The other two trenches exhibited similar levels of bioturbational disturbance; both displayed abundant in-filled burrows (krotovina) and, consequently, poorly differentiated horizons.

Mounds are composed of sand- and gravel-rich soils that are invariably thicker, darker, and better drained than intermound soils. For example, on mound A-M0028 (Figs. 7 and 8), the A and B horizons were collectively 68 cm thick at the center of the mound and only 40 cm at its edge bordering the intermound area. As indicated, mound soils are markedly bioturbated by rodents and their predators, especially near their centers. Buried horizon traces, mixed gravels, and in-filled animal burrows (krotovina) create a complex mound composition. Mound soils are thus notably easier for animals to penetrate than are the coarser gravel-rich intermound areas.

As a result of these compositional differences, probably largely created by the animals themselves, burrowing activities are concentrated in the mounds, making it easier for excavating burrows and nests. In addition, significant amounts of calcium carbonate ($CaCO_3$) have precipitated in B (3%–6%) and Ck (6%–8%) horizons, and give truncated or partly eroded mounds their light-toned appearance on air photos.

Mound Morphologies, Densities, and Spacings

Mound shapes, heights, and volumes differ between mound fields. Whether this is due to differences in the glacial deposits that comprise the soil parent materials or to the degree and perhaps longevity of bioturbation is not known. Most mounds are observed to have a distinctly higher central mound area, suggesting that mound centers represent the original mound nucleation

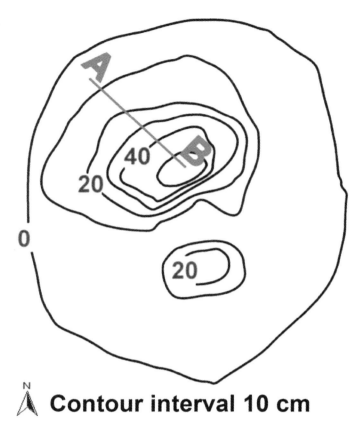

N
Contour interval 10 cm

Figure 7. Contour map of representative mound A-M0028 in the A-SE17 mound field ~20 m in diameter (cf. Fig. 8). The line A–B indicates the trenched area shown in Figure 8. Modified from Irvine (2005).

center that has been modified over time through a combination of both bioturbation and erosional processes. Given the fresh nature of rodent back dirt piles on mounds, it can be assumed that bioturbation processes are regularly reshaping the mounds, and that mound forming processes are still actively occurring. As a consequence of the coarser gravels that exist at the B-Angle Road mound field, bioturbators may have responded by creating smaller and more numerous mounds there, resulting in greater mound density and less spacing between mounds.

SASKATCHEWAN MOUNDS COMPARED WITH OTHER SELECTED NORTH AMERICAN MOUND SITES

Drawing on selected literature, Saskatchewan mounds are compared to the morphologies (heights, diameters, and volumes) and spacing of mounds described in over 30 mound sites elsewhere in North America (Fig. 9), summarized in the Appendix. Table 4 lists the mounds that are most similar in height, diameter, and spacing to the Saskatchewan mounds. These are the prairie mounds of Colorado (Vitek, 1978; Cox et al., 1987), Texas (Carty et al., 1988), and Wyoming (Spackman and Munn, 1984; Reider et al., 1999). These sites are in equivalent grassland environments that are likewise bioturbated by burrowing animals. In height, Saskatchewan mounds are most similar to those in the Laramie Basin (Wyoming) and the Texas Gulf Coast sites. In diameter, they are most similar to those along the Texas Gulf Coast, and in Colorado to both Blanca South and South-Central sites.

In North Dakota, no intact mounds were found by investigators there, only mound scars in the cultivated fields of former

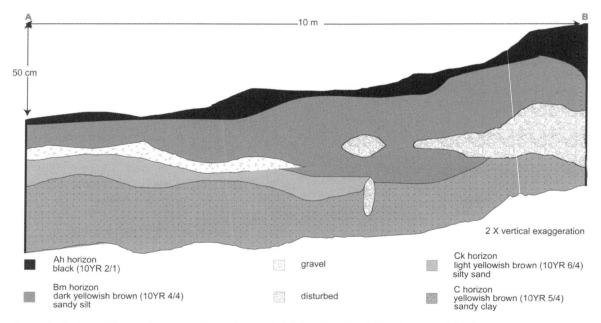

	Ah horizon black (10YR 2/1)		gravel		Ck horizon light yellowish brown (10YR 6/4) silty sand
	Bm horizon dark yellowish brown (10YR 4/4) sandy silt		disturbed		C horizon yellowish brown (10YR 5/4) sandy clay

2 X vertical exaggeration

Figure 8. Profile of the northwest quadrant of mound A-M0028 in the A-SE17 mound field, from mound edge A to mound center B (cf. Fig. 7). Note that soil horizon boundaries are gradual and blurred, not abrupt as shown. Modified from Irvine (2005).

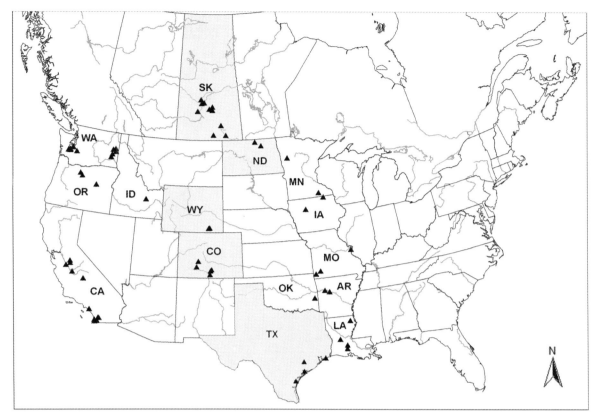

Figure 9. Mound sites in North America (triangles) based on selected literature. The states with mound morphologies and spacing similar to those in Saskatchewan are shown by gray highlights. Additional information on these sites is summarized in the Appendix. Modified from Irvine (2005).

prairies (Deal, 1972; Bluemle, 1983). The mounds were first observed as distinct spots on air photos, and referred to as "freckled land." Mound scars described at two such sites had diameters that ranged from 9.1 to 18.2 m. These are similar to the average 14.7 m diameter of intact mounds and mound scars in Saskatchewan.

The classic Mima mounds in Thurston County, Washington State are similar to Saskatchewan mounds in terms of diameter, gravelly compositions, and geologic contexts insofar as both formed from outwash gravels of approximately the same age (post-14,000 years) (Washburn, 1988). The western Washington mounds are, however, much more closely spaced, with a reported density of 86 mounds per ha, and have much greater heights—up to 2.1 m (Dalquest and Scheffer, 1942). They notably, however, lack the subsoil carbonates present in Saskatchewan mounds.

DISCUSSION AND INTERPRETATIONS

For well over a century, researchers across multiple disciplines have studied pimple-Mima-prairie mounds in Western North America, which has led to a plethora of genetic hypotheses and theories. These can be grouped into six genetic categories: (1) anthropomorphic; (2) deposition and/or erosion; (3) groundwater pressure; (4) seismic activity; (5) glacial/periglacial; and

(6) burrowing animals and discussed in Washburn (1988). In assessing these theories, the burrowing animal hypothesis best fits the results of this study for the following reasons.

The burrowing animal hypothesis was originally proposed by Gilbert (1875) to explain the abundant "prairie mounds" that dot the eastern flanks of the White Mountains in east-central Arizona. The animal burrowing model subsequently was endorsed by various workers (e.g., Campbell, 1906). The model was notably expanded and articulated in detail by Dalquest and Scheffer (1942, 1944), who produced two landmark studies that focused on multiple mound fields in the Olympia, Washington area, including the type locality at Mima Prairie. According to the Dalquest-Scheffer model, mounds form by the action of burrowing animals such as rodents, worms, insects, and other animals (see Johnson and Horwath Burnham, this volume, Introduction). Other workers have since further augmented and expanded the model with new information and supporting concepts (e.g., Price, 1949; Arkley and Brown, 1954; Cox, 1984, 1986, 1990; Johnson, 1989, 1999; Johnson et al., 1999, 2003; Horwath and Johnson, 2006, 2008; Reed and Amundson, 2007).

According to the expanded model, burrowing rodents and other mammals are attracted to soil that is deep and well-drained enough to allow foraging, to build their nests, reproduce, store

TABLE 4. LOCATION, MORPHOLOGY, AND SPACING OF OTHER NORTH AMERICAN
MOUND SITES MOST SIMILAR TO THE SASKATCHEWAN MOUND SITES

Location	Study	Height (m)	Diameter (m)	Volume (m³)	Density (mounds/ha)	Spacing (m)	Comments
Southern Saskatchewan	Irvine (2005)	0.36	14.7	9.8	2.25–16.96	22–48	Grasslands, gravelly soils within or near glacial spillways
Rolette County, North Dakota	Deal (1972)		9.1–18.3				Light-toned patches on air photos of lacustrine or wave-eroded glacial sediments
Towner County, North Dakota	Bluemle (1983)		9.1–18.3			Random spacing, some alignments	Light-toned patches on air photos of sandy, lacustrine sediments
Laramie Basin, Wyoming	Spackman and Munn (1984)	0.5	6–7			More regular than random spacing	Considerable bioturbation encountered in mound trenches but less in intermound areas
Laramie Basin, Wyoming	Reider et al. (1999)	0.15–0.65	10–20		26	21	Not located on the floodplain; composed of churned, unsorted materials (fines, pebbles, gravels)
South-central Colorado	Vitek (1978)	0.135	11.09 long and 10.18 wide	15		18.5, "regular spacing"	Mostly fine material with scattered pebbles; cobbles and boulders in intermound areas
Blanca South, Rocky Mtns., Colorado	Cox et al. (1987)	0.25	12.81				Soils contained an abundance of small stones
Gulf coast, Texas	Carty et al. (1988)	0.5	15.5				Prairie region

Note: Modified from Irvine (2005).

food, and to otherwise survive. However, in Saskatchewan, both intact mounds and mound scars occur in habitat-challenged areas where soils are thin, seasonally wet, and lie above a burrowing-resistant subsurface layer, notably densely bedded gravels. These mound sites tend to be associated with glacial spillways where the gravelly nature of the soils may preclude cultivation, thus insuring preservation of at least some intact mounds. In the Saskatchewan mound fields, burrowing animals, most notably rodents, appear to build their nests and store food in the highest (and driest) spots in a given area. Over time they augment these domiciles by displacing soil laterally toward their nests as they tunnel radially outwards, building their network of shallow feeding burrows. Vertical burrowing is restricted by the coarse nature of the spillway gravels, which requires them to excavate horizontally, thereby increasing the size and height of the mound at the expense of intermound soil.

Cox and Allen (1987) studied the centripetal (lateral and radial) translocation of soils by pocket gophers at Miramar Mounds National Landmark near San Diego, California. They found that over a one-year period, pocket gophers laterally moved soil plugs marked by iron pellets an average of 41 cm from intermounds toward their nest-centered mounds and 4.9 cm upward onto the mounds. Studies by Reed and Amundson (2007) in the extensively mounded-vernal pool Merced area of the Great Central Valley of California support the premise of biological activity in mound formation. In their studies, they calculated a median net soil erosion rate of 15 cm of soil per 1000 years offset by aboveground movements of soil by pocket gophers of 57 cm per 1000 years. The transport of soil by gophers is thus greater than erosion rates, suggesting that

gophers play a dominant role in the development and preservation of the mounds.

According to the animal burrowing hypothesis, mainly pebbles small enough to be moved by the dominant burrowers should be found in the mounds, while the maximum size of the pebbles moved varies with the species of rodent (i.e., the diameters of their burrows). Hansen and Morris (1968) found that pocket gophers in south-western Colorado typically moved pebbles less than 2.5 cm in diameter, whereas Johnson (1989) determined that gophers along the California coast typically move pebbles less than 6–7 cm in diameter, the maximum size of their burrows. In the Saskatchewan study, mound pebbles tend to be less than 5 cm, well within the range of the other studies. Larger particles (>7 cm) should bioturbationally sink to form a basal stonelayer (stone line) (Johnson, 1989; Johnson et al., 1999, 2002, 2003; Horwath and Johnson, 2006). In this regard, gravel concentrations, possibly stone lines, were encountered at approximate depths of 70 cm below mound centers in the excavated mounds A-M0028 (Fig. 8) and A-M0048 (Irvine, 2005). Occasional large pebbles and cobbles in and on the mounds are attributed to the larger, and abundant, ground squirrels, and especially to larger badgers and other predators.

Backhoe trenches confirm that Saskatchewan mounds have thicker, better-drained soils and display more evidence of bioturbation than do intermound areas. The latter have thinner and wetter soils, and in some cases a burrowing-resistant layer of glaciofluvial coarse gravels is encountered at shallow depth (less than 50 cm). The gravel layer extends under the mounds (>170 cm below mound centers).

As mounds bioturbationally develop and thicken, they presumably provide increasingly favorable nest sites because (1) the soil becomes increasingly deeper, better drained, and more loose for ease of subsequent burrowing; (2) mounded soils afford more protection from freezing, and thus provide more favorable overwintering sites; (3) mounds provide a refuge against periodically high water tables, and or incidences of surface flooding (cf. Koons, 1948); and (4) mounds provide occupants (e.g., ground squirrels, etc.) better visibility and protection from potential predators. Cox (1984) noted that all mound sites in North America lie within the range of the extensive pocket gopher (Geomyidae) family of rodents (40+ species).

Back dirt piles of the northern pocket gopher (*Thomomys talpoides*) are common in southern Saskatchewan and the mound fields described here. However, this cryptic animal throughout its range, including Saskatchewan, is rarely observed by researchers and others due to its fossorial nature. In our Saskatchewan study, we observed back dirt piles of northern pocket gophers associated with some mound areas. Other burrowing rodents in the Saskatchewan mound sites are the abundant Richardson's ground squirrel (*Spermophilus richardsonii*) and the thirteen-lined ground squirrel (*Spermophilus tridecemlineatus*). The regularity of mound spacing that we observed and determined to be present is likely due to the territorial nature of the burrowing rodents. Pocket gophers, for example, and presumably ground squirrels as well, tend to build their colonies to allow sufficient reproductive opportunities and food procurement space between their nests (Dalquest and Scheffer, 1942, 1944; Cox, 1984).

SUMMARY AND CONCLUSIONS

While pimple mounds have been studied in many areas of North America, this research represents the first time Saskatchewan mounds have been studied in any detail, and their regional distribution confirmed. Further, the Petrofka Bridge mound field, at 52° 33′ 30″ N, represents the northernmost occurrence of pimple-Mima-prairie mounds thus far documented in North America. In terms of morphology and spacing, Saskatchewan pimple mounds are most similar to certain grassland mounds of Wyoming, Colorado, and Texas. In terms of geologic (glaciogenic) context, internal composition, and age they are most similar to the temporally equivalent glacial spillway-outwash mounds of Mima and adjacent prairies in the Olympia area of western Washington.

Saskatchewan mounds are primarily distributed in, and genetically linked to, former glacial spillways. The coarse gravelly nature of spillway soils has limited cultivation in certain areas and has commensurately preserved certain mound fields. Further, coarse spillway gravels inhibit vertical burrowing by soil dwelling mammals, which forces lateral burrowing, a condition that has mound-forming consequences.

Based on nearest-neighbor statistical analyses, we conclude that the distribution of mounds within a given Saskatchewan mound field has a more regular than random pattern. This suggests a biological control, such as territoriality of a dominant species.

These and other observations, inferences, and results that converge from this study support the Gilbert-Dalquest-Scheffer hypothesis—that Saskatchewan mounds are primarily due to bioturbation by burrowing animals, mainly rodents (ground squirrels, gophers) and their predators. In addition to nearest-neighbor analyses, the hypothesis is further supported by (1) abundant evidence of small mammal bioturbation in mound soils; (2) the occurrence of thick soil on the higher mounds versus thin soil in the lower intermounds—suggesting soil movements from intermounds to mounds; and (3) pebbles in mound soils generally being less than 5 cm in diameter, and thus of a size commonly moved by small burrowing animals, such as ground squirrels and gophers.

We further conclude that as Saskatchewan mounds began forming, they increasingly represented favored, organic-rich, higher-lying and thus drier (less flood-prone) living sites for soil animals. As mounds form they would thus increasingly confer an adaptive living and overwintering advantage for many, if not most, soil animals.

Any positive relief landform or soil feature, however, must always be impacted to some degree by erosional and other processes that operate under gravity, as well as freeze-thaw, plant growth, and so on. Saskatchewan mounds are no exceptions. Further, vegetated mounds can and do trap eolian sediments, if only in miniscule amounts. In this collective sense, pimple-Mima-prairie mound formation, though animal-driven in the first instance, represents a complex, polygenetic set of processes that have operated in the past, that must vary in their rates with time, and that continue operating today.

ACKNOWLEDGMENTS

The research summarizes part of Lee-Ann Irvine's Master of Science thesis at the University of Regina, Saskatchewan, supervised by Dr. Janis Dale and Dr. David Sauchyn. Special thanks go to Tom Gentles, Dr. Jack Mollard (JD Mollard & Associates), Steve Beck and Don Turner (Prairie Farm Rehabilitation Administration), Brent Bitter (Saskatchewan Environment), Alan R. Smith (Canadian Wildlife Service), Frank McDougall (University of Saskatchewan), Dr. Don Johnson and Diana Johnson (University of Illinois), and Dr. Catherine Yansa (Michigan State University). In addition, technical and cartographic assistance was obtained from Nathan Friesen (Heritage Branch, Ministry of Tourism, Parks, Culture, and Sport) and Dr. Ray Poulin and Mike Benoit at the Royal Saskatchewan Museum, Ministry of Tourism, Parks, Culture, and Sport. The authors sincerely thank the various reviewers who contributed to the improvement of this paper.

APPENDIX

APPENDIX. LOCATION, MORPHOLOGY, AND SPACING OF NORTH AMERICAN MOUND SITES

Location	Latitude	Longitude	Researcher(s)	Height (m)	Diameter (m)	Volume (m³)	Density (mounds/ha)	Spacing (m)	Comments
Petrofka Bridge, Saskatchewan	52°33'30" N	106°50'00" W	Mollard (2000, personal commun.)						
Clarkboro Ferry, Saskatchewan	52°17'30" N	106°24'30" W	Mollard (2000, personal commun.)						
Saskatoon North, Saskatchewan	52°14'00" N	106°45'00" W	Mollard (2000, personal commun.)						
Wolverine, Saskatchewan	51°56'45" N	105°17'00" W	Mollard (1978, 1996)						
Young Erratic, Saskatchewan	51°45'00" N	105°42'30" W	Mollard (1978, 1996)						
Little Manitou Lake, Saskatchewan	51°38'57" N	105°17'12" W	Mollard (1978, 1996), Irvine (2005)	0.36	14.7	9.8	2.25–16.96	22–48	Grasslands, gravelly soils within or near glacial spillways
Bounty, Saskatchewan	51°33'00" N	107°16'00" W	Mollard (2000, personal commun.)						
Davin, Saskatchewan	50°19'30" N	104°09'00" W	Mollard (1978, 1996)						
Bengough, Saskatchewan	49°29'00" N	105°09'30" W	Mollard (1978, 1996)						
Halbrite, Saskatchewan	49°27'30" N	103°32'30" W	Mollard (1978, 1996)						
Rolette County, North Dakota	48°46'21" N	99°50'20" W	Deal (1972)		9.1–18.3				Light-toned patches on aerial photographs of lacustrine or wave-eroded glacial sediments
Towner County, North Dakota	48°25'22" N	99°06'20" W	Bluemle (1983)		9.1–18.3			Random, some alignments	Light-toned patches on aerial photographs of sandy, lacustrine sediments
Northwestern Minnesota	47°10' N	96° W	Ross et al. (1968)	0.15–1.3	4–40		2.35	70	Low, wet prairie area with mounds located on upland sites
Iron Horse Prairie, Dodge County, Minnesota	43°52'00" N	92°50'50" W	Johnson et al. (1999)	0.2–>1	0.4–10				Mesic tallgrass prairie, mounds contain gravels and evidence of current gopher activity
Eastern, channeled scablands, Washington	47°40' N	117°24' W	Freeman (1926)	1	10				Composed of loess-like material
Eastern, Medical Lake, Washington	47°34' N	117°41' W	Piper (1905)	0.6	4.6				Basalt mounds of the Columbia lava
Eastern, Spangle, Washington	47°25' N	117°19' W	Piper (1905)	0.6	4.6				Basalt mounds of the Columbia lava
Scabland Rock Lake, Washington	47°10' N	117°42' W	Berg (1990, 1991)						
Washington	47°8' N	122°34' W	Eakin (1932)						

(Continued)

APPENDIX. LOCATION, MORPHOLOGY, AND SPACING OF NORTH AMERICAN MOUND SITES (*Continued*)

Location	Latitude	Longitude	Researcher(s)	Height (m)	Diameter (m)	Volume (m³)	Density (mounds/ha)	Spacing (m)	Comments
Mima Prairie, Washington	46°56' N	122°52' W	Newcomb (1952)	2–2.5	12–15			Regular spacing	
Eastern, Winona, Washington	46°55' N	117°48' W	Piper (1905)	0.6	4.6				Basalt mounds of the Columbia lava
Thurston County area, Washington	46°55' N	122°53' W	del Moral and Deardorff (1976)	1.3	13.05				Single mound studied in detail
Puget Lowland prairies, Thurston County, Washington	46°53' N	123°3' W	Washburn (1988)	0.3–2	2.5–12				
Mima Prairie, Washington	46°53' N	123°3' W	Scheffer (1947)						
Southwest Washington	46°52' N	122°3' W	Ritchie (1953)		Up to 21				Boulder lag in intermound areas, and underlies mounds
Washington	46°52' N	122°3' W	Scheffer (1948)						
Tenino area, Washington	46°51' N	122°51' W	Horner (1930)	0.9–1.8	5.5–9		86	12	
Tenino area, Washington	46°51' N	122°53' W	Dalquest and Scheffer (1942)	0.3–2	2.5–12				
Mima Prairie, Washington	46°51' N	123°3' W	Scheffer and Kruckeberg (1966)						
Mima Prairie, Washington	46°51' N	123°4' W	Scheffer (1969, 1981)			38	20–25		
Columbia River Plateau, Washington and Oregon	45°9' N	121°5' W	Waters and Flagler (1929)	0.5–1.5	9–16				Mounds cover large areas of bare basalt
Lawrence Grasslands Preserve, Columbia Plateau, Oregon	44°57' N	120°48' W	Cox and Hunt (1990)	0.92	14.43			22–50	Surface of plateau are scablands, mounds in grasslands
Oregon	44°23' N	118°59' W	Le Conte (1877)						
Eastern Snake River Plain, Idaho	43°42'15" N	113°2'15" W	Tullis (1995)	<0.5	8–14			Varies from regular to random to strongly linear	8 mound sites studied on alluvial fans, river terraces, and loess over basalt; all mounds show current or past burrowing animal activity; only 8MFM4 site on alluvial fan mapped
Hayden Prairie, Howard County, Iowa	43°26'17" N	92°22'51" W	Johnson et al. (1999)	0.2–>1	0.4–10				Mounds contain gravels and evidence of current gopher activity

(Continued)

APPENDIX. LOCATION, MORPHOLOGY, AND SPACING OF NORTH AMERICAN MOUND SITES (*Continued*)

Location	Latitude	Longitude	Researcher(s)	Height (m)	Diameter (m)	Volume (m³)	Density (mounds/ha)	Spacing (m)	Comments
Kalsow Prairie, Pocahontas County, Iowa	42°34'12" N	94°33'55" W	Brotherson (1982); Ricks et al. (1997); Johnson et al. (1999)	0.2->1	0.4–10				Mounds contain gravels and evidence of current gopher activity
Laramie Basin, Wyoming	41°25' N	105°35' W	Spackman and Munn (1984)	0.5	6–7			More regular than random spacing	Considerable bioturbation encountered in mound trenches but less in intermound areas
Laramie Basin, Sabulsky site, Wyoming	41°23' N	105°42' W	Reider et al. (1999)	0.15–0.65	10–20		26	21	Not located on the floodplain; composed of churned, unsorted materials (fines, pebbles, gravels)
Missouri	38°38' N	90°19' W	Bushnell (1905)	0.9–1.22	12–19				
Southwest Missouri	37°7' N	93°49' W	Spillman (1905)	0.3–0.9	6–9				Limestone dominates parent material
Diamond Grove Prairie Natural Area, near Joplin, Missouri	36°54'26" N	94°19'54" W	Horwath et al. (2002); Horwath and Johnson (2006)	0.45	14		7.6		Active burrows visible on mound surface
Cochetopa Creek, Colorado	38°29' N	106°48' W	Hansen and Morris (1968)						
Black Mesa, Colorado	38° N	107° W	Hansen and Morris (1968)						
Mosca Flats, Colorado	37°46' N	105°23' W	Cox et al. (1987)						
Alpine Ridge, Colorado	37°39' N	105°29' W	Cox et al. (1987)						
South-central Colorado			Vitek (1978)	0.135	11.09 long and 10.18 wide	15		18.5, "regular spacing"	Mostly fine material with scattered pebbles; cobbles and boulders in intermound areas
California	37°35' N	120°27' W	Arkley and Brown (1954)	0.6–0.9	4.5–30.5		39	17	Mounds contain gravels and overly clay hardpan
Great Central Valley, near city of Merced, California	37°23'53" N	120°23'28" W	Reed and Amundson (2007)						Mima mound - vernal pool system
Blanca, South Colorado	37°20' N	105°33' W	Cox et al. (1987)	0.25	12.81				Soils contained an abundance of small stones
Merced County, California	37°10' N	120°47' W	Arkley (1948)						
California	36°41' N	120°3' W	Branner (1905)						
Tulare County, California	36°15' N	118°49' W	Whitney (1948)						

(*Continued*)

APPENDIX. LOCATION, MORPHOLOGY, AND SPACING OF NORTH AMERICAN MOUND SITES (Continued)

Location	Latitude	Longitude	Researcher(s)	Height (m)	Diameter (m)	Volume (m³)	Density (mounds/ha)	Spacing (m)	Comments
Orange County, California	33°42' N	117°47' W	Kelly (1948)	0.9–1.8	Up to 12				Mounds overlie clay hardpan
San Diego County, California	33°3' N	116°49' W	Kelly (1948)	0.9–1.8	Up to 12				Mounds overlie clay hardpan
San Diego County, California	33°2' N	116°45' W	Cox (1984)						
San Diego County, California	33°2' N	116°49' W	Cox (1986)						
California	32°51' N	117°10' W	Cox and Allen (1987)						Describes a single mound
California	32°46' N	116°54' W	Cleveland (1893)						
California	32°43' N	117°9' W	Barnes (1879)	0.3–1.22	3–15				
Northwest, along Arkansas River valley, Arkansas	35°25' N	93°38' W	Purdue (1905)						
Arkansas	35°15' N	93°10' W	Campbell (1906)	1.22	18.3				
Eastern Oklahoma	34°51' N	94°45' W	Knechtel (1952)	0.6–1.22				14–29	Mounds overlie clay hardpan
Epps, Louisiana	32°35' N	91°30' W	Cain (1974)	0.6				"evenly spaced"	Mound soil has a looser texture than surrounding and underlying soils
Calcasieu, Louisiana	31°5' N	92°42' W	Hilgard (1905)						
Opelousas, Louisiana	30°31' N	92°4' W	Hilgard (1905)						
Gulf Coast, Louisiana	30°13' N	92°6' W	Dietz (1945)						
Louisiana			Aten and Bollich (1981)	0.15–>1.5	2–>60				
Gulf coastal plain, Louisiana/Texas			Krinitzsky (1949)	0.6–0.9	10–15				Fine to sandy loam, area periodically floods
Coastal Texas	29°37' N	94°19' W	Aten and Bollich (1981)	0.15–1.5	2–>60				
Coastal plains, west of Garwood, Texas	29°27' N	96°25' W	Koons (1948)	Max. 0.6	9.1			Spacing shows no regularity	Flat, undrained regions that periodically flood
Gulf Coast, Texas	28°38' N	96°27' W	Udden (1906)	0.05–0.45	1.5–25				
Gulf Coast, Texas	27°50' N	97°20' W	Dietz (1945)	0.76	12.2		15	29	No gravel in or between mounds
Gulf Coast, Texas			Carty et al. (1988)	0.5	15.5				Prairie region
Gulf plain, Texas			Hobbs (1907)	0.9–1.2	6–30.5				

Note: Mound sites are sorted from north to south, and grouped by province or state. Modified from Irvine (2005).

REFERENCES CITED

Acton, D.F., and Ellis, J.G., 1978, The Soils of the Saskatoon Map Area, 73-B, Saskatchewan: Saskatchewan Institute of Pedology Publication S4, scale 1:126,720, 1 sheet.

Acton, D.F., Padbury, G.A., and Shields, J.A., 1990, Soil Landscapes of Canada, Saskatchewan Map: Agriculture Canada, Land Resource Research Centre (LRCC) Contribution No. 87-45, scale 1:1,000,000, 1 sheet.

Acton, D.F., Padbury, G.A., and Stushnoff, C.T., 1998, The Ecoregions of Saskatchewan: Regina, Saskatchewan, Canadian Plains Research Center, 205 p.

Arkley, R.J., 1948, The Mima mounds: The Scientific Monthly, v. 66, p. 175–176.

Arkley, R.J., and Brown, H.C., 1954, The origin of Mima mound (hog wallow) microrelief in the far western states: Soil Science Society of America Proceedings, v. 18, no. 2, p. 195–199, doi:10.2136/sssaj1954.03615995001800020021x.

Aten, L.E., and Bollich, C.N., 1981, Archeological evidence for pimple (prairie) mound genesis: Science, v. 213, no. 4514, p. 1375–1376, doi:10.1126/science.213.4514.1375.

Barnes, G.W., 1879, The Hillocks or mound-formations of San Diego, California: American Naturalist, v. 13, p. 565–571, doi:10.1086/272405.

Berg, A.W., 1990, Formation of Mima mounds: A seismic hypothesis: Geology, v. 18, no. 3, p. 281–284, doi:10.1130/0091-7613(1990)018<0281:FOMMAS>2.3.CO;2.

Berg, A.W., 1991, Reply on "Formation of Mima mounds: A seismic hypothesis": Geology, v. 19, no. 3, p. 284–285.

Bluemle, J.P., 1983, Freckled land: North Dakota Geological Society Newsletter, v. 12, p. 35–38.

Branner, J.C., 1905, Natural mounds or 'hog wallows': Science, v. 21, p. 514–516, doi:10.1126/science.21.535.514-b.

Brotherson, J.D., 1982, Vegetation of the Mima mounds of Kalsow Prairie, Iowa: The Great Basin Naturalist, v. 42, no. 2, p. 246–261.

Bushnell, D.I., 1905, The small mounds of the United States: Science, v. 22, p. 712–714, doi:10.1126/science.22.570.712.

Cain, R.H., 1974, Pimple mounds: A new viewpoint: Ecology, v. 55, no. 1, p. 178–182, doi:10.2307/1934633.

Campbell, M.R., 1906, Natural mounds: The Journal of Geology, v. 14, p. 708–717, doi:10.1086/621357.

Canada. Soil Classification Working Group, 1998, The Canadian System of Soil Classification: Agriculture and Agri-Food Canada Publication, 1646 (revised), 187 p.

Carty, D.J., Dixon, J.B., Wilding, L.P., and Turner, F.T., 1988, Characterization of a pimple mound-intermound soil complex in the Gulf Coast prairie region of Texas: Soil Science Society of America Journal, v. 52, no. 6, p. 1715–1721, doi:10.2136/sssaj1988.03615995005200060038x.

Clayton, L., and Moran, S.R., 1982, Chronology of Late Wisconsinan glaciation in middle North America: Quaternary Science Reviews, v. 1, p. 55–82, doi:10.1016/0277-3791(82)90019-1.

Cleveland, D., 1893, The hillock and mound formations of southern California: Science, v. 22, no. 544, p. 4, doi:10.1126/science.ns-22.544.4.

Cox, G.W., 1984, Mounds of mystery: Natural History, v. 93, no. 6, p. 36–45.

Cox, G.W., 1986, Mima mounds as an indicator of the presettlement grassland-chaparral boundary in San Diego County, California: American Midland Naturalist, v. 116, no. 1, p. 64–77, doi:10.2307/2425938.

Cox, G.W., 1990, Form and dispersion of Mima mounds in relation to slope steepness and aspect on the Columbia Plateau: The Great Basin Naturalist, v. 50, p. 21–31.

Cox, G.W., and Allen, D.W., 1987, Soil translocation by pocket gophers in a Mima moundfield: Oecologia, v. 72, p. 207–210, doi:10.1007/BF00379269.

Cox, G.W., and Hunt, J., 1990, Form of Mima mounds in relation to occupancy by pocket gophers: Journal of Mammalogy, v. 71, no. 1, p. 90–94, doi:10.2307/1381323.

Cox, G.W., Gakahu, C.G., and Allen, D.W., 1987, Small-stone content of Mima mounds of the Columbia Plateau and Rocky Mountain regions: Implications for mound origin: The Great Basin Naturalist, v. 47, no. 4, p. 609–619.

Dalquest, W.W., and Scheffer, V.B., 1942, The origin of the mounds of western Washington: The Journal of Geology, v. 50, no. 1, p. 68–84, doi:10.1086/625026.

Dalquest, W.W., and Scheffer, V.B., 1944, Distribution and variation in pocket gophers, *Thomomys talpoides*, in the State of Washington II: American Naturalist, v. 78, no. 778, p. 308–333, 423–450.

Deal, D.E., 1972, Areas with a freckled appearance on air photos, Geology of the Rolette County, North Dakota: Grand Forks, University of North Dakota Geology Department, p. 19–20.

del Moral, R., and Deardorff, D.C., 1976, Vegetation of the Mima mounds, Washington State: Ecology, v. 57, p. 520–530, doi:10.2307/1936436.

Dietz, R.S., 1945, The small mounds of the Gulf coastal plain: Science, v. 102, p. 596–597, doi:10.1126/science.102.2658.596.

Eakin, H.M., 1932, Periglacial phenomena in the Puget Sound region: Science, v. 75, p. 536–537, doi:10.1126/science.75.1951.536.

Eldridge, D.J., 2004, Mounds of the American badger (*Taxidea taxus*): Significant features of North American shrub-steppe ecosystems: Journal of Mammalogy, v. 85, no. 6, p. 1060–1067, doi:10.1644/BEH-105.1.

Freeman, O.W., 1926, Scabland mounds of Eastern Washington: Science, v. 64, p. 450–451, doi:10.1126/science.64.1662.450-a.

Fung, K., 1999, Atlas of Saskatchewan: Saskatoon, Saskatchewan, Printwest, 336 p.

Gilbert, G.K., 1875, Report on the geology of portions of Nevada, Utah, California, and Arizona, examined in the years 1871–72, *in* Volume III, Part I, U.S. Geographical and Geological Surveys West of the 100th Meridian, Lieut. George M. Wheeler, U.S. Corps of Engineers, in charge: Washington, D.C., Government Printing Office.

Hansen, R.M., and Morris, M.J., 1968, Movement of rocks by Northern Pocket Gophers: Journal of Mammalogy, v. 49, no. 3, p. 391–399, doi:10.2307/1378197.

Hilgard, E.W., 1905, The prairie mounds of Louisiana: Science, v. 21, p. 551–552, doi:10.1126/science.21.536.551-a.

Hobbs, W.H., 1907, Some topographic features formed at the time of earthquakes and the origin of the mounds of the Gulf Plain: American Journal of Science, v. 23, p. 245–256, doi:10.2475/ajs.s4-23.136.245.

Horner, J.B., 1930, The million mystery mounds of Tenino [unpublished paper]: Corvallis, Oregon State College, Oregon State University Archives, 13 p.

Horwath, J.L., and Johnson, D.L., 2006, Mima type mounds in southwest Missouri: Expressions of point-centered and locally thickened biomantles: Geomorphology, v. 77, p. 308–319, doi:10.1016/j.geomorph.2006.01.009.

Horwath, J.L., and Johnson, D.L., 2008, The biodynamic significance of double stonelayers at Diamond Grove Mima moundfield, southwest Missouri: Geological Society of America Abstracts with Programs, v. 40, no. 6, p. 208.

Horwath, J.L., Johnson, D.L., and Stumpf, A.J., 2002, Evolution of a gravelly Mima-type moundfield in southwestern Missouri: Geological Society of America Abstracts with Programs, v. 34, no. 6, p. 369.

Irvine, L.L., 2005, A study of pimple mounds in southern Saskatchewan [unpublished Master's thesis]: Regina, Saskatchewan, Department of Geography, University of Regina, 188 p.

Johnson, D.L., 1989, Subsurface stone lines, stone zones, artifact-manuport layers, and biomantles produced by bioturbation via pocket gophers (Thomomys Bottae): American Antiquity, v. 54, no. 2, p. 370–389, doi:10.2307/281712.

Johnson, D.L., 1999, Badgers as large clast translocators, petrocalcic rippers, and artifact movers and shakers, in Western Nearctica: Geological Society of America Abstracts with Programs, v. 31, no. 5, p. A-24.

Johnson, D.L., and Horwath Burnham, J.L., 2012, this volume, Introduction: Overview of concepts, definitions, and principles of soil mound studies, *in* Horwath Burnham, J.L., and Johnson, D.L., eds., Mima Mounds: The Case for Polygenesis and Bioturbation: Geological Society of America Special Paper 490, doi:10.1130/2012.2490(00).

Johnson, D.L., Johnson, D.N., and West, R.C., 1999, Pocket gopher origins of some midcontinental Mima-type mounds: Regional and interregional genetic implications: Geological Society of America Abstracts with Programs, v. 31, no. 7, p. A232.

Johnson, D.L., Johnson, D.N., and Horwath, J.L., 2002, In praise of the coarse fraction and bioturbation: Gravelly Mima mounds as two-layered biomantles: Geological Society of America Abstracts with Programs, v. 34, no. 6, p. 369.

Johnson, D.L., Horwath, J.L., and Johnson, D.N., 2003, Mima and other animal mounds as point-centered biomantles: Geological Society of America Abstracts with Programs, v. 34, no. 7, p. 258.

Kehew, A.E., and Teller, J.T., 1994, Glacial-lake spillway incision and deposition of a coarse-grained fan near Watrous, Saskatchewan: Canadian Journal of Earth Sciences, v. 31, p. 544–553, doi:10.1139/e94-048.

Kelly, A.O., 1948, The Mima mounds: The Scientific Monthly, v. 66, p. 174–176.

Knechtel, M.M., 1952, Pimpled plains of eastern Oklahoma: Geological Society of America Bulletin, v. 63, p. 689–700, doi:10.1130/0016-7606(1952)63 [689:PPOEO]2.0.CO;2.

Koons, F.C., 1948, The sand mounds of Louisiana and Texas: The Scientific Monthly, v. 66, p. 297–300.

Krinitzsky, E.L., 1949, Origin of pimple mounds: American Journal of Science, v. 247, no. 10, p. 706–714, doi:10.2475/ajs.247.10.706.

Le Conte, J., 1877, Hog wallows or prairie mounds (of California and Oregon): Nature, v. 15, p. 530–531, doi:10.1038/015530d0.

McDougall, F.H., 2000, A guide to some of the surface geological features of the Little Manitou Lake area, Saskatchewan: Saskatoon, Saskatchewan, Saskatchewan Archaeological Society Annual General Meeting, Field Trip Guidebook, 13 p.

Mollard, J.D., 1978, Landforms and Surface Materials of Canada: A Stereoscopic Airphoto Atlas and Glossary (sixth edition): Regina, Saskatchewan, JD Mollard & Associates, variously paginated.

Mollard, J.D., 1996, Landforms and Surface Materials of Canada: A Stereoscopic Airphoto Atlas and Glossary (eighth edition): Regina, Saskatchewan, Printwest, variously paginated.

Mollard, J.D., 1999, Locations, Projects Extraordinary Surface Features and Stratigraphy: 52nd Canadian Geotechnical Conference, Field Trip Guide Booklet: Regina, Saskatchewan, JD Mollard & Associates, 25 p.

Newcomb, R.C., 1952, Origin of the Mima mounds, Thurston County region, Washington: The Journal of Geology, v. 60, no. 5, p. 461–472, doi:10.1086/625998.

Padbury, G.A., and Acton, D.F., 1994, Ecoregions of Saskatchewan Poster Map: Regina, Saskatchewan, Minister of Supply and Services Canada and Saskatchewan Property Management Corporation, Canadian Plains Research Center, scale 1:2,000,000, 1 sheet.

Piper, C.V., 1905, The basalt mounds of the Columbia lava: Science, v. 21, no. 543, p. 824–825, doi:10.1126/science.21.543.824.

Price, W.A., 1949, Pocket gophers as architects of Mima (Pimple) mounds of the western United States: The Texas Journal of Science, v. 1, no. 1, p. 1–17.

Purdue, A.H., 1905, Concerning the natural mounds: Science, v. 21, p. 823–824, doi:10.1126/science.21.543.823.

Reed, S., and Amundson, R., 2007, Sediment, gophers, and time: A model for the origin and persistence of mima mound-vernal pool topography in the Great Central Valley, *in* Schlising, R.A., and Alexander, D.G., eds., Vernal Pool Landscapes: Studies from the Herbarium no. 14: Chico, California State University, p. 15–27.

Reider, R.G., Huss, J.M., and Miller, T.W., 1999, Stratigraphy, soils, and age relationships of mima-like mounds, Laramie Basin, Wyoming: Physical Geography, v. 20, no. 1, p. 83–96.

Ricks, D.K., Burras, L., Konen, M.E., and Bolender, A.J., 1997, Genesis and morphology of Mima mounds and associated soils at Kalsow Prairie, Iowa: Agronomy Abstracts, p. 255.

Ritchie, A.M., 1953, The erosional origin of the Mima mounds of southwest Washington: The Journal of Geology, v. 61, no. 1, p. 41–50, doi: 10.1086/626035.

Ross, B.A., Tester, J.R., and Breckenridge, W.J., 1968, Ecology of Mima-type mounds in northwestern Minnesota: Ecology, v. 49, no. 1, p. 172–177, doi:10.2307/1933579.

Saskatchewan Geological Society, 2002, Geological Highway Map of Saskatchewan: Saskatchewan Geological Society Special Publication Number 15, 1 sheet.

Saskatchewan Geological Survey, 2003, Geology, and Mineral and Petroleum Resources of Saskatchewan: Saskatchewan Geological Survey Miscellaneous Report 2003-7, 173 p.

Scheffer, V.B., 1947, The mystery of the Mima mounds: The Scientific Monthly, v. 65, no. 5, p. 283–294.

Scheffer, V.B., 1948, Reply on "Mima Mounds": The Journal of Geology, v. 56, no. 3, p. 231–234, doi:10.1086/625506.

Scheffer, V.B., 1969, Mima mounds: Their mysterious origin: Pacific Search (Seattle), v. 3, no. 5, p. 136–137.

Scheffer, V.B., 1981, Mima prairie's mystery mounds: Backpacker, v. 4/5, p. 96.

Scheffer, V.B., and Kruckeberg, A., 1966, The Mima mounds: Bioscience, v. 16, p. 800–801.

Spackman, L.K., and Munn, L.C., 1984, Genesis and morphology of soils associated with formation of Laramie Basin (mima-like) mounds in Wyoming: Soil Science Society of America Journal, v. 48, no. 6, p. 1384–1392, doi:10.2136/sssaj1984.03615995004800060038x.

Spillman, W.J., 1905, Natural mounds: Science, v. 21, p. 632, doi:10.1126/science.21.538.632-a.

Tullis, J.A., 1995, Characteristics and origin of earth-mounds on the eastern Snake River plain, Idaho [unpublished Master's thesis]: Pocatello, Department of Geology, Idaho State University, 164 p.

Udden, J.A., 1906, The origin of small sand mounds in the Gulf Coast country: Science, v. 23, p. 849–851, doi:10.1126/science.23.596.849.

Vitek, J.D., 1978, Morphology and pattern of earth mounds in south-central Colorado: Arctic and Alpine Research, v. 10, p. 701–714, doi:10.2307/1550738.

Washburn, A.L., 1988, Mima Mounds: An Evaluation of Proposed Origins with Special Reference to the Puget Lowlands: Washington Division of Geology and Earth Sciences, Report of Investigations, v. 29, 53 p.

Waters, A.C., and Flagler, C.W., 1929, Origin of small mounds on the Columbia River Plateau: American Journal of Science, v. 18, p. 209–224, doi:10.2475/ajs.s5-18.105.209.

Whitney, D.J., 1948, San Joaquin Valley hogwallows: The Scientific Monthly, v. 66, p. 356–357.

MANUSCRIPT ACCEPTED BY THE SOCIETY 5 MARCH 2012

The Geological Society of America
Special Paper 490
2012

Alpine and montane Mima mounds of the western United States

George W. Cox

Biosphere and Biosurvival, 108 Vuelta Maria, Santa Fe, New Mexico 87506, USA

ABSTRACT

Mima-type soil mounds result from the repeated outward tunneling of Geomyid pocket gophers from nest and food storage centers and the resultant backward displacement of soil and its accumulation near such centers. These mounds are widespread in alpine and montane grassland habitats in the western United States. Their abundance in highland areas is confirmed by Google Earth surveys, by published studies and other sources, and by personal fieldwork of the author and colleagues. Highland areas in Canada and Mexico were also surveyed by Google Earth, but Mima-type mounds have not yet been found in these locations. Almost all alpine mound sites surveyed are on ridge tops or south-facing slopes, with many best expressed just above timberline. In some northern locations alpine and montane mounds appear to have formed since the Pleistocene. The presence of mounds only on moraines of Illinoian age at one Wyoming site suggests that mounds there, and thus perhaps in more southern locations, may have begun forming much earlier.

INTRODUCTION

Mima mounds, named for the mounds on Mima Prairie, near Olympia, Washington, are widespread in many biomes and occur on different landforms, substrates, and soils of widely varying ages. Their locations are environmentally diverse, and range from coastal marine terraces and marshlands to plains, hills, plateaus, high mountain meadows, and alpine areas across the western United States (Cox and Scheffer, 1991). These mounds, composed of soil that may be gravelly or not depending on parent material and substrate, range from several to 25 or more meters in diameter and up to ~2 m in height. Many theories have been proposed for their origin (cf. Washburn, 1988; Johnson and Burnham, this volume, Introduction; Appendix D, this volume). Dalquest and Scheffer (1942) assembled evidence showing that, in the presence of rather specific local conditions (outlined below), Mima mounds can form by the tunneling action of pocket gophers, fossorial rodents of the family Geomyidae (widespread in western North America and endemic to the Western Hemisphere). Their hypothesized mechanism of mound formation is centripetal (radial) translocation of soil by repeated outward tunneling by the animals over many generations from activity centers favorable for nesting, food storage, predation avoidance, and survivability.

Local conditions favorable for this pattern of tunneling and soil translocation are created by (1) a shallow layer impervious to small mammal burrowing, such as bedrock, hardpan, claypan, or dense coarse gravels; (2) periodically wet or flood-prone soil conditions; or (3) some combination, the more usual case. The translocation results from the lateral movement of soil backward through tunnels made by the animals until it is deposited on or in the mound, either as surface heaps or packed into their abandoned tunnels. In time, the soil translocation process builds and maintains the mound. As the mound grows, it increasingly confers, as described, a living, reproductive, and survival advantage for the animals, while the soil between the mounds becomes progressively thinner.

Several lines of evidence support the pocket gopher hypothesis of Mima-type mound formation in North America. Perhaps the most notable is that such mounds occur only within the

Cox, G.W., 2012, Alpine and montane Mima mounds of the western United States, *in* Horwath Burnham, J.L., and Johnson, D.L., eds., Mima Mounds: The Case for Polygenesis and Bioturbation: Geological Society of America Special Paper 490, p. 63–70, doi:10.1130/2012.2490(03). For permission to copy, contact editing@geosociety.org.

geographical and altitudinal range of pocket gophers, with many or most mounds, and almost all moundfields, currently occupied by these animals (Cox and Scheffer, 1991; D.L. Johnson, personal observations). Another is that pocket gophers occupy an extremely eclectic range of environments in North America, in fact, the same wide-ranging eclectic environments in which Mima mounds occur. The appearance of pocket gopher heaps on most Mima mounds during the course of any given year is *prima facie* evidence that pocket gophers are inhabiting and bioturbating that mound. Moreover, the very habits and behaviors of the animals regarding their burrowing and food-harvesting styles, operating from point-centered nesting and food-storage sites (activity centers), support the biological model. Experimental field studies that confirm soil translocation by pocket gophers to Mima mounds in moundfield settings likewise support their role in mound development and maintenance (Cox and Allen, 1987). In fact, mounded landscapes similar to those in western North America occur in South America, South Africa, East Africa, and Eurasia where pocket gopher equivalents or near equivalents (mole rats, zokors, and other fossorial rodents) are implicated, sometimes in conjunction with other burrowers (Cox and Gakahu, 1983; Cox and Roig, 1986; Cox et al., 1987b; Milton and Dean, 1990).

Here I present information on the distribution of Mima terrain associated with populations of Geomyid pocket gophers in alpine and montane areas of the western United States. My purpose is to demonstrate that Mima mounds, presently known from widely diverse environments in North America, commonly form in montane and alpine environments. It is proposed that if pocket gophers live there, and if one or more of the specific local conditions outlined above are met, an environment for mound formation exists.

METHODS

Multiple alpine and high montane moundfields were located from the published and unpublished observations of other workers, and new ones were located by Google Earth aerial photographic surveys covering all of western North America. The Google Earth surveys led to the discovery of numerous new high-elevation moundfields. These moundfields may have been easier to recognize in southern mountains than in northern mountains, however, because of heavy snow cover in many images from the latter areas. Several high-elevation moundfields, including some localities discovered by the author, were visited on foot, and mound densities and dimensions determined. The association of mounds with pocket gopher populations was verified at all visited sites by checking for the presence of both their tunnels and their diagnostic, concentrically ridged surface heaps. The small stone content of mound and intermound soils was sampled at one alpine site in the Sangre de Cristo Mountains of Colorado (Cox et al., 1987a). Images of the sites visited on foot were compared to Google Earth images of the same sites to provide a basis for verification of sites seen in aerial photographs.

RESULTS

Moundfields occur at many alpine and high montane elevations in all western mountain states (Fig. 1). Mounds were not discovered at alpine and montane sites in Canada or Mexico, however, although they are present at some lower elevations (further work may reveal high elevation mounds in both countries).

In Washington, I personally examined Mima moundfields in two high montane locations on Mount Rainier and in the Olympic Mountains. On the northeast slope of Mount Rainier, Mima mounds occur in alpine meadows at timberline at Sunrise Visitor's Center (46° 54′ 54.57″ N, 121° 38′ 28.56″ W) at an elevation of 1945–2085 m. These mounds, dominated by the grass *Festuca viridula*, occur on south-facing meadow areas ranging from nearly level to over 15° in slope. The largest of these mounds range from 30 to 70 cm in height and 11.5–21.2 m in diameter. They were first noted by Henderson (1974), who attributed their origin to pocket gopher activity. The pocket gopher at this site, *Thomomys talpoides*, is a different species than those present in other western Washington moundfields.

In the Olympic Mountains of western Washington, I site-examined Mima mounds occupied by the pocket gopher *Thomomys mazama* in timberline meadows on south-to-southeast slopes on the High Divide bordering the southern rim of the Seven Lakes Basin (47° 54′ N, 123° 44′ W), at an elevation of 1610 m. The largest of these mounds are 20–75 cm high and 7.5–13.0 m in diameter. The pocket gopher here is a subspecies endemic to alpine meadows on ridges north of the Hoh River and west of the Elwah River, but is a very close relative of the pocket gophers that occupy moundfields in the Puget lowlands near Olympia.

Mima mounds are also present in abundance in central and eastern Washington (Kaatz, 1959): on Steptoe Butte (47° 01′ 39″ N, 117° 17′ 55″ W) essentially to its 1098 m summit; on Badger Mountain at many places and elevations (e.g., 47° 31′ 50″ N, 120° 03′ 41″ W, 655–1110 m elevation), and variously across Manastash Ridge at multiple elevations (e.g., 46° 54′ 36″ N, 120° 36′ 44″ W, 545–1220 m elevation). Pocket gophers are active at all of these sites (D.L. Johnson, 2011, personal commun.). At lower elevations, mounds are widespread in the Puget Lowlands and Channeled Scablands of eastern Washington. The pocket gopher present at these sites, and at most other high mountain areas eastward and southward in the western states is *Thomomys talpoides*.

Farther east, at Lemhi Pass on both sides of the Idaho/Montana border (44° 58′ 1.54″ N, 113° 27′ 0.22″ W) in the Bitterroot Range at an elevation of 2290 m, Mima mounds occur on east-, west-, and south-facing slopes. To the south, also on this border, near Grizzly Hill (44° 49′ 9.75″ N, 113° 21′ 20.97″ W), mounds straddle the S and SW slopes of ridges immediately above timberline at elevations of 2580–2640 m. Small moundfields occur at several other locations throughout this mountainous region. Mounds also occur in Beaverhead County, Montana northeast of Lima at 1920–2500 m (44° 39′ 24.33″ N, 112° 34′ 35.47″ W) and west of Wisdom (Thorp, 1949) at

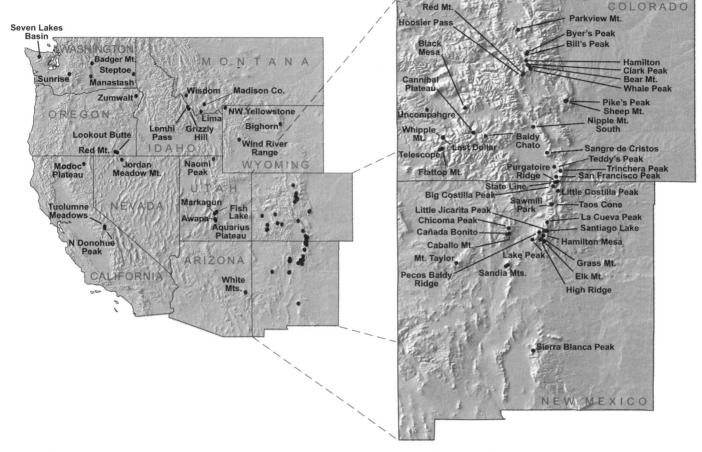

Figure 1. Locations of alpine and montane Mima moundfields in the western United States. Prepared by A. Gau and C. Gladish.

1860–1920 m (45° 37′ 12.28″ N, 113° 34′ 32.76″ W). This site is occupied by pocket gophers (D.L. Johnson, 2011, personal commun.). In Madison County, Montana mounds range over elevations of 1890–2165 m (45° 04′ 53.27″ N, 112° 17′ 14.96″ W).

In Wyoming, Mima-type mounds are widespread here and there across the state. For example, they are common throughout the Laramie Basin, notably at elevations of ~2150–2250 m (Reider, 1992; Reider et al., 1996; Spackman and Munn, 1984), as well as throughout much of the Great Divide Basin and Green River Valley farther west at similar elevations. At higher elevations to the northwest in the Wind River Range, they are developed on Illinoian glacial deposits such as the Bull Lake Moraine on the northwest flank of Willow Lake (43° 01′ 12.72″ N, 109° 52′ 59.13″ W). This westward sloping moraine appears to date from ca. 150,000 ^{14}C yr B.P. (Easterbrook, 2003). In this location they occur over an elevational range of ~2310–2750 m. Mima mounds appear to largely be absent on younger Pinedale moraines (40,000 ^{14}C yr B.P. and younger) of Wisconsin age. In northwestern Yellowstone National Park, mounds occur at an elevation of 2785 m in level meadow areas near tree line (44° 55′ 26.30″ N, 110° 50′ 39.47″ W). Farther east in the Bighorn Mountains of

Johnson County (43° 58′ 32.95″ N, 107° 02′ 08.23″ W), Mima mounds are widespread in loamy soils of open, gently rolling parklands at elevations of ~2500 m (USDA-SCS, 1978).

In northern Utah, mounds occur on south-facing slopes immediately above timberline at several locations on and near Naomi Peak (41° 53′ 30.49″ N, 111° 40′ 18.57″ W) in the Wasatch Range at elevations of 2670–2895 m. In the Dixie and Fish Lake National Forests in southern Utah, mounds also occur on the Aquarius, Awapa, Markagunk, Fish Lake, and other nearby plateaus (38° 03′ 48″ N, 111° 46′ 46″ W; 38° 07′ 42.52″ N, 111° 46.09′ 70″ W; 38° 33′ 41.09″ N, 111° 44′ 51.42″ W; 38° 41′ 38.50″ N, 111° 34′ 15.52″ W, respectively) at elevations ranging from 2800 to 3426 m. In northern Nevada, mounds occur at scattered locations on flats or south-facing slopes at elevations near 2040 m on Jordan Meadow Mountain (41° 50′ 48.87″ N, 118° 01′ 22.95″ W) in the Trout Creek Mountains. In southeastern Oregon, Google Earth images suggest that low mounds exist on N and S slopes at an elevation of 1690 m on Red Mountain (42° 17′ 21.47″ N, 118° 19′ 15.57″ W) and at 1850 m on nearby Lookout Butte (42° 15′ 15.59″ N, 118° 24′ 21.69″ W), also part of the Trout Creek Mountains. In this area mounds also occur at elevations as low as 1280 m. Farther north in Oregon, mounds occur on the Zumwalt Plateau (45° 28′ 41″ N, 117° 04′ 07″ W) at an elevation of 1158 m. At lower elevations, Mima mounds occur in parts of the Willamette Valley and over extensive mesas bordering the Deschutes River in north-central Oregon.

In California, although Mima mounds are widespread at lower elevations in coastal areas and in the Central Valley, few have been reported in the Sierra Nevada. An exception is to the north on the Modoc Plateau (41° 33′ 57″ N, 120° 41′ 41″ W), where mounds are present at an elevation of 1525 m. Weakly developed mounds are present in some wet meadows in the Sierra Nevadas, as in parts of Tuolumne Meadows (37° 52′ 34.76″ N, 119° 21′ 38.76″ W) at an elevation of 2615 m, and in a lakeside meadow (37° 47′ 40.73″ N, 119° 13′ 44.28″ W) north of Donohue Peak at an elevation of 3300 m. The species present at these California sites is *Thomomys monticola*.

Mima-type mounds are known from many alpine and subalpine locations in Colorado (Table 1). Features on Niwot Ridge, Boulder County, termed "mounds" by Litaor et al. (1996), appear to be surface heaps of newly mined soil deposited by pocket gophers. The northern pocket gopher, *Thomomys talpoides*, is abundant at this site, with a density of ~18 animals per hectare in moist meadow habitat (Thorn, 1982). Areas of concentrated activity by the northern pocket gopher, *Thomomys talpoides*, on Niwot Ridge result in large patches rich in gravel and small stones, but not in earth mounds (T. Seastedt, 2009, personal commun.). In northernmost Colorado, mounds occur on south-facing slopes just above tree line on Clark Peak and Parkview Mountain (Table 1). Farther south, moundfields are present on Byer's Peak, Bill's Peak, above Hamilton Gulch, and on Bear Mountain, Whale Peak, Red Moun-

tain, and Hoosier Pass on the border of Summit and Lake Counties (Table 1). Smaller moundfields occur in similar sites throughout this region. Some moundfields occur on moderately steep slopes. Those on Bill's Peak and above Hamilton Gulch, for example, are located on slopes of ~6.4°–9.7°. Farther east, moundfields occur at Windy Point on the south side of Pike's Peak and on Sheep Mountain, ~8 km to the southwest (Table 1). The Windy Point moundfield is located on a south-facing slope of ~6.2°.

In southern Colorado, Hansen and Morris (1968) examined alpine mounds on Baldy Chato, Saguache County (Table 1), and subalpine mounds nearby in meadows bordering the headwaters of Pauline Creek at an elevation of 3500 m. At these sites, surface heaps and snow tunnels backfilled with mined soil contained numerous small stones up to ~2.5 cm in diameter. Mima-type mounds at these sites are low, measuring ~10–15 cm in height, and occupied by *Thomomys talpoides*. In addition to these sites, Mima mounds were studied in a montane grassland on Black Mesa, Gunnison County, Colorado (Table 1) by Scheffer (1958) and Hansen (1962). Mounds at this site measured 25–50 cm high and 9.1–12.2 m in diameter, and were also occupied by *Thomomys talpoides*. McGinnies (1960) carried out studies on a Mima-mound site on the Uncompahgre Plateau of western Colorado (Table 1). Mounds at this site are occupied by pocket gophers (D.L. Johnson, 2011, personal commun.).

In the San Juan Mountain region of western Colorado, moundfields occur mainly on south-facing slopes at elevations of 3350–3678 m at Whipple Mountain North and South, Last Dollar Mountain, Flattop Mountain, and Telescope Mountain (Table 1).

In southernmost Colorado, Mima-type mounds also occur in alpine tundra areas of the Sangre de Cristo Mountains

TABLE 1. ALPINE AND SUBALPINE MIMA MOUND SITES IN COLORADO

Site name	Latitude and longitude	Elevation (m)	Aspect
Clark Peak	39° 35′ 35.09″ N, 105° 56′ 03.90″ W	3500	S facing
Parkview Mountain	40° 19′ 53.30″ N, 106° 09′ 00.67″ W	3490	S facing
Byer's Peak	39° 51′ 57.71″ N, 105° 56′ 26.09″ W	3585	S facing
Bill's Peak	39° 49′ 52.53″ N, 105° 57′ 31.31″ W	3645	S facing
Hamilton	39° 41′ 22.39″ N, 105° 58′ 20.07″ W	3565–3770	S facing
Bear Mountain	39° 34′ 24.42″ N, 105° 53′ 33.40″ W	3630–3765	S facing
Whale Peak	39° 30′ 43.47″ N, 105° 51′ 43.39″ W	3830	S facing
Red Mountain	39° 23′ 24.47″ N, 106° 01′ 34.03″ W	3544–3770	S facing
Hoosier Pass East	39° 21′ 39.69″ N, 106° 02′ 59.86″ W	3630–3675	S facing
Pike's Peak	38° 48′ 51.53″ N, 105° 01′ 23.17″ W	3605–3665	S facing
Sheep Mountain	38° 47′ 13.09″ N, 105° 02′ 59.65″ W	3600–3905	S facing
Black Mesa	38° 37′ 43.27″ N, 107° 24′ 09.39″ W	3140	Level
Uncompahgre	38° 29′ 15.62″ N, 108° 21′ 43.11″ W	2535–2555	Level
Nipple Mountain S	38° 14′ 29.77″ N, 105° 48′ 23.93″ W	3750	W facing
Cannibal Plateau	38° 06′ 31.78″ N, 107° 13′ 29.55″ W	3140	Level
Baldy Chato	38° 02′ N, 106° 56′ W	3900	S facing
Whipple Mountain N	38° 00′ 09.38″ N, 107° 56′ 14.07″ W	3460	S facing
Last Dollar Mountain	37° 59′ 59.13″ N, 107° 57′ 35.35″ W	3350	S facing
Whipple Mountain S	37° 59′ 39.84″ N, 107° 56′ 03.63″ W	3632	N-S facing
Flattop Mountain	37° 44′ 35.29″ N, 107° 56′ 49.02″ W	3650	S facing
Telescope Mountain	37° 42′ 55.64″ N, 108° 00′ 20.38″ W	3678	S facing
Sangre de Cristos	37° 39′ 38.82″ N, 105° 28′ 53.37″ W	3637–3714	W facing
Teddy's Peak	37° 20′ 94.44″ N, 105° 19′ 04.21″ W	3630–3770	S-W facing
Trinchera Peak	37° 16′ 19.05″ N, 105° 10′ 15.64″ W	3630	S facing
San Francisco Peak	37° 08′ 11.70″ N, 105° 16′ 05.78″ W	3075–3200	S facing
Purgatoire Ridge	37° 02′ 03.55″ N, 105° 14′ 07.10″ W	3640–3745	S facing

(Table 1), as well as at lower elevations in the San Luis Valley and Sangre de Cristo foothills. In alpine fellfields, Mima mounds occur in a saddle of the main ridge of the Sangre de Cristos on the border of Alamosa and Huerfano Counties (Vitek, 1978; Cox et al., 1987b). These mounds tend to be somewhat elliptical in shape and average 12.6–17.9 m in minimum and maximum diameters, respectively, and from 12.2 to 61.0 cm in height (Frederking, 1973), and are occupied by the northern pocket gopher, *Thomomys talpoides*. At this alpine site Cox et al. (1987b) sampled the small stone composition of mound top, mound edge, and intermound soils to determine if gravel (8–15 mm diameter) and small stones (15–50 mm diameter) were concentrated in mound soils. This result is expected if pocket gophers mine and move soil, gravel, and small pebbles moundward, and erosion selectively returns soil fines to intermound areas. Small stone samples from four large mounds were screened in the field and transported to our base camp for counting and weighing. The results showed that gravel and pebbles in the size range that pocket gophers can easily mine and transport are concentrated in the mounds relative to intermound soils (Cox et al., 1987b). These results are similar to those obtained at several other Mima mound locations in western North America (Cox et al., 1987a).

On the east slope of the Sangre de Cristo Mountains in southern Colorado, mounds also occur in lower montane grasslands over an extensive area at elevations of 2700–2950 m between 37° 39′ 43″ N and 37° 47′ 45″ N. Vitek's (1973) Mosca Flats study area is located in this mound region. This site is also occupied by *Thomomys talpoides* (Cox et al., 1987a). Analysis of small stone content of mound and intermound sites at this location showed concentration patterns of these elements in mound soils similar to those seen at the Sangre de Cristo alpine site (Cox et al., 1987a). In southernmost Colorado, several other mound sites exist on south- to west-facing slopes above timberline in the Sangre de Cristo Mountains (Table 1).

Mounds are present in many locations in the mountains of New Mexico (Table 2). On three peaks just south of the Colorado state line, mounds occur on south-facing slopes immediately above timberline. On Little Costilla Peak, at an elevation of 3600–3700 m on the border of Taos and Colfax Counties, moundfields occur on areas ranging from nearly level to 7° in slope. Mounds I examined at this site range from 8 to 19 m in diameter and from 0.3 to 1.5 m in height, and are occupied by the northern pocket gopher, *Thomomys talpoides*. Mound soils here also contain gravel and pebbles. Farther south, east of Wheeler Peak, mounds occur on Taos Cone and in Sawmill Park on the border of Taos and Colfax Counties at elevations of 3320–3680 m.

Mima-type mounds occur on several alpine sites in the Pecos Wilderness of Mora and San Miguel Counties, New Mexico, at elevations of 3000–3785 m (Table 2). These sites vary in aspect, and some are on mesa-like alpine summits. Still farther south, I have examined Mima mounds in the alpine zone on Lake Peak, Santa Fe County. The northern pocket gopher, *Thomomys talpoides*, is abundant at this site. I have also examined well-developed mounds in the alpine zone of Elk Mountain, San Miguel County, New Mexico, east of the upper Pecos River valley. The mounds, formed on bedrock, are scattered throughout a rich alpine meadow just above timberline on a west-facing 2°–5° slope (Fig. 2). These mounds range from ~20 cm to 1.4 m in height, and from slightly less than 5 m to slightly more than 19 m in mean diameter, with the larger mounds being elongate up- and downslope. The vegetation of the mounds is dominated by shrubby cinquefoil (*Dasiphora fruticosa*). Mound soils contain dense concentrations of pebbles up to ~9 cm in greatest diameter, whereas intermound areas have very thin soil with large

TABLE 2. ALPINE AND SUBALPINE MIMA MOUND SITES IN NEW MEXICO

Site name	Latitude and longitude	Elevation (m)	Aspect
State Line Peak	36° 59′ 22.34″ N, 105° 17′ 48.47″ W	3655–3860	S facing
Big Costilla Peak	36° 57′ 11.70″ N, 105° 19′ 53.35″ W	3885	S facing
Little Costilla Peak	36° 49′ 42.34″ N, 105° 13′ 18.44″ W	3600–3700	S facing
Taos Cone	36° 33′ 12.05″ N, 105° 22′ 27.18″ W	3650–3680	S facing
Sawmill Park	36° 34′ 04.89″ N, 105° 21′ 26.43″ W	3320	E-W facing
La Cueva Peak	36° 10′ 00.66″ N, 105° 29′ 19.42″ W	3180–3275	S-E facing
Little Jicarita Peak	35° 59′ 01.03″ N, 105° 33′ 47.05″ W	3785	SW facing
Santiago Lake	35° 57′ 41.78″ N, 105° 29′ 53.36″ W	3559	L-W facing
Pecos Baldy Ridge	35° 55′ 38.63″ N, 105° 38′ 14.28″ W	3590–3640	W facing
Hamilton Mesa N	35° 53′ 47.70″ N, 105° 35′ 04.72″ W	3190	L-WSE facing
Hamilton Mesa S	35° 52′ 04.82″ N, 105° 35′ 36.81″ W	3145	L-NWSE facing
Grass Mountain	35° 49′ 08.19″ N, 105° 37′ 58.86″ W	3000	L-S facing
High Ridge	35° 48′ 47.81″ N, 105° 35′ 42.03″ W	3167	S facing
Lake Peak	35° 47′ 36.82″ N, 105° 46′ 25.81″ W	3780	W facing
Elk Mountain	35° 46′ 15.44″ N, 105° 33′ 09.43″ W	3506	W facing
Mount Taylor	35° 13′ 56.11″ N, 107° 36′ 27.05″ W	3225	S facing
Sandia Mountains	35° 12′ 33.47″ N, 106° 26′ 31.59″ W	3143	NE facing
Chicoma Peak	35° 59′ 53.94″ N, 106° 22′ 44.22″ W	3195	S facing
Caballo Mountain	35° 57′ 01.72″ N, 106° 21′ 58.66″ W	3145	S facing
Cañada Bonito	35° 54′ 52.68″ N, 106° 23′ 43.98″ W	2915	S facing
Sandia Mountains	35° 12′ 33.47″ N, 106° 26′ 31.59″ W	3143	NE facing
Sierra Blanca Peak	33° 21′ 45.42″ N, 105° 49′ 08.63″ W	3095–3635	NWSE facing

Figure 2. (A–B) Elk Mountain, New Mexico. Google Earth images of Mima biotopography (arrow) near summit (elevation 3500 m) in thin soil over bedrock (road is 3 m wide). (C) Bedrock mounds of photo B being measured by author and colleague. (D) Weathered rolls of soil and small stones deposited in snow tunnels by pocket gophers at arrow, typical indicators of pocket gopher bioturbation in and under snow cover. (E) Maximum size of stones moved by pocket gophers taken from gopher-produced heaps and backfilled snow tunnels. (F) Intermound bedrock from which soil is moved centripetally (laterally) to Mima mounds by repeated seasonal, multigenerational, burrowing by pocket gophers outward from their activity centers in mounds. As mound-intermound biotopography evolves intermounds become lower and wetter, and on slopes often function as runoff channels which can remove any remaining residual soil. Elk Mountain is probably representative of many of the alpine moundfields in Table 2.

exposed angular cobbles. The northern pocket gopher is also abundant at this site.

West of the Rio Grande in New Mexico, high-elevation mounds occur in mountain grasslands on south-facing slopes at elevations of 2915–3195 m in the Jemez Mountains of Los Alamos and Sandoval Counties (Table 2). At Cañada Bonito, Los Alamos County, mounds ranging from ~8.5 to11.0 m in diameter and up to ~45 cm in height occur in dense grassland with a 7° slope. Mounds at this site are evident in an aerial photograph presented by Mohlenbrock (1988). Mounds at all the Jemez Mountain localities are occupied by the northern pocket gopher, *Thomomys talpoides*. Farther west, mounds are weakly developed on the south slope of Mount Taylor in Cibola County. In the Sandia Mountains, Bernalillo County, New Mexico, a few low Mima mounds occupied by northern pocket gophers also occur in meadows within coniferous forest (Table 2). The pocket gopher at this site is *Thomomys bottae*.

The southernmost high-elevation mounds in New Mexico are found in Otero County, at several locations on south-facing slopes of Sierra Blanca and adjacent lower peaks (Table 2). These moundfields span an elevational range of 3095–3635 m, and on some steep slopes are greatly elongated parallel to the slope, like those at lower elevations on the Columbia Plateau (Cox, 1990). Some of these mounds also occur on north-facing grassy slopes. The pocket gopher at this site is also *Thomomys bottae*.

In Arizona, Mima mounds occur over large areas of montane grassland in the White Mountains near Alpine, Apache County (33° 57′ 37.44″ N, 109° 22′ 59.67″ W), at elevations of ~2675–2870 m. The species here is *Thomomys bottae*.

DISCUSSION AND CONCLUSIONS

Thus, in North America, Mima terrain is well developed in many alpine tundra, mountain meadows, and montane grasslands. On the other hand, some mountain areas, such as the front range of the Rocky Mountains in Montana, seem to lack mound sites. Moundfields do not seem to occur in the Sierra Nevada of California, except in several wet meadows. All the sites identified in this survey, however, probably represent only a fraction of those alpine and montane areas with some development of Mima-type mounds.

Thomomys pocket gophers occur throughout the mountains of the western United States and are present in all mound sites that have been visited on foot. The high concentration of small stones in all Mima mounds examined, as well as experimental studies of soil translocation by pocket gophers in California (e.g., Cox and Allen, 1987), show that tunneling patterns of pocket gophers are adequate to build and maintain these mounds.

Several general patterns emerge from these observations and surveys. First, alpine Mima mounds appear to be restricted to situations immediately above timberline where herbaceous vegetation is well developed (e.g., Fig. 1). Second, most alpine and subalpine localities are on relatively steep south-facing slopes.

Warmer temperatures and good drainage may be important for permanent residence of pocket gophers at these elevations. Third, the northernmost alpine mounds also appear to be low in profile, usually less than 0.5 m in maximum height. In some northern localities, such as Niwot Ridge, Colorado, pocket gophers are present in alpine tundra, but their tunneling activity results only in areas of concentration of surface heaps rich in small stones. Farther south, in alpine areas of New Mexico, mounds often exceed 1.0 m in height. This suggests that alpine mounds in more northern areas may be younger in age, or that thin soils there limit the size of mounds that are formed.

Ages of these high-elevation moundfields are difficult to estimate. On Mount Rainier, the mounded areas have probably developed within the past 4000 years (Henderson, 1974). Mounds of similar size in alpine areas in central Colorado are likely also to have developed in post-Wisconsin time. At lower elevations and in more southern areas, mounds may be much older. The presence of mounds on moraines of Illinoian age and their absence on neighboring Wisconsin-age moraines in Wyoming hint that some moundfields may have been initiated during the Sangamon interglacial period.

Mima-type mounds exist in alpine and high montane environments in other parts of the world, where fossorial rodents of different families, but similar tunneling behaviors, occur. In Kenya and Ethiopia in East Africa, for example, mole rats of the genus *Tachyoryctes*, belonging to the family Rhizomyidae, are associated with Mima-type mounds (Cox and Gakahu, 1987). In Kenya, Mima-type mounds occupied by *Tachyoryctes splendens* occur at elevations of 3400–3600 m on Mount Kenya, above the elevation at which fungus-gardening termites are associated with large mounds. In Ethiopia, Mima mounds occupied by *T. macrocephalus* are widespread in the Bale Mountains at elevations of 4000–4100 m. In Chile, South America, the cururo, *Spalacopus cyaneus* (family Octodontidae), is a widespread fossorial rodent that occurs from near sea level to alpine elevations. Near Farellones, at ~2900 m in the Andes east of Santiago, I have examined low mounds formed by cururos in damp meadows. Similar mound features also occur at high elevations in Gansu and Ningxia Provinces, China, within the range of mole rats of the family Spalacidae.

Thus, it may be that throughout the world, high mountain environments occupied by fossorial rodents will exhibit Mima mounds or Mima-like terrain. This phenomenon represents a remarkable example of convergence among diverse rodent families in morphology, behavior, and geomorphological impact.

ACKNOWLEDGMENTS

I thank Darla Cox, Chris Gakahu, Mark Gregory, Diana and Don Johnson, and John Vitek for assistance in studies of mounds at various subalpine and alpine sites. I am also indebted to Alexandra Gau and Callie Gladish for preparing the map (Fig. 1) of mound sites in the western states and to Don Johnson for many valuable suggestions on the manuscript.

REFERENCES CITED

Cox, G.W., 1990, Form and dispersion of Mima mounds in relation to slope steepness and aspect on the Columbia Plateau: The Great Basin Naturalist, v. 50, p. 21–31.

Cox, G.W., and Allen, D.W., 1987, Soil translocation by pocket gophers in a Mima moundfield: Oecologia, v. 72, p. 207–210, doi:10.1007/BF00379269.

Cox, G.W., and Gakahu, C.G., 1983, Mima mounds in the Kenya highlands: Significance for the DalquestScheffer Hypothesis: Oecologia, v. 57, p. 170–174, doi:10.1007/BF00379577.

Cox, G.W., and Gakahu, C.G., 1987, Biogeographical relationships of mole rats with Mima mound terrain in the Kenya highlands: Pedobiologia, v. 30, p. 263–276.

Cox, G.W., and Roig, V.G., 1986, Argentinian Mima mounds occupied by ctenomyid rodents: Journal of Mammalogy, v. 67, p. 428–432, doi:10.2307/1380907.

Cox, G.W., and Scheffer, V.B., 1991, Pocket gophers and Mima terrain in North America: Natural Areas Journal, v. 11, p. 193–198.

Cox, G.W., Gakahu, C.G., and Allen, D.W., 1987a, The small stone content of Mima mounds of the Columbia Plateau and Rocky Mountain regions: Implications for mound origin: The Great Basin Naturalist, v. 47, p. 609–619.

Cox, G.W., Lovegrove, B.G., and Siegfried, W.R., 1987b, The small rock content of mima-like mounds in the South African Cape region: Implications for mound origin: Catena, v. 14, p. 165–176, doi:10.1016/S0341-8162(87)80015-2.

Dalquest, W.W., and Scheffer, V.B., 1942, The origin of the Mima mounds of Western Washington: The Journal of Geology, v. 50, p. 68–84, doi:10.1086/625026.

Easterbrook, D.J., ed., 2003, Quaternary Geology of the United States: INQUA 2003 Field Guide Volume: Reno, Nevada, The Desert Research Institute, 438 p.

Frederking, R.L., 1973, Spatial variation of the presence and form of earth mounds on a selected alp surface, Sangre de Cristo Mountains, Colorado [Ph.D. thesis]: Iowa City, University of Iowa, 201 p.

Hansen, R.M., 1962, Movements and survival of *Thomomys talpoides* in a Mima-mound habitat: Ecology, v. 43, p. 151–154, doi:10.2307/1932058.

Hansen, R.M., and Morris, M.J., 1968, Movement of rocks by northern pocket gophers: Journal of Mammalogy, v. 49, p. 391–399, doi:10.2307/1378197.

Henderson, J.A., 1974, Composition, distribution, and succession of subalpine meadows in Mount Rainier National Park [Ph.D. thesis]: Corvallis, Oregon State University, 150 p.

Kaatz, M.R., 1959, Patterned ground in central Washington—A preliminary report: Northwest Science, v. 33, p. 145–156.

Litaor, M.I., Mancinelli, R., and Halfpenny, J.C., 1996, The influence of pocket gophers on the status of nutrients in Alpine soils: Geoderma, v. 70, p. 37–48, doi:10.1016/0016-7061(95)00069-0.

McGinnies, W.J., 1960, Effect of mima-type microrelief on herbage production of five seeded grasses in western Colorado: Journal of Range Management, v. 13, p. 231–234, doi:10.2307/3895047.

Milton, S.J., and Dean, W.R.J., 1990, Mima-like mounds [heuweltjies, kraaltjies] in the southern and western Cape: Are the origins so mysterious?: South African Journal of Science, v. 86, p. 207–208.

Mohlenbrock, R.H., 1988, Cañada Bonito, New Mexico: Natural History, v. 97, no. 11, p. 27–32.

Reider, R.G., 1992, Mima-like mounds as salt-movement structures, Laramie Basin, Wyoming, USA: American Quaternary Association 12th biennial meeting: Geological Association of Canada Program with Abstracts, v. 12, p. 74.

Reider, R.G., Huss, J.M., and Miller, T.W., 1996, A groundwater vortex hypothesis for mima-like mounds, Laramie Basin, Wyoming: Geomorphology, v. 16, p. 295–317, doi:10.1016/0169-555X(95)00142-R.

Scheffer, V.B., 1958, Do fossorial rodents originate Mima-type relief?: American Midland Naturalist, v. 59, p. 505–510, doi:10.2307/2422495.

Spackman, L.K., and Munn, L.C., 1984, Genesis and morphology of soils associated with formation of Laramie Basin (mima-like) mounds in Wyoming: Soil Science Society of America Journal, v. 48, p. 1384–1392, doi:10.2136/sssaj1984.03615995004800060038x.

Thorn, C.E., 1982, Gopher disturbance: Its variability by Braun-Blanquet vegetation units in the Niwot Ridge alpine tundra zone, Colorado Front Range, U.S.: Arctic and Alpine Research, v. 14, p. 45–61, doi:10.2307/1550814.

Thorp, J., 1949, Effects of certain animals that live in soil: The Scientific Monthly, v. 68, p. 180–191.

USDA-SCS (U.S. Department of Agriculture–Soil Conservation Service), 1978, Soil Survey Data and Descriptions for Some Soils of Wyoming: Soil Survey Investigations Report No. 32: Wyoming Agricultural Experiment Station, 145 p.

Vitek, J.D., 1973, The mounds of south-central Colorado: An investigation of geographic and geomorphic characteristics [Ph.D. thesis]: Ames, University of Iowa, 229 p.

Vitek, J.D., 1978, Morphology and pattern of earth mounds in south-central Colorado: Arctic and Alpine Research, v. 10, p. 701–714, doi:10.2307/1550738.

Washburn, A.L., 1988, Mima mounds, an evaluation of proposed origins with special reference to the Puget Lowlands: Report of Investigations, State of Washington Department of Natural Resources, Division of Geology and Earth Resources Report no. 29, Olympia, Washington, 53 p.

Manuscript Accepted by the Society 5 March 2012

The Geological Society of America
Special Paper 490
2012

The biodynamic significance of double stone layers in Mima mounds

Jennifer L. Horwath Burnham
Department of Geography, Augustana College, Rock Island, Illinois 61201, USA

Donald L. Johnson
*Department of Geography, University of Illinois, Urbana, Illinois 61801, USA, and
Geoscience Consultants, Champaign, Illinois 61820, USA*

Diana N. Johnson
Geoscience Consultants, Champaign, Illinois 61820, USA

ABSTRACT

In hopes of shedding light on their genesis, Mima-type soil mounds were investigated at two environmentally and geologically disparate gravelly prairies, Diamond Grove Prairie in southwestern Missouri, and Mima Prairie in the southern Puget Sound of Washington. Mound soils were described, with large volume samples collected at narrow depth increments and laboratory analyses conducted. Results reveal, as predicted, that the soils contain small gravels (≤6 cm) scattered throughout the mound above a basal stone layer of large clasts (>6 cm). The stone layer is exposed across the intermound areas as a pavement.

Biomantle principles predicted that mound soils would be texturally biostratified by small burrowing vertebrates into locally thickened, two-layered biomantles, and they are. What was not expected, but might have been predicted by the principles, was the revelation from laboratory data of a second, upper, weakly expressed stone layer of pebbles (≤6 cm). To explain the secondary stone layer we introduce the concept of *dominant bioturbator*. At both prairies the dominant bioturbator was probably, until geologically recently, a species of the Geomyidae family of rodents, pocket gophers. This animal does not presently inhabit either prairie, but is present nearby. We attribute the secondary, apparently incipient stone layer to new dominants, almost certainly invertebrates, whose textural effects were regularly erased by the earlier (gopher) dominants. These effects are evident and expressed as a new upper biomantle that is now being superimposed upon the old.

INTRODUCTION

Mima-type and Mima-like natural prairie mounds occur intermittently from the Mississippi River to the west coast of the United States, from southern Canada to northern Mexico; and from sea level to above tree line in widely different climates, elevations, plant communities, geologic substrates, landforms, soils, on level-to-steep slopes, and slope aspects. Much has been written on their origin, with many theories offered. All, however, can be reduced to either biological or physical theories, or some combination. Evidence has significantly accumulated, exemplified by chapters in this volume and cited works within, that a combination of both fits essentially all the information known thus far, and that a polygenetic explanation is clearly indicated.

Horwath Burnham, J.L., Johnson, D.L., and Johnson, D.N., 2012, The biodynamic significance of double stone layers in Mima mounds, *in* Horwath Burnham, J.L., and Johnson, D.L., eds., Mima Mounds: The Case for Polygenesis and Bioturbation: Geological Society of America Special Paper 490, p. 71–84, doi:10.1130/2012.2490(04). For permission to copy, contact editing@geosociety.org. © 2012 The Geological Society of America. All rights reserved.

However, within this polygenetic mix evidence also suggests that one process invariably is a common denominator. That process is animal bioturbation.

A Spectrum of Mound Attributes

Prairie mounds, or more specifically Mima-type and Mima-like mounds, may be viewed as undefined and unquantified *activity centers* across a microtopographic spectrum. Mounds form when burrowing animals: (1) take advantage of natural micro-highs such as coppices, dunes, boles of uprooted trees, rill divides, low levees, anastomosed flood channel highs, and high spots on meander scrolls; (2) create their own in situ raised loci through repeated backspoil movements due to centripetal (radially outward) tunnelings from nesting-reproductive centers (Dalquest and Scheffer, 1942; Cox, 1984); or (3) occupy narrow zones of deep soil, as may occur above bedrock fracture traces that are bordered by thin or no soil (Shlemon et al., 1973; personal observations).

In the first case, animals simply occupy preexistent micro-highs to create a nesting center, which may confer a survival advantage, for example against flooding. Such mounds are called "hybrid mima-type mounds" (Johnson and Johnson, 2008). In the second case mounds may originate as self-induced raised activity centers anywhere that burrowing animals live, den, and/or nest. If the site confers a living and/or survival advantage over nearby areas, it will persist and grow, up to a point, over multiple animal generations. As it grows and gains organic matter, the low-density looseness of an incipient mound may in itself increase living advantages. Other animals and plants will invariably also occupy it for similar reasons. In the third case, as in thin soil or eroded bedrock areas with fractures, deep soil suitable for nesting and survival may exist only in bedrock fractures. Hence mounds form above the fractures during nest-centered burrowing activities. (Field confirmed examples are Upper Table Rock near Medford, Oregon [42° 28' 47" N, 122° 54' 43" W], and North Table Mountain near Oroville, California [39° 34' 33" N, 121° 33' 32" W].)

Across the microrelief spectrum, mounds vary in height, diameter, shape, density, soil type, particle-size composition, and other external-internal attributes. At one end of the spectrum are the usually unspectacular and indistinct *Mima-like* mounds that are activity centers (for denning, reproduction, food storage, etc.) of burrowing animals. These can develop essentially anywhere in pristine (unplowed-undeveloped) landscapes, or in abandoned or fallow agricultural tracts, and invariably form in deep soil where constraints to vertical burrowing are absent. Such activity centers are usually short-lived, ephemeral, and impermanent because survival advantages to the animals that made them are weak or neutral.

At the other end of the spectrum are the more distinct, prominent, and invariably better-expressed *Mima-type* mounds. These are longer lasting because survival advantages for their multigenerational rehabitation by the species that produced them are strong. These fixed, more permanent activity centers commonly occur in areas unsuitable, or marginally suitable, for agriculture, which insures their longevity in the modern world where pressure for crop production is intense. They usually share common characteristics of having formed in thin soil over some barrier or limitation to vertical burrowing, often areas of limited cultivation value. Barriers include hard bedrock, hardpan, dense gravels, heavy clay, areas of episodic wetness, or some combination (Cox, 1984; personal observations). Most prairie mounds lie between these endmembers, and all are subject to the myriad environmental forces that impact landforms in general, which renders them polygenetic.

Gravelly Soils and Mounds, Clast Sizes, Dominant Bioturbators, and Stone Layers

Mounds formed in gravelly soils that contain a range of particle sizes—pebbles, cobbles, boulders—usually display an internal morphology that is quite different from mounds formed in nongravelly soils. Gravelly mounds, like gravelly soils in general, commonly possess a stone layer of large clasts expressed as a continuous or discontinuous layer (Johnson et al., 2002). They form as a result of soil being removed from around and under large stones by the dominant bioturbating animals. Undermined stones settle downward, ultimately to the maximum depth of bioturbation. Discussed in several of the chapters in this volume (Reed and Amundson, Chapter 1; Irvine and Dale, Chapter 2; Johnson and Johnson, Chapter 6), the stone layer is a common internal attribute of mixed-clast gravelly soil mounds in western North America (e.g., Arkley and Brown, 1954; Cox, 1984a; Dalquest and Scheffer, 1942; Fosberg, 1963; Gentry, 2006; Johnson et al., 2002, 2005, 2008b; Kienzle, 2007; Larrison, 1942; McFaul, 1979; Mielke, 1977; Ritchie, 1953; Page et al., 1977; Wilkes, 1845; Rice and Watson, 1912). The principle requirement is that some percentage of gravel sizes must be *larger* than the burrow sizes of the dominant bioturbators that inhabit—or did inhabit—the soil or mound. If all gravels are *smaller* than burrow diameters, then basal stone layers cannot form, and the gravels likely will be bioturbationally spread throughout the mound. If large clasts are sparsely represented in a soil, the mound that forms will likely display a sparse or discontinuous stone layer. If no clasts larger than burrow diameters are present then a stone layer will not form.

The basal stone layer is not only under mounds, but also often forms a gravelly stony continuum across some or most of the moundfield, being exposed between mounds as a stony pavement (cf. Arkley and Brown, 1954; Cox, 1984b; Cox and Allen, 1987; Dalquest and Scheffer, 1942; Finney, this volume; Fosberg, 1963; Gentry, 2006; Irvine and Dale, this volume; Johnson and Johnson, this volume; Kienzle, 2007; Larrison, 1942; McFaul, 1979; Reed and Amundson, this volume; Ritchie, 1953; Page et al., 1977; Wilkes, 1845). Where the large clast size requirement is met, the presence of a basal stone layer under essentially all gravelly mounds, and a stone pavement between them, strongly supports the model that animal bioturbation is the common process denominator in polygenetic Mima mound formation.

By definition, a soil stone layer, or stone line, is a three-dimensional layer, or soil horizon, dominated by coarse particles (>2 mm) that generally follows the surface topography (Sharpe, 1938, p. 24). However, in the case of prairie mounds, the basal stone layer may be planar or concave. Stone layers form only in soils with mixed fine and coarse particles, and often involve pebble size (4–64 mm) or larger clasts. Stone layers may be one stone thick and appear in profile as a "stone line," or they may be several stones thick and appear in profile as a "stone zone" (Johnson, 1989). Most stone layers are pedogenic in origin (exceptions discussed below), are produced through animal bioturbation-biosorting-biostratification activities, and form the basal layer of two-layered soil biomantles (Horwath and Johnson, 2006; Johnson, 1990; Paton et al., 1995; Schaetzl and Anderson, 2005; Wilkinson et al., 2009). To best understand how two-layered biomantles form it is necessary to appreciate the concept of *dominant bioturbator*.

The dominant bioturbator refers to the bioturbating organism most responsible for particle distributions in a given soil biomantle, including the stone layer, during some time period. In western North America, dominant bioturbators include ants, moles, such rodents as kangaroo rats, pocket gophers, and ground squirrels, and vegetation (root growth, tree uprooting). Dominant bioturbators can vary in population density, can change from one organism or species to another over short periods, and/or may involve several organisms as environmental conditions change, and as species numbers and ranges change.

Two-layered biomantles consist of an upper layer or zone of biosorted particles smaller than the mean burrow diameter of the dominant bioturbator, which often—at least for foraging pocket gophers—approximates their body diameters. Larger particles that are too big for the animals to move settle downward to form the lower, or basal, biomantle—the *stone layer* (cf. figure 11, Johnson and Horwath Burnham, this volume, Introduction). If the dominant bioturbator is one of 40+ species of pocket gophers, for example—as is the case for many western North American soils—the largest pebbles in the upper biomantle will be <6 cm, the mean body diameter of most adult members of that group (Geomyidae). The clasts in the stone layer (lower, basal biomantle) will be >6 cm in diameter. If, on the other hand, the dominant bioturbator is an ant or termite, whose body diameters are ≤2 mm, the largest particles in the upper biomantle will be <2 mm, and those in the lower biomantle—the basal stone layer—will be >2 mm.

In the midlatitudes, where bioturbation is usually shallow and seasonal, basal stone layers tend to occur at shallow depth, often ≤1–2 m. Small burrowing vertebrates are predominantly responsible for producing them, most being active only seasonally, and rarely all-year. Because of shallow depth, midlatitude stone layers can be easily destroyed or disrupted, and the biosorting process is reversed and/or interrupted. This happens when tree uprooting occurs, or when large burrowing predators (badgers, bears, and canids) in search of prey or while den-making return stone layer clasts to the surface.

In subtropical and tropical latitudes, where bioturbation is generally much deeper, stone layers tend to reside at greater depths, often ≥2–6 m or more. The invertebrates (e.g., ants, termites, worms, etc.) mainly responsible for producing them are represented in great numbers and species. Unlike seasonal midlatitudes, bioturbation is year-round, and so great that its effects on soil thickness appear to offset or balance normal landscape denudation and erosional downwasting. Because tropical stone layers lie more deeply, they are less susceptible to tree uprooting and large mammal disturbance than are the shallower stone layers in the midlatitudes. Hence, in the subtropics and tropics stone layers are comparatively more preserved as continuous and stable, unbroken three-dimensional entities, and thus more commonly encountered.

Imbricated "stone lines" that formed from geologic processes and are now visible in fluvial and glacial sedimentary sections, and that can be unequivocally attributed to surface erosion followed by geological burial (Ruhe, 1959), appear far less common than stone layers ("stone lines") produced by normal pedogenic-biogenic processes (Johnson, 1988, 1990). However, in some volcanic landscapes a stone layer is present that appears to be equally erosional and pedogenic-bioturbational in origin. An example is the Pacific Northwest volcanic region of North America. Here Miocene and post-Miocene basalt flows have, to varying degrees, experienced multiple cycles of comparatively rapid loess-ash accumulations followed by soil formation and erosion (Easterbrook and Rahm, 1970). Such ash- and loess-rich soils are particularly susceptible to erosion (Blong and Enright, 2011), especially in montane and hilly regions where soils tend naturally to be thin. Owing to repeated eolian deposition-soil formation-erosion cycles, a polygenetic stone layer presently underlies the current (latest) ash-loess soil mantle. Where these mantles are thin (≤1 m), prairie mounds commonly are formed that are separated by broad, largely erosionally produced stone pavements. Hence, the stone layer that underlies the mounds and mantles in this region, and the pavement exposed between the mounds, are polygenetic equivalents. Graphic stills from a computer animation of erosion-deposition cycles where stone layers become pavements and vice versa linked to bioturbation is displayed in Johnson et al. (2005).

Purpose, Caveats, Beginnings, and Future Studies

In this chapter, we report our findings of double stone layers in three gravelly Mima mounds at two mounded prairies in Missouri and Washington State. We then touch on their self-evident biogenic implications, and how the stone layers in both prairies fulfill a process-morphologic prediction of the dynamic denudation approach. We end by suggesting how biodynamic principles applied to both field observations and laboratory data can predict, reveal, and confirm some obvious, but also some very subtle, soil-landform topographical and morphological signatures of life.

Three mounds were laboratory analyzed for this study, one mound at Diamond Grove Prairie involving 12 cross-sectional profiles, and two mounds at Mima Prairie involving one center-point profile per mound. Although broad generalizations to other mounds of each moundfield and to other moundfields can be made, there remains uncertainty in doing so, and all generalizations should be done cautiously and judiciously. Nevertheless, our data and conclusions establish working hypotheses that can be tested and modified against future studies.

MOUND STUDY SITES

The location, age, substrate contexts, and geologic histories of the two mounded study sites could not be more different. Mima Prairie occupies relatively unweathered loess-ash veneered late Pleistocene glacial outwash gravels on an ~30-m-high terrace above the Black River, near Littlerock, just south of Olympia in Washington. It is one of many mounded gravelly outwash tracts in the immediate area. Diamond Grove Prairie conversely occupies weathered residuum on the ancient Springfield Plateau underlain by mid-Paleozoic rocks near Joplin in southwestern Missouri (Figs. 1 and 2). The Missouri Department of Conservation manages the Diamond Grove Prairie Conservation Area, and the Washington Department of Natural Resources manages the Mima Mound Natural Area Preserve.

Diamond Grove Prairie

The climate of Diamond Grove Prairie is warm moist temperate continental, with warm to hot summers and moderately cool to cold winters punctuated by severe cold snaps (Aldrich, 1989). Precipitation is 104 cm (41 in.) annually that falls mainly as rain in most months, with a June maximum. Snow falls in winter months, but is normally light and of short duration. The soils are described as Oxyaquic Fragiudalfs, and mapped as Keeno very cherty silt loam and Hoberg silt loam (Aldrich, 1989; USDA-NRCS, 2011). They are developed in highly weathered, partly saprolitized ancient cherty residuum of Mississippian-aged limestone, sandstone, and dolomite of the Springfield Plateau (Unklesbay and Vineyard, 1992). The plateau lies 200 km south of the maximum extent of Pleistocene glaciations in a region of gently rolling topography. Mounds are low and broadly shield-like in form (Fig. 2C), average 45 cm in height, 14 m in diameter, and have densities that approximate 7.6 mounds per hectare (Horwath, 2002; Horwath and Johnson, 2006). The larger mounds approach 1.3 m in height and 23 m in diameter (Horwath, 2002).

Vegetation on the mounds consists of native tallgrass, herbs and forbs, more moisture tolerant species occupying the wetter

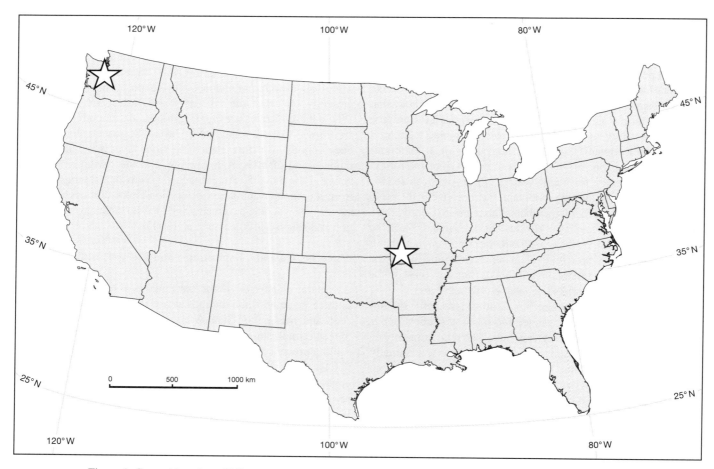

Figure 1. General location of Mima-type moundfield study sites in southwest Missouri and western Washington.

intermound areas (Horwath, 2002). Shrub invasion and weed competition of Diamond Grove Prairie are managed with periodic fire and mowing, and historical records indicate that the area has never been plowed. The unplowed private field abutting Diamond Grove Preserve that was used for the mound analysis presented here is similarly vegetated, but is annually mowed for hay production and thus not fire-managed.

Mima Prairie

The climate of Mima Prairie is Mediterranean-like, with marine-influenced cool wet winters, and warm, usually dry summers (Pringle, 1990). Annual precipitation is 130 cm (51 in.) at nearby Olympia, most (~80%) falling as rain from October through April. The blackish-colored soils are described as Andic Xerumbrepts and Vitrandic Dystroxerepts, and mapped as a complex of Spanaway gravelly sandy loam and Nisqually ashy loamy fine sand (Pringle, 1990; USDA-NRCS, 2011). They are developed in braided terrace outwash gravels derived from the wasting Vashon Lobe of the Cordilleran Ice Sheet. The area underwent rapid deglaciation ca. 17,500 radiocarbon years ago, which discharged torrents of gravel-bearing, anastomosing floodwaters (Porter and Swanson, 1998; Lea, 1984; Thorson, 1980; Washburn, 1988; Bretz, 1913). The anastomosing process, whose braided patterns and flowing symmetries are discernible on light detection and ranging (LIDAR) imagery of these mounded prairies (TGDC, 2011), left micro-highs on which the mounds have biodynamically formed. (Our use of the term *biodynamic* and its derivatives [*biodynamics*, *biodynamically*, etc.] is in the context of principles established in Johnson [1990] and Johnson et al. [2005].) Loess from the wasting Vashon glacier, and later ash from nearby Mount Rainier, contributed postglacial fine fractions (silt, fine sand) to the initial gravelly mantled terraces. The anastomosed gravelly outwash microrelief with subsequent ash-loess overprints, processed by bioturbation and other soil processes in a prairie-like hypsothermal environment that experienced frequent fires (Storm, 2004; Ugolini and Schlichte, 1973; Schlichte, 2010), led to the mound patterns and soil textures that we see today. Water tables in the mounded prairies were lowered by postglacial and Holocene stream incision, and by historic water well draw-downs and drainage practices (Hirschey and Sinclair, 1992; Lea, 1984; Mundorff et al., 1955; Noble and Wallace, 1966; Wallace and Molenaar, 1961). Hence, the geologic, hydrologic, and soil parent material histories of Mima Prairie are fundamentally different and the soils eminently younger than the ancient, highly complex and weathered residuum of the Paleozoic Springfield Plateau in Missouri.

Mounds at Mima Prairie are notably more dome-like in form (Fig. 2A), average 1–2 m in height and 10–20 m in diameter, with greater densities that approach 34 mounds per hectare (Bretz, 1913, Plate 8; TGDC, 2011). Prairie-like vegetation on the mounds is composed of grasses, ferns, herbs, lichens, and forbs under constant pressure of invasion by surrounding Douglas fir-dominated forest (del Moral and Deardorff, 1976; Giles, 1970; Evans et al., 1975). Recurrent Holocene fires are indicated

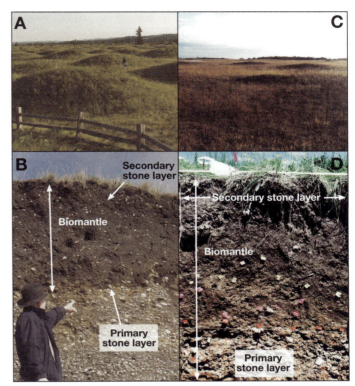

Figure 2. Above- and below-ground images of study site moundfields. (A) Mima Mounds Natural Area Preserve, Washington (person for scale). (B) Internal structure of bisected mound at Mima Prairie. The approximate location of the secondary stone layer, not obvious in the field, is marked by the arrow. The lower primary stone layer is visually masked by dark organic colors. (C) Mounds of Diamond Grove Prairie, Missouri. (D) Internal structure of mounds at Diamond Grove Prairie (surface to base of photo is 125 cm; small square tags mark soil features used in soil descriptions). As in photo B, an arrow marks the approximate location of the secondary stone layer.

by finely comminuted charcoal dispersed through the black ashy-sooty gravelly soils (cf. Ugolini and Schlichte, 1973; Storm, 2004; Schlichte, 2010). Mima Prairie, like Diamond Grove Prairie, also has not been plowed.

Research strategies employed at the two study areas were similar in some ways, and different in others. The Diamond Grove Prairie research was comprehensive in scope and executed during a prescribed time frame (2000–2002). The Mima Prairie research had less comprehensive goals and was executed over a much longer and open time frame (1998–present).

METHODS

Soil Analytical Methods

Mound soil of Diamond Grove Prairie was described and analyzed in a south-facing wall of a backhoe trench cut through the center of one mound on an unplowed private tract adjacent to Diamond Grove Prairie. The 21-m-long trench was bulk (kg) sampled from the surface to 2 m (at deepest) at 10 cm intervals in 12 pedons established at 1 m intervals across the trench wall. Samples were oven-dried and the pipette method was used to determine the fine soil fraction (≤2 mm). The coarse fraction (>2 mm) was analyzed with nested sieves (31.5, 16, 8, and 4 mm) with each clast then hand counted, weighed, and hand measured for long axis diameter (detailed methodologies and results are in Horwath, 2002).

Soils at Mima Prairie were described and analyzed from the south-facing walls of two mounds cross-sectionally exposed in a gravel borrow pit at the south end of Mima Prairie Preserve (here-after "Mima borrow pit"). Bulk (kg) samples were collected from one (center) pedon of each mound at 10 cm intervals from mound crests (apices) down into subjacent stratified outwash gravels. Fine and coarse soil fractions were analyzed in the laboratory following the methods of Horwath (2002). Intermittent multiyear photographic and foot surveys of Mima Prairie and other Puget Sound prairies, both mounded and unmounded, were also made for comparative soil-ecologic assessments.

Ecological Methods

At Diamond Grove Prairie, six 1 ha plots were established to study the mound density and vegetation composition of the mounds. The burning of one of the plots (for prairie management) in the month prior to fieldwork provided an excellent opportunity for observing and mapping of all visible animal burrows. The exposed ground surface, free of vegetation, provided an unobscured view of animal burrows on the prairie. The plot was evaluated in both north-south and east-west transects at 10 m intervals during which burrows were located and flagged. The size of each burrow was approximated and placed in one of three categories: small (0.1–1 cm), medium (1–5 cm), and large (6+ cm). Occasional observations of animals in burrows were also recorded (summaries in Horwath, 2002).

At Mima Prairie, soil-ecological studies began in 1998 with a photographic overflight, followed by nonintrusive, multiyear (11) observations and photographic ground surveys at different times and seasons, the last in October 2011. While the focus was on Mima Prairie, comparative ground surveys were also conducted at other mounded and nonmounded prairies, both locally and regionally. This work included comparative analysis of literature, topographic maps, ground photographs, multiyear airphotos, LIDAR, and Google Earth and other online resources (TGDC, 2011; limited summaries in Johnson et al., 2006, 2008b).

RESULTS

Although the moundfields of Diamond Grove and Mima Prairies differ widely in terms of location (Fig. 1), mound geometries, densities (Fig. 2), heights, diameters, ages, soils, geology, parent materials, local elevations, and regional climates, the mounds within them share several internal attributes that shed light on their common, predominantly biogenic origins.

What has not been observed until this study—although in hindsight might have been predicted by the theory we employed—is the presence of a secondary, somewhat diffuse stone layer found in the uppermost bioturbationally active portion of some Mima-type mounded biomantles (~10–30 cm depth). This secondary, poorly expressed near-surface stone layer, which appears to be incipient (i.e., recently formed), follows the mound topography unlike the primary stone layer that is planar or concave. It is present in each of the three mounds studied at both moundfields.

Diamond Grove Prairie

At Diamond Grove Prairie a thick A horizon, and two poorly expressed EBg and EBtg horizons that are apparently incipiently formed, lie above three exceptionally well-developed Btg, Bt/Btg, and Bt horizons (Fig. 3; details of soil horizon characteristics are in Horwath, 2002). The upper younger horizons are generally more continuous and thickest near the mound center, and less continuous and thinner toward mound edges. Of particular note is the strongly expressed and thick primary lower, basal stone layer of the Btg horizon, with coarse fractions (>2 mm) averaging 75% by weight (Fig. 3) and long-axis diameters of the largest stones averaging 8.0 cm (Horwath and Johnson, 2006). As mentioned, an unanticipated discovery was the presence of a second, weakly expressed near-surface stone layer that, while not obvious in the field, was noted in the laboratory data. This layer of coarse material, which mimics the mound topography, averages 10–30 cm depth and was present in 9 of the 12 pedons analyzed (Horwath, 2002). The largest stones range from 2.7 to 8.2 cm in long-axis diameter, with an average of 4.7 cm (Horwath and Johnson, 2006; the 8.2 cm anomalously large clast possibly reflects predator activity). The coarse fraction of the upper stone layer ranges from 13% to 34% (by weight) with an average of 21.3% (Horwath and Johnson, 2006). A slight coarsening of the layer was noted near the center of the mound.

The results of mapping animal burrows on the burned section (plot 6) of the Missouri study site are shown in Figure 4. In addition to showing individual animal burrows and Mima-type mounds, this figure also displays small heaps of burrow spoil deposited on the surface under the winter snowpack of 2000–2001 by the animals that inhabit the mounds. Observing such heaps was fortuitous in timing because the presence of snowpack over an extended period is unusual for southwest Missouri. (Snowpack-protected heaps quickly disappear following rains, with such evidence lost.) The majority of burrows on plot 6 are of medium size (1–5 cm diameter) and were observed both on mounds and intermound areas. Numerous clusters of these medium-sized burrows were found throughout the plot and likely reflect adjacent entrance and exit burrows of small burrowing mammals (rodents, Table 1). At least 10 small rodents have been documented on Diamond Grove Prairie (Table 1). Crayfish burrows and associated chimneys are almost certainly responsible for many of the nonclustered medium-sized burrows (Fig. 4B) in

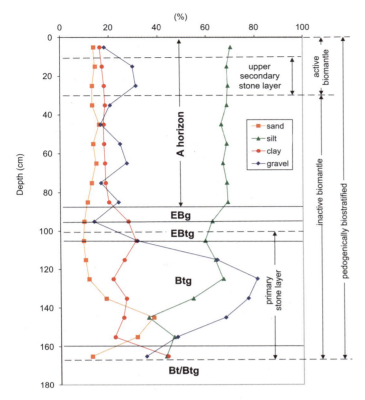

Figure 3. Fine and coarse fraction depth distributions for mound soils at Diamond Grove Prairie. This profile is from the center (thickest) portion of the mound. Other profiles have similar particle-size distributions albeit in thinner horizons and less overall depths (Horwath, 2002). Gravel curve is a ratio of coarse (≥2 mm) to fine (<2 mm) fractions of the bulk sample.

the wetter intermound areas (several crayfish exoskeletons were observed in intermounds).

Of the eight documented large burrows (6+ cm diameter) on the plot, five were located on mounds, and were presumably made by badgers, coyotes, foxes, armadillos, and/or box turtles, all of which inhabit the area. The smallest burrows (0.1–1 cm) are presumed to be due to insects, arachnids, and small invertebrates (Fig. 4A). The vast majority of all the documented burrows were of small and medium size (<5 cm).

Mima Prairie

At Mima Prairie, the two mounds analyzed, plus other mounds cross-sectionally exposed in Mima borrow pit, are 1–2+-m-thick accumulations of black Spanaway gravelly and very gravelly sandy loam above a biogenically differentiated ~20–30 cm thick, extremely gravelly basal stone layer (Fig. 2B; Table 2). The large clasts that compose the primary (basal) stone layers of both mounds are effectively masked by the black soil colors and were not perceived until the soil profile was physically sampled. Only then did the large clasts that compose the basal stone layer become obvious.

As at Diamond Grove Prairie, the secondary near-surface stone layer at Mima Prairie is generally 10–30 cm deep, and it likewise was not initially seen in the field, but became apparent from laboratory data (Fig. 5, Table 2). The largest stones of the upper stone layer of mound 1 range from 4.5 to 4.7 cm in long-axis diameter, with an average of 4.6 cm. The coarse fraction of the upper stone layer is comparatively coarser than at Diamond Grove Prairie, with a range of 64.3%–54.0% (by weight) with an average of 59.2%. The lower stone layer contains larger stones (average 9.4 cm) and a higher coarse fraction percentage (average 71.6%) compared to the upper stone layer. The clasts that make up the stone layers of the Mima Prairie mounds and inter-mound pavements are also essentially unweathered (Fig. 6D) in contrast to the highly weathered, saprolitized, dense, thick, and extremely gravelly stone layer of Diamond Grove Prairie. Since only one pedon in each of the two mounds at Mima Prairie was analyzed, we could not determine whether the secondary stone layer mimics mound topography, as it does at Diamond Grove Prairie where multiple pedons were analyzed. Careful examinations of bisected mounds during subsequent visits to Mima borrow pit, however, suggest that it does.

DISCUSSION

What Happens when Dominant Bioturbators Disappear and Smaller Ones Take Over?

While the Diamond Grove Prairie and Mima Prairie mound-fields formed in completely different environments, evidence suggests that the dominant bioturbators of each—until recent local extirpation—were in both cases members of the Geomyidae family of burrowing rodents. These were the Plains Pocket Gopher (*Geomys bursarius*) at Diamond Grove Prairie, and the Mazama Pocket Gopher (*Thomomys mazama*) at Mima Prairie. Although pocket gophers do not presently occupy either prairie now, they do occupy nearby and/or regional prairies in both areas. The pebbly nature of the mounds at both prairies and their co-evolved basal stone layers, are strong and predicted indicators of small mammal bioturbation in gravelly soils (Horwath and Johnson, 2006; Johnson, 1990; cf. Cox, 1984). Moreover, the gopher-size krotovina at the base of the Mima Prairie mounds (Figs. 6F and 6G), and similar krotovina within the Diamond Grove Prairie mounds (Horwath, 2002), all suggest that these animals did occupy the soils of both sites in recent geologic time, probably during very late Holocene or protohistoric time (cf. Faunmap Working Group, 1994; Dalquest and Scheffer, 1942, 1944). Further, paleontological records show that certain beetles that are obligate inhabitants of pocket gopher burrows are present as subfossil remains at Mima Prairie (Nelson, 1997). Schwartz and Schwartz (2001, p. 178) stated that the current range of pocket gophers in Missouri is not known with certainty, and that they may be abundant for 6–8 years in some Missouri sites, then become rare: "Frequently there are areas near active colonies that are unoccupied, although formerly pocket gophers

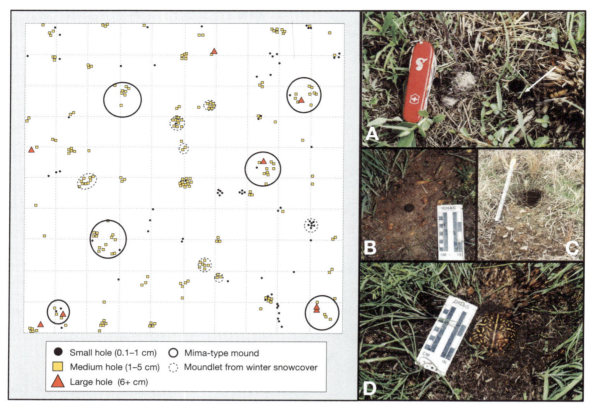

Figure 4. Distribution of animal burrows observed at the 1 ha burned plot of Diamond Grove Prairie. Note that burrows of various sizes occur both on mounds and intermound areas. (A) Example of a small (0.1–1 cm) burrow, likely of an insect or arachnid (knife is 9 cm long). (B) Example of a medium (1–5 cm) burrow, likely of a crayfish. (C) Example of a large (6+ cm) burrow occupied by a rabbit (as evidenced by scat). Tape measure units are marked in decimeters. (D) One of many box turtles observed digging at Diamond Grove Prairie.

TABLE 1. SMALL RODENT SPECIES DOCUMENTED AT DIAMOND GROVE PRAIRIE AND THEIR BURROW-ING BEHAVIOR (MODIFIED FROM ROBBINS AND HADLEY, 1998; CF. SCHWARTZ AND SCHWARTZ, 2001)

Species	Common name	Burrow behavior
Cryptotis parva	Least shrew	Burrower
Microtis ochrogaster	Prairie vole	Burrower
Microtis pinetorum	Pine vole	Burrower
Mus musculus	House mouse	Burrow occupier
Peromyscus leucopus	White-footed mouse	Occasional burrower
Peromyscus maniculatus	Deer mouse	Occasional burrower
Reithrodontomys fulvescens	Fulvous harvest mouse	Rare burrow occupier
Reithrodontomys megalotis	Western harvest mouse	Rare burrow occupier
Sigmodon hispidus	Cotton rat	Occasional burrower
Zapus hudsonius	Meadow jumping mouse	Burrow occupier

inhabited them." In this regard, Howell (1923) documented the rapidity with which populations and ranges of small animals, like gophers, can fluctuate and change.

Moles, either Townsend's (*Scapanus townsendii*) or Coast mole (*Scapanus ovarius*), or possibly both, presently inhabit and actively bioturbate many of the mounds at Mima Prairie (Figs. 6A and 6B). Unlike Diamond Grove Prairie, however, a small mammal survey has not been done, nor is a mole-specific study available. Hence, whether both species of moles inhabit and bioturbationally impact Mima Prairie is unclear. What is clear,

however, based on multiple intermittent personal observations (D.L.J. and D.N.J.) over many years is that while some mounds and sectors of Mima Prairie display abundant fresh to wasted mole heaps (Figs. 6A and 6B), others display none. The burrowing-heaping-mounding capacities of these two Pacific Northwest moles are remarkably similar, and may match or even exceed the profound impacts of pocket gophers on soil formation (Carraway et al., 1993; Grinnell, 1923; Hartman and Yates, 1985; cf. Mapes, 2011). Whether gophers and moles were co-dominant bioturbators at Mima Prairie during mound formation is uncertain, but

TABLE 2A. PARTICLE-SIZE DISTRIBUTION DATA FROM MOUND 1 AT MIMA BORROW PIT, SOUTH END OF MIMA PRAIRIE, THURSTON COUNTY, WASHINGTON

Horizon	Depth (cm)	Horizon thickness (cm)	Bulk sample (g)	Percent gravels (by wt)	Percent fines (by wt)	Total wt three largest stones (g)	Average wt three largest stones (g)	Wt largest stone (g)	Average length three largest stones (mm)	Length largest stone (mm)	Fines (<2 mm)			Texture class
											Percent sand	Percent silt	Percent clay	
Biomantle														
A	0–10	0–130	654.6	41.8	58.2	75	25	44	37	39	53.5	39.9	6.7	gsl*
A/SL	10–20	0–130	1150.6	64.3	35.7	83	28	33	37	45	60.2	34.9	5.0	vgsl†
A/SL	20–30	0–130	1145.4	54.0	46.0	141	47	68	38	47	58.4	36.8	4.8	vgsl†
A/SL	30–40	0–130	1122.2	54.9	45.1	134	45	71	42	51	55.5	41.2	3.3	vgsl†
A	40–50	0–130	957.9	49.6	50.4	92	31	34	39	42	53.9	42.6	3.6	vgsl†
A	50–60	0–130	1437.1	56.5	43.5	183	61	76	52	57	55.4	40.4	4.3	vgsl†
A	60–70	0–130	1255.0	58.3	41.7	160	53	61	47	49	54.6	41.6	3.8	vgsl†
A	70–80	0–130	1070.6	44.9	55.1	58	19	30	33	38	52.2	44.1	3.8	vgsl†
A	80–90	0–130	1179.6	57.4	42.6	174	58	86	47	51	57.0	39.6	3.5	vgsl†
A	90–100	0–130	1214.2	50.8	49.2	133	44	67	39	43	53.7	42.7	3.6	vgsl†
A/SL	100–110	0–130	1317.9	68.0	32.0	594	198	522	59	98	58.4	38.6	3.0	egsl§
A/SL	110–120	0–130	1483.5	68.6	31.4	585	195	275	66	73	57.7	39.9	2.4	egsl§
A/SL	120–130	0–130	1895.1	78.3	21.7	1036	345	515	89	110	59.8	37.9	2.3	egsl§
Subsoil														
Bw/C	130–140	130+?	2124.8	90.3	9.7	732.0	244.0	299.0	76.0	83.0	93.4	4.6	2.0	egs#

TABLE 2B. PARTICLE-SIZE DISTRIBUTION DATA FROM MOUND 2 AT MIMA BORROW PIT, SOUTH END OF MIMA PRAIRIE, THURSTON COUNTY, WASHINGTON

Horizon	Depth (cm)	Horizon thickness (cm)	Bulk sample (g)	Percent gravels (by wt)	Percent fines (by wt)	Total wt three largest stones (g)	Average wt three largest stones (g)	Wt largest stone (g)	Average length three largest stones (mm)	Length largest stone (mm)	Fines (<2 mm)			Texture class
											Percent sand	Percent silt	Percent clay	
Biomantle														
A	0–10	0–110	1135.2	49.2	50.8	106	35	50	39	45	56.8	38.0	5.3	gsl*
A/SL	10–20	0–110	1255.5	59.4	40.6	207	69	95	53	63	58.7	37.1	4.2	vgsl†
A/SL	20–30	0–110	1611.2	59.0	41.0	128	43	51	45	55	56.4	39.5	4.1	vgsl†
A	30–40	0–110	1288.1	53.1	46.9	86	29	52	37	51	54.3	42.0	3.7	vgsl†
A	40–50	0–110	1345.2	54.5	45.5	80	27	33	41	44	54.8	42.9	2.3	vgsl†
A	50–60	0–110	1380.5	55.0	45.0	166	65	66	43	46	59.1	37.3	3.7	vgsl†
A	60–70	0–110	1529.2	64.4	35.6	85	28	31	39	45	54.0	43.1	2.9	vgsl†
A/SL	70–80	0–110	1598.0	64.1	35.9	771	257	342	76	85	55.1	41.1	3.9	egsl§
A/SL	80–90	0–110	1829.4	67.3	32.7	1114	371	805	77	109	53.7	43.0	3.4	egsl§
A/SL	90–100	0–110	3045.5	83.0	17.0	391	130	157	69	83	61.4	45.4	3.2	egsl§
A/SL	100–110	0–110	2087.1	83.3	16.7	324	108	144	54	58	84.2	14.3	1.5	egsl§
Subsoil														
Bw/C	110–120	110–150	2283.3	82.7	17.3	163	64	66	46	52	88.7	10.3	1.0	egs#
Bw/C	120–130	110–150	2309.8	69.4	30.6	309	103	140	56	65	95.0	4.3	0.8	egs#
Bw/C	130–140	110–150	2429.1	81.4	18.6	580	193	224	68	77	91.8	7.1	1.1	egs#
Bw/C	140–150	110–150	2335.9	84.1	15.9	422	141	296	61	85	91.5	7.4	1.1	egs#
C	150–160	150–170+	2239.0	80.9	19.1	315	105	128	58	61	94.7	4.3	0.9	egs#
C	160–170	150–170+	2180.7	81.1	18.9	209	70	84	51	55	95.8	3.6	0.6	egs#
Krotovina	110–120		1318.8	59.7	40.3	122	41	61	40	45	68.0	30.0	2.0	vgsl†

Note: Percent total gravels and fines (columns 5 and 6) relative to bulk sample weights (column 4), and fines expressed as percent sands, silts, and clays (columns 12–14), were plotted together in Figure 5 for comparative purposes. Data of other columns (7–11) provide supplementary and comparative information on representative sizes and weights of stones distributed within these two mounds.

*Gravelly sandy loam.
†Very gravelly sandy loam.
§Extremely gravelly sandy loam.
#Extremely gravelly sand.

moles are active on some mounds now, and indirect evidence (krotovina at base of mounds) is strong that gophers recently were active on most or all of them. It is also uncertain whether the two mounds studied at Mima Prairie were historically bioturbated by either of these animals. If they were, it was long enough ago that their textural legacies in the upper 30 cm are now erased

and palimpsestically overprinted at Mima Prairie, as at Diamond Grove Prairie, by a new incipient two-layered biomantle.

As pocket gopher dominants vacated Diamond Grove and Mima Prairies, soils of these mounded landscapes came under a new set of soil bioturbations that, as before, impacted their textural signatures. Specifically, new texturally layered biomantles began

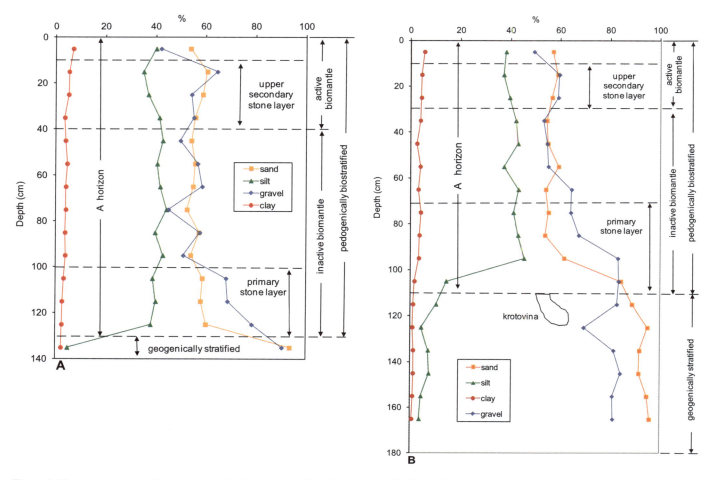

Figure 5. Fine and coarse fraction depth distributions for mound 1 (A) and mound 2 (B) at Mima borrow pit (cf. Tables 2A and 2B). The A horizons of both mounds constitute their biomantles. Gravel curve is a ratio of coarse (≥2 mm) to fine (<2 mm) fractions of the bulk sample. Sand, silt, and clay curves are plotted independently of the coarse-to-fine ratio in order to display them on the same chart. The "active biomantle" is the focus of most biodynamic activity in terms of present particle-size differentiation in the mound. The krotovina (one krotovine shown) common at the base of the primary stone layer in B compositionally mirrors the pebbly soil between 0 and 70 cm depth (above the primary stone layer; cf. Table 2B).

forming in the upper reaches of the mounds. Stated otherwise, the newer incipient upper and shallower "active biomantle" (Figs. 3 and 5) is owed to a new set of dominant bioturbators, whose effects on soil texture and biostratification are now being superimposed onto mounds previously biostratified by small mammals. The new dominants likely include insects, arachnids, crustaceans (pill bugs), mollusks (snails), annelids (worms), other "bugs," and/or possibly involving such smaller vertebrates as lizards, voles, and shrews (cf. Nardi, 2003, 2007; Wallwork, 1976).

The mound soil *below* the uppermost shallow, now active biomantle is conceptually and functionally termed the "inactive biomantle" (Figs. 3 and 5). This does not mean that lower mound areas are free of bioturbation, but conveys the point that biodynamic activity is now focused primarily in the upper reaches of the mounds, specifically the upper 30 cm of soil. Hence, the reworking of soil previously biosorted by small mammals is again being biosorted, this time by new dominants, which we believe are invertebrates.

From our limited data we can only hypothesize and speculate which species might now be playing the hierarchically dominant roles in these upper active and new biomantles. We suspect that a range of invertebrate soil infauna is involved, as suggested by the microbioturbation displayed at Diamond Grove Prairie (cf. Fig. 4).

Stone Sizes and Quantitative Dominant Bioturbator Estimations

As indicated, three mounds were laboratory analyzed for this study among the thousands present in the two moundfields of Mima and Diamond Grove Prairies. Hence any generalizations must be made with such limitations in mind. But also as indicated, our data and conclusions represent initial working hypotheses that can be tested in future studies.

Drawing partly on the biodynamic principles employed in this chapter and on common logic, most burrowing dominants, like pocket gophers, biotransfer soil material that is smaller than

Figure 6. Collage of typical mounds and examples of bioturbation signatures, Mima Mounds Natural Area Preserve, Littlerock, Washington. (A) Numerous recent and old *Scapanus* mole heaps on Mima mounds, both on mound in front of man (as per arrow), and on mound behind him. (B) Close-up of mole heaps (keys, at arrow, for scale): a—recent *Scapanus* heap; b—older, more weather-wasted heap; c—oldest, most wasted *Scapanus* heap. (C) Intermound cobble-boulder pavement, largely obscured by grass, that continues as a stone layer under mounds (boulders in photos are larger than most in intermounds). (D) Sample of larger pebbles collected from several mole heaps. (E) Small mammal (vole?) runnels and burrows common on mounds and intermounds. (F) South-facing cross section of a bisected mound immediately east of Mima borrow pit; arrows at krotovina. Unlike mounds 1 and 2, this mound was field-determined to possess an intermittent discontinuous basal (primary) stone layer. (G) Close-up of F, showing multiple gopher- and mole-size krotovina (arrows) that are typical at the base of essentially all bisected mounds in Mima Prairie and other nearby mounded prairies.

the diameters of their foraging tunnels. Hence, comparing the sizes of stones dispersed through the mound soils (main biomantle)—stones that presumably have been repeatedly recycled through foraging tunnels—to the larger stones in the primary stone layers that could not be moved by such animals, allows for size and identity estimations of the animals involved. Forage

tunnel diameters often are only slightly larger than, or approximate, the body size diameters of the animals that make them—at least as repeatedly personally observed in the case of gophers. Hence, as mentioned earlier, stones passing through these tunnels must have diameters smaller than tunnel diameters. The average maximum long-axis diameters of pebbles dispersed through the mounds at both prairies is 5.0 cm for Diamond Grove Prairie and 4.6 cm for Mima Prairie (Horwath and Johnson, 2006; Table 2). These data suggest that the average diameter of the responsible animals at both prairies approximates this or a slightly larger size.

These data, again in conjunction with the biodynamic theory employed, also suggest that all larger stones have settled downward due to bioturbational undermining, ultimately to form the basal (primary) mound stone layers at both prairies. The average diameter of large stones in the lower (basal) primary stone layers is 8.0 cm for Diamond Grove Prairie and 9.4 cm for Mima Prairie (Horwath and Johnson, 2006; Table 2). They are, of course, significantly larger than the stone diameters in the upper (secondary, incipient) stone layer, which average 4.7 cm for Diamond Grove Prairie and 4.6 cm for Mima Prairie (Horwath and Johnson, 2006; Table 2).

These data, added to information outlined earlier, converge to provide strong evidence that the origin of the basal mound stone layer at both prairies was through biosorting and biostratification by small fossorial mammals, such as pocket gophers (and/or possibly moles at Mima Prairie). Stones in the upper stone layer were small enough for the former dominants (gophers, moles) to move, mix, and disperse them throughout, but these stones were larger than could be moved by the much smaller new dominants, like insects, worms, snails, and arachnids. These data strongly suggest that the incipient upper secondary stone layer now being formed is likely by just such soil invertebrates. The textural impacts of the latter doubtless had always operated, but were previously regularly obliterated by the larger bioturbators. Textural soil signatures of larger dominants invariably mask the signatures of smaller animals—until the larger dominants leave. Only then can the textural expressions and roles of the smaller bioturbators be assessed in the formation of soil.

The Meaning of Other Gravel Peaks

In addition to the gravel peaks that define the primary and secondary stone layers of the three mounds studied, other peaks are also evident in the data of the particle size charts (Figs. 3 and 5) and tables (Table 2). Determining whether or not such peaks have genetic significance, however, is challenging. For example, Figure 3 represents the central (thickest) pedon of the 12 pedons analyzed in a cross-sectional cut across the Diamond Grove Prairie mound. The peak at 50–70 cm depth in the chart is almost as prominent as the upper second stone layer peak at 10–30 cm. But, unlike the latter, which occurs across 9 of the 12 pedons, no other peaks are present at that level in the other 11 pedons (Horwath, 2002). In addition, other peaks are present in some of the other 11 pedons that appear somewhat random. Such a comparative assessment of peaks cannot be made in the two mounds studied at Mima Prairie because only one pedon each was analyzed. Hence the reasonably prominent peak at 50–70 cm depth of mound 1 in Figure 5A may or may not have any genetic significance.

These data, augmented by observations made on many other mounded prairies, strongly suggest that bioturbation of Mima mounds can be a heterogeneous process. Dominant bioturbators come and go, and some mounds may be dotted with their heaps, as in Figure 6A, at one point in time, while others in the same moundfield, as at Mima Prairie, display none.

CONCLUSIONS

Gravelly Mima-type soil mounds at Diamond Grove Prairie in Missouri and Mima Prairie in Washington State both express textural signatures of relatively large formerly dominant bioturbators, and relatively small currently dominant bioturbators. Signatures left by the former dominant bioturbators, *G. bursarius* at Diamond Grove Prairie and *T. mazama* at Mima Prairie, are the biostratified and deep, two-layered Mima-type mounds themselves. As predicted by the theory employed, the base of each mound is defined by a strongly expressed primary coarse clast stone layer. The new signatures in the three mounds studied are presently being imparted by invertebrates (insects, worms, pill bugs, and snails), and are expressed at both prairies as shallow, bioturbationally active ~30-cm-thick miniature biomantles. These now occupy and have formed in the upper parts of the mounds. Also as predicted, each new biomantle is defined by a diffuse, incompletely formed secondary stone layer at a depth of 20–30 cm that consists of small (<5 cm diameter) pebbles.

The primary stone layer at the base of Mima-type mounds that dot both prairies is the legacy of the former dominant bioturbators, pocket gophers—and possibly also moles in the case of Mima Prairie. The upper secondary stone layer in the mound profiles at both study sites is indirectly a reflection of the local disappearance of the two still regionally pervasive and dominant soil bioturbators, namely pocket gophers: *G. bursarius* in the case of Diamond Grove Prairie, and *T. mazama* in the case of Mima Prairie. We argue that a preexistent suite of invertebrate bioturbators is responsible for the now forming and weakly expressed secondary biomantle and stone layer. We assume that these new dominant invertebrate bioturbators had always played important, though subordinate, biomixing-biosorting roles—that is, subordinate to Geomyidae bioturbation until it ceased.

We conclude and hypothesize that the same textural biodynamic principles, where a change in dominant bioturbators promotes new textural profiles, apply to gravelly soils in general, and to gravelly paleosols of Quaternary, Tertiary, Mesozoic, and earlier ages (Meysman et al., 2006; Hasiotis and Halfen, 2010). If a small dominant bioturbator (worm, insect) replaces a larger one (rodent, mole), a double stone layer should form in any gravelly soil. Moreover, because Mima mounds are thicker than typical midlatitude soil biomantles, double stone layers are more easily discerned in them than in normally much shallower (nonmounded) soils, where they are likely less to be perceived. However, this would not be the case in the humid tropics where biomantles are notably thicker and soils deeper. Hence, field observations and laboratory data paired with biodynamic principals can predict, and reveal, some rather subtle biotic-soil-landform attributes that would otherwise likely be overlooked.

The discovery of these secondary stone layers was by accident, and to our knowledge appears to lack precedence. Nevertheless, biodynamic principles predict their occurrence whenever dominant bioturbators change to a smaller size in gravelly soils. This study represents a small first step. Those who study soils and paleosols are challenged to seek similar effects in gravelly *and* nongravelly soils, mounded or otherwise, in the search to further identify topographical and soil morphological signatures of life.

ACKNOWLEDGMENTS

Special thanks to referees Dick Berg, Barbara Lewis, Tony Lewis, Jonathan Phillips, and Ken Schlichte who improved this chapter with their suggestions. For Diamond Grove Prairie support special thanks goes to Bob and Mary Jean Hockman, Mike and Joan Banks, Steve and Missy Taylor, Melissa Georgas, Mike Skinner, Mark Hutchings, and Don Kurz. For Mima Prairie support special thanks goes to Lee Walkling, Connie Manson, Leslie Durham, Roberta Davenport, Ken Schlichte, Jeff Bradley, Lois Richards, Claudia Westbrook, Fred and Katherine Colvin, Stephanie Zurenko, Victor Scheffer, and Andrew Deffobis. Funding for Diamond Grove Prairie research was graciously provided by the Missouri Prairie Foundation, the Missouri Department of Conservation, the Geological Society of America (grant no. 6922-01), University of Illinois Graduate College, and the University of Illinois, Department of Geography.

REFERENCES CITED

Aldrich, M.W., 1989, Soil Survey of Newton County, Missouri: United States Department of Agriculture, Natural Resources Conservation Service: Washington, D.C., U.S. Government Printing Office, 123 p.

Arkley, R.J., and Brown, H.C., 1954, The origin of Mima mound (hogwallow) microrelief in the far western states: Proceedings of the Soil Science Society of America, v. 18, no. 2, p. 195–199, doi:10.2136/sssaj1954.03615995001800020021x.

Blong, R.J., and Enright, N., 2011, Preservation of thin tephras: Unpublished manuscript on file: Perth, Australia, Murdoch University Library, 28 p., http://researchrepository.murdoch.edu.au/5785/ (accessed 19 June 2012).

Bretz, J.H., 1913, The mounds of the Vashon Outwash: Chapter 5, *in* Glaciation of the Puget Sound Region: Washington Geological Society Bulletin No. 8: Olympia, Frank M. Lamborn, Printer, p. 81–108.

Carraway, L.N., Alexander, L.F., and Verts, B.J., 1993, Scapanus townsendii: Mammalian Species, no. 434, p. 1–5, doi:10.2307/3504286.

Cox, G.W., 1984a, Mounds of mystery: Natural History, v. 93, no. 6, p. 36–45.

Cox, G.W., 1984b, Soil transport by pocket gophers in Mima mound and vernal pool microterrain, *in* Jain, S., and Moyle, P., eds., Vernal Pools and Intermittent Streams: Institute of Ecology Publication no. 28: Davis, University of California, p. 37–45.

Cox, G.W., and Allen, D.W., 1987, Sorted stone nets and circles of the Columbia Plateau: A hypothesis: Northwest Science, v. 61, no. 3, p. 179–185.

Dalquest, W.W., and Scheffer, V.B., 1942, The origin of the mounds of western Washington: The Journal of Geology, v. 50, no. 1, p. 68–84, doi:10.1086/625026.

Dalquest, W.W., and Scheffer, V.B., 1944, Distribution and variation in pocket gophers, *Thomomys talpoides*, in the state of Washington: American Naturalist, v. 78, no. 777, p. 308–333, 423–450, doi:10.1086/281203.

del Moral, R., and Deardorff, D.C., 1976, Vegetation of the Mima mounds, Washington State: Ecology, v. 57, p. 520–530, doi:10.2307/1936436.

Easterbrook, D.J., and Rahm, D.A., 1970, Landforms of Washington—The Geologic Environment: Bellingham, Washington, Union Printing, 155 p.

Evans, S.A., Gilbert, M.E., Johnson, C.S., and Schuett, J.P., 1975, The Pseudotsuga menziesii invasion on Mima Prairie: A study of prairie-forest dynamics, *in* Herman, S.G., and Wiedemann, A.M., eds., Contributions to the Natural History of the Southern Puget Sound Region, Washington: Olympia, Washington, The Evergreen State College, p. 63–108.

Faunmap Working Group, 1994, Faunmap: A database documenting late Quaternary distributions of mammal species in the United States: Springfield, Illinois State Museum Scientific Papers 25, no. 1–2, 288 p.

Finney, F.A., 2012, this volume, The forgotten natural prairie mounds of the Upper Midwest: Their abundance, distribution, origin, and archaeological implications, *in* Horwath Burnham, J.L., and Johnson, D.L., eds., Mima Mounds: The Case for Polygenesis and Bioturbation: Geological Society of America Special Paper 490, doi:10.1130/2012.2490(05).

Fosberg, M.A., 1963, Genesis of some soils associated with low and big sagebrush complexes in the Brown, Chestnut and Chernozem-prairie zones in southcentral and southwestern Idaho [unpublished Ph.D. thesis]: Madison, University of Wisconsin, 308 p.

Gentry, H.R., 2006, Soil Survey of Yakima Training Center, Parts of Kittitas and Yakima Counties, Washington: U.S. Department of Agriculture–National Resources Conservation Service, in cooperation with U.S. Department of Army, Yakima Training Center, U.S. Army Corps of Engineers, and Washington State University Agricultural Research Center: Washington, D.C., U.S. Government Printing Office.

Giles, L.J., 1970, The ecology of the mounds on Mima Prairie with special reference to Douglas-fir invasion [unpublished Master's thesis]: Seattle, University of Washington, 99 p.

Grinnell, J., 1923, The burrowing rodents of California as agents in soil formation: Journal of Mammalogy, v. 4, no. 3, p. 137–149, doi:10.2307/1373562.

Hartman, G.D., and Yates, T.L., 1985, *Scapanus orarius*: Mammalian Species, no. 253, p. 1–5, doi:10.2307/3504000.

Hasiotis, S.T., and Halfen, A.F., 2010, The story of O: The dominance of organisms as a soil-forming factor from an integrated geologic perspective and modern field and experimental studies: Proceedings of the 19th World Congress of Soil Science, Soil Solutions for a Changing World, 1–6 August 2010; published on DVD, http://www.iuss.org; Brisbane, Australia, p. 1–4.

Hirschey, S.J., and Sinclair, K.A., 1992, A hydrologic investigation of the Scatter Creek-Black River area, southern Thurston County, Washington State [unpublished Master's thesis]: Olympia, Washington, The Evergreen State College, 126 p.

Horwath, J.L., 2002, An assessment of Mima type mounds, their soils, and associated vegetation, Newton County, Missouri [unpublished Master's thesis]: Champaign-Urbana, University of Illinois, 265 p.

Horwath, J.L., and Johnson, D.L., 2006, Mima-type mounds in southwest Missouri: Expressions of point-centered and locally thickened biomantles: Geomorphology, v. 77, no. 3–4, p. 308–319, doi:10.1016/j.geomorph.2006.01.009.

Howell, A.B., 1923, Periodic fluctuations in the numbers of small mammals: Journal of Mammalogy, v. 4, no. 3, p. 149–155, doi:10.2307/1373563.

Irvine, L.L.-A., and Dale, J.E., 2012, this volume, "Pimple" mound microrelief in southern Saskatchewan, Canada, *in* Horwath Burnham, J.L., and Johnson, D.L., eds., Mima Mounds: The Case for Polygenesis and Bioturbation: Geological Society of America Special Paper 490, doi:10.1130/2012.2490(02).

Johnson, D.L., 1988, A genetic classification of stone lines and stone zones in soils: Geological Society of America Abstracts with Programs, v. 20, p. 284–285.

Johnson, D.L., 1989, Subsurface stone lines, stone zones, artifact-manuport layers, and biomantles produced by bioturbation via pocket gophers (*Thomomys bottae*): American Antiquity, v. 54, no. 2, p. 370–389, doi:10.2307/281712.

Johnson, D.L., 1990, Biomantle evolution and the redistribution of earth materials and artifacts: Soil Science, v. 149, no. 2, p. 84–102, doi:10.1097/00010694-199002000-00004.

Johnson, D.L., and Horwath Burnham, J.L., 2012, this volume, Introduction: Overview of concepts, definitions, and principles of soil mound studies, *in* Horwath Burnham, J.L., and Johnson, D.L., eds., Mima Mounds: The Case for Polygenesis and Bioturbation: Geological Society of America Special Paper 490, doi:10.1130/2012.2490(00).

Johnson, D.L., and Johnson, D.N., 2008, White paper: Stonelayers, mima mounds and hybrid mounds of Texas and Louisiana—Logical alternative 'biomantle working hypotheses' (appendix 3), *in* Johnson, D.L., Mandel, R.D., and Frederick, C.D., eds., The Origin of the Sandy Mantle and Mima Mounds of the East Texas Gulf Coastal Plains: Geomorphological, Pedological, and Geoarchaeological Perspectives: Pre-meeting field trip guidebook, Geological Society of America Annual Meeting, Houston, Texas, Oct. 4–9, 2008, 57 p., 30 plates, 3 appendices.

Johnson, D.L., and Johnson, D.N., 2012, this volume, The polygenetic origin of prairie mounds in northeastern California, *in* Horwath Burnham, J.L., and Johnson, D.L., eds., Mima Mounds: The Case for Polygenesis and Bioturbation: Geological Society of America Special Paper 490, doi:10.1130/2012.2490(06).

Johnson, D.L., Horwath, J., and Johnson, D.N., 2002, In praise of the coarse fraction and bioturbation: Gravelly Mima mounds as two-layered biomantles: Geological Society of America Abstracts with Programs, v. 34, no. 6, p. 369.

Johnson, D.L., Domier, J.E.J., and Johnson, D.N., 2005, Animating the biodynamics of soil thickness using process vector analysis: A dynamic denudation approach to soil formation: Geomorphology, v. 67, no. 1–2, p. 23–46, doi:10.1016/j.geomorph.2004.08.014.

Johnson, D.L., Johnson, D.N., Horwath, J.L., Wang, H., Hackley, K.C., and Cahill, R.A., 2006, Mima mounds as upper soil biomantles: What happens when the dominant bioturbators leave and the invertebrates take over?: Poster, 18th World Congress of Soil Science, 9–15 July 2006, Philadelphia, Pennsylvania.

Johnson, D.L., Johnson, D.N., Benn, D.W., and Bettis, E.A., 2008a, Deciphering complex soil-site formation in sands: Geomorphology, v. 101, no. 3, p. 484–496, doi:10.1016/j.geomorph.2007.04.033.

Johnson, D.L., Johnson, D.N., Horwath, J.L., Wang, H., Hackley, K.C., and Cahill, R.A., 2008b, Predictive biodynamic principles resolve two long-standing topographic-landform-soil issues: Mima mounds and soil stone-layers: Geological Society of America Abstracts with Programs, v. 40, no. 6, p. 209.

Kienzle, J., 2007, Soil Survey of Wallowa County Area, Oregon: U.S. Department of Agriculture–National Resources Conservation Service, in Cooperation with U.S. Forest Service, U.S. Department of Interior-Bureau of Land Management, Oregon State University Agricultural Experiment Station, and Wallowa Soil and Water Conservation District: Washington, D.C., U.S. Government Printing Office.

Larrison, E.J., 1942, Pocket gophers and ecological succession in the Wenas region of Washington: The Murrelet, v. 23, no. 2, p. 34–41, doi:10.2307/3535581.

Lea, P.D., 1984, Pleistocene glaciation at the southern margin of the Puget lobe, western Washington [unpublished Master's thesis]: Seattle, University of Washington, 96 p.

Mapes, L.V., 2011, Moles: Nature's rototillers are mightier than a machine: Seattle Times, 7 April 2011, http://seattletimes.nwsource.com/html/localnews/2014158720_moles08m.html (accessed 14 October 2011).

McFaul, M., 1979, A geomorphic and pedological interpretation of the mima-mounded prairies, South Puget Lowland, Washington State [unpublished Master's thesis]: Laramie, University of Wyoming, 77 p.

Meysman, F.J.R., Middleburg, J.J., and Heip, C.H.R., 2006, Bioturbation: A fresh look at Darwin's last idea: Trends in Ecology & Evolution, v. 21, no. 12, p. 688–695, doi:10.1016/j.tree.2006.08.002.

Mielke, H.W., 1977, Mound building by pocket gophers (*Geomyidae*)—Their impact on soils and vegetation in North America: Journal of Biogeography, v. 4, p. 171–180, doi:10.2307/3038161.

Mundorff, M.J., Weigle, J.M., and Holmberg, G.D., 1955, Ground water in the Yelm area, Thurston and Pierce Counties, Washington: U.S. Geological Survey Circular 356, 58 p., 2 pl.

Nardi, J.B., 2003, The World beneath Our Feet: A Guide to Life in the Soil: New York, Oxford University Press, 223 p.

Nardi, J.B., 2007, Life in the Soil: A Guide for Naturalists and Gardeners: Chicago, University of Chicago Press, 293 p.

Nelson, R.E., 1997, Implications of subfossil Coleoptera for the evolution of the Mima mounds of southwestern Puget Lowland, Washington: Quaternary Research, v. 47, p. 356–358, doi:10.1006/qres.1997.1891.

Noble, J.B., and Wallace, E.F., 1966, Geology and ground water resources of Thurston County, Washington: Washington Division of Water Resources Water Supply Bulletin 10, v. 2, 141 p.

Page, W.D., Swan, F.H., III, Hanson, K.L., Muller, D., and Blum, R.L., 1977, Prairie mounds (mima mounds, hog-wallows) in the Central Valley, *in* Singer, M.J., ed., Soil Development, Geomorphology, and Cenozoic History of the Northeastern San Joaquin Valley and Adjacent Areas, California: A Guidebook for the Joint Field Session of the American Society of Agronomy, Soil Science Society of America, and the Geological Society of America (10–13 November 1977, Modesto, California, compiled by Huntington, G.L., Begg, E.L., Harden, J.W., and Marchand, D.E.): Davis, California, Department of Land, Air and Water Resources, University of California, p. 247–266.

Paton, T.R., Humphreys, G.S., and Mitchell, P.B., 1995, Soils: A New Global View: New Haven, Connecticut, Yale University Press, 213 p.

Porter, S.C., and Swanson, T.W., 1998, Radiocarbon age constraints on rates of advance and retreat of the Puget Lobe of the Cordilleran Ice Sheet during the last Glaciation: Quaternary Research, v. 50, no. 3, p. 205–213, doi:10.1006/qres.1998.2004.

Pringle, R.F., 1990, Soil Survey of Thurston County, Washington: U.S. Department of Agriculture, Soil Conservation Service, in cooperation with Washington State Department of Natural Resources, and Washington State University, Agriculture Research Center: Washington, D.C., U.S. Government Printing Office.

Reed, S., and Amundson, R., 2012, this volume, Using LIDAR to model Mima mound evolution and regional energy balances in the Great Central Valley, California, *in* Horwath Burnham, J.L., and Johnson, D.L., eds., Mima Mounds: The Case for Polygenesis and Bioturbation: Geological Society of America Special Paper 490, doi:10.1130/2012.2490(01).

Rice, T.D., and Watson, E.B., 1912, Soil survey of Titus County, Texas: U.S. Department of Agriculture, Bureau of Soils, Field Operations Bureau of Soils for 1909, 11th Report, p. 1005–1027.

Ritchie, A.M., 1953, The erosional origin of the Mima mounds of southwest Washington: The Journal of Geology, v. 61, no. 1, p. 41–50, doi:10.1086/626035.

Robbins, L.W., and Hadley, B., 1998, 1997 Small mammal survey of Diamond Grove Prairie conservation area in Newton County, Missouri: Unpublished report to the Missouri Department of Conservation, on file at Duane G. Meyer Library, Southwest Missouri State University, Springfield, Missouri, 9 p.

Ruhe, R.V., 1959, Stone lines in soils: Soil Science, v. 87, p. 223–231, doi:10.1097/00010694-195904000-00006.

Schaetzl, R.J., and Anderson, S., 2005, Soils—Genesis and Geomorphology: Cambridge, UK, Cambridge University Press, 832 p.

Schlichte, K., 2010, Climate changes and their effects on northwest forests: Northwest Woodlands, v. 26, no. 2, p. 22–23.

Schwartz, C.W., and Schwartz, E.R., 2001, The Wild Mammals of Missouri (second revised edition): Columbia, University of Missouri Press, 368 p.

Sharpe, C.F.S., 1938, Landslides and Related Phenomena: New York, Columbia University Press, 137 p.

Shlemon, R.J., Begg, E.L., and Huntington, G.L., 1973, Fracture traces: Pacific Discovery, v. 26, p. 31–32.

Storm, L., 2004, Prairie fires and earth mounds: The ethnoecology of Upper Chehalis prairies: Douglasia, v. 28, no. 3, p. 69.

TGDC (Thurston GeoData Center), 2011, http://www.geodata.org/website/cadastral/viewer.htm: Olympia, Washington, Thurston County Planning Department (accessed 14 October 2011).

Thorson, R.M., 1980, Ice-sheet glaciation of Puget Lowland, Washington, during the Vashon Stade (late Pleistocene): Quaternary Research, v. 13, p. 303–321, doi:10.1016/0033-5894(80)90059-9.

Ugolini, F.C., and Schlichte, A.K., 1973, The effect of Holocene environmental changes on selected western Washington soils: Soil Science, v. 116, no. 3, p. 218–227, doi:10.1097/00010694-197309000-00010.

Unklesbay, A.G., and Vineyard, J.D., 1992, Missouri Geology: Three Billion Years of Volcanoes, Seas, Sediments, and Erosion: Columbia, University of Missouri Press, 189 p.

USDA-NRCS (U.S. Department of Agriculture–Natural Resources Conservation Services), 2011, Official Series Descriptions, updated 29 September 2010, http://soils.usda.gov/technical/classification/osd/index.html (accessed 3 June 2011).

Wallace, E.F., and Molenaar, D., 1961, Geology and Ground-Water Resources of Thurston County, Washington: Washington Division of Water Resources Water Supply Bulletin 10, v. 1, 254 p., 2 pl.

Wallwork, J.A., 1976, The Distribution and Diversity of Soil Fauna: London, Academic Press, 355 p.

Washburn, A.L., 1988, Mima mounds: An Evaluation of Proposed Origins with Special Reference to the Puget Lowland: Olympia, Washington Division of Geology and Earth Resources, Report of Investigations, v. 29, 53 p.

Wilkes, C., 1845, Narrative of the United States Exploring Expedition During the Years 1838–42, v. 4: Philadelphia, Lea and Blanchard.

Wilkinson, M.T., Richards, J.P., and Humphreys, G.S., 2009, Breaking ground: Pedological, geological, and ecological implications of soil bioturbation: Earth-Science Reviews, v. 97, no. 1–4, p. 257–272, doi:10.1016/j.earscirev.2009.09.005.

Manuscript Accepted by the Society 5 March 2012

The Geological Society of America
Special Paper 490
2012

The forgotten natural prairie mounds of the Upper Midwest: Their abundance, distribution, origin, and archaeological implications

Fred A. Finney*

Upper Midwest Archaeology, P.O. Box 106, St. Joseph, Illinois 61873, USA

ABSTRACT

Mima mounds in North America are primarily known from the western states of Washington, Oregon, and California; the Rocky Mountains; the mid-lower Mississippi Basin; and Louisiana-Texas Gulf Coast. By contrast, their former extent and abundance across the Upper Midwest prairie belt has never been systematically established due to their destruction by agriculture and historic confusion as to whether they were natural or anthropic mounds. Recent maps showing their distribution identify only two small moundfields, one centered on Waubun Prairie in western Minnesota, the other on Kalsow Prairie in north-central Iowa. But in fact, natural mounds were once a common feature of many Upper Midwest prairies, having extended from Kansas, Missouri, and Illinois north into Wisconsin, Iowa, Minnesota, the Dakotas, and across the prairies and parklands of Canada. Several remnant tracts, intact and preserved, bear witness to their former much greater extent. This chapter documents the original distribution across the prairie belt, which has implications for their origin insofar as it falls more or less entirely within the range of the Geomyidae (pocket gopher) family of fossorial rodents. Natural prairie mounds in the Upper Midwest invariably are found where limitations to vertical burrowing occur, or did occur, which leaves lateral burrowing as the only option to these and other soil animals. Owing to extensive overlaps between natural mounds and morphologically similar prehistoric "Moundbuilder" mounds, the idea is advanced that prairie mounds were opportunistically used for prehistoric interments, and later as ideation templates for prehistoric burial, effigy, and other mounds and utilitarian structures.

INTRODUCTION

This chapter documents the former distribution of natural prairie mounds in the Upper Midwest of North America that were once commonly present. It thus corrects the impression that this broad prairie region has, and had, a dearth of Mima mounds. W. Armstrong Price (1949) produced a first approximation map of Mima mound distributions in the United States west of the Mississippi River. It included the mid-lower Mississippi Valley Basin and Louisiana-Texas Gulf Coast, the central Rocky Mountains, the broad Columbia Plateau basalt region, the Olympia, Washington, mounded prairies, and the Great Central Valley and coastal strip of California. George Cox (1984) subsequently produced a similar somewhat expanded and updated map that displayed additional sites in the United States, Canada, and Mexico. This map included two small areas in the Upper Midwest, one in Iowa (Kalsow Prairie), and one in Minnesota (Waubun Prairie). Beyond these two areas, this broad region is shown as lacking mound distributions. Cox's map, reproduced by Washburn (1988) in his Mima mound monograph, has since reappeared in various

*fafinney@aol.com

Finney, F.A., 2012, The forgotten natural prairie mounds of the Upper Midwest: Their abundance, distribution, origin, and archaeological implications, *in* Horwath Burnham, J.L., and Johnson, D.L., eds., Mima Mounds: The Case for Polygenesis and Bioturbation: Geological Society of America Special Paper 490, p. 85–133, doi:10.1130/2012.2490(05). For permission to copy, contact editing@geosociety.org. © 2012 The Geological Society of America. All rights reserved.

subsequent mound studies. The Price-Cox maps are a measure of how little is known—and how much is forgotten—about the once numerous and extensive Upper Midwest natural prairie mounds.

To correct the record, and fill in the blank, this chapter documents the former natural prairie mound distributions in the Upper Midwest. The record shows that natural prairie mounds were, before settlement and agriculture, far more numerous and greater in extent than is generally known. But it also shows that the region, at least north of the Wisconsin glacial boundary, lacked the high density and extensive mounded terrains and "pimpled plains" so representative of the lower unglaciated Mississippi Valley and Gulf Coast, and far western states. Mounds were present in smaller but still sizeable numbers in the Upper Midwest, and except for several notable tracts, were generally less densely clustered. In addition, Mima mounds in the prairie heartland were often confused with, and in many cases mistaken for, the small morphologically similar American Indian mounds that once occurred by the scores across the prairie belt. Natural prairie mounds of the Upper Midwest slowly disappeared under the plow and European settlement, and while they were known to earlier nineteenth-century generations, they are now largely gone and forgotten, but preserved in long-ignored primary and secondary document sources. Consequently, no comprehensive accounting of the extent and distribution of former natural prairie mounds exists. In this chapter they are accounted for with reference to their attributes, distribution, and certain environmental factors that correlate with and largely explain their distribution. Information sources are primarily older accounts written by archaeologists, geologists, soil scientists, and other researchers. In addition, modern observations on those few scattered remnants of unplowed prairie tracts are also included. Collectively these sources converge and allow the construction of a reasonably accurate Mima mound distribution map for the Upper Midwest that provides an updated perspective (Fig. 1).

BACKGROUND

As indicated, in the Upper Midwest the majority of natural prairie mounds vanished during the natural to cultural transformation of the landscape, and were thus forgotten by recent generations. The principal sources for their former distribution are observations made by early geologists and archaeologists. Sources for those few extant prairie remnants with mounds are by contemporary archaeologists, plus fieldwork by colleagues and me (Figs. 2 and 3). Aerial imagery was not used in discovering Mima mounds for this chapter, although this technique can be useful for extensive unplowed tracts elsewhere (Bragg and Weih, 2007; Vogel, 2005). The Upper Midwest exhibits a near ubiquitous landscape modification of the original prairie setting caused by repeated plowing and cultivation.

A subtle but important regional distinction exists in the early prairie mound literature that should be noted. In the far west and lower Mississippi Basin–Gulf Coast regions, most mound references reside in the earth science, soil, and ecological literature.

But, in the Upper Midwest early mound sources reside mainly within the domain of archaeology—which in the nineteenth and early twentieth centuries was unequally influenced by "Moundbuilder" concerns. Every mound encountered in the Upper Midwest during this early period was viewed as a potential burial mound that might contain artifacts, and thus a potential antiquities repository to be exploited. As a result, archaeologists discovered thousands of presumed Indian mounds in the Upper Midwest, many of which proved to be natural prairie mounds. They were conventionally referred to as "conical mounds," or just "conicals," even though most—though not all—were dome-shaped and often morphologically indistinguishable from natural prairie mounds. (The terms "dome-shaped mounds" and "conical-shaped mounds" are used interchangeably in this chapter.)

During this early emphasis on burial mound archaeology, field methods and quality of data recordation often varied greatly between investigators, which proved problematic. To further complicate matters, natural prairie mounds were often considered unimportant both by early geologists intent primarily on identifying glacial features, and by contemporary archaeologists seeking spectacular burial finds. Even the better-regarded field archaeologists typically wrote little about natural prairie mounds, for once they were shown to lack artifacts or burials they were dismissed as objects of interest. Negative evidence was not in vogue at that time, or at least there was no explicit recognition of its value to cumulative research results. A notable exception was in the early twentieth century when botanist Edward Schmidt investigated several thousand natural prairie mounds over a sustained 40-year effort that spanned three counties in southeastern Minnesota (see multiple Schmidt references at the end of this chapter). This work, which resides in obscure archaeological and historical publications, never crossed over into the geological-soil and natural prairie mound literature because Schmidt and his contemporaries were convinced they were human-made, a conclusion not substantiated by modern research (Anfinson, 1984, 1999, 2007; Arzigian and Stevenson, 2003; Finney, 2006).

CAVEATS

A few explanations concerning terms are necessary at this point. First, Mima mounds, prairie mounds, natural prairie mounds, natural mounds, pimple mounds, hog-wallows, biscuit mounds, prairie blisters, and freckled land are but a few of the many names applied to soil mounds. Certain regional preferences exist, with Mima mounds, hog-wallows, and biscuit mounds used in the far west and mountains; pimple mounds in the south-central Gulf Coast and Saskatchewan; and natural prairie mounds, freckled land, and prairie blisters in the Upper Midwest (Bluemle, 1983). In his extensive field notes and letters, archaeologist T.H. Lewis, who made numerous observations of natural prairie mounds between 1882 and 1895 across the midcontinental region, referred to them variously as marsh mounds, meadow mounds, prairie mounds, swamp mounds, tumuli,

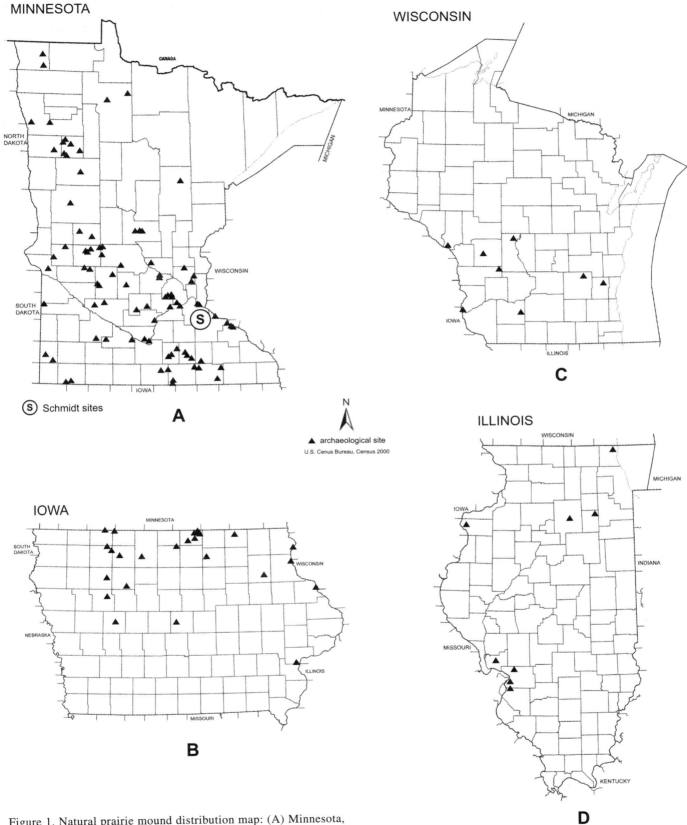

Figure 1. Natural prairie mound distribution map: (A) Minnesota, (B) Iowa, (C) Wisconsin, and (D) Illinois.

Figure 2. Photograph of natural prairie mounds in Minnesota. This view depicts a Mahnomen Mound at Waubun Prairie (courtesy Donald L. Johnson).

Figure 3. Photograph showing natural prairie mounds and vegetation differences near Booneville, Logan County, Arkansas (courtesy Donald L. Johnson).

and natural mounds. Other writers not uncommonly misidentified them as garden mounds and house mounds. E.W. Schmidt called them lowland mounds, but erroneously assumed they were anthropic mounds. Several researchers even confused natural prairie mounds with earth lodges made by the Great Plains Indians (explained below). This chapter uses the terms "Mima mounds" and "natural prairie mounds" as synonyms. (Note: Nothing in this chapter should be used to minimize protection of any potential human burial site, for the professional archaeological practice and site preservation ethic in the Upper Midwest is to treat all known or possible mounds as mortuary features until proven otherwise.)

MOUNDS OF CONFUSION

Some considerations are necessary for untangling "mounds of confusion"—separating those that are natural (i.e., nonhuman origin) from anthropic mounds, considerations that can vary from one author to another. Many kinds of natural mounds are recognized in the literature, with Mima-pimple-natural prairie mounds being the more frequent type west of the Mississippi River. East of the Mississippi, however, with the exception of parts of Wisconsin and Illinois where examples are directly coincident with pocket gopher distributions and wettish landscape settings, the likelihood of finding natural prairie mounds decreases significantly. In preparing this chapter, the author reviewed hundreds of older mound accounts found in a broadly eclectic and scattered literature in an attempt to discern known or probable natural prairie mounds in the Upper Midwest. Many historically notable late-nineteenth- and early-twentieth-century archaeologists and geologists recorded their findings and opinions on natural prairie mounds in this region. Among others, the list includes such prominent names—distinguished and respected individuals in their time—as Clement Webster, William McGee, Theodore Lewis, Stephen Peet, William McAdams, David Bushnell,

Edward Schmidt, Jacob Brower, Newton Winchell, Warren Upham, Gerard Fowke, and Louis Houck (Appendices 1 and 2). These individuals worked as archaeologists, geologists, or historians, or in a few cases a mix of each. In Appendices 1 and 2 many of their site descriptions are quoted verbatim and reproduced from the original documents, descriptions which clarify the site settings of known and probable natural prairie mounds.

The new distribution map (Fig. 1) is based on this literature review followed by individual site assessments. By careful reading, it became obvious that some early observers could clearly distinguish natural prairie mounds from anthropic ones. This ability was characteristic of several prominent researchers with considerable field experience and a discerning eye. An examination of their body of professional work was necessary to ascertain individual bases for distinguishing natural from cultural mounds. For other less experienced and less perspicacious researchers, identifying instances of known or probable Mima mounds versus cultural burial mounds involved some degree of uncertainty (Appendix 1).

The possible natural prairie mounds initially identified were then subjected to individual critical reviews. For this site assessment it bears repeating that a major objective of this chapter was to identify and assess small, round, or possibly oval, dome- and conical-shaped mounds that lacked burial, artifact, or other anthropogenic signatures (Table 1). As noted throughout this chapter (and in appendices), distinguishing natural prairie mounds from those culturally produced is often difficult. It requires more than just noting the absence of burials and artifacts. An extensive site list—much longer than Table 2—can be constructed for the mounds that only possess traces of cultural materials and those apparently void of artifacts and human remains. Notably, several researchers regularly recognized the existence of a class of sterile, artifact-free mounds but steadfastly believed that they and all other tumuli must be of American Indian origin. Some researchers did not speculate on the function

TABLE 1. NATURAL PRAIRIE VERSUS ANTHROPIC MOUNDS

Natural prairie mounds	Anthropic mounds
1. Height is 2 m or less	1. Height often greater than 2 m
2. Shape is domed, conical, or elongated if on slopes	2. Shape is conical, linear, effigy, platform
3. Contents—none or rare; can have surface materials	3. Contents—human burials and primary and secondary artifacts
4. Not used for buildings	4. Can be used as platform for buildings
5. Fill—massive, homogeneous biomantle (one or two layered); develops soil horizons when abandoned	5. Fill—variable, can exhibit either massive or loaded appearance in profile; stratiform (i.e., anthropogenic stratified); prepared surfaces may be present; can exhibit special or ritual sediments (i.e., different colors and textures) and earthen features; develops soil horizons when abandoned
6. Mounds often have vegetation differences with surrounding soil	6. Mounds could have vegetation differences with surrounding soil
7. Found in prairies but rarely in woods	7. Found in both forests and prairies
8. Typically occurs in wet settings or poorly drained areas	8. Typically found in well-drained areas and rarely in wet settings; may be in riverine settings or along bluff lines
9. Sites often found in thin soils over 'hardpan,' gravel layer, near surface bedrock, seasonally wet, or permanent wet conditions	9. Sites are located in elevated areas with scenic viewshed of the surrounding countryside
10. Rarely on upland bluff lines	10. Locations include upland bluff lines
11. Rarely found in well-drained areas	11. May contain post-construction rodent bioturbation or krotovinas
12. Surface artifacts are very rare in Upper Midwest	12. Surface artifacts may be present

of small, sterile mounds, while for others the possible types of small mound-like human constructions for habitation or subsistence functions included house mounds, earthlodges, hut rings, garden mounds, and corn hills. The sometimes bizarre interpretations advanced for some Mima mounds in other regions (see Reed and Amundson, this volume, Chapter 1) are absent from the Upper Midwest literature. For this region, mounds remain a question of being either anthropic or natural. The possible functions listed in the literature include examples that could be evaluated by the artifact content (house mounds, earthlodges, hut rings, and burial mounds) and those that can be assessed by their size (garden mounds and corn hills) (Fowke, 1922, 1928; Lapham, 1855; Melton, 1929; Sasso, 2003; Thoburn, 1937; Wilson, 1917). In addition, any human construction should possess a distinct anthropic sedimentological signature (Appendix 2).

NATURAL PRAIRIE MOUND DISTRIBUTION

The literature review initially divided mounds into cultural and noncultural sites depending upon the presence (cultural) or absence (noncultural) of human burials, artifacts, and anthropic sediments (Table 1). A further consideration of the noncultural mounds allowed separation by their attributes into natural prairie mounds and other kinds of noncultural mounds (see previous section). The natural prairie mound category includes the sites possessing Mima or Mima-like attributes (Table 2; Appendix 2). Since this chapter documents a substantial reappraisal of the natural prairie mound distribution in the Upper Midwest (Fig. 1), the site-specific data necessary for reaching this conclusion are presented in Appendix 2. Some of the possible Mima mounds remain poorly documented. This category includes instances where a knowledgeable researcher described a mound or mounds with apparent noncultural attributes and pointedly does not state any disqualifying attributes. For example, a rock mound would immediately be disqualified as it is not composed of unconsolidated sediments. Nearly all possible Mima mound sites found

in the literature were rejected since their status was considered inconclusive and thus could not sufficiently support the chapter's objective. For the purposes of this chapter, the site listing includes those sites considered definite, probable, and possible natural prairie mounds (Table 2; Appendix 2). The possible sites actually used in this chapter typically have little documentation but exhibited other Mima mound attributes such as a relatively low landscape position. Overall these sites comprise a sufficient body of reliable evidence to demonstrate that a wider distribution existed in the Upper Midwest than that previously published in North American continental perspective (Fig. 1).

The largest portion of the natural prairie mound database derives from Minnesota. A total of 144 known, probable, or possible sites are scattered across the state (Table 2; Appendix 2). This distribution covers 42 counties, principally in southern and western Minnesota. Many counties have two or three Mima mound examples. The largest concentration is not Waubun Prairie (*n* = ~200 [Ross et al., 1968]), but rather at the Schmidt Moundfield located south of the Twin Cities in southern Dakota; northern, central, and eastern Rice; and western Goodhue counties where Schmidt (1937) recorded more than 2700 "lowland mounds" herein defined as the Schmidt Moundfield (Fig. 4; Appendix 2). The total extent falls within an area ~25 miles in diameter with Northfield near the center. These mounds were concentrated along the valleys of the Cannon River between Cannon Falls and Bridgewater, Chub Creek, and Prairie Creek. Outlier concentrations occurred north of this corridor with a few more mounds to the south. The salient characteristics of Schmidt's lowland mounds are their site settings at low positions in the landscape, their typical location near water, their near complete absence of artifacts, their lack of human burials, and few were found in forests. In fact he commented on "the presence of much underground water" (Schmidt, 1937, p. 14) and that "All of the mounds are in the neighborhood of water" (Schmidt, 1937, p. 15) in reference to the mounds. These low settings are dominated by broad river valleys with lakes, marshes, meadows, and glacial outwash plains from the Des Moines Lobe.

TABLE 2. NONCULTURAL MOUND SITES

Site (or location)	County	State	Mounds (n)	Status Noncultural (?)	Comments
21AK28	Aitkin	MN	59	Noncultural (?)	Either noncultural Mima, ant hills, or historic Indian corn hills
Cedar Creek Ecosystem Science Reserve	Anoka	MN	n/a	Mima	Huntly and Inouye, 1988
Mounds near Centerville Lake	Anoka	MN	9	Probable	Upham, 1888a
21AN159	Anoka	MN	1	Probable	One mound in a wetland setting
Tamarac National Wildlife Refuge	Becker	MN	n/a	Possible	Low setting
Narrows at Upper and Lower Red Lake	Beltrami	MN	n/a	Probable	T.H. Lewis reported mounds in low setting
Waskish vicinity	Beltrami	MN	1	Possible	L.A. Wilford reported sand mound
Rothi WPA	Big Stone	MN	n/a	Mima	Reported by USFWS personnel
Springfield vicinity	Brown	MN	n/a	Mima	T.H. Lewis
New Ulm vicinity	Brown	MN	1	Mima	T.H. Lewis
21DKab	Dakota	MN	2	Mima	Schmidt, 1928
21DKac	Dakota	MN	2	Mima	Schmidt, 1928
21DKad	Dakota	MN	6	Mima	Schmidt, 1928
21DKae	Dakota	MN	12	Mima	Schmidt, 1928
21DKaf	Dakota	MN	6	Mima	Schmidt, 1928
21DK8	Dakota	MN	118	Mima	Schmidt, 1928; no evidence of burials or artifacts
Empire Township locality	Dakota	MN	36	Mima	Schmidt, 1928
Lakeville Township locality	Dakota	MN	9	Mima	Schmidt, 1928
Lakeville Township - Sec. 11	Dakota	MN	3	Mima	Schmidt, 1928
Castle Rock Township - Secs. 11–12	Dakota	MN	20	Mima	Schmidt, 1928
Castle Rock Township - NW	Dakota	MN	6	Mima	Schmidt, 1928
Greenvale and Waterford Townships	Dakota	MN	330	Mima	Schmidt, 1928
Greenvale Township - Sec. 11	Dakota	MN	12	Mima	Schmidt, 1928; Locality 1
Greenvale Township - Sec. 10	Dakota	MN	14	Mima	Schmidt, 1928; Locality 2; Thomas Moore farm
Greenvale Township - Sec. 11	Dakota	MN	14	Mima	Schmidt, 1928
Eureka Township - Secs. 2–33	Dakota	MN	74	Mima	Mounds follow old slough from Section 2 to Section 33
Eureka Township - Sec. 7	Dakota	MN	7	Mima	Schmidt, 1928
Eureka Township - Sec. 16	Dakota	MN	12	Mima	Schmidt, 1928
Eureka Township – Sec. 34	Dakota	MN	26	Mima	Schmidt, 1928; Locality 3
Eureka Sec. 33 to Greenvale Sec. 4	Dakota	MN	20	Mima	Schmidt, 1928
Greenvale Township - Sec. 4	Dakota	MN	11	Mima	Schmidt, 1928
Greenvale Township - Sec. 18	Dakota	MN	18	Mima	Schmidt, 1928
Waterford Township - Sec. 4	Dakota	MN	18	Mima	Schmidt, 1928; Locality 4; possibly a continuation of 21DK23
Waterford Township - Sec. 12	Dakota	MN	8	Mima	Schmidt, 1928
Waterford Township - Sec. 21	Dakota	MN	13	Mima	Schmidt, 1928
Waterford Township - Sec. 18	Dakota	MN	8	Mima	Schmidt, 1928
Waterford Township - Sec. 7	Dakota	MN	6	Mima	Schmidt, 1928
Waterford Township - Sec. 6	Dakota	MN	6	Mima	Schmidt, 1928
Waterford Township - Sec. 8	Dakota	MN	n/a	Mima	Schmidt, 1928; between railroad and wagon road
Greenvale Township - Sec. 3	Dakota	MN	7	Mima	Schmidt, 1928; Excavation found a horse skeleton; possibly a continuation of 21DK23
21DKj	Dakota	MN	3	Mima	Schmidt, 1928; Locality 5
21DKi	Dakota	MN	8	Mima	Schmidt, 1928; Locality 6; Markley, 1948
Greenvale Township - Sec. 9	Dakota	MN	1	Mima	Schmidt, 1928; Markley, 1948
Sciota Township - Secs. 18–19	Dakota	MN	2	Mima	Schmidt, 1928; Martin Elstad farm
21DK21	Dakota	MN	2	Mima	Schmidt, 1928, 1937; Simpson pasture
21DK22	Dakota	MN	n/a	Mima	Schmidt, 1928; Boudreau Mounds or Chub Creek Creek Mounds I
21DK23	Dakota	MN	40	Mima	Schmidt, 1928; Castle Rock Slough Mounds; east side of former lake; John Street Mounds; ~40 mounds

(Continued)

TABLE 2. NONCULTURAL MOUND SITES (Continued)

Site (or location)	County	State	Mounds (n)	Status	Comments
21DKd	Dakota	MN	72	Mima	Schmidt, 1928; south of Castle Rock station on north shore of former lake
21DKa	Dakota	MN	n/a	Mima	Schmidt, 1928; Brotzler Mounds
Iron Horse Prairie	Dodge	MN	9	Mima	D.L. Johnson
Evansville locality	Douglas	MN	n/a	Probable	Either Mima or glacial hummocky topography
Lake Geneva locality	Douglas	MN	n/a	Mima	Schmidt, 1928
Between Freeborn and Wells	Fairbault	MN	n/a	Mima	Upham, 1884a; mounds extend for ~3 miles between Freeborn and Wells
South of Wells	Fairbault	MN	n/a	Mima	Upham, 1884a
Mansfield Township - Sec. 13	Freeborn	MN	3	Mima	T.H. Lewis
Mansfield Township - Sec. 34	Freeborn	MN	3	Mima	T.H. Lewis
South of Blooming Prairie	Freeborn	MN	5	Mima	Schmidt, 1928
Warsaw Township - Sec. 8	Goodhue	MN	10	Mima	Schmidt, 1928
Kenyon Township - Sec. 4	Goodhue	MN	10	Mima	Schmidt, 1928
Florence Township	Goodhue	MN	n/a	Probable	Winchell, 1888b
Prairie Island	Goodhue	MN	n/a	Possible	Small mounds with no burials or artifacts as well as anthropic mounds
21HE27	Hennepin	MN	50	Mima	Winchell, 1911
21HE47	Hennepin	MN	27	Mima	Marsh mounds; anthropic mounds present at large site totaling 98 mounds
21HE3	Hennepin	MN	2	Mima	Anthropic mounds also present at large site
21HE30	Hennepin	MN	1	Mima	Anthropic mounds also present at large site
21HE19	Hennepin	MN	20	Probable	Either noncultural mounds or soil covered historic features
21KH8	Kandiyohi	MN	1	Mima	Anthropic mounds also present at large site totaling 39 mounds
SW of Pennock	Kandiyohi	MN	n/a	Mima	Marsh setting
21KH9	Kandiyohi	MN	n/a	Mima	L.A. Wilford
W.C. Dayton Conservation and Wildlife Area	Kittson	MN	n/a	Mima	Wet prairie setting
21KT2	Kittson	MN	1	Mima	Anthropic mounds are also present
Schaefer Prairie	McLeod	MN	n/a	Mima	Listed in site guide
Glencoe vicinity	McLeod	MN	2	Probable	L.A. Wilford
Waubun Prairie Research Area	Mahnomen	MN	200	Mima	Ross et al., 1968
STH 113 and US 59 in Waubun	Mahnomen	MN	1	Mima	Mahnomen mound
S. of County Highway 1	Mahnomen	MN	1	Mima	Mahnomen mound
W. of Bejou and Santee Prairie	Mahnomen	MN	1	Mima	Mahnomen mound
Santee Prairie	Mahnomen	MN	n/a	Mima	D.L. Johnson
21MH6	Mahnomen	MN	n/a	Probable	Mahnomen mound(s) may exist among the nine reported 21MH6 mounds
Litchfield Township	Meeker	MN	n/a	Possible	
21MO2, 21MO3, and 21MOx	Morrison	MN	55	Probable	Winchell, 1911
21MOad	Morrison	MN	n/a	Noncultural (?)	Noncultural mounds caused by tree falls
Between Grand Meadow and Leroy	Mower	MN	500	Mima	Winchell, 1911
Between Auston and Fairbault	Mower	MN	n/a	Mima	Winchell, 1911
Racine Prairie	Mower	MN	n/a	Mima	D.L. Johnson
21NL119	Nicollet	MN	3	Mima	Bruce Koenen
Nicollet vicinity	Nicollet	MN	15	Mima	Determined noncultural by OSA
21NO1	Nobles	MN	n/a	Mima	Upham, 1884b; T.H. Lewis
Little Rock Township	Nobles	MN	n/a	Probable	Upham, 1884b
Twin Valley vicinity	Norman	MN	n/a	Mima	D.L. Johnson; noncultural pseudo-mounds
Pease Prairie near Henning	Ottertail	MN	n/a	Possible	T.H. Lewis; natural hillocks
Otter Tail River vicinity	Ottertail	MN	n/a	Possible	T.H. Lewis; either Mima mounds, natural hillocks, or both
Pipestone vicinity	Pipestone	MN	n/a	Possible	Rock debris mounds near quarry and possible Mima mounds near marsh on creek

(Continued)

TABLE 2. NONCULTURAL MOUND SITES (*Continued*)

Site (or location)	County	State	Mounds (n)	Status	Comments
Edgerton vicinity	Pipestone	MN	n/a	Possible	L.A. Wilford; noncultural mounds
Malmberg Prairie	Polk	MN	n/a	Mima	D.L. Johnson
Pembina Trail Preserve	Polk	MN	n/a	Mima	D.L. Johnson
Ashley Creek Valley	Pope	MN	n/a	Mima	Winchell, 1885
North shore Lake Minnewaska	Pope	MN	n/a	Probable	Mounds in low land near lake shore (but not 21PO1 or 21PO4)
Uplands north of Lake Minnewaska	Pope	MN	n/a	Probable	Mounds in upland prairie (but not 21PO1 or 21PO4)
21POi	Pope	MN	25	Mima	L.A. Wilford
Lamberton vicinity	Redwood	MN	n/a	Mima	T.H. Lewis
Prairie west of Renville	Renville	MN	n/a	Mima	Reported by USFWS personnel
21RN13	Renville	MN	n/a	Mima	L.A. Wilford; marsh setting
Wheatland Township - Sec. 16	Rice	MN	6	Mima	Winchell, 1911
Webster Township - Sec. 17	Rice	MN	n/a	Mima	Winchell, 1911; 1/8 mile north of McFadden
Bridgewater Township - Secs. 12–14	Rice	MN	120	Mima	Schmidt, 1928; Locality 1
Bridgewater Township - Sec. 14	Rice	MN	13	Mima	Schmidt, 1928; Locality 2
Thielbar - Revier farm vicinity	Rice	MN	54	Mima	Schmidt, 1928; Locality 3; in 1916 Schmidt excavated 4 and tested 14 more mounds
Cannon City Township - Sec. 4	Rice	MN	10	Mima	Schmidt, 1928; Locality 4
Webster Township - Sec. 9	Rice	MN	6	Mima	Schmidt, 1928; Locality 5
Webster Township - Secs. 16–17	Rice	MN	12	Mima	Schmidt, 1928
Webster Township - Secs. 29 and 31	Rice	MN	6	Mima	Schmidt, 1928
Northfield Township et al.	Rice	MN	577	Mima	Schmidt, 1928; Locality 6; mounds along Prairie Creek
Wheeling Township - Sec. 14	Rice	MN	25	Mima	Schmidt, 1928; Locality 7
21RC12	Rice	MN	10	Mima	Schmidt, 1937
Carl Kester farm et al.	Rice	MN	45	Mima	Schmidt, 1937
21RC13	Rice	MN	8	Mima	Schmidt, 1937
21RC14	Rice	MN	59	Mima	Schmidt, 1937
Shakopee vicinity	Scott	MN	n/a	Mima	T.H. Lewis; mounds extend for at least a mile
Savage vicinity	Scott	MN	6	Mima	Determined noncultural by OSA
21SH12	Sherburne	MN	49	Mima	Upham, 1888e
Sherburne National Wildlife Refuge	Sherburne	MN	n/a	Mima	Blair et al., 2001
Henderson vicinity	Sibley	MN	n/a	Mima	T.H. Lewis
Ashley Creek Valley	Stearns	MN	n/a	Mima	Winchell, 1885
Raymond Township	Stearns	MN	n/a	Mima	Reported by USFWS personnel
Roscoe Prairie	Stearns	MN	n/a	Mima	D.L. Johnson
E. of Aurora	Steele	MN	n/a	Mima	Harrington, 1884; T.H. Lewis
S. of Aurora	Steele	MN	n/a	Mima	Harrington, 1884; T.H. Lewis
S. of Owatonna	Steele	MN	n/a	Mima	Schmidt, 1928
Verlyn Marth Memorial Prairie	Stevens	MN	n/a	Possible	D.L. Johnson; small mounds of uncertain origin
21SEm	Stevens	MN	11	Probable	Wilford
Svor WPA	Swift	MN	n/a	Mima	Reported by USFWS personnel
Welker WPA	Swift	MN	n/a	Mima	Reported by USFWS personnel
21WB6	Wabasha	MN	100	Probable	Site had one anthropic mound
S. of Miller Creek and 21WB35	Wabasha	MN	n/a	Mima	Marsh mounds (Winchell, 1911)
21WB35	Wabasha	MN	57	Probable	May have included anthropic mounds
Wilton Township - Sec. 10	Waseca	MN	3	Mima	Upham, 1884c
Wilton Township - Sec. 30	Waseca	MN	3	Mima	Upham, 1884c
Woodville Township - Sec. 3	Waseca	MN	n/a	Possible	21WE1 and 21WE2 have anthropic mounds in this area, and possibly Mima mounds
21WA1	Washington	MN	1	Possible	Anthropic mounds present at large site
21WA2	Washington	MN	1	Possible	Anthropic mounds present at large site

(Continued)

TABLE 2. NONCULTURAL MOUND SITES (Continued)

Site (or location)	County	State	Mounds (n)	Status	Comments
21WRh	Wright	MN	3	Mima	Upham, 1888f; in a low closed depression setting
21WR29	Wright	MN	1	Probable	Upham, 1888f; at least one anthropic mound existed among the 14 mounds
Mound Springs Prairie Scientific and Natural Area	Yellow Medicine	MN	n/a	Possible	Low wetland has domed features
13AM79	Allamakee	IA	800	Mima	Finney and Johnson, 2010; T.H. Lewis
13AM260	Allamakee	IA	2	Mima	Fokken, 1980
East half of county	Buena Vista	IA	n/a	Possible	MacBride, 1902
Wolters Prairie Preserve	Butler	IA	n/a	Possible	Low setting
Leeper Prairie Preserve	Butler	IA	n/a	Possible	Low setting
Carrollton vicinity	Carroll	IA	n/a	Probable	White, 1870; large closed depression
Peat swamp north of Clear Lake	Cerro Gordo	IA	n/a	Probable	White, 1870
Hoffman Prairie	Cerro Gordo	IA	n/a	Possible	Low setting
Dewey's Pasture	Clay	IA	n/a	Possible	Low setting
Kirchner Prairie Wildlife Management Area	Clay	IA	n/a	Possible	Low setting
13CT18	Clayton	IA	1	Possible	Whittaker and Storey, 2008; 18 possible noncultural mounds in the 128 mounds
Near Bakers Point on Spirit Lake	Dickinson	IA	n/a	Probable	Upham, 1884d
13DB363	Dubuque	IA	70	Probable	Woodman, 1873
Anderson Prairie State Preserve	Emmet	IA	n/a	Mima	Klass et al., 2000; Wolfe-Bellin and Moloney, 2001
Smithfield Township	Fayette	IA	n/a	Mima	Savage, 1905
S. of Turkey River	Fayette	IA	1	Mima	Natural prairie mound illustrated by McGee (1891)
Webster farm NE of Rockford	Floyd	IA	10	Mima	Natural prairie mounds illustrated by Webster (n.d.)
13HK10	Hancock	IA	16	Mima	Benn, 1976; A.C. Johnson, 1974–1979; Mima mounds extend for ~7 miles west of 13HK10
Hayden Prairie State Preserve	Howard	IA	n/a	Mima	Benn, 1976; D.L. Johnson
Stinson Prairie	Kossuth	IA	n/a	Possible	
Muscatine Island	Muscatine	IA	n/a	Possible	Various (see Appendix 2)
Central and western parts of county	Palo Alto	IA	n/a	Possible	MacBride, 1905
Kalsow Prairie State Preserve	Pocahontas	IA	128	Mima	Brotherson, 1982
Sac City	Sac	IA	3	Probable	White, 1870
Ames vicinity	Story	IA	n/a	Possible	Noncultural mounds or glacial hillocks
Banks of Shell Rock at Northwood	Worth	IA	n/a	Mima	Schmidt, 1928
N. of Kensett	Worth	IA	n/a	Mima	Schmidt, 1928
NE of Hanlontown	Worth	IA	n/a	Mima	Schmidt, 1928
13WT4	Worth	IA	1	Mima	Benn and Petersen, 1976
13WT6	Worth	IA	1	Mima	Benn and Petersen, 1976
13WT8	Worth	IA	2	Mima	Benn and Petersen, 1976
Prairie du Chien vicinity	Crawford	WI	n/a	Probable	Various (see Appendix 2)
Horicon vicinity	Dodge	WI	n/a	Possible	Lapham, 1855
Shea Prairie	Iowa	WI	n/a	Possible	
Cranberry Creek vicinity	Juneau	WI	n/a	Possible	
47MO7	Monroe	WI	19	Probable	Mier et al., 1996
Trempealeau vicinity	Trempealeau	WI	n/a	Possible	Gale, 1867
Millard Prairie	Vernon	WI	13	Possible	Cole and Flint, 1913
West Bend vicinity	Washington	WI	6	Possible	
Goose Lake Prairie State Natural Area	Grundy	IL	n/a	Mima	D.L. Johnson
Otter Creek in Jerseyville vicinity	Jersey	IL	n/a	Probable	McAdams, 1919
Upper Des Plaines River	Lake	IL	n/a	Mima	T.H. Lewis
Ottawa vicinity	La Salle	IL	n/a	Possible	Kett, 1877
Northern American Bottom	Madison	IL	n/a	Probable	McAdams, 1881
11MS30	Madison	IL	2	Possible	Porter, 1974
11S706	St. Clair	IL	n/a	Possible	Brackenridge, 1814
New Boston vicinity	Mercer	IL	n/a	Possible	McWhorter, 1875
Grand total (minimum estimate for documented Mima mounds)			4229		

Abbreviations: IA—Iowa; IL—Illinois; MN—Minnesota; WI—Wisconsin; USFWS—U.S. Fish and Wildlife Service.

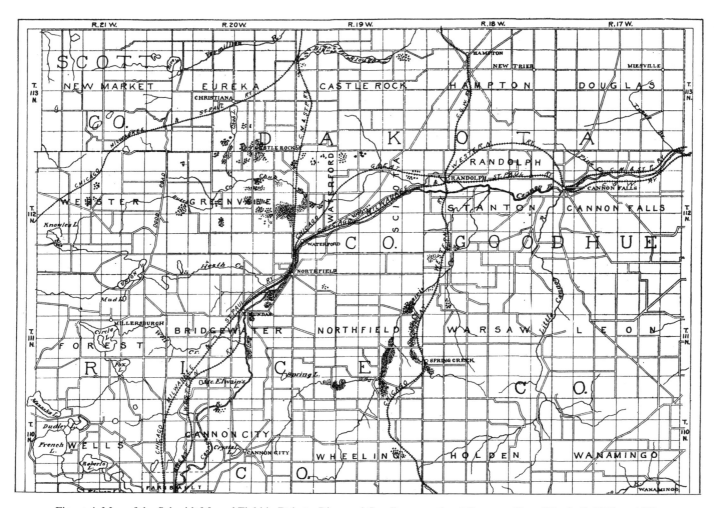

Figure 4. Map of the Schmidt Mound Field in Dakota, Rice, and Goodhue counties, Minnesota (from Winchell, 1911, p. 129).

Since they were not reported for the uplands Schmidt believed that his lowland mounds comprised a specialized class of mounds. In fact in an early article Schmidt (1910) referred to anthropic mounds as the "highland mounds." Winchell contributed a description of another large moundfield in southern Minnesota. In the eastern half of Mower County he found a natural prairie mound group with more than 500 mounds that extended more than three miles across a wet prairie.

In Iowa a total of 31 possible Mima mound sites were noted on the Des Moines Lobe and the Iowan Erosion Surface (Table 2). This quantity is remarkable considering the near total Euro-American destruction of the original Iowa prairie environment (Rosburg, 2001; Smith, 1998; White and Glenn-Lewin, 1984). The best known sites are from 1976 investigations in Hancock and Worth counties (Benn, 1976; Benn and Petersen, 1976) and Kalsow Prairie with 128 Mima mounds (Brotherson, 1982). The largest natural prairie mound sites in Iowa were in Hancock and Allamakee counties.

East of the Mississippi River in Wisconsin ($n = 8$) and Illinois ($n = 8$) possible Mima mound sites are rare. The best known

sites are the West Prairie Mound Group in Wisconsin and the Goose Lake Prairie in Illinois. In addition, natural prairie mounds are believed to have existed in a series of localities along the Mississippi River where special environmental conditions, i.e., near surface wetness, existed in the past (Fig. 1). These sites began southeast of the Twin Cities and extended downstream at Lake City, Minnesota; Trempealeau vicinity, Wisconsin; Harpers Ferry, Iowa; Prairie du Chien vicinity, Wisconsin; Dubuque, Iowa; Muscatine vicinity, Iowa; and New Boston vicinity, Illinois (Table 2).

Although outside the Upper Midwest, a brief comment is necessary concerning the Mima mound distribution in the Eastern Plains Border and Ozark Regions of Missouri. In this state, a mound survey made a century ago illustrates the effects of modern changes to the landscape. Historian Louis Houck hired two informants to canvass the state to list its archaeological sites including Indian mounds. Houck's (1908) industrious informants were Lewis M. Bean and D.L. Hoffman, who spent two years traversing the state gathering data and logged a total of over 28,000 Indian mounds in Missouri. Archaeologists know that this incredible total is not possible (Lynott, 1982; O'Brien, 1996). In retrospect it is easy to discern that Bean and Hoffman had included all mounds, both Indian and natural, in their observations. The Ozark Plateau and the southeast Missouri lowlands are known to possess numerous Mima mounds (Berry et al., 1940; Blankinship, 1889; Bushnell, 1904, 1905; Hoy, 1865; Lewis, 1881–1895a, 1881–1895b; Lynott, 1982; O'Brien and Wood, 1998; O'Brien et al., 1989; Price, 1977; Price et al., 1975; Snyder, 1890). Two observations can be made from the Missouri data: (1) throughout the Midwest there is tremendous potential for confusing cultural and noncultural mounds, and (2) for the majority of all mounds their aboveground superstructure is no longer visible because of damage inflicted by the implements of modern agriculture. This situation is not unique to natural prairie mounds or Missouri. An estimated 80% of the anthropic mounds in Iowa and Wisconsin had been plowed away by 1979 (Petersen, 1984). These numbers underscore the remarkable presence of any natural prairie mounds observed on the modern landscape, which includes only miniscule quantities of the original or unplowed prairie. Such small prairie remnants can be found scattered across Minnesota, Iowa, Missouri, and a few locations in Illinois and Wisconsin.

DISCUSSION

Natural prairie mounds are a long ignored feature of the Upper Midwest. During the Moundbuilder emphasis of the late nineteenth century, the trend was to not write about negative finds or empty mounds. Since the emphasis was on positive finds—spectacular finds if possible (McKusick, 1991; Meltzer, 1985, 1998; Silverberg, 1968; Smith, 1985; Squier, 1850; Squier and Davis, 1848)—in several cases negative results were mentioned only because it was an unexpected consequence of the fieldwork. However, since the beginning of the nineteenth century there have been various reports of culturally sterile mounds devoid of human remains or artifacts. Because of similar sizes between anthropic and natural prairie mounds the problem of distinguishing them continued into the twentieth century in the Upper Midwest. The potential clues provided by the environmental setting were often ignored; and Schmidt's work was the leading example. In general Mima mounds tend to form in periodically wet settings, either low terraces on floodplains, outwash plains, or wettish tracts in the uplands. By contrast one finds anthropic mounds on the higher floodplains, bluff edges, and dry uplands with a viewshed.

Various theories, most of them advanced by persons who are but slightly, if at all, familiar with the country, have been propounded to account for mounds of this character. Their vast number has led some writers to believe that they cannot be artificial but must be due to natural phenomena.... (Fowke, 1922, p. 163)

In archaeology most problems are resolved, or at least better questions can be asked, once the investigators have excavated the site, mound, or feature in question. This was not the case with natural prairie mounds, where certain investigators insisted on interpreting all small mounds as human-made in origin despite repeated evidence to the contrary. These mounds were initially assumed to be Indian burial mounds. Surprisingly, the excavations that produced no burials and no or very few artifacts did not alter this opinion. Instead the negative excavation results only meant that a cultural interpretation changed from a mortuary function to the domestic sphere. This rigid instance for anthropic origin derives from the persuasive influence of the theoretical anthropologist Lewis H. Morgan and the existence of house mounds elsewhere in the New World, as well as an oral tradition recorded in Minnesota that discussed house mounds used by earlier inhabitants. Yet again, interpretation ignored the fact than an absence of artifacts cannot support any habitation function. Of the five researchers, Winchell, Upham, Bushnell, Brower, and Schmidt, who strongly held this idea, none was a trained archaeologist. Only one of these individuals had excavated earthlodges on the Great Plains similar to the study of site formation processes, taphonomic assessment, and ethnoarchaeological research conducted by Lewis (1883b). Immediately before his death Brower (1904) had excavated earthlodges in North Dakota. Almost certainly his purpose was to gather comparative data. The Missouri River earthlodges were sometimes called "Mandan house mounds." The oral traditions of the Mandan state that they originally came from Minnesota and had migrated to the Missouri River in North Dakota. Based on this tradition Winchell (1911) believed that earthlodges should be found in Minnesota. While excavations never supported this hypothesis, such results were ignored. These two Mandan and Dakota traditions provide examples of simplistic attempts to match folklore, i.e., oral stories such as traditions, myths, and legends, to the prehistoric archaeological record. Such a blind adherence to all types of folklore as a data source was rejected as being unreliable, often conflicting, and nonscientific during the late-nineteenth-century professionalization of archaeology (Gazin-Schwartz and Holtorf, 1999; Hall, 2008; Mason, 2006; Nabokov, 2002; Trigger, 2006). A modern summary of the house mound hypothesis is provided by Anfinson (1984):

As to mounds representing the remains of collapsed earth lodges, Winchell (1911:380–398) speculated at length on this topic with respect to the flat-topped mounds in Minnesota. Subsequent excavations of some of these mounds indicated that they were indeed burial mounds and not houses. (Anfinson, 1984, p. 4)

The natural prairie mound distribution data indicate that any given mound group can be either anthropic, natural, or a mix of both in origin. The human-made mounds retain evidence of loading (and sometimes massive fill), burials, and various kinds of artifacts. Such mounds often possess attributes of soil engineering, as Woodland mounds have sod blocks, whereas Mississippian mounds use clay basal platforms, clay edge buttresses, buckshot fill, and clay caps. Cultural interpretations for color symbolism exist for the Woodland period sod blocks, and the Mississippian practice of alternating light and dark fill units (Fortier and Finney, 2007; Pauketat, 1993, 2004, 2009; Perttula, 1996; Sherwood and

Kidder, 2011; Van Nest, 2006; Van Nest et al., 2001). Natural prairie mounds invariably exhibit a relatively homogeneous biomantle without burials or artifacts—unless an artifact (manuport) was dropped on a natural mound, or a natural mound was opportunistically used for burial, which is known to have occurred (Smith, 1910; Aten and Bollich, 1981; Galm and Keene, 2006; Greene, 1975).

The biomantle concept provides a useful approach to explain the origin of natural prairie mounds in the Upper Midwest (Johnson, 1990, 2002; Johnson et al., 2002, 2003, 2005, 2006). Point-centered, locally thickened biomantles invariably occur under certain environmental parameters (see below). Application to Mima mounds is the outcome of a taphonomic assessment outlining the most likely or parsimonious natural origin (Cox, 1984; Cox and Scheffer, 1991; D.L. Johnson et al., 1999, 2005, 2008; Martin, 1999; Nelson, 1997; Rick et al., 2006). This approach seems to explain the data better than other hypotheses, and thus may be considered a reasonable resolution to the problem (Fogelin, 2007; Hodder, 1999; Kelley and Hanen, 1988). While the biomantle concept represents a significant formation process, polygenesis and hybrid Mima mounds have been recognized in different regions (Collins, 1975; Davids, 1968; Horwath Burnham et al., this volume, Chapter 4; Johnson and Burnham, this volume, Introduction; D.L. Johnson et al., 2008). There are, however, potential overlaps in the appearance of natural mounds and those culturally produced with massive fill. For mound groups of mixed origin (i.e., natural mounds, versus those impacted by humans, versus mounds made entirely by humans), interpretations should be based on excavation data and comparisons of mound fills and profiles to expectations of cultural versus natural prairie mounds. In this context it should be noted that natural prairie mounds are hypothesized to have been prehuman templates for anthropic mounds for the midcontinental region (Finney and Johnson, 2010).

An overview of the Upper Midwest natural prairie mounds documented in this chapter would focus on their position on the landscape, often where anthropic mounds do not occur. Mima mounds represent soil mounds produced from unconsolidated deposits, consisting of a biomantle that exhibits, usually, relatively homogeneous soil. In the Upper Midwest an expectation for their occurrence is a local environment with a shallow soil that lies above some barrier to vertical burrowing for soil animals. The barrier may be caused by either wet conditions, such as a high water table, or dense subsoil (Bt horizon) or dense stratum (coarse gravels, a pan, bedrock) beneath the topsoil (Ellery et al., 1998; Horwath and Johnson, 2006; Johnson and Johnson, 2008; Schaetzl and Anderson 2005). Upper Midwest natural prairie mound sizes generally range from 10 to 120 feet (3–36 m) in diameter and 1 to 6 feet (0.3–2 m) in height. Mima mound densities of more than 100 per acre (>50/ha) have been recorded for the far west and lower Mississippi Valley regions, where mounds can number in the thousands up to millions (cf. Mielke, 1977; Bragg, 2003; D.L. Johnson et al., 2008; Reed and Amundson, this volume, Chapter 1). In contrast, densities are less in the Upper

Midwest where the historical data suggest much lower numbers, in some situations only one to three mounds per acre (2.5–7 mounds/ha). For some local Upper Midwest situations, Webster (1900) noted a maximum of eight per acre (20/ha). Mound density distinctions between the Upper Midwest and the other regions may be related to local and regional environmental settings, and perhaps other variables. The largest areal moundfield documented in this study constitutes the series of lowland mound sites in Dakota, Rice, and Goodhue counties in southeastern Minnesota recorded by Schmidt (1937). The 2.6 square mile (6.7 km²) terrace at Harpers Ferry, Iowa, had 900 mounds, which equates to 346 mounds per square mile (134/km²), the bulk of which were natural prairie mounds (Finney and Johnson, 2010). Another high density mound site was in eastern Mower County, Minnesota, where Winchell (1884a, 1911) recorded more than 500 mounds.

After observations by the early archaeologists and geologists in this region, the natural prairie mounds were subsequently noted by early soil surveyors (e.g., Bennet et al., 1911). However, during the beginning of the twentieth century nearly all of the Upper Midwest landscape came under cultivation with the mechanization of agriculture and the installation of field tile for draining wet localities (Deiss, 1992; Montgomery, 2007). Schmidt (1937) commented on the continuing destruction of wetland tracts in Minnesota when he surveyed areas in the mid-1930s that were unavailable before 1910. Smith (1998) used the agricultural census schedules from the Iowa Department of Agriculture to document the rapid demise of the prairie hay fields. Based on the number of acres devoted to "wild hay" fields the census data indicate that only ~5.5% of the Iowa pre–Euro-American settlement prairie remained in 1896. At present the quantity is less than 0.1%. As a result, the modern landscape in the Upper Midwest is dominated by corn and soybeans where every available field has been drained (if necessary), plowed, and planted. During this process the natural prairie mounds had become a distant memory from past generations. The natural prairie mounds are no longer mentioned in the county soil surveys because they are gone. Even state prairie guides are inconsistent in stating whether or not natural mounds are present for undistorbed prairie tracts. In some cases their presence appears to be unknown to state preserve managers, interpreters, rangers, and law enforcement personnel. Why are these facts important? The presence of natural prairie mounds, apart from examples in north-central Illinois and west-central Wisconsin, extends in an arc from southeastern Minnesota and northern Iowa to southern Alberta on the northwest. What are the common factors for the occurrence of the Upper Midwest natural prairie mounds over this extensive distribution from Iowa to Alberta? They typically occur where wetness is a seasonal or semipermanent condition. In Iowa the Des Moines Lobe is described as

a relatively level, poorly drained landscape. This resulted in an immense complex of prairie interspersed with potholes, marshes, wet meadows, meandering rivers and streams, shallow lakes and four deeper lakes. In a wet spring, it is likely that much of the area was a continuous complex of wet prairies, sloughs, sedge meadows, and marshes. (Smith, 1998, p. 95)

A virtually identical description, with only the addition of hundreds if not thousands of lakes, would be appropriate for the Des Moines Lobe across Minnesota (Patterson et al., 2003). In the past, potholes (lakes, ponds, etc.) covered ~20% of the prairie pothole region. Of these potholes 60% were temporary and another 35% were seasonal in nature (R.R. Johnson et al., 2008). In addition, the identified prairie types in Iowa include four examples with thin soil: hill prairies on the Paleozoic Plateau of northeast Iowa, gravel prairies on the Des Moines Lobe, limestone prairies on the Iowan Erosion Surface, and Sioux Quartzite prairies on the Northwest Iowa Plains (Schennum, 1986). There are examples of incipient or juvenile mounds, apparently forming today, at a number of preserves and natural areas in Minnesota and northern Iowa. These include Hayden Prairie State Preserve, Howard County, Iowa; Anderson Prairie State Preserve, Emmet County, Iowa; Racine Prairie Scientific and Natural Area, Mower County, Minnesota; and Cedar Creek Ecosystem Science Reserve, Anoka County, Minnesota.

Of equal importance is the negative data generated by this chapter for the absence of natural prairie mounds on the Northwest Iowa Plains, the Southern Iowa Drift Plain, and the northern Missouri glaciated plain. These physiographic regions have deep loess soils. For this reason the dense barriers or wet conditions found in various locations in the Prairie Pothole region are minimal or absent.

Various explanations for Mima mounds also exist among American Indians (e.g., Curtis, 1924; Krinitzsky, 1949). These stories are not mischievous coyote or trickster tales. Of interest is the world creation myth of the Klamath Indians of southern Oregon featuring a pocket gopher who builds up the entire dry landscape from a water-covered world (Curtis, 1924). This choice of animal is a variant on the widespread earth diver myth among North American Indians where an animal, e.g., duck, turtle, beaver, badger, or crayfish, dives through the omnipresent water to retrieve the piece of earth that creates terra firma (Hall, 1979, 1997). The Klamath Indian use of a pocket gopher for a world creation myth indicates their familiarity with this species. While the earth diver myth serves a different purpose, this story also reflects the knowledge that pocket gophers built mounds.

SUMMARY AND CONCLUSIONS

Natural prairie mounds (i.e., Mima mounds, pimple mounds, etc.), once common across much of western and central North America, have long piqued the interest of scientists and lay public alike. Most mound investigators, however, have focused principally on their southern occurrences in Arkansas, Louisiana, Texas, and Oklahoma, and western occurrences in California, Oregon, Washington, Idaho, Colorado, and Wyoming. Insofar as a consensus on the nature and origin of natural prairie mounds, or Mima mounds, has been lacking, a deep literature on them has accumulated, especially in the earth sciences, covering nearly two centuries. Notably few contemporary or former researchers are, or were, aware that natural prairie mounds were once fairly common in the Upper Midwest—albeit in smaller overall numbers than those in the southern and western sectors mentioned.

Because natural mounds in the Upper Midwest were often confused with a class of human burial mounds long associated with "Moundbuilder" traditions of American Indians, which are and were similar in morphology and dimensions, reference to natural mounds is found principally in the archaeological literature of the nineteenth and early twentieth centuries. However, there was limited discussion of the "confusion issue" in this older literature, because most archaeologists of the time reasonably considered their prime directive accomplished (i.e., to find burials and/or artifacts) whenever an investigated mound was found to be noncultural. Once a noncultural determination was made, such mounds were no longer considered important. Nevertheless, during and since this early period, a few archaeologists and geologists, including noted mound investigators T.H. Lewis, C. Webster, W. McGee, and G. Fowke, did in fact recognize the noncultural nature of numerous Upper Midwest mounds.

Conversely, other prominent mound researchers during this period, including D. Bushnell, E.W. Schmidt, J. Brower, and N. Winchell, adamantly held to a cultural origin despite the near complete absence in such mounds of artifacts or other indicators. Despite the low profile of natural (Mima) mounds scattered in this older literature, before the prairies were drained and plowed, they had a wide spatial distribution, especially in the prairie belt of southern Minnesota and north-central Iowa. Their former distribution across these two states, as noted, is associated primarily with the Prairie Pothole region, including both the Des Moines Lobe and the Iowan Erosion Surface. They extended into eastern South and North Dakota and through the prairie parts of Manitoba, Saskatchewan, and Alberta. In addition, they were present at scattered locations in Wisconsin and Illinois.

The majority of the natural prairie mound sites reflect the presence of local environmental conditions that involve either a dense substratum (hardpan, claypan, coarse gravels, and bedrock), periodic to semipermanent soil wetness, or some combination of these conditions. Notably, natural prairie mounds are and were absent—or at least have never been reported—from the thick loess-covered Northwest Iowa Plains, the Southern Iowa Drift Plain, and the northern Missouri Drift Plain—all north of the Missouri River. These deep loess soils by and large lack impediments to vertical burrowing. In contrast, nearly all the unglaciated Missouri counties south of the Missouri River—which is the approximate glacial-nonglacial boundary—have, or formerly had, Mima mounds. They were, for example, very common before draining and plowing activities began, in both the Ozark Plateau and the Mississippi River Alluvial Plain. (They are still common in the Ozarks; cf. Horwath, 2002; Horwath and Johnson, 2006, and references therein.) As outlined above, they form and did form invariably where impediments to vertical burrowing exists, or existed.

In summary, this chapter documents the former abundance and wide distribution of natural prairie mounds in the Upper Midwest. Their distribution coincides with that of the suite of burrowing animals and their predators that have been implicated as being primarily responsible for producing them, most notably members of the Geomyidae (pocket gopher) family of animals.

ACKNOWLEDGMENTS

This chapter would not have been possible without the interest and assistance given by the editors in this endeavor. Many colleagues responded to my information requests, but alas often did not have any new site leads. Those who provided data used in this chapter include Scott Anfinson (Minnesota OSA), Bruce Koenen (Minnesota OSA), Craig Johnson (MnDOT), David Mather (Minnesota SHPO), Kim Davis (Savage, Minnesota), Thomas Thiessen (MWAC), William Green (Logan Museum, Beloit College), Shirley Schermer (Iowa OSA), David Kluth (USFWS), Scott Glup (USFWS), Sara Vacek (USFWS), and Amy Rosebrough (Wisconsin Historical Society). Thomas Emerson and Michael Lewis (Illinois State Archaeological Survey) at the University of Illinois at Urbana-Champaign produced Figure 1 and scanned the slide used for Figure A2-1.

APPENDIX 1. PROMINENT INFORMATION SOURCES

During this literature search, notes were compiled for the various nineteenth-century researchers in order to discern their perspectives on cultural and natural mounds. The thumbnail biographical sketches comprise the historical primer necessary to identify the individuals who made relevant observations, including the geologists and archaeologists Webster, McAdams, McGee, Lewis, Winchell, Upham, Brower, Schmidt, Bushnell, and Fowke. Their written descriptions are particularly useful for documenting Mima mounds in the Upper Midwest. In his history of Missouri archaeology Michael O'Brien (1996) distinguishes three types of late-nineteenth-century archaeologists: (1) those who investigated big questions not restricted to one locality or state, (2) antiquarians who examined their own local areas, and (3) those who worked in multiple states and locations on multiple problems. Researchers at the Smithsonian Institution, e.g., Cyrus Thomas, William H. Holmes, and Ales Hrdlicka, are prominent examples of the first type. The individuals who examined and reported the natural prairie mounds comprised the second and third types. The following paragraphs are restricted to relevant background material for this chapter concerning the prominent members of the third type of late-nineteenth-century archaeologists.

Clement Webster

Clement L. Webster (1859–1930) was a geologist and mining engineer from Charles City, Iowa, who also conducted archaeological investigations in Iowa, New Mexico, and Washington. Today Webster is best remembered for the lithographic limestone quarried in the early twentieth century at Lithograph City (renamed Devonia), Iowa. His observation about the "Mounds of the Western Prairies" bears repeating:

… a widely prevalent feature … is the great number of isolated or grouped mounds which are seen over the surface and which are often denominated by the inhabitants as Indian mounds. These mounds are generally circular and have an oval to flattened top with a diameter at the base of from 4 to 20 feet [1.2–6.1 m], and commonly rise to a height of from 1 to 3 feet [0.3–0.9 m].

Although the marginal outline of these mounds is generally circular, still at times some of them are oblong or have a gently flowing contour. These mounds … [can be] closely and irregularly grouped … . In some instances, as many as sixteen of these mounds have been counted in an area containing ~2 acres

[0.8 ha]. The location of these mounds is almost exclusively in the prairie regions and may be found on high and dry or low and rather moist land. In Iowa and southern Minnesota, where these mounds have been most studied by me, they may be seen for many miles over the level prairies of these regions. …

Although the external appearance of these mounds is analogous to that of some of the Indian mounds of Iowa and other States, still they may be distinguished from those of the mound builders by their relative position and the region occupied, as well as by their greater irregularity of contour.

For more than twenty-five years I have resided in the prairie regions of the West, and have thus been afforded a fine opportunity to study the origin and development of these "singular" mounds. By far the greater number of them owe their origin to the … gopher.

Webster also mentions badger and wolf mounds of similar dimensions. The description left by Webster is typical of late-nineteenth-century accounts of natural prairie mounds. Once interpreted as nonhuman in origin, the site locations were not considered important and not precisely given. An unpublished book manuscript that collected an 1889 article on natural prairie mounds in north-central Iowa included an illustration not published in the original source (Webster, 1889, 1900). It showed at least 10 Mima mounds on the Webster family farm in a numerous mound group (Fig. A1-1).

William McAdams

Professor William McAdams, Jr., (1834–1895) was a farmer, geologist, archaeologist, and state representative who lived at Otterville and Alton in western Illinois. He is best known for his archaeological interests and publications on mounds and village sites of the Illinois River Valley and the American Bottom, including Cahokia, in the late nineteenth century (Farnsworth, 2004; McAdams, 1880, 1881, 1887, 1919). McAdams served as the superintendent of the Illinois archaeological exhibit at the 1893 Chicago World Fair (Chappell, 2002). He also worked on sites in Missouri, Kansas, and the Dakotas, a fact that almost certainly influenced his views about possible house mounds.

William McGee

William J. McGee (1853–1912) was a geologist and native of Farley in northeast Iowa who also worked as an archaeologist and ethnologist. He worked for the U.S. Geological Survey between 1883 and 1893 and subsequently for the Bureau of Ethnology (Hinsley, 1981; Meltzer, 1985; Woodbury and Woodbury, 1999). Of particular interest is his publication titled *The Pleistocene History of Northeastern Iowa*, which included an illustration of Mima mounds in that region (Fig. A1-2). This excellent drawing is one of the best natural prairie mound illustrations of its era. Unfortunately McGee (1891) did not give a site position beyond stating the upland prairies south of the Turkey River. This location is believed to be northern Fayette County, Iowa.

Theodore Lewis

Theodore H. Lewis (1856–1930) was arguably the most experienced archaeological mound surveyor of the late nineteenth century in eastern North America. He worked throughout the Midwest, Plains, and Southeast between 1876 and 1895. Lewis did the fieldwork for Alfred J. Hill's Northwestern Archaeological Survey which recorded over 17,000 Indian mounds and earthworks at more than 2000 sites (Dobbs, 1991; Finney, 2006; Lewis, 1898). The mound measurements, site descriptions, and maps from his work are housed at the Minnesota Historical Society. Lewis published 60 articles on his work. These articles presented a significant body of research for prehistoric archaeology that emphasized mounds, rock art, the exposure of Moundbuilder fantasies and other fraudulent claims, and a pioneering approach that combined field surveys with historic documents in ethnohistorical studies of the De Soto route (Finney, 2001, 2004, 2005, 2006, 2008, 2010).

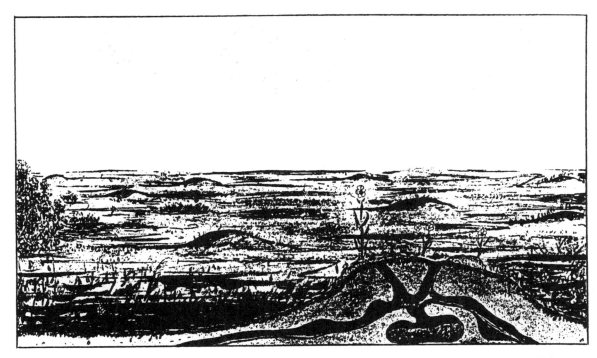

Figure A1-1. Natural prairie mounds on the Webster farm, Floyd County, Iowa (from Webster, 1899, Plate 20 between pages 39 and 40). Drawing by Clement L. Webster, whose caption stated, "Mounds of the prairies and sectional view, as seen near our early farm home two miles northeast from Rockford, Iowa."

PRAIRIE LANDSCAPE, WITH NATURAL (?) MOUND IN FOREGROUND.

Figure A1-2. Natural prairie mounds in northeast Iowa (from McGee, 1891, Plate 12 between pages 124 and 125). The location is only specified in a general sense as the upland prairies south of the Turkey River and may be from northern Fayette County (McGee, 1891, p. 108). Drawing by Hobart Nichols.

Among Lewis' publications an 1883 article on natural prairie mounds is of particular relevance. It appeared as a brief note in the correspondence section of the *American Antiquarian* entitled "Swamp Mounds" (Lewis, 1883a). He also used the terms marsh mounds, meadow mounds, natural mounds, and prairie mounds for small mound-shaped hillocks of nonhuman origin. In his 1883 article Lewis asked what interpretations other researchers had for these swamp mounds that appeared to him as natural phenomena. While Lewis was not certain of their origin, he should be credited with recognizing their noncultural nature. In a comment appended to the Lewis note, *American Antiquarian* editor Stephen D. Peet (1883a) stated that he believed the swamp mounds were created by burrowing rodents.

Note.—The opinion has been expressed that these mounds are formed by gophers. They are frequently found in the midst of swampy or low lands, separate from any artificial works.—Editor. (Peet, 1883a, p. 331)

Soon after Peet (1883b) noted their presence in a wet environmental context on the flat, low portions of the Minnesota prairie.

Swamp Mounds.—On the way to Minneapolis through Minnesota, mounds of the character described by Mr. Lewis were noticed from the [railroad] car window. They resemble Indian mounds, but more likely were erected by gophers. This is the opinion of those that have explored them. Indian mounds are found in Minnesota but always near lakes or rivers; these are remote from streams, in the low places of the prairie—in sinks or swamps in the midst of the prairies. (Peet 1883b, p. 337)

A careful perusal of the Lewis field notebooks and correspondence provides a background for the swamp mounds article that reveals a significant finding. In addition to recording thousands of Indian mounds, he encountered Mima mounds on numerous occasions. Discussions about confusing and distinguishing natural with artificial mounds are further revealed in the Lewis correspondence with his patron, Alfred J. Hill, who supported and funded his extensive mound work. Lewis wrote about examples from Minnesota, Wisconsin, Illinois, Missouri, Arkansas, and Kansas. These close encounters with Mima mounds began with a survey late in the 1882 field season when Lewis found hundreds of mounds near Shakopee in Scott County, Minnesota. Not liking their appearance and landscape position, Lewis did not believe these mounds were worthy of the time and expense necessary for the detailed measurements that were his standard procedure for mapping Indian mounds.

There are some rises here that look like mds[.] but I cannot tell whether they are or not, but think they were built up of turf or peat from the swamps. They extend above here for at least a mile. ... There are also some of the same kind south of the track. They are all built alike. Occasionally one is round and nice while most of them are uneven. (Lewis, 1881–1895b, Lewis to Hill dated October 11, 1882, from Shakopee, Minnesota)

This survey near Shakopee appears to be the epiphany that prompted Lewis to realize this potential problem for completing his fieldwork. Afterwards Hill and Lewis continued to ponder natural mounds as the December 18, 1882, letter from Jonesboro, Arkansas, is devoted to this problem. This letter is the ultimate "smoking gun" that outlines Lewis' field observations and opinions on the subject. The key sentence—and a statement made very early in the history of his surveys—that summarizes his philosophy and approach to natural versus anthropic mounds is: "If in the future I survey any more mounds, it will be only those that I am positive are artificial ..." (Lewis, 1881–1895b, Lewis to Hill dated December 18, 1882, from Jonesboro, Arkansas). Further the tone implies he is seeking to convince Hill that refusing to make a formal survey of natural mounds is the correct policy for conducting their fieldwork.

Your letter of the 10th just received also the "Prescott" letter and contents noted. Instead of adding to the number of mounds, I rather think I have been to[o] cautious in not adding enough for I am satisfied that I would have had at least

100 more mounds surveyed being well and satisfied in a number of instances that mounds were artificial, but on account of ill shapes they were passed by for I would not survey them unless I was positive. It is more than likely that the mounds surveyed if ever they are placed in printed form will be examined and a few mistakes whether accidental or otherwise would destroy its effect and make the book worthless, for that reason I have thought it best to let doubtful mounds remain unsurveyed, and if they are examined in future and turnout to be artificial nothing can be said [p. 2] against the work here. I have always had hopes that something would turn up by the time the work is completed—to have the whole embodied in a good standard work and if it cannot be done right I do not want anything to do with it. The mounds passed as doubtful would cost to[o] much to excavate them to prove their character and I should prefer to leave them and expend an equal sum in excavating some of the larger mounds, for they are more liable to contain relics than the smaller ones, and to make complete book illustrations of the relics found in the state should accompany it as these relics prove conclusively that the moundbuilders of other states were of the same class and people as those in Minn. If in the future I survey any more mounds, it will be only those that I am positive are artificial leaving the remainder to future explorers—in case I cannot make the examination. I am satisfied that I could have added from 75 to 100 in the large group below Shakopee and while I think of it I will mention the fact that there are several of the same class [p. 3] of mounds in Shakopee but I will never survey them unless they are first opened, but I think it will cost too much money and time to do anything in this line. (Lewis, 1881–1895b, Lewis to Hill dated December 18, 1882, from Jonesboro, Arkansas)

The December 18th letter and Hill's unknown reply prompted the subsequent swamp mounds query in the May 1883 issue of the *American Antiquarian* that solicited wider opinions on the subject (Lewis 1883a). Four months after publication, Lewis wrote Hill:

My trip to Springfield [Brown County, Minnesota] and Lamberton [Redwood County, Minnesota] did not amt. to anything as I anticipated. There are plenty of mds[.] in that vicinity but all were natural. (Lewis, 1881–1895b, Lewis to Hill dated September 13, 1883, from Watertown, Dakota Territory)

It is worth noting that Lewis (1883b) made a visit to the Missouri River in North Dakota for research to determine the appearance of earthlodges in the archaeological record. He excavated a portion of an abandoned earthlodge and as a result developed objective expectations for assessing the function of different kinds of archaeological finds. This work was another step toward Lewis becoming satisfied that natural prairie mounds had a nonhuman origin.

In the summer of 1884 Lewis made a brief statement about Mima mounds. On June 11, 1884, from Prairie du Chien, Wisconsin, he wrote: "marsh mds[.] are not wanting in this region" (Lewis, 1881–1895b, Lewis to Hill dated June 11, 1884, from Prairie du Chien, Wisconsin). This comment might be a reference to Mima mounds on the nearby Mississippi River terrace at Harpers Ferry, Iowa. In any event, Lewis noted the presence of marsh mounds in the region around Prairie du Chien, which is ~10 miles south of Harpers Ferry.

Two fall 1884 surveys in Steele and Wabasha counties, Minnesota, document Lewis making distinctions in his field notes between different kinds of natural mounds.

Yesterday afternoon I examined the country for 2 or 3 miles S of Aurora and found nothing but sand hills and marsh mounds. This morning went down Strait River on E side as far as Medford and returned on W side and saw only marsh mounds. (Lewis, 1881–1895b, letter dated October 12, 1884, from Owatonna, Minnesota)

Went to Plainview but the famous mds there are natural. Stone is being taken out of one. (Lewis, 1881–1895b, letter dated October 13, 1884, from Owatonna, Minnesota)

Three later Lewis survey trips yielded clues to his perception of prairie mounds in the midcontinent. In field notes and letters of his 1893 surveys near Des Arc and Lonoke in east-central Arkansas, near Kansas City, Greenfield, and Warrensburg in western Missouri and adjacent Kansas, and north of Chicago, Illinois, Lewis noted the presence of natural mounds. He explicitly used the terms "prairie mounds," "marsh mounds," and "meadow mounds" for describing them in his letters. These terms refer to the distinctions Lewis made based on the associated vegetation and landscape position of the natural prairie mounds (e.g., Figs. 3 and A1-3).

There are no large mounds in this vicinity, the largest not being over three feet [0.9 m] in height. But there are probably thousands of these small mounds in this county. There is probably 100 on this townsite not counting those that have been destroyed. They are not so numerous on the high lands—covered with woods and brush—near the river as they are further back on prairies where they are almost countless. (Lewis, 1881–1895b, Lewis to Hill dated February 12, 1893, from Des Arc, Arkansas)

In the letter dated February 19, 1893, Lewis briefly mentioned visiting prairie mounds in the vicinity of Lonoke, Arkansas. Both Des Arc and Lonoke are in the former Grand Prairie of east-central Arkansas.

The several thousand mounds in Johnson and neighboring counties both in Kansas and Mo[.] are not "Marsh mounds" but "Prairie Mounds" similar to those in Ark[.], Texas and Louisiana. (Lewis, 1881–1895b, Lewis to Hill dated May 9, 1893, from Greenfield, Missouri)

Saturday I drove out to the mounds north of here but did not find any platforms or earthworks, but did find quite a number of round mounds, most of which were like the Minnesota marsh mounds, but a few extended along up the slope to the top of the hill, but nothing worth surveying. (Lewis, 1881–1895b, Lewis to Hill dated May 22, 1893, from Warrensburg, Missouri)

Lewis reiterated his views on Kansas City and Johnson County, Missouri, in his field notebook during June:

Missouri. The small round mounds extending south along the different lines of RR from Kansas City—those bordering the prairies and edges of creek bottoms are similar to the "Prairie mounds" of Arkansas, Indian Territory [Oklahoma], Texas and Louisiana. June 8th 93. (Lewis, 1881–1895a, Field Notebook 41, p. 13)

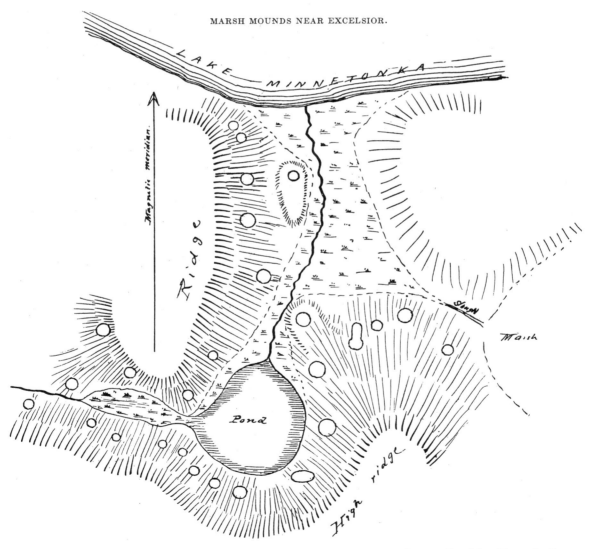

Figure A1-3. A marsh mound group found on the south shore of Lake Minnetonka near Excelsior, Hennepin County, Minnesota (from Winchell, 1911, p. 260).

Yesterday afternoon I looked up some of the mounds on the west side of Des Plaines river in Lake Co[.,] Ill[.], and found them to be Meadow mounds. (Lewis, 1881–1895b, Lewis to Hill dated September 8, 1893, from Burlington, Wisconsin)

Another 1893 field notebook entry from Johnson County, Missouri, reports mounds that almost certainly were dominated by natural mounds because of the tremendous site extent.

NW[¼] – [Section] 15 – [T]47[N] – [R]25W. There are several mounds on a side hill and near the edge on the top of the hill, all under cultivation. These mounds seem to follow Blackwater Creek for miles. May 20th 1893 (Lewis, 1881–1895a, Field Notebook 41, p. 5)

In summary, beginning in the fall of 1882 Lewis expressed skepticism that the numerous mounds observed near Shakopee in Scott County, Minnesota, were entirely anthropic in origin. Subsequent field notebook entries and letters clearly indicated that he recognized prairie mounds as natural entities—as did many of his contemporaries. Some of these sites, such as Harpers Ferry, Iowa, apparently represented an unusual case where both Indian and prairie mounds shared the same landform. At Harpers Ferry Lewis noted a group consisting of an estimated 900 mounds, most being small round mounds. The observation was not repeated by subsequent visitors, though Lewis is considered to have been a competent field archaeologist. Finney and Johnson (2010) believe that natural prairie mounds existed on the Harpers Ferry terrace in 1892. This position is based on Lewis' field notebooks and unpublished correspondence, the then (before drainage) wet prairie, nondunal setting of the alluvial Harpers Ferry terrace, application of new soil-geomorphic explanatory theory, and synthesis of recent, wide-ranging prairie mound research and observations (Finney and Johnson, 2010). In all, Lewis noted prairie mounds in Scott, Brown, Redwood, Sibley, Waseca, Steele, Wabasha, Stern, Nobles, and Freeborn counties, Minnesota; Allamakee County, Iowa; Crawford County, Wisconsin; Lake County, Illinois; Johnson, Dade, Jackson, and Cass counties, Missouri; and Johnson County, Kansas. His field notes also acknowledge an understanding of the pimple mounds in Texas, Louisiana, Arkansas, and Oklahoma.

Newton Winchell

Other early observers who reported natural prairie mounds in the Upper Midwest and the eastern Plains border included those working with the state geological surveys. A native of New York State, Newton H. Winchell (1839–1914) graduated from the University of Michigan with bachelor's (1866) and master's (1867) degrees. In 1872 Winchell started the Minnesota Geological Survey. He also taught at the University of Minnesota and is one of the scholars who helped establish its reputation as a major institution (Upham, 1916). During the geological fieldwork Winchell and his assistants appended a list of Indian mounds to the end of the county reports. Such a practice was a common occurrence for the nineteenth-century state geological surveys. In Minnesota the quantity of mounds was far less than those observed by Lewis for the same counties, and appears to have been dominated by those easily observed from roads and rail lines. In a number of counties Winchell et al. reported natural prairie mounds as well as anthropic mounds. In particular they saw probable or definite Mima mounds in Anoka, Dakota, Fairbault, Goodhue, Hennepin, Morrison, Mower, Nobles, Pope, Rice, Sherburne, Stearns, Steele, Wabasha, Waseca, and Wright counties (Winchell, 1873, 1877, 1878, 1884a, 1884b, 1885, 1888a–1888c; Winchell and Upham, 1884). Unfortunately these researchers held the basic thesis that all mounds were human-made constructions. Only the observation of more than 500 mounds in eastern Mower County gave Winchell (1911) pause to reconsider this hypothesis. After completing his final reports for the geological survey, Winchell became the archaeologist at the Minnesota Historical Society and

compiled the Northwestern Archaeological Survey, the Brower surveys, and ethnographic notes on the Dakota and Chippewa into the *Aborigines of Minnesota* (Winchell, 1911).

Warren Upham

The geologist, librarian, historian, and archaeologist Warren Upham (1850–1934) was a native of Amherst, New Hampshire, who graduated from Dartmouth College in 1871. He worked on the New Hampshire Geological Survey before joining Winchell on the Minnesota Geological Survey. In the course of his western and central Minnesota research he noted the presence of mounds in his county reports (Upham, 1884a–1884d, 1888a–1888f). Next, Upham (1887, 1890, 1895) continued the studies on tracing glacial Lake Agassiz for the Geological Survey of Canada and then the U.S. Geological Survey. He returned to St. Paul as the librarian for the Minnesota Historical Society (MHS), then the administrative secretary, and eventually became the archaeologist after Winchell passed away. While the MHS librarian Upham published important contributions on various historical topics (e.g., Upham, 1920). In his geological endeavors Upham supported positions espoused by Winchell and Brower concerning the reputed glacial age human sites at Little Falls, Minnesota, and Lansing, Kansas, as well as the belief that all mounds were anthropic in origin (Schmidt, 1935a).

Jacob Brower

Jacob V. Brower (1844–1905) was the significant personality in the establishment of the Minnesota state park system. He was the first and only state employee who opened Itasca Lake State Park in the northwoods wilderness during the early 1890s. Brower had no formal training in archaeology. After Civil War service, he served as the Todd County auditor, then made his living as an attorney, and was a member of the state legislature. As an archaeologist, Brower held a reputation that began with his Itasca Lake research, then at Mille Lacs, at Red Wing, and other Minnesota localities. He was also known for geographical exploring finding the ultimate source of the Missouri River at the Montana-Idaho border and tracing the Coronado expedition in Kansas (Finney, 2006). Brower published eight volumes between 1898 and 1904 in a series titled *Memoirs and Explorations in the Basin of the Mississippi*. Five of these books were about Minnesota, two on Kansas research, and one for North Dakota sites. Except at Itasca Lake, Brower's Minnesota investigations were completed without any knowledge of where Lewis had surveyed and the Northwestern Archaeological Survey results. Of particular interest to this chapter are his surveys near Red Wing (Brower 1903) and his collaboration with Schmidt (1937).

Edward Schmidt

Edward W. Schmidt (1866–1955) graduated from the University of Wisconsin in 1887 and started teaching botany at St. Olaf College in Northfield, Minnesota. He retired from St. Olaf in 1933. Schmidt also served as the president and a faculty member at the Red Wing Seminary from 1910 to 1918 (Withrow, 2005). As summarized by Russ Withrow,

Through the years this versatile man taught many subjects, including ancient languages, religion, philosophy, mathematics and science, focusing on botany and biology in later life. Though he was principally a botanist and had no formal training in archaeology, he regarded his hobby of Indian archaeology and the publication of local geology as his primary work. (Withrow, 2005, p. 1)

According to Schmidt, he first noticed the lowland mounds around Northfield in 1887. Not until the 1901 and 1902 interval when he visited Brower and W.M. Sweeney at Prairie Island and Red Wing did Schmidt become satisfied with the importance of investigating the lowland mounds (Brower, 1903; Schmidt, 1928, 1937). There are three primary subsequent fieldwork episodes documented for Schmidt. The first period was between 1906 and 1908, with the last year occurring in conjunction with Winchell (1911) excavating mounds with negative results. In this work Schmidt (1908, 1909, 1910, 1928) initially amassed a database of 1575 lowland mounds. His distribution map, selected sites, and conclusions appeared in *Aborigines of Minnesota* (Winchell, 1911) while the lowland mound paper was placed in histories for Goodhue and Rice counties and the *Northfield Independent* newspaper (Schmidt, 1909, 1910, 1928). Next, Schmidt conducted additional fieldwork excavating 26 mounds for the Minnesota Historical Society in 1916 (MHS, 1917; Schmidt, 1928). The third period of investigations took place in the mid-1930s after Schmidt retired from St. Olaf, beginning with limited fieldwork digging several mounds in 1933–1934 with more intensive excavations following in 1935–1936. The final lowland mound count for his investigations was estimated to be more than 2700 in total mounds (Schmidt, 1937). There are occasional inconsistencies for the sites (e.g., names, mound count) between the earlier and later accounts. The earlier sites were not repeated, at least not in full detail, in his final paper. The 1930s fieldwork that produced another ~1125 mounds is only partially mapped and described site by site outlining the location and mound counts enumerated for the earlier surveys. Mounds found from 1933 to 1936 often occurred in wetlands not drained or accessible during the earlier surveys. In the past decade Donald and Diana Johnson attempted to relocate Schmidt mound groups but discovered that nearly all have been destroyed by plowing. A few remaining sites might still exist in inaccessible areas dominated by water and dense vegetation (D.L. Johnson, 2010, personal commun.).

At several points in time Schmidt helped excavate anthropic mounds near Red Wing, Minnesota, and across the Mississippi River at the Diamond Bluff and Adams sites in Pierce County, Wisconsin (Schmidt, 1937). In addition to his numerous lowland mound contributions (Schmidt, 1908, 1909, 1910, 1928, 1934a, 1934b, 1934c, 1935a, 1937), Schmidt had a few publications that involved prehistoric artifacts and sites (Schmidt, 1935b, 1941), Indian trails (Schmidt, 1909, 1910, 1910), and beaver dams (Schmidt, 1909, 1910, 1935c, 1937). For an interpretation of the lowland mound problem Schmidt remained undaunted by the lack of human burials and a near universal absence of any artifacts. In his writings all mounds were attributed to human construction, with the presumed function of lowland mounds serving as a small basal platform for individual habitation units. Schmidt (1937) sincerely believed that the inhabitants had policed their sites and completely removed all debris evidence of their presence and that he could discern "load marks" in the profiles of certain mounds. This identification of basketloading is suspect, and more likely refers to backfilled rodent disturbances with fills exhibiting a high contrast in color, texture, or both. In various writings he attributed the lowland mounds to historic tribes as well as to the very distant past in a belief that the burials and buildings had disintegrated leaving no traces. Despite changing the lowland mound hypothesis to accommodate the repeated negative findings, Schmidt never abandoned a belief in their human construction. His writings indicated a "hope springs eternal" perspective (Shermer, 1997). Schmidt explained away the gopher hypothesis with the following ideas: (1) the sites were too wet for gophers, (2) gophers live at numerous localities that do not exhibit mounds, (3) he never met anyone who could claim to have seen gophers create large mounds, (4) an absence of innermound cavities large enough to contain the soil from the mound superstructure, (5) animal burrows did not penetrate the clay or gravel underneath the mounds, and (6) an overriding certainty of a human origin (Schmidt, 1928, 1937).

David Bushnell

David I. Bushnell, Jr., (1875–1941) was a native of St. Louis. He had wide research interests in ethnology, archaeology, and photography despite never receiving formal training in anthropology. One of his earliest archaeological fieldwork experiences was with Brower at Mille Lacs, Minnesota. Later Bushnell was associated with Harvard's Peabody Museum from 1901 to 1904 and then with the Smithsonian Institution (Swanton, 1942). His writings concern prehistoric salt manufacture, the direct historic approach, Indian villages and burial sites, the Choctaw, prehistoric Virginia Piedmont sites, and frontier artists who depicted American Indians. Some of his early articles attributed natural prairie mounds to the Indians as functioning for elevated habitation sites (Bushnell, 1904, 1905). He acknowledged the possibility of multiple origins for the small natural mounds from different regions of the country. Bushnell (1905, p. 713) had excavated small conical mounds of anthropic origin in St. Louis that he interpreted as "ruined habitations." While Bushnell (1905) recognized that this interpretation did not seem to fit the small conical mounds that lacked artifacts or burials, he still persisted in the belief that all small conical mounds were human-made.

Probably if these small mounds were not so numerous the question of their origin would never have been raised and they would have been considered, together with the larger mounds, as having been made by man, but the question of number should not influence the decision. It is doubtful if the combined bulk of all these small mounds in the Mississippi Valley is more than equal to that of the one great mound [Monks Mound] of the Cahokia group.

Without conclusive proof to the contrary, I feel that the most plausible theory of the origin of these small mounds, in Missouri and in other localities where they occur under similar conditions, is that they were made by man, probably to serve as elevated sites for habitations. (Bushnell, 1905, p. 714)

Gerard Fowke

Born in Mason County, Kentucky, as Charles Mitchell Smith and orphaned at age 15, Gerard Fowke (1855–1933) legally changed his name in 1887 to that of a famous ancestor from colonial Maryland (Fowke, 1929). His academic background consisted of one year at Ohio State University. Fowke was a regular field assistant from 1887 to 1890 for the Bureau of Ethnology mound survey who subsequently worked for the Smithsonian Institution and other organizations across North America and in Siberia (Fowke, 1929; Smith, 1985; Thomas, 1894). During his career Fowke authored numerous archaeological, and a few geological, papers. In several publications Fowke (1910, 1922, 1928) reported numerous Mima mounds in Missouri and Arkansas. He is notable for rejecting the house mound hypothesis (Fowke, 1922, 1928; Melton, 1929). This is significant since Fowke enjoyed a considerable reputation as an experienced field archaeologist, a skeptic, and for exposing incorrect ideas and frauds. As explained by Fowke:

The small, low, flattened mounds of the lower Mississippi Valley are a problem to archaeologists. (Fowke, 1922, p. 161)

They are found everywhere ... (Fowke, 1922, p. 163)

But they were not "residence mounds" or "house sites" in the sense that they furnished a base or foundation for structures that were used for dwellings; for there has never been found on their surface or in the earth immediately around them any of the debris invariably accompanying Indian huts or houses, such as fireplaces, ash beds, burned rocks, broken implements, or fragments of bones or pottery. These considerations also interfere with a full acceptance of the hypothesis that they are the remains of houses built of wood and covered with earth. It is true that such evidence is very frequently found in other localities; but to establish the fact that they were residence sites, refuse of this kind should be found wherever the mounds occur. (Fowke, 1922, p. 164)

In addition Fowke (1922, p. 161–165) rejected all other existing theories for the origin of natural prairie mounds leading him to speculate that more than one method of origin might be present. He summarized the problem: "A few of these mounds have been explored by the writer, but no discoveries were made upon which can be based a definite statement as to their probable purpose" (Fowke, 1922, p. 165).

APPENDIX 2. NATURAL PRAIRIE MOUNDS IN MINNESOTA, IOWA, WISCONSIN, AND ILLINOIS

In the Upper Midwest, the natural prairie mounds disappeared during the anthropogenic transformation of the prairie into the modern agricultural landscape. Similar to large-scale deforestation, little was recorded about this process ("progress") and details of the prairie landscape on which it occurred (e.g., Bogue, 1963; Cronon, 1983; Kaplan et al., 2009; Montgomery, 2007; Ramankutty and Foley, 1999; Urban, 2005; Williams, 2008). Noncultural mounds of various types are found in the Upper Midwest. Those found west of the Mississippi River in the glaciated areas are post–Des Moines Lobe in age. The following site descriptions are not comprehensive overviews of all previous investigations and interpretations, but rather concentrate on the salient aspects of the natural prairie mounds. Many of these sites are unpublished, or minimally reported in the literature, and (if recorded) the state archaeological site file comprised the primary source for their description. Sites listed in this appendix consist of a natural prairie mounds database that inasmuch possible lists their distinctive characteristics and attributes previously enumerated in the body of the paper (Table 2). The assessment made from the combination of these multiple lines of evidence is comparable to a "thick description" that conveys interpretation. In contrast merely stating that a mound is noncultural would be a "thin description" (Geertz, 1973).

Literature Review Findings

Minnesota House Mound Hypothesis

In Minnesota researchers at the beginning of the twentieth century pondered the hundreds of so-called swamp or lowland mounds found in Rice, Dakota, Goodhue, and Scott counties that Lewis had seemingly ignored (Finney, 2006; Winchell, 1911). Since these mounds lacked human burials, another possible cultural function was sought from the anthropological literature and American Indian oral traditions. Following the ideas expressed by Lewis H. Morgan (1881), the Marquis de Nadaillac (1884), and those recorded in ethnographic descriptions (Pond, 1908; Winchell, 1908, 1911) these researchers considered the lowland mounds as the probable basal platforms for dwellings, i.e., low house mounds (Brower, 1903; Bushnell, 1904, 1905; Schmidt, 1908, 1909, 1910, 1928, 1934a, 1934b, 1934c, 1935a, 1937; Winchell, 1908, 1911). Superficially similar house mounds, with habitation debris, exist in other portions of the New World (Morgan, 1881; de Nadaillac, 1884).

Morgan suggested that these mounds were house sites. Present-day opinion is that they were mortuary mounds. There were structures on them [temple mounds], but it is unlikely that they were houses; there were no middens, and evidence of row after row of houses has been found on the regular ground surface, nearby but separated from the mounds. (Bohannan, 1965, p. xi)

Unfortunately Morgan's house mound hypothesis was not based on excavation data and consisted of mere speculation as noted by Lewis and many other subsequent investigators (e.g., Bohannan, 1965). In Minnesota this idea was expressed in several Dakota ethnographic accounts where the lowland mounds were attributed to earlier historic peoples, the Iowa and Omaha tribes, who had been driven out of the state. Pond (1908) was told that numerous small mounds along the Minnesota River near Bloomington and West St. Paul were made by the Iowa Indians while Col. William Colville's informant stated that the Omaha lived at the mounds near Red Wing (Winchell, 1888b, 1911).

The small mounds, which may be seen on the left bank of the Minnesota at Eden Prairie and Bloomington, and perhaps at other places, are, the Dakota say, the ruins of dwelling houses built by the Iowas. These mounds are in rows or groups, on the bluff of the northeast side of the river. They are circular and of various sizes. I never measured them, and it long since I have seen them; but

I think they were three or four feet [0.9–1.2 m] high and fifteen to twenty feet [4.6–6.1 m] in diameter at the base. Their situation on the north side of the river, if there are none on the south side, indicates that the Iowas were more apprehensive of an attack from the south than from the north. (Pond, 1908, p. 174–175)

In addition, Winchell (1911) believed that the Mandan and Hidatsa had moved to the Missouri River during the historic period, and at that time, their purported point of origin was Minnesota. Using this logic suggested by this oral tradition, Winchell believed that earthlodges or house mounds should be found in Minnesota. Such an uncritical acceptance for features superficially similar in morphology or site specific aspects of oral traditions invariably resulted in disappointments when the myth description did not match the sites found in the archaeological record. This problem was especially compounded when the informants came from a different ethnic group. For these reasons the development of professional archaeology at the end of the nineteenth century necessarily resulted in a rejection of folklore (Gazin-Schwartz and Holtorf, 1999; Hall, 2008; Mason, 2006; Nabokov, 2002; Trigger, 2006). Winchell should have been aware of the divergence of folklore and archaeology, but along with several colleagues retained a firm belief in the house mound hypothesis and accepted the oral traditions that agreed with their hypothesis. Further, Winchell noticed that an earthlodge-building tribe living on the Missouri River had originated from Minnesota before being driven west. For this reason he decided that earthlodges would have been used by the tribe when resident in Minnesota and that these were represented by the lowland mounds. An early twentieth-century summary of American Indian dwelling types made this observation on earthlodges: "The structures are always covered over with earth, so much so that they look almost like natural mounds, whence the common term 'earth lodge'" (Waterman, 1924, p. 13). As noted elsewhere in this chapter, Winchell, Schmidt, and others persisted in this superficial similarity identification and ignored the overwhelming archaeological evidence to the contrary. Bushnell (1904, 1905) who worked briefly in Minnesota was a particular advocate of the house mound hypothesis. He had worked in Missouri prior to his May to June 1900 fieldwork at Mille Lacs in Minnesota and appears to have influenced Schmidt, Brower, Winchell, and Upham (Finney, 2006). Schmidt (1928, 1937) in particular carried the hypothesis for house mounds despite not finding habitation debris—or virtually any artifacts—in his prairie mound excavations. Such interpretations were made despite an absence of credible cultural materials and contemporary warnings to the contrary (e.g., Fowke, 1922; Lewis, 1883a, 1886; Melton, 1929; Peet, 1883a, 1883b; Thoburn, 1929, 1937; Thomas, 1884; Webster, 1889). Winchell (1911) had the final word on this subject in the early twentieth century while compiling *The Aborigines of Minnesota* where he cast dispersions on the completeness of the Lewis surveys because he had skipped numerous natural prairie mound groups. On the other hand, E.S. Macgowan (1948) noted that he interviewed Winchell in 1911 about prehistoric house types in Minnesota. According to Macgowan their discussion revealed that Winchell was not personally familiar with the archaeological expression of various house types. Further, Winchell believed that earthlodges had not only occurred but were widespread in Minnesota because peoples who came from the state later built this house type on the Great Plains. As stated by Winchell (1911, p. 385): "The Mandan house [earthlodge] may therefore be taken as the type of the early Dakota house in the places of their sedentary residence." It is no wonder that Winchell and his colleagues persisted in their belief of Minnesota house mounds.

Other Types of Noncultural Mounds

A variety of noncultural mound types exist in the Upper Midwest that can be potentially confused with Mima-pimple-natural prairie mounds and anthropic conical mounds (Arzigian and Stevenson, 2003). They are the result of various bioturbation, i.e., faunalturbation and floralturbation, and geological, mostly periglacial, processes and can be readily identified with careful investigation (Hole, 1981; Schiffer, 1987; Stein, 2001; Wood and Johnson, 1978). The following list is designed for the Upper Midwest region.

Two kinds of faunalturbation that create mounds are highlighted by burrowing animals (Cox, this volume, Chapter 3; Horwath Burnham et al., this volume, Chapter 4; Johnson and Horwath Burnham, this volume, Introduction) and ants. Some of the ant hills situated in unplowed fields from the Chicago region (Gregg, 1944; Lane and BassiriRad, 2005) form dome to conical mound-like structures that reach sizes large enough (1.5 m [5 feet] in diameter by 0.9 m [3 feet] in height) for potential confusion with smaller Mima mounds. Ants are another species that can be a significant factor in soil modification (Baxter and Hole, 1967; Curtis, 1959; Folgarait, 1998). The ant mounds share characteristics with Mima mounds where the cumulative effect of grazing animals in pastures reduces the mound height while plowing removes and destroys the superstructure (King, 1981).

In wooded areas the floralturbation process that produces potential mounds and moundfield-like topography is tree falls or uprooting typically caused by high wind events (Embleton-Hamann, 2004; Schaetzl et al., 1989). Tree falls are more common where the soil is shallow and roots cannot extend to their normal depth. They are marked by small soil mounds, often with an inverted A-B soil profile, and an adjacent pit. Surface extent of the pits can be 12–16 m² (Schaetzl et al., 1989). The tree-tip pit represents the hole caused by the tree root ball being torn from the earth and the adjacent mound results from the remaining soil adhering to the root ball. Pits are usually shallow and fill in before the adjacent mound erodes away. The mound sizes exhibit a positive correlation with the tree dimensions, with larger trees possessing greater root ball uplift and producing mounds large enough for potential confusion with smaller Mima mounds (Gallaway et al., 2009; Norman et al., 1995). A recent study of tree falls in upper Michigan and Wisconsin demonstrated the continuing presence of mounds 0.25–0.50 m in height that were hundreds of years in age and in several cases more than 2000 years old (Schaetzl and Follmer, 1990). A careful examination of a possible tree fall should reveal the simple or complex uprooting. Another possible attribute is the inversion of the root ball marked by the B horizon resting on the ground surface. Mounds are found on slopes as well as level surfaces. Windstorms can topple a series of trees in a catastrophic event resulting in a widespread pit/mound (a.k.a. cradle knoll) microtopography consisting of multiple synchronous pit/mounds (Schaetzl et al., 1989). In a woodlot setting the presence of pit/mound microtopography can increase the visual illusion of a mound or even a moundfield. In this regard it should be noted that Increase Lapham specifically recognized tree fall mounds as distinct from the mounds he mapped as prehistoric in origin (Dorney, 1983).

A number of possible historic disturbance types can create soil mounds or earth-covered debris piles in the possible size range of anthropic and natural prairie mounds as well as much larger in extent. Large-scale landscaping can create numerous mound look-a-likes. A well-known example of rock debris piles with a thin soil veneer seemingly appearing as mounds was found near the quarry pits from the historic catlinite quarry (21PP2) at Pipestone, Minnesota (Arzigian and Stevenson, 2003; Scott et al., 2006). Unfinished construction sites are another potential source of grassy mound-like features. In addition, the appearance of a heavily disturbed area can mimic pit/mound microtopography. The recent nature of these historic disturbances should be readily apparent to the careful observer.

The geological noncultural mounds are dominated by features of glacial origin that could appear similar to anthropic and natural prairie mounds. Fortunately, it is often only the small end of their possible size range that could be a "mound of confusion" and most of these features are not composed of soil. Natural landscape features small enough to be mound-like are various types of dunes and bedrock domes. These

features can appear to be in apparent association with anthropic mound groups. Sand dunes should (but do not always) exhibit crescent and linear shapes. This absence of a round plan view should eliminate any potential confusion about their origin. However, archaeologists have found a few mound-like dunes. In Minnesota Lloyd Wilford excavated and interpreted a low mound (<0.9 m [1 feet]) at the McAloon Mound Group (21CW3) as a natural sand dune (Arzigian and Stevenson, 2003). Small landscape features, whether composed of soil or with a rock core, are expected to be the most frequent natural mounds encountered east of the Mississippi River. In addition there are small domes of volcanic ash in Washington that were later used for human burial mounds by American Indians (Smith, 1910).

The idea that mounds form at the base of a tree by wind deposited sediments, i.e., nebkha dunes or plant obstacle dunes, is a competing hypothesis for Mima mound origins. Despite its popularity among some researchers, this hypothesis fails to explain all Mima mound situations. For example, in the southwest Lee (1986) noted that the mound fill around creosote bushes was the same as the underlying soil and thus concluded that burrowing rodents caused the mound. Nebkha dunes also fail to explain the round shape of natural prairie mounds. An early twentieth-century account of anthropic mounds in the Dakotas contains a pertinent observation.

All burial mounds of class A [ordinary round anthropic mounds] examined bore evidence of having been blown or washed out toward the southeast, as might be expected where the prevailing winds are from the northwest. (Montgomery, 1906, p. 641)

This cautionary tale should be remembered by those advocating an aeolian hypothesis for Mima mound formation as the evidence for this activity might actually be a secondary site formation process (Schiffer, 1987) rather than the primary mode of formation, i.e., the polygenetic reworking of an existing anthropic or natural prairie mound. Fowke added his perspective:

Others are convinced that they are formed by the piling up of earth around a bush, clump of grass, stone, or other object acting as a nucleus around which wind-borne material may accumulate—overlooking the fact that clay, gravel, or gumbo soil cannot be carried by wind, and that lighter soil or sand will form elongated instead of circular masses. (Fowke, 1922, p. 163)

In the Upper Midwest there are annual examples of wind-created mounds, ridges, and dunes caused by an unconsolidated deposit of drifting snow. The patterns of these snow features, even when formed around vertical obstacles such as tree trunks or fence posts, represent an infinite array of shapes and sizes. This problem is best addressed with the viewing and sampling of an entire cross-section profile wall across a Mima mound (Horwath and Johnson, 2006; Horwath Burnham et al., this volume, Chapter 4). In contrast an approach using small diameter coring and impressions based on surface morphological similarities provides a limited perspective.

Sand blows in eastern Arkansas and southeast Missouri represent earthquake-related features that experienced observers can easily differentiate from natural prairie mounds. The identifying attributes of sand blows are their sandy composition, a sand dike, and a concave shape. Mima mound sediments mimic the topsoil and have a convex (dome/conical) shape (Fuller, 1912).

The literature has infrequent instances of "natural mounds" used as a synonym for bedrock outcrops. In geology the term "mound" is applied to isolated hills regardless of their size as well as various elevated hill, knob, bluff line, or bedrock exposure landforms. During their Minnesota mound excavations from the 1930s to the 1960s, Lloyd Wilford and Elden Johnson occasionally encountered mound-like earth-covered bedrock domes in surveys and excavations. In Washington County, Minnesota, the informant H.A. Haskell reported mounds that proved to be earth-covered limestone domes (Wilford,

1949). At the Steele Mound Group (21SC24) in Scott County, Minnesota, Elden Johnson identified one mound (unnumbered by Lewis) as a soil-covered limestone dome. This feature was one of several noncultural mounds at an anthropic mound group (Arzigian and Stevenson, 2003). Natural features (with a sufficiently deep soil profile) in the uplands are likely to express well-developed soils. In contrast anthropic mounds have Inceptisols and Entisols (Bettis, 1988; Parsons, 1962). The process of soil formation in natural prairie mounds begins after the bioturbators leave (possibly occurring more than once); thus if this happens within the last two millennia the soils are recent, and if mid-Holocene or older the soils will be well developed.

A number of features found in glacial topography have the potential for confusion with natural prairie mounds. In each case these features have a size range in excess of the dimensions of Mima mounds, and only the small specimens could pose an interpretive dilemma. The hummocky and semiclosed depression topography of the Des Moines Lobe exhibits hillocks, knolls, and kames as well as other glacial features that are mostly greater in size including ridges, low escarpments, and isolated moraine deposits (Bettis et al., 1996). Low elevation examples of glacial hillocks and knolls are readily identifiable in profile as the base of the A horizon follows the undulations in the ground surface (Eyles et al., 1999) instead of having a thickened expression at a surface "bump" for a mound. Closed depressions on the Des Moines Lobe are less frequent (Bettis et al., 1996), and Mima mounds can be associated with this landform (Buol et al., 2003). Isolated segments of low morainal ridges and eskers may include mound-shaped examples. At the Steele Mound Group (21SC24) in Scott County, Minnesota, Elden Johnson identified three mounds (Mounds 8, 11, and 18) as "moraine nodules." These mounds were between 0.3 and 0.75 m (1–2.5 feet) in height (Arzigian and Stevenson, 2003).

The name "prairie mounds" has been applied to at least four distinct phenomena—Mima mounds (but not in this chapter, which uses "natural prairie mounds"); mound-like glacial till features of glacial origin; natural rises, knobs, knolls, and ridges; and prehistoric burial mounds. The piles of glacial till that collapsed into cone- or doughnut-shaped piles through wasting ice are called "prairie mounds"" in the Dakotas and the adjacent Canadian prairie (Bik, 1967; Gravenor, 1955). The majority of the Gravenor prairie mounds are too extensive—up to 15.2 m (50 feet) tall—to be mistaken for Mima mounds. In addition, for the small Gravenor prairie mounds their glacial till composition would readily remove them from consideration. Frost mounds are both permanent and seasonal (perennial) in duration. Two types of periglacial frost mounds, pingos and palsas, once existed in the Upper Midwest. Pingos represent earth mantled ice blocks in a periglacial environment and collapsed fossil pingos are marked by a ring of soil. Their distinctive shape ensures that pingos are recognized as glacial features (Flemal, 1976). The second periglacial frost feature, palsas, consist of peat-covered ice lenses with a low dome shape caused by the ice expansion. Fossil palsas are ponds (French, 1993). The fourth use of "prairie mounds" in the literature occurs as a synonym for prehistoric burial mounds (e.g., Blackmar, 1912; Peet, 1892).

Finally, another periglacial phenomenon occasionally confused with Mima mounds is the patterned ground features associated with cold climates. In some situations gophers can expose stone lines (or stone pavement) between Mima mounds (Cox and Hunt, 1990). In periglacial environments ice frost wedges can form polygons and circles on roughly level terrain (as well as frost mounds) and stone stripes on hillslopes (Kessler and Werner, 2003; Walker et al., 2008). Various periglacial processes have been dismissed as the creator of Mima mounds because they also occur on the Gulf Coast and in southern and Baja California beyond the range of past glacial events. Although Mima mound researchers never achieved a consensus concerning their origin, it has been assumed that a common ancestry existed, and thus arbitrarily ruled out equifinality (i.e., parallel evolution) of a similar

morphological form by unrelated causal mechanisms. More recent investigations have recognized polygenesis and hybrid Mima mounds in different regions (Johnson and Horwath Burnham, this volume, Introduction; Vogel, 2005).

Identifying Natural Prairie Mounds

When a researcher wrote that Mima mounds, natural prairie mounds, or a closely related variant term are present, then that statement is almost always conclusive. For other names the site description must be subjected to a thoughtful assessment. Application of the term "small mound" was never considered a sufficient identification in the absence of supporting evidence. However, small mounds reported in the literature that are described as being culturally sterile (no artifacts or human remains), when combined with other lines of evidence can be possible or probable natural prairie mounds. The site setting can be indicative when the landscape position is low relative to the surrounding topography, has a wet setting, or both (Table 1). The following quotation is from a book notice where finding the detailed descriptions in the original source clarified the situation, thus revealing the presence of probable natural prairie mounds.

Some space is also given to descriptions of Indian mounds, usually circular in form, but thus far found to be barren of implements or other remains, and occupying the most elevated and picturesque elevations. No conjecture is made in respect to their character or the purpose for which they were intended by their ancient builders. (Anonymous, 1868, p. 208)

On the other hand, the following examples reflect an antiquarian objective of museum-quality artifact acquisition. The first example does not mean that 80% of the excavated mounds were culturally sterile. Because the excavator's goal was retrieving fancy grave goods, neither description, or similar ones, contains sufficient information that could permit these sites to be included in this chapter.

... the Rev, Mr. Gass, who has spared no time or labor, and who has recently presented his report of the exploration of 75 mounds within the year, only one fifth of which afforded any relics for the museum, though the investigations are always instructive, and many facts are thus learned. (Pratt, 1883, p. 155)

The country [Louisa County, Iowa] is here everywhere dotted thickly with ancient mounds. Many of them have already been explored by parties of the neighborhood and by strangers, as I learned, comparatively few relics have been discovered. (Blumer, 1883, p. 132)

In the Upper Midwest natural prairie mounds can be isolates or occur in groups numbering a few or many mounds. Natural prairie mounds can appear either as unpatterned or arranged in linear patterns (D.L. Johnson et al., 2008). At two mapped Iowa sites, Kalsow Prairie and the Hovey Mounds (13HK10), the mounds are arranged according to the local topography and thus appear to be without a pattern. When the site topography permits, the natural prairie mounds appear in regular patterns including apparent rows. When occurring in large moundfields, natural prairie mounds typically are patterned ground microfeatures. Stated another way, the presence of a row, or rows, of mounds does not automatically mean that the site can be deemed anthropic in origin. By contrast Mima moundfields in the far west and south-central regions number in the hundreds and often in the thousands as patterned ground microfeatures. Published claims in the Upper Midwest literature noting the existence of numerous mounds (e.g., $n > 100$) is indicative of prairie mounds but not always conclusive. With a few exceptions, e.g., the Bryan, Silvernale, and Diamond Bluff sites in the Red Wing locality at the head of Lake Pepin on the Mississippi River, the upper threshold of counts for anthropic mounds does not appear to exceed 100–120 per site. Sites with more than 120 mounds, particularly when small low mounds exist, should be considered probable sites for containing natural prairie mounds even when known anthropic mounds are present. Descriptions where the mounds continue "for miles" along a waterway or across the prairie refer to Mima mounds (e.g., Lewis, 1881–1895a, Field Notebook 41, p. 5; Snyder, 1890). The following two examples illustrate the abundance of natural prairie mounds seen by nineteenth-century observers. These examples differ only in that one posits a natural origin and the other incorrectly ascribes a human origin.

In passing over the prairies of this county [Floyd County, Iowa], especially during the earlier days, the attention is arrested by the great number of isolated or grouped mounds which are seen over the surface. The origin of these "peculiar structures" is a mystery to most people; they believing them to be "Indian mounds" or even "ancient muskrat houses." By far the greater number of these mounds owe their origin to the pocket gopher (*Geomys bursarius*), which year by year made additions to them by the dirt brought forth in the extension of their underground channels, until they finally assumed the proportions now seen. The pocket gopher is abundant with us and is usually considered a great pest. (Webster, 1897, p. 15)

There are evidences of tilling the soil, of quite a novel character, which still exist in prodigious numbers, not only in Missouri but also in other regions west of the Mississippi. I have heard of very few east of that river. These works consist of low circular elevations, generally two or three feet [0.6–0.9 m] above the level of the natural surface of the soil, with diameters varying from two to sixty feet [0.6–18.3 m]; all are round, or nearly so, sloping off gently around the edges. All that I have seen among the Ozark hills are composed of black alluvial soil, and disclosed, when excavated, no implement or relic of any sort. Their presence may always be detected in cultivated fields when covered with growing crops, by the more luxuriant growth and deeper green of the vegetation. ... Residence sites they could not have been, or they would have contained some relic of stone or bone, or fragment of pottery, or at least the ashes of the family fire. ...

From all that can be learned about them, I see no reason to doubt that they were erected for agricultural purposes, and have therefore presumed to name them Garden Mounds. (Conant, 1879, p. 65–66)

Another source of potential confusion arises from the occasional presence of prehistoric artifact surface scatters on the ground surface of natural prairie mounds. This phenomenon is essentially unknown in the Upper Midwest. In the following examples, the use of Mima mounds for small-scale surface sites placed on an existing natural landform appears related to subsistence activities focused on the adjacent wetlands. This activity occurs in the alluvial bottom of the Mississippi River in southeast Missouri where the pre–Euro-American settlement floodplain was often swampy in nature. During the Late Woodland Barnes phase these elevated locations were frequently utilized for small temporary surface sites (O'Brien et al., 1989; Price, 1977; Price et al., 1975). The Western Lowlands and Grand Prairie of eastern Arkansas have prairie bumps with some surface evidence of use as sites (Morse and Morse, 1983). Use of pimple mounds by prehistoric peoples in the Gulf Coast region is known in several localities (D.L. Johnson et al., 2008). In their excavation of three pimple mounds, Aten and Bollich (1981) discovered artifacts in the biomantle as well as recovering prehistoric ceramic and lithic scatters on the mound surface. Based on the earlier ages for the diagnostic artifacts in the biomantle, they estimated that the mound aggradations took 300–500 years each to accumulate and further observed that not all mounds at the site had formed at the same time (Aten and Bollich, 1981).

Archaeological and Sedimentological Mound Signatures

The literature review revealed a continuing problem for potentially confusing the small dome- and conical-shaped anthropic and natural prairie mounds in the Upper Midwest. This predicament is not common on the West Coast or in the south central Gulf Coast regions. What factors are different in the Upper Midwest? The greatest distinction is that the Upper Midwest has thousands of small anthropic burial mounds. By contrast on the West Coast prehistoric burial mounds are less common and the south central Gulf Coast anthropic mounds are

typically larger and found at an obvious archaeological site. Thus the potential for mound confusion remained a continuing problem in the Upper Midwest.

In the Upper Midwest, human burial mounds have a two millennium time depth (Arzigian and Stevenson, 2003; Birmingham and Eisenberg, 2000). Their shapes include conical, linear, effigy, and platform. Of these, only the small conical mounds possess the potential for confusion with natural prairie mounds. Writings about known Mima mounds indicate that these sites share a set of diagnostic characteristics that differ from those of anthropic mounds (Table 1). When a number of these attributes occur in combination then a probable assessment of natural prairie mounds can be made for the site in question. A few researchers wrote about the soil matrix and mound shape, particularly if the matrix or shape was out of the ordinary. When these observations are recorded, the site in question might be assignable to a natural or cultural origin. A surprising number of the accounts readily explain the landscape position that was typically low and often wet or seasonally wet. A mound's small size is not by itself conclusive and must instead be considered in the context of its other attributes (Table 1). Those numerous archaeological articles and reports written about obvious, and probable, anthropic cultural mounds were ignored for the purposes of this chapter. Human-made mounds were identified by the presence of artifacts, burials, and a soil profile indicating loaded fill (defined below) for the construction method. The presence of artifacts encompasses both exotic grave goods as well as ordinary refuse consisting of ceramic sherds, lithic debitage, floral remains, and faunal items. The presence of ceramic vessel fragments and debitage in a mound, particularly if the sherds exhibit weathering on their broken surfaces, should represent a secondary deposit (Schiffer, 1987) occurring in a redepositional event as they were likely incidental materials incorporated into the fill. Conversely the occurrence of only one or a few items, particularly if surface finds, would not be sufficient to conclusively identify an anthropic mound in the Upper Midwest.

Occasionally an archaeological report expressed surprise when the mound excavations produced no human burials or grave goods. For these cases, such a statement is not sufficient proof whether artifacts may or may not have been present. Some early writers only discussed spectacular finds, e.g., whole pots, copper, ear spools, projectile points, cache blades, and marine shells. In general, reading more of that researchers work will indicate whether they typically mentioned all artifact classes when present in an excavation. On the other hand, reports noting no finds were present often mean just that. Meager burial mound finds were not published as reliably or consistently as those that yielded interesting materials. In the literature search certain papers written by experienced archaeologists were found that simply commented on the inexplicable absence of artifacts in a group of small mounds. Again, such a statement by itself is not conclusive; however, an absence of an associated artifact assemblage—or present only in a mundane, small quantity—certainly makes these sites candidates for natural prairie mounds. When occurring in combination with other attributes, such as landscape position and small size, this leads to a reasonable hypothesis of Mima mounds for the site in question (Table 1).

Each human-made mound usually exhibits one or more interments, although this is not always the case (e.g., McKern, 1930; Salkin, 1976). In conical mounds, the burial location is either centrally located on the former sod surface, or in a submound floor pit extending below the original ground surface. In the literature a documented absence of human burials is the best and most reasonable indicator of natural prairie mounds. Only 5% of the documented mounds in Minnesota conclusively failed to yield human remains from excavation contexts. Stated another way, ~78% of excavated mounds are known to have possessed burials. The remaining 17% of the mound sample fell into an indeterminate category due to either the small size of the excavation, a lack of adequate existing documentation, or both possibilities

(Arzigian and Stevenson, 2003). An examination of 496 reported mounds from Wisconsin suggests a similar figure—71%—for mounds having yielded human burials (Riggs, 1989). If better documentation existed for all mound excavations in both states, the number yielding human remains would likely approach 90% or greater. Another potential problem is the poor preservation of bone in certain soil and environmental conditions for some sites. Also, human burials in mounds could include intrusive inhumations made during the Historic period. Not all nineteenth-century investigators would have recognized the intrusive nature of such burials placed into older tumuli.

The identification of anthropic mounds involves more than the presence of burials and artifacts (Table 1). There is a distinctive sedimentological signature of construction activities representing a liturgical sequence in human-made mounds (Buikstra et al., 1998; Sherwood and Kidder, 2011; Van Nest, 2006). An experienced excavator of mounds in 15 states commented:

During the past forty-two years my field parties have examined some hundreds of mounds. Instead of uniform construction, there are variations not merely in the whole river valley, but even in the mounds of one group. (Moorehead, 1929a, p. 547)

Variation, and not uniformity, remains the predominant theme for anthropic mounds. The total possible range of mound strata variation remains beyond the scope of this chapter where the potential confusion exists only for the small conical mounds. For this reason, the following discussion of mound fills is oriented toward the anthropic small conicals of the Upper Midwest typically built in only one or two episodes. The construction technique for creating a small conical mound superstructure (i.e., aboveground portion) is more elaborate than a least-effort principle of scooping up the surrounding surface sediments in a container and dumping the contents at the proposed burial location. There is limited evidence for the utilization of the surrounding soils. Mound structural elements can begin with the removal of the sod layer prior to building the dome-shaped superstructure followed by designating the ritual space with a thin layer of sand or loess spread across the exposed surface. In a mound, the dedicated offerings may include sediment deposits exhibiting a particular color, texture, or both, or a special source, prepared earthen surfaces and features, as well as the expected inhumations with burial pits and grave goods (Greber, 1996; Hall, 1979, 1997; Henriksen, 1965; Osborne, 2004; Owoc, 2002; Perttula, 1996). Several of these "special earths" are attributed to a symbolic recreation of the earth diver myth (Charles and Buikstra, 2002; Charles et al., 2004; Hall, 1979, 1997; Van Nest, 2006). In addition to utilizing nearby borrow pits (Bettis, 1988; Parsons, 1962) the superstructure can incorporate allogenic fills derived from outside the mound vicinity (Van Nest et al., 2001). This situation is obvious when no borrow pits are present or the mound sediments include foreign soil bodies of various colors and textures. Such a mix, mostly notably differences in colors expressed as alternating fill units (placed horizontal, oblique, or vertical) or earthen features, can have symbolic meanings within the mound (Byers, 2004) or merely represent the basketloading construction method. The fill units and earthen features interpreted as special earths and color symbolism can occur in small conical mounds as well as the Mississippian platform mounds (Barrett and Hawkes, 1919; Buikstra et al., 1998; Bullington, 1988; Byers, 2004; Greber, 1996; Hall, 1979, 1997; Henriksen, 1965; Pauketat, 1993; Sherwood and Kidder, 2011; Van Nest, 2006; Van Nest et al., 2001). This basketloading technique is typical of small human-made mounds and consequently is readily apparent in the variegated multiple color and texture appearance of a vertical cross-section view (Fig. A2-1). A caveat necessary for reviewing earlier literature is that not all excavators recognized basketloading or would have considered it a physical attribute worthy of recordation. A more telling statement was made when a report specifically mentioned the absence of basketloading in a mound.

The types of mound fill recognized for conical Middle Woodland mounds in Illinois are (1) loaded, (2) massive, and (3) stratiform (Van Nest et al., 2001). The loaded fills represent those mound construction stages exhibiting various heterogeneous fills. Three methods are suggested for creating this result, compositional loading (i.e., basketloading), sod blocks, and soil blocks. Basketloading is widely known in the existing literature because of its frequent occurrence in the archaeological record (Buikstra et al., 1998) (Fig. A2-1). By contrast the gathering and deposition of sod blocks and soil blocks as a compositional loading fill type, which conveys a basketloading perspective in profile, remains poorly recognized (Byers, 2004; Sherwood and Kidder, 2011; Van Nest et al., 2001). Van Nest et al. (2001) argued for the symbolic value of sod blocks in Illinois Hopewell mounds, perhaps for world renewal rites, as a greater consideration over their undoubted engineering properties for soil stability. Massive fills are defined by their homogeneous appearance. Examples of prehistoric Woodland period mounds built with massive fills (e.g., Bettis, 1988) are less frequent than those with loaded fills. For anthropic tumuli the massive fills remain best known for a mound cap construction stage. In addition the soil pedogenic process could eventually have masked basketloading (Van Nest et al., 2001). Examples of anthropic mounds with a homogeneous fill are uncommon compared to those that exhibit loaded fills. The types of stratiform fills are typically thin bedded and laminated fill lens. This category could be either geogenic, anthropic, or both in origin. If geogenic the lens could represent a precipitation event on a recently constructed mound surface, and if anthropic is the result of a ritual application of a sediment layer. In summary, while examples of loaded fills are best expressed in the literature, anthropic Woodland period mounds can exhibit either loaded or massive fills from one or more vertical accretion stages of construction (Arzigian and Stevenson, 2003; Bettis, 1988; Buikstra et al., 1998; Van Nest, 2006; Van Nest et al., 2001). In addition, subsurface features (i.e., burials, prepared surfaces, and hearths) can remain intact and persist long after plowing has removed the mound superstructure. Finally, certain soil fill characteristics (zoned fills, veneers) of Mississippian platform mounds are not reviewed here because their size, shape, and anthropic natures make them unlikely to be confused with natural prairie mounds (Pauketat, 1993; Sherwood and Kidder, 2011).

In contrast to the anthropic mounds, the natural prairie mounds result from pedogenic processes. This fact is expressed in the absence of human burials, artifact assemblage, ritual sediments, and often in the landscape setting (Table 1). Published examples of Mima mound profiles are notable for the overall uniformity of the mound sediments that are similar to the A horizon of the surrounding soil (e.g., Baisden et al., 2002; Borst, 1975; Horwath and Johnson, 2006). This homogeneous nature conveys the appearance of an overthickened A horizon at the mound location. Vegetation differences exist between the mound and the surrounding soil (e.g., Fig. 3). In the Van Nest et al. (2001) terminology the natural prairie mounds represent massive fills. The biomantle concept (Johnson, 1990, 2002), which emphasizes faunalturbation, serves as the best approach for explaining this phenomenon. In this approach each Mima mound consists of a point-centered locally thickened biomantle, which, when bioturbationally active, consists of mixed homogeneous sediments (Horwath and Johnson, 2006, 2007; Horwath et al., 2002; Johnson et al., 1999, 2003, 2005; Lee and Carter, 2010; Lee et al., 2004). If a gravelly soil is present, then a two-layered biomantle develops (Johnson et al., 2002). The causal mechanism is burrowing animals (Cox, 1984; Dalquest and Scheffer, 1942; Horwath and Johnson, 2006; Johnson et al., 1999, 2003, 2005; Schaetzl and Anderson, 2005). Vertical size sorting of pebbles and stones normally occurs in gravelly soils (Johnson et al., 2002). The same two-layered signature represents faunalturbation in the sediments at archaeological sites (Bocek, 1986; Erlandson, 1984; Johnson, 1989; Pierce, 1992; Schiffer, 1987). The significant role of burrowing animals and bioturbation in soils is underestimated (Darwin, 1881; Johnson, 2002; John-

Figure A2-1. Example of the basketloading type of loaded anthropic mound fill from salvage excavation at the Adler site (11S64) located in the American Bottom, Illinois (photograph from the University of Illinois at Urbana-Champaign, used with permission of the Illinois State Archaeological Survey).

son and Johnson, 2008; Meysman et al., 2006), while their potential for the destruction of crops, tree plantations, and manicured lawns is better known (Engeman and Witmer 2000). A frequent concern is the past and present distribution of burrowing animals vis-à-vis natural prairie mounds. This ignores significant possible variation in local environmental conditions. It is the local presence of a dense or wet barrier to vertical burrowing by fossorial rodents that causes mounding to occur. Examples of local environmental conditions that prevent vertical burrows are thin soils, hardpans, near surface bedrock, a high water table, and other near surface wetness situations. Under these conditions burrowing animals must dig laterally with the resulting soil accumulation growing near the nest (Ellery et al., 1998; Finney and Johnson, 2010; Horwath and Johnson, 2006; Johnson and Johnson, 2008; Johnson et al., 2003, 2005; Schaetzel and Anderson, 2005). This requirement would explain the discontinuous distribution of natural prairie mounds in the Midwestern prairies. In contrast anthropic mounds in Minnesota occur in greatest quantity along rivers and the deciduous forest (Anfinson, 1984). Further, natural prairie mounds can occur with anthropic mounds instead of always being segregated in separate groups. In the lower Mississippi River Valley pimple mounds may be found in clusters or randomly distributed across the landscape (Smith, 1996).

Natural Mound Distribution in the Upper Midwest

If an accurate distribution map had existed for the Upper Midwest, then the seismic hypothesis for Mima mound origins (Berg, 1990a, 1990b, 1991; Cox, 1990; Saucier, 1991, 1994) could not have been proposed for the entire continent. At the time of the Berg (1990a) paper the known Mima mounds in the Upper Midwest were mapped as two isolated patches centered on Waubun Prairie, Minnesota, and Kal-

sow Prairie, Iowa. The research presented in the paper demonstrates that a much wider distribution once existed for this region. More importantly, one can now discuss where natural prairie mounds occur and where they did not. The negative evidence is of equal importance for the interpretation. The southern extent of this distribution follows the Des Moines Lobe and several adjacent landforms, e.g., Iowan Erosion Surface, Prairie Coteau, from northern Iowa through Minnesota, the eastern Dakotas, southern Manitoba, southern Saskatchewan, and southern Alberta following the prairie pothole region. This pattern is sufficient to indicate that the natural prairie mound extent is not coterminus with the areas of greatest seismic activity in North America.

Minnesota

Information on Mima mounds in Minnesota derives from four principal sources: (1) a review of archaeological, geological, botanical, and biological literature; (2) Minnesota archaeological site file; (3) Minnesota Office of the State Archaeologist (OSA); and (4) modern prairie observations. In the Upper Midwest, Minnesota is notable for being where modern archaeological research most frequently encounters natural prairie mounds (Fig. 1A). In fact, the Minnesota State Archaeologist, who is charged with the protection of Indian mounds as locations of human burials, receives inquiries about a Schmidt lowland mound site every year. The Schmidt Moundfield encompasses portions of Dakota, Rice, and Goodhue counties. It represents the largest concentration of Mima mounds in the Upper Midwest. Noncultural mounds are not intentionally recorded in the Minnesota archaeological site file. Thus the majority of Schmidt's lowland mounds were never formally assigned site numbers, even though some that appeared in *Aborigines of Minnesota* (Winchell, 1911) were recorded. After the advent of cultural resources management (CRM) archaeology the conventional wisdom by the mid-1980s among archaeologists working in Minnesota was that the lowland mounds were noncultural in origin (Anfinson, 1984, 1999). For this reason, Scott Anfinson (1984) ignored the Schmidt sites when compiling a statewide summary of Minnesota mound distributions, where anthropic mound locations are dominated by sites along the major rivers and in the deciduous forests. Kenneth Wedding (1985) reviewed the lowland mound problem and concluded that the Cox (1984) hypothesis for gopher origin appeared to fit the lowland mound data set. Natural prairie mounds are found throughout western and southern Minnesota and at least one locality along the Mississippi River where special environmental conditions, i.e., near surface wetness, existed in the past.

An explanation is necessary for certain Minnesota archaeological site numbers that contain a lowercase letter in place of a number in the Smithsonian trinomial system. In Minnesota the State Archaeologist assigns the formal site numbers. Additional sites were found in the State Historic Preservation Office (SHPO) site files. The latter are informally known as "alpha sites" and have lowercase-letter designations in place of the number (e.g., 21ANg). They were recorded by Anfinson in the SHPO files only, and not in the official OSA site file. During his tenure as Municipal-County Highway Archaeologist, Anfinson collected archaeological site location information from numerous printed sources including atlases, county histories, and other historic accounts. In 1990 this information was systematically added to the SHPO site files (Anfinson, 1991, 2008, 2010). Since this information had not been field checked for either site location or current status, they were not assigned site numbers in the usual manner for reporting known archaeological resources. They represent unverified site information derived from document research not yet subjected to a ground truth survey by professional archaeologists. In particular, Schmidt's lowland mounds are deemed unverified sites.

Aitkin County

Aitkin County has one site with possible noncultural mounds. The Sandy Lake Mounds (21AK28) was initially defined as a group of 32 possible mounds in Section 31, T50N, R23W. These small dome-shaped mounds are ~0.9–3 m (~3–10 feet) in diameter and ~0.3–0.45 m (~1–1.5 feet) tall. None of the mounds were positioned on the 3.6 m (12 feet) high Sandy Lake bank edge, but some existed within 7 m (23 feet). Artifacts were not found in the lake cut bank or in a brief excavation. During the 1977 survey one mound was tested with a 0.2 × 0.5 m trench that extended 0.15 m deep. The trench was positioned on the mound edge so that 0.3 m was outside and 0.2 m was inside. The profile showed that the mound consisted of a thickened soil because the base of the topsoil continued across as a horizontal deposit and did not follow the surface contour of the dome-shape mound (Hudak and Ready, 1979; Ready, 1977). A subsequent site evaluation in 1983 enlarged the mound count to 59 at site 21AK28. The investigators cored all of these possible mounds. Of these, 30 mounds were deemed noncultural and the status of the other 29 mounds was considered unknown. The investigators favored the hypothesis that the mounds were created by ants. Alternatively these mounds were noncultural or represent historic Indian agricultural features such as corn hills (Salkin and Hendrickson, 1983). Based on negative results for artifacts and burials the Minnesota OSA interpreted these features as possible burial mounds, corn hills, or ant hills (Arzigian and Stevenson, 2003).

Anoka County

One known and two possible natural prairie mound sites are in Anoka County. Small Mima mounds are reported to exist in the Cedar Creek Ecosystem Science Reserve. This facility is owned by the University of Minnesota (Huntly and Inouye, 1988; Inouye et al., 1997).

Upham (1888a) mentioned possible natural mounds near Centerville Lake in Section 15 or 16, T31N, R22W.

Within an eighth of a mile westward from this [anthropic mound] are nine low mounds, not so definite at the edge as usual, probably from having been plowed over, each being three or four feet [0.9–1.2 m] high and about fifty feet [15.2 m] across. (Upham, 1888a, p. 425)

Site 21AN159 is a possible Mima mound discovered in a wetland setting during archaeological investigations at nearby sites prior to construction of the Anoka County Harness Track.

During the archaeological work, a possible burial mound was noted. Because the possible mound was a lone mound, had a relatively low profile, and was in a somewhat unusual location, authentication would have required extensive fieldwork including trenching or shovel testing the feature. The State Archaeologist recommended avoidance and the developer concurred. (Anfinson, 2008, p. 27)

Becker County

The Tamarac National Wildlife Refuge has possible Mima mounds in T140N, R39W (Boyle, 2010).

Beltrami County

A possible site location is represented by the mounds reported at the narrows between Upper Red Lake and Lower Red Lake by Lewis (1886) but never recorded in the Minnesota site file (Finney, 2006). Their absence from the site file might be an accidental oversight or taken as evidence for inferring a noncultural status.

According to Wilford (1956) there is an irregular sand mound on the lake shore near Waskish in the SE¼ of Section 17, T154N, R30W. This site is judged to be a possible noncultural mound. Unfortunately the area in question has been developed in the half century since Wilford's visit.

Big Stone County

Mima mounds are reported by U.S. Fish and Wildlife Service personnel for the Rothi Waterfowl Production Area (WPA) located in the S½, Section 34, T122N, R45W (Sara Vacek, 2010, personal commun.).

Brown County

T.H. Lewis twice reported natural prairie mounds here. First, in 1883 he noted their presence in the Springfield vicinity (Lewis, 1881–1895b, letter dated September 13, 1883, from Watertown, Dakota Territory). This location is roughly west of Springfield near the intersection of U.S. 14 and County Road 2.

The next year Lewis gave a specific description for a location in the New Ulm vicinity. This site is listed as Section 21, T109N, R30W (Lewis, 1881–1895b, letter dated October 2, 1884, from New Ulm, Minnesota):

It quit raining early this morning, so that ~8 oclock I started out to hunt up the mds on 20–109–30, on the Little Cottonwood. There are no artificial mds there. The one referred to in [the] note is on 21 and natural. (Lewis, 1881–1895b, letter dated October 2, 1884, from New Ulm, Minnesota)

Dakota County

There are numerous Mima mound sites reported here with site concentrations in southern Dakota; northern, central, and eastern Rice; and western Goodhue counties representing the Schmidt Moundfield (Fig. 4) mapped in the early twentieth century (Schmidt, 1928, 1937; Winchell, 1911). Some of these sites, but not all, entered the Minnesota site file. In Dakota County a few natural prairie mounds are in the northern portion of the county near the Minnesota and Mississippi rivers. The vast majority are from the south half of Dakota County. Listed in approximate order of mound frequency from south to north these are the townships of Greenvale, Waterford, Sciota, Eureka, Castle Rock, Lakeville, and Empire. The following list of sites is arranged by the rough chronological order of the surveys; however, certain lowland mound sites were visited by multiple investigations.

Five sites from Winchell's (1888a) Dakota County geology report represent probable Mima mound groups. None of these sites were subsequently included in *The Aborigines of Minnesota* (Winchell, 1911). The first three sites are in Greenvale Township, the last in Eureka Township, and the fourth in both Greenvale and Eureka. Site 21DKab consisted of "several" mounds surrounding a marsh in the E½, Section 23, T112, R20W (Winchell, 1888a, p. 100). At site 21DKac "several" more mounds were in the SW¼, Section 24, T112N, R20W (Winchell, 1888a, p. 100). Two mounds are listed in the OSA database for both 21DKab and 21DKac (Arzigian and Stevenson, 2003). The Dutch Creek Mounds (21DKad) had six mounds in the SE¼, Section 18, T112N, R20W (Winchell, 1888a). At the Border Mounds (21DKae) the site consisted of about a dozen mounds ~0.75 m (~2.5 feet) high in S½, S½, Section 35, T113N, R20W and N½, N½, Section 2, T112N, R20W (Winchell, 1888a). Finally, Winchell noted six mounds at site 21Dkaf that were 0.3–0.6 m (1–2 feet) high situated on low, moist land in S½, S½, Section 34, T113N, R20W despite much higher land being nearby (Winchell, 1888a).

The Black Dog Mound Group (21DK8) located on a Minnesota River terrace was initially published by Winchell (1888a) in his report on Dakota County geology. The George Scott farm is now the location of the intersection of STH 13 and STH 77.

A great many artificial [*sic*] mounds are found on the east side of the Minnesota in Eagan township. They are on the great-river terrace which has been described, and near the western margin of the same. They are abundant on sec. 19, Eagan [T27N, R23W], on the farms of Mr. George Scott and Mr. Thornton. By long cultivation they have been flattened out. They are of all sizes, from ten feet [3 m] to forty feet [12.2 m] across, rising from two to five feet [0.6–1.5 m] above the surrounding surface. They differ from the surrounding soil in being pebbly and sandy, and drying quicker after showers. When thus dry they are contrasted with the black color of the rest of the field. Twenty or thirty are visible on the farms of Messrs. Scott and Thornton. They are very uniform in outward appearance; occasionally they show a stone six inches in diameter, the soil generally being stoneless. They show nothing artificial, outwardly, and suggest an artificial origin only by their existence, and their resemblance to other mounds known to be artificial. They are uniform in slope, sub-circular, and are only on the highest parts of the terrace-flat, the general surface descending from them slightly east and west. They are regarded by the residents as "Indian mounds," but, according to Mr. James Slater [another landowner in Section 19], no Indian relics, such as arrow-points or pipes or hammers, were ever found in them, nor in the surrounding country. They extend along the terrace for a distance estimated as at least two miles.

This tract is well known as Black Dog, by the residents in this part of the county, and is probably the place at which major [Stephen] Long made a short stop in 1823, and which professor Keating made mention of under the name Oanoska. The Indian village which he mentions, about six miles further up, as Tetanketane, must have been not far from Hamilton. Of these mounds professor Keating makes the following note: "On the right bank major Long observed numerous ancient tumuli or artificial mounds, some of which are of large size. They occupy a considerable extent of the prairie upon which they are situated. In one part they formed a line of about half a mile in a direction parallel with the river, from which they were distant about three hundred yards. The mounds were erected at a distance of from twelve to fifteen yards asunder, and when observed from one end of the line, presented the appearance of a ridge or parapet." (Winchell, 1888a, p. 100)

The summary of site 21DK8 in *Aborigines of Minnesota* (Winchell, 1911) listed 118 mounds consisting of 101 round, 3 elongated, and 14 damaged. In 1881 Lewis mapped a total of 39 round mounds with diameters in the 4.6–6.1 m (15–20 feet) size range. Almost four decades later Schmidt (1928) reported only 24 mounds from site 21DK8.

For the lowland mounds Schmidt (1928) arranged his sites by locality within the county. The first series of sites was inadvertently listed under Goodhue County, but was actually from Dakota County. In west-central Dakota County Schmidt began finding scattered natural prairie mounds. The mound numbers increased to the south toward the

center of the Schmidt Moundfield. The first mound locality was between the Vermillion River and South Branch of the Vermillion River east of Farmington where ~50 mounds were counted. Schmidt (1928) noted that ~36 of these mounds were at the junction of Sections 25, 26, 35, and 36, Empire Township (T114N, R19W). In Lakeville Township (T114N, R20W) he found nine mounds located between Farmington and Lakeville and three mounds in Section 11. In northeast Castle Rock Township (T113, R19W) there were 20 mounds in Sections 11 and 12, and in the northwest part of the township another six mounds approximately three miles south of Farmington (Schmidt, 1928).

For Dakota County Locality 1 Schmidt grouped ~330 mounds found in six sections from Greenvale and Waterford Townships (Schmidt, 1928, 1934c). It included Sections 13, 14, 23, and 24 in Greenvale (T112N, R20W) and Sections 18 and 19 in Waterford (W½, T112N, R19W). From September 19–21, 1908, Winchell and Schmidt dug mounds in Sections 13, 21, and 24 in Greenvale Township (Section 21 is listed only in Schmidt, 1937), while the Sections 18 and 19 Waterford Township mounds were the subject of the August 25, 1916, excavations at the Simpson pasture, and Section 23 mounds on the C.A. Bierman farm were examined during 1935 (Schmidt, 1934c, 1937). Photographs of two Simpson lowland mounds were illustrated in his final publication (Schmidt, 1937).

Locality 2 consisted of six sites in Greenvale (T112N, R20W) and Eureka (T113N, R20W) townships. There were 12 mounds in Section 11, T112N, R20W, on the Thomas Moore farm and five sites on the south bank of Chub Creek (Schmidt, 1928). These five sites were (1) 14 mounds in Section 10, T112N, R20W, located 0.5 mile south of Thomas Moore mounds; (2) 14 mounds in Section 11, T112N, R20W; (3) 74 mounds positioned along the slough running from Section 2 to Section 33, T113N, R20W, between the Vermillion River and Chub Lake; (4) 7 mounds in Section 7, T113N, R20W; and (5) 12 mounds in Section 16, T113N, R20W (Schmidt, 1928).

In Locality 3, Schmidt (1928) listed four sites from Greenvale (T112N, R20W) and Eureka (T113N, R20W) townships. They were (1) 26 mounds in Section 34, T113N, R20W; (2) 20 mounds extending along Chub Creek from Section 33, T113N, R20W, to the adjacent Section 4, T112N, R20W; (3) a continuation of the previous group with 11 mounds in Section 4, T112N, R20W; and (4) 18 mounds in Section 18, T112N, R20W (Schmidt, 1928).

Six Waterford Township (W½, T112N, R19W) sites represent Schmidt's (1928) Dakota County Locality 4. They were (1) 18 mounds in Section 4 (possibly including site 21DK23); (2) 8 mounds in Section 12; (3) 13 mounds in Section 21; (4) 8 mounds in Section 18; (5) 6 mounds in Section 7; and (6) 6 mounds in Section 6 between the railroad and the wagon road (Schmidt, 1928). In a subsequent paper Schmidt (1937) mentioned mounds on the Jens Bundlie farm in the SE¼ of Section 8, Waterford Township, as a continuation of site 21DK23. The excavation of a mound at Bundlie's began with great anticipation when bones began to appear in the units. The crew uncovered a horse skeleton and Schmidt displayed a good sense of humor in his description of this find among the results.

One Greenvale Township (T112N, R20W) site was Locality 5. It consisted of 7 mounds in Section 3 (Schmidt, 1928).

There were two Sciota Township (E½, T112N, R19W) mound groups in Schmidt's Locality 6. The first was composed of 3 mounds in Section 3 [21DKj], which Markley (1948) listed in Section 2, and the second had 8 mounds in Section 12 [21DKi] (Schmidt, 1928).

In his final report Schmidt (1937) stated he had recently tested one lowland mound on the Martin Elstad farm along Chub Creek in Section 9 of Greenvale Township (T112N, R20W). Schmidt (1937, frontispiece) illustrated two mound examples from the Simpson pasture in Waterford Township. Mounds in the Simpson pasture were excavated on August 25, 1916 (Schmidt, 1934c). This location is in Sections 18 and 19, T112N, R19W (Pinkney, 1896; Schmidt, 1934c; Webb

Publishing, 1916). In addition to mounds, Schmidt (1935c) observed the remains of numerous beaver dams on Chub Creek. He also noted that lowland mounds did not occur in the beaver meadows, i.e., former ponds. This position on Chub Creek has a positive correlation with the greatest concentrations of lowland mounds in Dakota County.

Wilford's (1939a) memorandum on Dakota County dated May 16, 1939, indicated that he visited the following four sites with Schmidt. Site 21DK21, named the Boudreau Mounds or Chub Creek Mounds I, is located in a low meadow bordering Chub Creek in Section 12, Greenvale Township (Schmidt, 1937; Wilford, 1939a; Winchell, 1911). One of the mounds was measured as being 11.9 × 11.4 × 0.9 m (39 × 37.5 × 3 feet) high (Schmidt, 1937).

Group of two mounds on Chub Creek, two miles south of Castle Rock on the Everett Boudreau farm, Sec 12, R20, T112. One dug by Schmidt ([s]ee p. 11 of his 1937 report) was sterile. (Wilford, 1939a, p. 1)

The Castle Rock Slough Mounds (21DK22) are located in Waterford Township on the east shore of Castle Rock Slough immediately west of STH 65 in Section 5, T112N, R19W (Schmidt, 1937; Wilford, 1939a; Winchell, 1911). Schmidt did not provide a mound count estimate. The site 21DK22 setting is described by Wilford:

This slough is a peat bed, the bed of a former lake now drained and very acid. The mounds would probably yield nothing, as they are but little elevated above the slough. (Wilford, 1939a, p. 1)

The John Street Mounds (21DK23) in Sections 4–5, T112N, R19W, Waterford Township were reported by Winchell (1911), Schmidt (1937, p. 10), and Wilford (1939a). Mound dimensions recorded during the 1935–1936 excavations were (1) 9.1 × 8.2 × <1.2 m (30 × 27 × <4 feet) high; (2) 9.1 m (30 feet) diameter and 0.75 m (2.5 feet) high; and (3) 5.5 m (18 feet) diameter by 0.6 m (2 feet) high (Schmidt, 1937). Based on the negative excavation results site 21DK23 represented Mima mounds.

Mounds on John Street farm, Secs. 4 and 5, R19, T112. These are directly east of previous group, across highway 65 and up over a hill where the Cox cottage is. Schmidt dug some of these (p. 10 of 1937 report) and found nothing. (Wilford, 1939a, p. 1)

Site 21DKd is a large mound group positioned in NE, NW, Section 6, T112N, R19W, on a level plain adjacent to a wetland. As explained by Wilford (1939a, p. 1), "Immediately south of depot at Castle Rock station is a group of 70 mounds on the north shore of Castle Rock Slough." Schmidt (1937) stated 72 mounds existed at this site.

Two Schmidt sites (21DKi and 21DKj) were visited ca. 1948 by Max Markley, a biochemist in the Twin Cities, who had a strong interest in archaeology. His field observations remain pertinent. For site 21DKi Markley observed:

One group of eight mounds [21DKi] is located in the center of section 12 in Sciota Township about two miles due west of the village of Randolph. These mounds occupy an area of twenty acres on the north side of Chub creek, three miles above its junction with the Cannon river. The land is in grass and has never been plowed. The soil is classified as Wabash Silty Clay Loam, which is rated as one of the most productive of all corn lands. On the south side of the creek the land is a little higher, with a sparse gravel subsoil. This land, while it has a good top soil, is droughty. There are no mounds on this side of the creek, even though it is above overflow. Lower down the valley the Wabash is replaced by Cass Sand and here also mounds are lacking. To the west the Wabash is replaced by the heavier boggy Clyde soil. No mounds have been found in the Clyde area.

The mounds of Section 12 appear to be undisturbed, but Chub creek has cut through one of them leaving a vertical profile. The mound appears to have been made from the local top soil and shows no load marks. No ash appears in the profile. No artifacts are found on the top of the ground or in the body of the mound, which is not surprising in-as-much as floods each year deposit silt over the whole area. Between the main group of mounds and the creek the author

found a firehearth about one foot below the surface. This had a number of fire-blackened glacial stones. Two of these showed flattening on one side typical of the mano or hand stone. (Markley, 1948, p. 63)

A second group of mounds [21DKj] is in the southwest quarter of Section 2, Sciota Township, on the east side of Spring creek, a few rods above where it empties into Chub creek. Here are three on about four acres of Wabash soil. This area forms a low ridge along the creek with the swampy Clyde soil to the south and east and the lighter, but mucky, Kato Soil to the west. This past summer the writer trenched the largest of these mounds, which was one foot [0.3 m] high and thirty feet [9.1 m] in diameter. The mound was composed of uniform black silty clay loam to a depth of three feet [0.9 m] at the center. Below this was undisturbed blue clay saturated with water. No artifacts, ashes or load marks were found. The mound was restored. (Markley, 1948, p. 63–64)

Use of the soil survey data is an improvement in archaeological research methods over his predecessors. However, Markley stated that the mound fill was three feet (0.9 m) thick. This thickness figure must have included the mollic epipedon with the mound height and emphasizes that the mound superstructure was composed of sediments derived from the underlying solum. The Markley interpretations are problematic as he too persisted in projecting all mounds as cultural in origin. Although he admitted that Winchell's earthlodge hypothesis was not likely, use as a seasonal garden or field house was proposed as the mound function that would account for the absence of an artifact assemblage (Markley, 1948).

The Brotzler Mounds (21DKa) in SW, NW, Section 16, T28N, R22W, were recorded in 1972 as a group of plowed down burial mounds (Nystuen, 1973). The subsequent 1973 Phase 2 site evaluation consisted of test unit excavations followed by topsoil stripping, both of which produced negative results. Since the mound excavations yielded no artifacts, burials, or archaeological features, they were considered natural prairie mounds (Peterson, 1974).

Dodge County

Mima mounds occur at the Iron Horse Prairie located two miles south of Hayfield in Section 27, T105N, R17W. Measurements for the known natural prairie mounds are three large mounds ~10 × 8 m in diameter and 0.6–0.8 m in height, and four to six incipient mounds with a 2 m diameter and ~0.4 m tall. These mounds are near the west edge of the central portion of the preserve (Donald L. Johnson, 2009, personal commun.).

Douglas County

Two notices of noncultural mounds were discovered for Douglas County. Lewis wrote that the low "hills" in Sections 17, 20, 30, and 31, T130N, R39W, near Evansville were natural (Lewis, 1881–1895b, letter dated September 19, 1886, from Sauk Centre, Minnesota). These sections have a glacial hummocky and wetland topography. Low topographic features in this area could be glacial or Mima in origin.

There were lowland mounds observed at Lake Geneva (Schmidt, 1928).

Fairbault County

The Minnesota Geological Survey found two Mima mound sites in Fairbault County. According to Upham (1884a; Thomas, 1891; Winchell, 1911, p. 100) the first site consisted of numerous round mounds with diameters of 4.6–6.1 m (15–20 feet) and elevations of 0.3–0.45 m (1–1.5 feet). The total site extent continued for ~4.8 km (~3 miles) along the road across the prairie. They were positioned approximately in the middle between the villages of Freeborn and Wells. This site location would be near the Fairbault-Freeborn county line. A second mound group was observed to be 3.2 or 4.8 km (2 or 3 miles) west of Wells (Upham, 1884a). Upham believed that both sites represented artificial mounds. Instead the tremendous extent of the first site identified its natural prairie mound status, while the second one was a probable natural prairie site.

Freeborn County

According to Winchell (1911; Thomas, 1891),

Mounds in Mansfield [Township] have been noted by Upham, N.W. ¼, sec. 13 and at the south side of sec. 34 close to the [Iowa] State line. In the former place there are "two or three," and at the latter "two or more," small mounds. (Winchell, 1911, p. 96)

During fieldwork in southern Minnesota, Lewis investigated the mounds reported for Section 13, T101N, R23W, Mansfield Township, Freeborn County. His background research included checking the Government Land Office survey plat maps. This location appears to be part of a former glacial lake bed that is positioned a short distance north of the Iowa border. Based on the maps and an informant interview Lewis concluded that meadow mounds existed at the reported location.

I looked up the government plats of 101 – 23 this morning and am satisfied that the reported mds are of the "meadow" order, the country being flat and marshy. I also saw one of the owners of the land described in 13 and he states that the land is marsh. (Lewis, 1881–1895b, letter dated October 21, 1889, from Albert Lea, Minnesota)

Schmidt observed five lowland mounds positioned five miles south of Blooming Prairie. This location is in Freeborn County, or possibly Mower County, Minnesota (Schmidt, 1928).

Goodhue County

The westernmost townships of Goodhue County, along with adjacent southern Dakota and northern, central, and eastern Rice counties, comprise the Schmidt Moundfield (Fig. 4) (Schmidt, 1909, 1928, 1937; Brower, 1903; Winchell, 1888b, 1911). In particular Stanton Township had numerous small mounds in the low areas near Prairie Creek and the Stanton Flats (Schmidt, 1928; Winchell, 1911). This area was the major concentration for the Schmidt Moundfield in Goodhue County. Because the vast majority of the Prairie Creek drainage is in the adjacent county the numbers for this mound concentration (*n* = 577) are reported for Rice County. For the rest of Goodhue County Schmidt (1928) noted Warsaw Township had 10 mounds in Section 8 and Kenyon Township had another 10 mounds along the Zumbro River in Section 4. A few of the lowland mounds found their way into the Minnesota site file from Schmidt (1928, 1937; Winchell, 1911).

In Florence Township for the southeast corner of the county Col. William Colville mentioned the existence of small mounds 3 m (10 feet) in diameter and 0.6 m (2 feet) high (Winchell, 1888b). This might be a reference to the 21WB6 mounds at Lake City in adjacent Wabasha County.

Finally, at Prairie Island (named treeless or bald island by French traders), which is adjacent to the Mississippi River north of the Cannon River confluence and the Red Wing area, there are several hundred reported mounds. Prairie Island is ~10.5 miles long and 1–3 miles wide. It is (was) low and marshy on the west-northwest side. The survey of Jacob Brower (1903) reported 260 mounds and this island has numerous prehistoric habitation sites as well as anthropic mounds. The Bartron site (21GD2) excavated by Wilford is at the south end. Prairie Island also has small mounds with few or no artifacts (Brower, 1903; Schmidt, 1928, 1937). The majority of mounds on Prairie Island were clearly anthropic in origin, but given the number of small examples some could have been natural prairie mounds.

Hennepin County

Records exist for one probable natural prairie mound group (21HE27) and three cases of probable Mima mounds at an anthropic mound group (21HE3, 21HE30, and 21HE47), all associated with Lake Minnetonka in Hennepin County. The Lake Minnetonka locality possessed the largest concentration of mounds in the county. Most of these disappeared without a trace during the twentieth-century development

of suburbs on the west side of Minneapolis. In addition there is another possible natural prairie mound site (21HE19) in Hennepin County.

The information recorded by Winchell (1911) includes a probable natural prairie mound group on the south shore of Lake Minnetonka known as the Shaver Mounds (21HE27). The smallest size class at 3 m (10 feet) in diameter is of particular interest. The Minnetonka locality had numerous mounds, many of which disappeared without investigations during the urban development of the area. According to the 1873 county atlas, James Shaver, Jr., owned the W½ of the NW¼ in Section 17, the adjacent portion of Section 8 along the lake shore, and the E½, NE¼ of Section 18 (Wright, 1873).

On the land of James Shaver, NW. ¼, Sec. 17, Minnetonka, are a great many mounds. In the summer of 1875 a number of these were located by chain and compass by a party from the Minnesota Academy of Sciences. They were found to lie on the bluff and knolls overlooking the water of the lake, following the higher land without regard to direction or relative position. No plan or order was discernable, though 20 were carefully surveyed. They vary in height from two or three feet to five or six [0.6 or 0.9–1.5 or 1.8 m], and from ten feet [3 m] in diameter to forty [12.2 m]. There are in that neighborhood fifty or more within the area of a quarter section of land. (Winchell, 1877, p. 201)

At present no aboveground mounds exist at site 21HE27. A portion of the site area was tested in 2008 for a county road project. No features or artifacts were found (Anfinson, 2010).

In *Aborigines of Minnesota* Winchell (1911) identified the southwest portion of the Gideon Bay Mound Group (21HE47) found along Lake Minnetonka near Excelsior as being marsh mounds. A total of 98 mounds were reported for site 21HE47. This site is another example of anthropic and natural prairie mounds occurring in a single group. The marsh mounds were in the NE¼ of Section 33, T117N, R23W, City of Excelsior, and consisted of 27 mounds found on low or sloping surfaces on the southwest shore of Gideon Bay (Fig. A1-3). The site description indicates the existence of other comparable mound groups in the region.

Marsh mounds near Excelsior. There is a class of mounds, which differ in location from all the foregoing [anthropic mounds], in Hennepin County, being found on low ground. They are not confined to Hennepin County. Outwardly they resemble the ordinary tumulus, and are in a similar manner elongated and extended or "approached." Of these the following groups have been surveyed near Excelsior, on lake Minnetonka. (Winchell, 1911, p. 259)

Mound 10 at the Halpin Mound Group (21HE3) was excavated in July 1947 by Wilford. This mound is one of 13 round and 7 elongated mounds at the site. Two of the elongated mounds possess a lengthy linear shape. This site is also known as the Starvation Point or the Orono Point Mound Group surveyed by Hill, Robert Sweeny, and Lewis. The setting is on a narrow peninsula jutting into Lake Minnetonka between Smith Bay and Browns Bay in Section 11, T117N, R23W (Winchell, 1911). Mound 10 was 8.5 m (28 feet) in diameter and 0.45 m (1.5 feet) in height. Wilford's field crew dug a circle 6.1 m (20 feet) in diameter from the mound center. According to Wilford et al. (1969, p. 7), "The mound proved to be sterile. Nothing was found in the mound fill and it contained no burials. The only feature was the presence of a group of twenty rocks in the southeast part of the excavations at the top of the subsoil." These rocks do not represent a hearth or a cultural feature and the extent of the excavations was sufficient to deem this mound devoid of burials (Arzigian and Stevenson, 2003). Mound 10 can be considered a probable natural prairie mound found with anthropic mounds, i.e., Mound 11 (Wilford et al., 1969), at site 21HE3. In addition one of the 21HE3 mounds excavated by Sweeny in 1867 produced negative results. As summarized by Sweeny (quoted in Winchell, 1911, p. 230): "That they are not all burial mounds we are aware, since one of the largest two away from the main group contained no indication of burials, or fires, or anything suggestive of its past history or uses." This mound is a second probable natural prairie mound found with anthropic mounds at 21HE3.

Also in 1947, Wilford dug the Mound 4 at Preston Haglin (21HE30) located in Section 22, T117N, R23W, near the Lake Minnetonka shore. This area had numerous mounds (Arzigian and Stevenson, 2003; Winchell, 1911). As described by Wilford et al. (1969, p. 47) his excavation examined a "Single, low, conical mound 20 feet [6.1 m] in diameter and 1.5 feet [0.45 m] in height. No evidence of burials; no cultural materials; modern trash in the mound center." At many other sites his investigations revealed that previously looted mounds yielded at least a few jumbled human bones and artifacts (Arzigian and Stevenson, 2003; Wilford, 1970; Wilford et al., 1969). For this reason Mound 4 at site 21HE30 can be considered a possible natural prairie mound found with a presumed anthropic mound group.

Arzigian and Stevenson (2003) provide an excellent summary for site 21HE19. The location is approximately Sections 34–35, T116N, R22W.

Lewis mapped 20 mounds at this site, including 11 small mounds in the southern subgroup. In 1998, the southern mounds were no longer visible, and geomorphological work did not encounter soil features that would be expected of mound fill or mound construction. The mound loci were interpreted as either natural landforms or the remnants of recent historic activity (evidence of historic pits, debris, and intensive cultivation were present) (Kolb 1998; Madigan and Bailey 1998). Subsequent archival research indicated that this had been the original location of the town of Hennepin, platted in the 1850s, abandoned in the 1870s, and returned to cultivation by the time Lewis mapped it in 1882. Lewis noted not historic features or disturbances. The landscape features Lewis mapped might have been historic or natural features. (Arzigian and Stevenson, 2003, p. 394)

Kandiyohi County

The Green Lake Mounds (21KH8) is an anthropic group mapped by Upham (1888b; Winchell, 1911) that likely includes noncultural mounds. This site is on the east side of Green Lake in Sections 29 and 30, T121N, R33W. According to Upham it consisted of 31 round mounds, 6 oval mounds, and 2 linear mounds. In contrast the OSA database lists 69 mounds for 21KH8 (Arzigian and Stevenson, 2003). The sons of the landowner William Taggart dug in two of the larger mounds. The first was:

… the northernmost of the two mounds which are marked on the map [refers to Upham site map] as six feet [1.8 m] in height, and found within it a hollow chamber, dome-shaped, about three feet [0.9 m] high, with a flat floor, which was on a level with the base of the mound. The mound marked with the height of 5 feet [1.5 m] has also been dug into, but it appears that no bones nor implements were found in either. (Upham, 1888b, p. 242)

During his archaeological investigations Wilford encountered two Mima mound groups in Kandiyohi County. The first was an informant reported site that he visited while excavating at a nearby human burial mound. This one is a known noncultural site.

In July [1951], I dug the [anthropic] mound at the north east side of Nest Lake [21KH2], camping at Hillman's. I was visited by [Ray] Svobodny two or three times. On July 12 he took me to see a group of mounds in a low marshy area SW of Pennock. Very difficult to reach because most of the roads leading to it were blocked by water. The mounds [no site number] look like low hummocks to me, and were certainly poorly situated for burial mounds. Later Ray tested these and found nothing. These are the mounds he had reported in his letter of Dec. 6, 1951, as in Sec. 8, T.116[N]-[R]36[W]. (Wilford, 1951a)

At site 21KH9 Wilford's excavation identified marsh mounds in Section 8 of T119N, R36W. He tested the site in 1952 and found the mounds to be culturally sterile (Wilford, 1951a).

Kittson County

A recent prairie guide lists the presence of natural prairie mounds at the 14,500 acre Wallace C. Dayton Conservation and Wildlife Area. They occur at a density of about one Mima mound per one acre in the wet prairies (Jones and Cushman, 2004). This park is a unit of the Tallgrass Aspen Parklands in northwestern Minnesota. The tallgrass-aspen parklands ecosystem is a mixture of tall grass prairie, marshes, and deciduous forest in adjacent portions of northwest Minnesota and southern Manitoba.

A second Kittson County site is represented by a small oval mound excavated in 1961 by Gary Hume from the University of Minnesota. Mound 2 at site 21KT2 was described as being 3 × 1.5 m (9.8 × 5 feet) in plan view and 0.3 m (1 foot) tall. According to the profile map the fill was homogeneous black sandy humus. No burials or artifacts were mentioned in the field notes. Hume interpreted Mound 2 as a natural bump on the ground surface and indicated that others were present at the site. Mound 2 is a possible natural prairie mound. It should be noted that Mound 1 at site 21KT2 was an anthropic mound (Arzigian and Stevenson, 2003).

McLeod County

One known and one possible Mima mounds sites are in McLeod County. There are Mima mounds listed among the natural features at the Schaefer Prairie owned by The Nature Conservancy. This location is seven miles southwest of Glencoe. The Mima mounds are listed as being part of the wetland complex at Schaefer Prairie (Anonymous, 2009). For the second site, Wilford noticed a different set of probable natural prairie mounds near Glencoe.

On Mar. 19, 1938 riding on the C.M. &St.P. [railroad] I noticed two distinct low mounds in a cultivated field adjoining the trunk on the northside. This was 5 min before reaching Glencoe, hence 3 or 4 miles east of that point. (Wilford, 1938)

Mahnomen County

Located in west-central Minnesota, Mahnomen County is the center of a Mima mound concentration. The 640 acre Waubun Prairie Research Area is an unplowed prairie tract used for scientific research by the University of Minnesota. The site is famous for the Canadian toads (Bufo hemiophrys) that burrow and overwinter in the Mima mounds. At Waubun Prairie the ~200 Mima mounds consist of black silt loam over dense yellowish clay subsoil, range from 3 to 40 m in diameter, and from 0.15 to 1.3 m in height (Davids, 1968; Ross et al., 1968; Tester, 1995; Tester and Breckenridge, 1964).

A number of individual or isolated natural prairie mounds scattered throughout the county are locally known as "Mahnomen mounds." Examples of their locations are (1) at the corner of STH 113 and U.S. 59 in Waubun; (2) on the south side of County Highway 1, a road that extends east and west from Bejou; and (3) west of Bejou and Santee Prairie.

A 1999 visit resulted in the verification of intermediate-sized Mima mounds at Santee Prairie in Sections 6 and 7, T145N, R41W (Anonymous, 1979; Donald L. Johnson, 2009, personal commun.).

The Naytahwaush Mounds (21MH6) in the NW¼, Section 28, T144N, R39W, were reported by Elden Johnson in 1970 as consisting of six conicals, two elongated, and one more mound located one-quarter mile away. Johnson found no artifacts in the numerous gopher holes and backdirt piles around the mounds or on the ground surface between the mounds and North Twin Lake. This site is located on the White Earth Reservation. A Phase I archaeological survey for a sewer line project in Naytahwaush produced no cultural materials or evidence of burials in several disturbed mounds at site 21MH6. The report noted that local residents had impacted one or more mounds as the village expanded (Goltz, 1993). The absence of any reported artifacts or burials at 21MH6, combined with an apparent lack of concern shown by the American Indian residents of Naytahwaush, suggests that at least some noncultural mounds are present.

Meeker County

A possible noncultural mound site was reported during the late-nineteenth-century Euro-American settlement of Meeker County. It was located in Litchfield Township, T119N, R31W. An examination of the USGS 7.5 min quadrangles reveals the presence of marshes dominating six square miles west of the city of Litchfield with more marshlands scattered through the township south of the municipal limits. The primary source is a centennial county history that referred to Meeker County as part of the Big Prairie located west of the Big Woods (Smith, 1877). According to Smith (1877, p. 17–18): "There are a number of small mounds simulating Indian mounds in this town[ship], mostly in the timber, and evidently of great age. None have ever been explored." A notice of this site appeared in the Smithsonian annual reports (Cramy, 1880; Winchell, 1911).

Morrison County

According to Winchell (1878) mounds were found around Frieze Lake east of Little Falls. This site setting has an extensive marsh on three sides of the lake.

About fifty-five earthworks or mounds are found one the border of a small lake on Sec. 35, Belle Prairie [Township] and Sec. 9, Little Falls [Township] six miles east of the village of Little Falls. They follow round the shore of the lake, which is known by the Indians as "The Lake between the Hills." (Winchell, 1878, p. 59)

The final report on Morrison County by Upham (1888c) revised the site location to Section 35 in T41N, R31W, which would indicate that either Frieze Lake or Pelkey Lake was the "Lake between the Hills." He also mentioned additional mound groups one to two miles distant near Rice Creek that included probable anthropic as well as additional probable natural prairie mounds. Upham's informant was Nathan Richardson (Upham, 1888c; Winchell, 1911). These mound groups might be sites 21MO2, 21MO3, and 21MOx.

Winchell (1878; repeated in Upham, 1888c) also identified numerous small mounds on the east bank of the Mississippi River at Little Falls. This location is part of the famous Little Falls site that was the subject of a long discussion in the literature whether the human occupation was preglacial in age, i.e., Paleolithic, or not; and a small portion of the controversial debate on this topic for North America (Finney, 2006; Meltzer, 2009). That debate does not concern the present chapter. Winchell's Little Falls mounds, which he recognized as postdating the artifacts for the Paleolithic debate, are worthy of further attention.

The mounds themselves are somewhat different from those seen elsewhere, inasmuch as they consist of low, circular ridges, from eight to twelve feet [2.4–3.6 m] across, rising but two to three feet [0.6–0.9 m] above the general level. These are scattered over a small distance on the east bank of the [Mississippi] river … . They may have been designed for habitation, having been formed at first by slightly excavating the surface of the ground, and then building rude arched coverings supported by wooden branches and enclosed by earth. As these decayed and fell in, the resulting forms would be exactly what are now seen. (Winchell, 1878, p. 58)

When William H. Holmes of the Smithsonian Institution visited the Little Falls site, he recognized Winchell's mounds at 21MOad as being tree falls, i.e., floralturbation. Holmes was far ahead of his late-nineteenth-century contemporaries in understanding archaeological site formation processes and chipped stone artifact reduction sequences (Finney, 2006; Holmes, 1893; Meltzer and Dunnell, 1992; Schiffer, 1987). Despite the Holmes (1893) identification of the mounded tree falls, Winchell (1911) persisted in considering these features to be mounds. Further, Winchell stated that he had visited the site twice since the original observation in 1877 and could no longer see the mounds, a fact which was attributed to the development of Little Falls (Winchell, 1911).

Mower County

There are two separate documented Mima mounds occurrences in the eastern half and one in the western half of Mower County. Despite believing that all small mounds were the work of humans Winchell

(1873, 1884a, 1911) encountered a natural prairie mound group so extensive that it caused a pause to reconsider, at least for this site. It contained more than an estimated 500 mounds extending for more than three miles across the prairie. This location is in eastern Mower County, probably in T102N, R14–15W.

There is a multitude of mounds on the high prairies between Grand Meadow and Le Roy, which, were it not for their great number, would be unhesitatingly pronounced artificial. They are first seen surrounding a marsh about a quarter of a mile across, about two miles and a quarter south of Grand Meadow. About twenty are here visible, rising each about two feet [0.6 m] above the surface. Farther south they increase in number, extending three or more miles toward the south and southwest. Probably five hundred could be counted, some being five feet [1.5 m] high. They are scattered promiscuously over the upper prairie. The surface has the appearance of having been poorly drained formerly, and was perhaps covered with shallow water till late into the summer season. These mounds have the popular reputation of being "gopher knolls." It is thought that they occur where the ground is wet and the clay near the surface. Yet, south of the region designated they do not exist, though there is no apparent difference in the prairie. The material of which they consist is the ordinary loam of the surface soil. Several of them have been removed, when near the highway, and the material hauled into the street for grading. There is no record or knowledge of any human bones or other relics having been found in them. (Winchell, 1884a, p. 365)

Cyrus Thomas (1891, p. 120) mentioned this site in his compendium of mounds in the Eastern United States and added the comment "probably natural." In *Aborigines of Minnesota* Winchell (1911) added a postscript to the above site description:

Since then nothing further has been learned of these mounds, but the writer has been inclined to regard them as not of aboriginal origin. (See under Rice County.)

It is the opinion of Mr. C. L. Webster, as published in the *Charles City Intelligencer*, May 7, 1889, that mounds of this character, which are common in Iowa and southern Minnesota, and which have been denominated "prairie mounds," owe their origin to the pouched gopher (*Geomys bursarius*) which, by the extension of its underground channels, year by year brought forth the dirt until the mounds assumed the shapes and sizes which they now possess. When they were abandoned by the removal or death of the gophers, the indigenous plants of the region took possession, and here maintain a ranker growth than is usual under other conditions.

These mounds are not so located as to be characteristic of the mound-builders. (Winchell, 1911, p. 95)

Another Mima site reported for Mower County was the existence of lowland mounds along the road between Austin and Faribault (Curtiss-Wedge, 1911). This location would be in the northwest portion of the county.

In the northeast corner of Mower County a portion of the Racine Prairie is extant in a Minnesota scientific and natural area along a former railroad line parallel to U.S. Route 63. Small Mima mounds are visible in Sections 14 and 23, T104N, R14W (Donald L. Johnson, 2009, personal commun.).

Nicollet County

Three mound-like features at 21NL119 were mapped in 1997 by Bruce Koenen for the OSA. The site setting is on the upland plateau near the bluff edge on the west side of the Minnesota River Valley in the SE, NE, SE, Section 11, T109N, R27W. According to the site form, Mound 1 was 9.75 m (32 feet) in diameter and 0.45 m (1.5 feet) in height. Mounds 2 and 3 were measured as being 7.6 m (25 feet) in diameter and 0.3 m (1 foot) high. When Koenen cored these mound-like features the soil profile revealed an overthickened topsoil above the subsoil. For this reason it was concluded that the 21NL119 mounds have a probable noncultural origin (Bruce Koenen, 2009, personal commun.).

A recent survey for a proposed Trunk Highway 14 bypass discovered three groups of Mima mounds from prairie hay fields in the NE¼, Section 9, T109N, R28W. These sites are in a level upland setting near a wetland southwest of Nicollet. The mounds are grouped in clusters of two, nine, and four mounds, each separated by more than 300 m (Terrell, 2005). After site visits by Mark Dudzik and Scott Anfinson, their noncultural nature was sufficiently established that no site number was assigned (Dudzik, 2005).

Nobles County

In his report on the county geology, Upham (1884b) reported two sites containing what he believed were artificial mounds. Site 21NO1 consisted of circular mounds that were 0.45–0.9 m (1.5–3 feet) tall from the NW¼ of Section 18 in Ransom Township, T101N, R41W (Thomas, 1891). A second site was noted for the south half of Little Rock Township (T101N, R42W). The Little Rock River drains the level to slightly undulating uplands in these townships. This locality was subsequently visited by Lewis in summer 1889 and he identified the first (21NO1) of the two Upham sites for Nobles County as being noncultural in origin.

Yesterday I got 1 mds S of Rushmore but not a "Winchell md," his being meadow mds on NE and SW of [Section] 18. (Lewis, 1881–1895b, letter dated August 23, 1889, from Heron Lake, Minnesota)

No specific comment was made about the second Upham site, but a reported extent encompassing the south half of a township, rather than listing individual sections, strongly implies the presence of an extensive natural prairie moundfield.

Norman County

Donald and Diana Johnson (2009, personal commun.) have observed noncultural "pseudo-mounds" near Twin Valley in the eastern half of the county. This locality is adjacent to Mahnomen County.

Ottertail County

Hundreds of mounds, usually known or considered to be anthropic, were found around and near the numerous lakes and wetlands in Ottertail County. Upham (1888d) noted many mound groups but began his description with the following observation: "In the townships of Everts, Girard, Amor, Perham and Star Lake, numberless mounds are found of every shape and size together with many sites of prehistoric villages" (Upham, 1888d, p. 560). It remains probable that among the "numberless mounds" are noncultural mounds.

During an 1886 survey in eastern Ottertail County, Lewis found probable noncultural mounds on the prairie. He told Hill: "There are no mds on Pease prairie near Henning station. They are only natural hillocks" (Lewis, 1881–1895b, letter dated September 6, 1886, from Wadena, Minnesota). Near the center of the county he reported: "There are plenty of natural mds on 28 & 29 – 134 – 41, but I could not find any others" (Lewis, 1881–1895b, letter dated September 14, 1886, from Fergus Falls, Minnesota). Section 28, T134N, R41W, has a huge marsh along the Otter Tail River, while Section 29 contains West Lost Lake, glacial ridges and knobs, and small wetlands. It is probable that the Section 28 mounds are of the Mima variety but those in Section 29 could be Mima, glacial, or both.

Pipestone County

Reportedly, small mounds occurred in the vicinity of Pipestone National Monument as related by various nineteenth-century observers, and subsequently were not mentioned by numerous later visitors. As summarized in a park archaeology overview, "… mounds are so poorly known at Pipestone national Monument as to be regarded as unconfirmed and dubious features" (Scott et al., 2006, p. 198). Any of these "mounds" that occurred near the pipestone quarry most likely represent soil and grass-covered piles of rock derived from quarrying activities (Arzigian and Stevenson, 2003). As for other mounds positioned away from the diggings, the geologist Charles White (1869) mapped a wide

marsh along Pipestone Creek approximately a mile from the quarry. If they occurred in the marsh setting, the presence of natural prairie mounds can be considered a probable interpretation for these reported mounds that later disappeared (Scott et al., 2006, p. 203).

Wilford found a possible Mima mound site in Pipestone County. He examined a site reported to be southwest of Edgerton and on the east side of the Rock River at the Tinklenberg farm. In May 1954 Wilford went to verify a reported mound group but wrote that they were not mounds—meaning he saw something other than human burial mounds. In the absence of any statement about bedrock outcrops or glacial landscape features then natural prairie mounds remains the most likely explanation (Wilford, 1954a).

Polk County

Two natural prairie mound locations are known in Polk County. Mima mounds occur at the Malmberg Prairie in Section 16, T149N, R48W, and the Pembina Trail Preserve in T149N, R44–45W (Donald L. Johnson, 2009, personal commun.).

Pope County

A total of four probable or possible natural prairie mound sites existed in Pope County. During a geological reconnaissance survey made to Pope County in May 1884 Winchell (1885) noted several mound groups, all of which he believed to be artificial. Of particular interest to this chapter are those found along Ashley Creek in the northeast corner of the county and mounds from the Glenwood vicinity near the county center. The Ashley Creek valley mounds are probable natural prairie mounds and more of the same occur among the cultural, natural, and glacial features described for the Glenwood mounds locality.

Near the county line between Stearns and Pope counties, along the valley of Ashley Creek, are a great many artificial mounds of earth. They are the north side of the railroad accompanying a marshy tract. (Winchell, 1885, p. 13)

Twenty or more other artificial mounds [site 21PO1] are on the land of Dan. F. Bartke, SW 1/4, sec. 2, T. 125, 38, a short distance west of Glenwood. On in this vicinity is known as White Bear Mound [21PO4]. This rises ~200 feet [61 m] above the lake [Minnewaska], but is situated on a natural conical hill. This is on the north side of the lake, about three miles from Glenwood. Numerous other mounds are on the low land [no site number], southwest of the White Bear mound, on the north side of Pelican lake; also north of White Bear mound, and northwesterly, scattered over the upland prairie [no site number]. (Winchell 1885, p. 13–14)

In 1939 Wilford tested two mounds from the Hutchins Mounds (21POi) in Section 22, T126N, R37W. The published report (Wilford et al., 1969) listed the site as the Lake Villard Mounds (21PO10), but this identification was corrected by the OSA to a nearby site using Wilford's field notes (Arzigian and Stevenson, 2003). Site 21POi was positioned in a slough hay field at the southwest end of Rice Lake. This group consisted of

25 low, conical mounds, all undisturbed, comprise this group located in a low, wet area. Two circular mounds, one 40 feet [12.2 m diameter by 0.5 m high] and the other 30 feet [9.1 m] in diameter [by 0.45 m high], were excavated. No complete site map exists. Neither mound showed any evidence of burials or cultural materials. Wilford questioned whether they were artificial or natural features. (Wilford et al., 1969, p. 47)

According to the field notes the Mound 1 excavation consisted of a 5.5 m (18 feet) diameter circle while the Mound 2 dig consisted of a 4.3 m (14 feet) diameter circle. Both excavation circles were placed over the mound center. The contour map made by Wilford's crew gives the mound shape as a roughly round instead of a symmetrical circular appearance. The notes further state that rodent bones were found in both mounds and a "flint chip" from Mound 2 (Arzigian and Stevenson, 2003; Wilford, 1939b).

Redwood County

Lewis noted that natural prairie mounds existed in the Lamberton vicinity (Lewis, 1881–1895b, letter dated September 13, 1883, from Watertown, Dakota Territory).

Renville County

Two Mima mound locations are reported for Renville County. The first is positioned two miles west of Renville in a tallgrass prairie easement tract managed by the U.S. Fish and Wildlife Service (Scott Gulp, 2010, personal commun.).

Site 21RN13 has marsh mounds located in Section 22, T116N, R35W, near the north line of the section. Mrs. Tom Svobodny reported to Wilford that Indian mounds existed on the island in Long Lake during a 1944 interview (Wilford, 1944). When Wilford finally saw the site he recognized the mounds as being noncultural in origin. Long Lake is a marsh on the West Fork of Beaver Creek. It is a former glacial meltwater channel.

With Ray Svobodny visited the village site on the island in Long Lake, mentioned in the memos of 1944 and 1950. Long Lake is a drained lake and the island is now of ridge in lower land. In wet seasons the trail to it leads thru a marshy area and a car cannot be driven to it. The site is not very promising. There are no true mounds, Ray Svobodny and his boys have dug at various points and have unearthed animal bones. It looks like a kitchen midden. (Wilford, 1951b)

Rice County

A lengthy list of Mima mound sites is available for Rice County, as it is part of the extensive Schmidt Moundfield (Fig. 4) in northern, central, and eastern townships. The following lowland mound groups are arranged in the approximate chronological order of the investigations, with some overlap for subsequent projects at the identical sites (Anfinson, 2007; Schmidt, 1910, 1928, 1937; Wedding, 1985; Winchell, 1884b, 1911). Some of these mounds, or at least many of those included in Winchell (1911), became formally recorded in the Minnesota archaeological site file. Winchell used the Rice County earthworks entry for announcing his interpretation on the subject.

The writer concurs with the [lowland mounds] opinion expressed by Prof. Schmidt. Owing to the great number of these tumuli found in that part of the state within accessible limits, he spent several days, in company with Prof. Schmidt and a couple of laborers, in digging into several of them, and in making a general reconnaissance round about Northfield. Although he found no human bones or artifacts, there are other evidences which conspire to convince him that these earthworks cannot be excluded from the category of "Indian mounds" which are common in the state. Their numbers are so startling, and their having not been observed by any other archaeologist is so remarkable, that it seemed necessary to make some examination before incorporating them in this report. (Winchell, 1911, p. 130)

The NW¼ of Section 16, Wheatland Township (T112N, R22W), contains a probable Mima mound group (Thomas, 1891; Winchell, 1911):

They lie along a small lake which is on the west side of the north-and-south road. They are rather small, not exceeding two feet [0.6 m] in height. Five or six are visible from the road. There are probably others. (Winchell, 1884b, p. 673)

In Section 17 of Webster Township (T112N, R21W) Winchell (1911; Thomas, 1891) had a site located:

… an eighth of a mile north of Edward McFadden's, on the highest ground, but yet surrounded by a marsh, may be seen a number of mounds rising two and a half or three feet [0.75 or 0.9 m]. (Winchell, 1884b, p. 673)

For his Rice County investigations, Schmidt (1928) identified seven lowland mound localities. The first consisted of 120 mounds scattered across Sections 12, 13, and 14 of Bridgewater Township (T111N, R20W). These mounds were found on the east side of Cannon River south of Northfield. A second group had 13 mounds in Section

14 of Bridgewater Township (T111, R20W) east of Dundas. Third was a site listed as having 54 mounds and being south of Dundas at Thielbar's place and the adjoining land. This site description is virtually identical with the ~50 mounds reported for the George Revier farm on the east side of the Cannon River in Section 22. The 1915 county atlas shows that the Revier farm (within SE, NE and NE, SE of Section 22) comprised the north half of the A.E. Thielbar farm shown on the 1900 county atlas (Northwest Publishing, 1900; Webb Publishing, 1915). On August 24, 1916, Schmidt, his brother, and others tested mounds in the Revier hay meadow. The excavated mound fill was fine sandy sediments over peat below the level surface of the field. The recorded dimensions for mounds extensively tested were (1) 7.9 m (26 feet) diameter by 0.6 m (2 feet) high, (2) 7.9 m (26 feet) diameter by 0.9 m (3 feet) tall; (3) 6.4 m (21 feet) diameter but height not recorded; and (4) 8.8 m (29 feet) diameter by 0.9 m (3 feet) high. Smaller excavations were made in an additional 14 mounds. The goal was to determine if these mounds might have been "tenting places." All results were negative (Schmidt, 1934c). The fourth site had 10 mounds south of the Cannon River in Section 4 of Cannon City Township (T110N, R20W). A fifth locality consisted of three lowland mound groups in Webster Township (T112N, R21W). They were reported as being (1) 6 mounds in Section 9; (2) 12 mounds from Sections 16 and 17 (see the last Winchell site above); and (3) 6 mounds listed for Sections 29 and 31. The sixth locality was a mega-grouping of 577 mounds located between Stanton and the headwaters of Prairie Creek and its tributaries. Stanton is actually in Goodhue County while most of the Prairie Creek Valley falls in Northfield Township (T111N, R19W) of Rice County. For this reason the mega-mound group could have extended for 6–8 miles along Prairie Creek whose uppermost reaches are in Cannon City and Wheeling Townships (Schmidt, 1928). The Dennison Flats along Prairie Creek had hundreds of the mounds in this mega-grouping. Schmidt dug three mounds at the Arndt farm in Section 23 of Northfield Township on August 25, 1916. The same day the crew dug one mound in the Slee farm meadow in Section 24 of Northfield Township. The results of all mound excavations were negative (Schmidt, 1934c). The seventh Rice County locality was for ~25 mounds in Section 14, Wheeling Township (T110, R19W) (Schmidt, 1928).

On May 16, 1939, Wilford and Schmidt toured four lowland mound sites. At site 21RC12 Wilford received permission to excavate but did not believe the effort would prove productive (Wilford, 1939c). This group of eight mounds is believed to represent natural prairie mounds (Arzigian and Stevenson, 2003).

Farm of Wm. F. Bickel ~1½ mi. south of Northfield on Highway 65. Farm is on east side of Cannon River, between the highway and the river, in the N½ of Sec 11 and the S½ of Sec 2, R. 20, T 111. Mounds are near river and back in fields to the east, a group of 8 or 10 mounds [21RC12]. These are all low-lying mounds, where skeletal material is very doubtful. They are entirely similar to the mounds excavated by Schmidt which have all proved sterile. Mr. Bickel is willing to have me excavate these. (Wilford, 1939c)

A Schmidt site not previously reported to Winchell was the subject of the second visit (No. 2 in Wilford, 1939c).

A large group of mounds [no site number], ~45, on property belonging to Carl Kester, E.B. Anderson, or -- Jones. May belong to a man named Hollis. Difficulty due to uncertainty as to exact boundaries, and to legal status of estates. Carl Kester operates one of these farms. These are also low-lying mounds, though two on the bank of the river from a small rivulet runs in, seem better drained and may be better preserved. (Wilford, 1939c)

Site 21RC13 near Dundas in Bridgewater Township is another mound group where Wilford received permission for excavation (Wilford, 1939c). This group of eight mounds is believed to represent natural prairie mounds. Only one mound is in the Office of the State Archaeologist database for 21RC13 (Arzigian and Stevenson, 2003).

A group due east of Dundas, south of Highway 65 belongs to Ed Chester, Dundas. This group [21RC13] runs N and S but is now back from the river. Also low-lying, S¼ of Sec. 11, R. 20, T. 111. Can be excavated. (Wilford, 1939c)

The fourth site Wilford visited in 1939 was 21RC14 (Wilford, 1939c). According to the Office of the State Archaeologist database 21RC14 consisted of 59 natural prairie mounds (Arzigian and Stevenson, 2003). Wilford gave this site a much lower mound count.

A group of ~20 mounds [21RC14] on the farm of Clarence Komoll, P. O. Dundas. These are on the east side of the Cannon River ~1 mi. south and east of Dundas, Probably E½ of Sec. 22 or W½ of Sec. 23, R. 20, T. 111. These also are low lying mounds. (Wilford, 1939c)

A recent proposed residential development in Dundas threatened recorded archaeological site 21RC14 (Anfinson, 2007). As explained by Anfinson:

In October 2005, an attorney contacted the State Archaeologist regarding a proposed development near Dundas where a burial mound group had previously been reported. The site in question, 21RC14, had originally been recorded in the early 1900s by Edward Schmidt, a history professor at St. Olaf College. Schmidt had taken University of Minnesota archaeologist Lloyd Wilford to the site in 1939, but Wilford had been non-committal as to the origin of the mounds. In May of 1985, a local avocational archaeologist, Ken Wedding, reported that the mounds had been "destroyed by tiling (tilling?)." A Carleton College geology professor, Connie Sansome, who had examined the possible mounds just prior to this, thought they were of natural rather than cultural origin.
    The State Archaeologist visited the site on 10/21/05 and walked the harvested agricultural fields where the mounds had been reported. No mounds were evident. In a letter to the attorney and MIAC dated 10/24/05, the State Archaeologist declined to authenticate the location as a cemetery.
    Schmidt had recorded hundreds of mounds in southern Dakota County and northern Rice County in the early 20th century, but none of his mounds have ever been confirmed as burial sites. Schmidt himself dug into almost a hundred of these mounds and did not find any artifacts or human remains. They were reported in Winchell's Aborigines of Minnesota (1911), however, and the locations were therefore given site numbers by the first State Archaeologist in the early 1960s. All of Schmidt's mounds appear to natural features known as "mima mounds." (Anfinson, 2007, p. 27)

In his memorandum summarizing a 1945 site visit, Wilford (1945) noted that the Mazaska Lake Mounds (21RC3) had a lower setting than the nearby site 21RC2. The position of site 21RC3 at near lake level suggests a possible Mima mound site. This site was previously surveyed by Lewis and reported in Winchell (1911).

Scott County

In 1882 Lewis identified an extensive natural prairie moundfield in the Shakopee vicinity (Finney, 2006). This report contained the diagnostic prairie mound description that the site extended "for at least a mile" in T115N, R23W (Lewis, 1881–1895b, letter dated October 11, 1882, from Shakopee, Minnesota).

There are some rises here that look like mds[.] but I cannot tell whether they are or not, but think they were built up of turf or peat from the swamps. They extend above here for at least a mile. ... There are also some of the same kind south of the track. They are all built alike. Occasionally one is round and nice while most of them are uneven. (Lewis, 1881–1895b, Lewis to Hill dated October 11, 1882, from Shakopee, Minnesota)

In 2007 five or six conical mounds were reported in a wetland from the vicinity of Savage. These mounds were examined by Scott Anfinson and Bruce Koenen from the Office of the State Archaeologist and deemed noncultural Mima mounds because of the low, wet floodplain setting on the south side of the Minnesota River. Since this type of landscape is not an expected location for anthropic mounds, a site number was not assigned to the Savage mounds. The site location is near the Savage Fen and existing subdivisions in the NE, NW, Section

15, T115N, R21W (Anfinson, 2009; Kim Davis, 2007, personal commun.). The easternmost mound appears to lie in Dakota County and the site is accessed via a small park at the southeast corner of Savage. This site is unplowed as there are numerous boulders (up to 4 × 5 feet in size) completely and partially exposed on the ground surface. Trees have invaded the mound area in the past two to three decades.

In July 2007, the State Archaeologist received several reports from the public about a proposed housing development threatening mounds in Savage. This project was extremely controversial in the local community. The State Archaeologist examined the location on 8/13/07 and located a number of earthen mounds. A more detailed examination of these features in November 2007 by OSA personnel included soil coring of several of the mounds. Based on the results of the soil cores and the location in a wetland, it was determined that the features were not burial mounds. Mounds are almost always located in upland areas and have a view of the surrounding terrain. While expedient or accidental burials are occasionally found in lowland areas, formal cemeteries (mound groups included), never are. Mounds found in lowland areas are often referred to as "mima" mounds and can be constructed by burrowing toads, sedges, and other natural processes. The State Archaeologist is attempting to have a soil scientist examine the Savage mima mounds in more detail to better understand the natural process that built them. (Anfinson, 2009, p. 29)

Sherburne County

In his report on the geology of Sherburne County, Upham (1888e; Winchell, 1911) noticed anthropic mounds as well as probable natural prairie mounds. For the probable noncultural site (21SH12) Upham wrote:

On the southeast side of this [Elk] lake, northward from its outlet, in the east [half of] section 3 of Clear Lake [T34N, R29W] township, is a series of many round mounds two to four feet high, extending about a quarter of a mile. (Upham, 1888e, p. 444)

This site (21SH12) was recorded in 1886 as consisting of 49 mounds dominated by small tumuli (Winchell, 1911). Nearly a century later Barbara Lass (1981) measured extant 21SH12 mounds in two fields. First, she found 10 conicals that were 0.6 m (2 feet) high with the largest diameter at 10–15 m (33–49 feet). At the second location there were two mounds, both 0.6 m (2 feet) high with plan view dimensions of 3 × 3 m (10 × 10 feet) in diameter and 5 × 2 m (16.5 × 6.5 feet), respectively (Lass, 1981).
    Mahnomen mounds have been reported for T35N, R27W, at the Sherburne National Wildlife Refuge in Sherburne County (Blair et al., 2001).

Sibley County

Writing from Henderson, Minnesota, Lewis related his experience at trying to reach a location approximately five miles to the west-northwest,

Yesterday I went west of here to near the center of the N line of [Section] 6 and did not get beyond. S[ection] 4 [T112, R26W] is mostly marsh. As near as I can learn the hill is somewhere on W half of [Section 31] or still W of that on [Section] 36. Don't know whether I will try again or not because the highest natural md has a house on it and there are many of them on the prairie. (Lewis, 1881–1895b, letter dated October 5, 1884, from Henderson, Minnesota)

Stearns County

There are three reported Mima mound locations in Stearns County. Winchell (1885, 1911) observed numerous mounds in a lengthy wetland in the valley of Ashley Creek near the Stearns and Pope counties line. "They are on the north side of the railroad accompanying a marshy tract. The railroads passes up an old valley of glacial drainage, abundantly strewn with gravel, and these mounds are frequent along this valley" (Winchell, 1885, p. 13). This probable natural prairie mound description appeared with a report of a geological reconnaissance made to Pope County. However, an examination of the

7.5 minute topographic map indicates that the wetland in question is mostly in Stearns County. In addition Winchell's (1885) volume index indicates a Stearns County location for the mounds.

Mima mounds are reported in a waterfowl production area approximately eight miles north of the village of Brooten according to U.S. Fish and Wildlife Service personnel (Scott Glup, 2010, personal commun.). This location is Section 19 of Raymond Township (T125N, R35W), near the Pope County line, and approximately seven miles south of Ashley Creek.

Mima mounds also occur at Roscoe Prairie in Section 35, T123N, R32W (Donald L. Johnson, 2009, personal commun.).

Steele County

A total of three natural prairie mound sightings occurred in Steele County. The geologist M.W. Harrington (1884; Thomas, 1891; Winchell, 1911) included a note about two mound groups near Aurora at the end of his county geology report. He incorrectly considered these mounds to have been human made.

A series of large mounds, which have much the appearance of being artificial, are situated on the east side of the slough at Aurora station; and several others lie near the railroad a few miles farther south. (Harrington, 1884, p. 403)

In contrast, Lewis found marsh mounds both north and south of Owatonna in Steele County. His field examination included the Medford and Aurora localities. The Aurora findings involved at least one of the two Harrington sites.

Yesterday afternoon I examined the country for 2 or 3 miles S of Aurora and found nothing but sand hills and marsh mounds. This morning went down Strait [Straight] river on E side as far as Medford and returned on W side and saw only marsh mounds. (Lewis, 1881–1895b, letter dated October 12, 1884, from Owatonna, Minnesota)

The third Steele County report was Schmidt's observation of lowland mounds located eight miles south of Owatonna (Schmidt, 1910, 1928).

Stevens County

There are two possible natural prairie sites in Stevens County. The Verlyn Marth Memorial Prairie contains small mounds of uncertain origin (Donald L. Johnson, 2009, personal commun.). This possible natural prairie mound site is located in Section 6, T126N, R42W. A second site, 21SEm, consisted of ~11 small mounds approximately one foot in height that a local informant reported to Wilford. This site was located southwest of Gravel Lake in Section 18, T144N, R44W. However Wilford found only one low, flat mound in his 1954 site visit (Wilford, 1953, 1954b).

Swift County

Mima mounds are reported by U.S. Fish and Wildlife Service personnel at two locations: the Svor WPA in the NW¼, Section 21, T122N, R38W, and at the Welker WPA in the NE ¼ Section 5, T122N, R39W (Sara Vacek, 2010, personal commun.).

Wabasha County

Two sites in the Lake City locality represent natural prairie mounds found in wet environments along Lake Pepin and the Mississippi River. The first is site 21WB6 in Sections 4–5, T111N, R12W. Dr. L.C. Estes (1867) reported a mid-nineteenth-century observation that Lake City "was at first thickly strewn with mounds" (Winchell, 1888c, p. 19).

A portion of the village of Lake City is built upon and over these remains [mounds]. Between my residence and the lake [Pepin] there seems to have been a regularly laid out town or city. The streets are regular and the mounds equi-distant from each other. In the center of this city there was a very large mound, much larger than any of the others, which was located in the center of the widest street, and the only one out of line. It was very probably the "headquarters," the residence of the chief, or it might have been the town hall. Nine years ago I sketched and counted these mounds. There were about one hundred of them, occupying, perhaps, a space of thirty acres.

I am also as well satisfied that the very many round mounds found standing separate from each other were simply turf houses, in which once dwelt a people far above, in point of intelligence, the present race of savages. The common idea that these mounds were receptacles of the dead, or that each one is now or has been a sepulchre, is a theory which none have been able to maintain. I have dug into and seen many of them levelled, but never succeeded in finding human bones but in one, and those in the large mound standing in the center of the city of mounds before described, and I am convinced that these were of more recent interment, and that they belonged to the existing race of Indians. It was a natural and convenient place in which to bury their dead.

Every investigation proves beyond a doubt, in my mind, that these works were built of turf, and that they are always composed wholly of the upper strata of soil, and that there is no perceptible depression of the earth around their base. Again, the soil in the vicinity is not found as deep as in other places, proving that it had once been removed. If these mounds were originally sepulchral in their design, then in all we should find human bones. (Estes, 1867, p. 366)

Winchell (1888c, 1911) repeated the idea for house mounds at site 21WB6.

These mounds, which generally were not large, may have been simply the ruins of habitations made of, or at least liberally covered with, mud and turf, of the nature, perhaps, of the more modern adobe of the west and southwest. (Winchell, 1888c, p. 19)

A short distance south of Lake City on the Mississippi River floodplain at Lake Pepin is another probable natural prairie mound group (no site number). The location is a marsh along the south side of Miller Creek in the SW ¼ of Section 9, T111N, R12W, where Winchell's (1911, p. 139) map has the label "marsh mounds" without showing the individual mounds. In this vicinity on the north side of the creek is site 21WB35, a group of 57 mounds (48 round and 9 elongated) first mentioned by Estes (1867) for the Dwelle farm. The round mounds at 21WB35 include examples as small as 17 feet in diameter by 1 foot in height. Burial authentication investigations in 1998 failed to relocate the 21WB35 mounds. Only two possible cultural features (not excavated) were found at the site location (Arzigian and Stevenson, 2003).

Waseca County

For Waseca County, Upham (1884c; Thomas, 1891; Winchell, 1911, p. 98) reported two purported artificial mound sites (no site numbers):

…two or three low, circular and dome-like heaps of earth 20 or 30 feet in diameter but only 1 or 2 feet in height, seen in and beside the road that runs Wilton southwest to Vivian, occurring nearly at the south line of section 10, and again in the northeast quarter of section 30, Wilton. (Upham, 1884c, p. 414)

In the south half of Waseca County Lewis tracked down the mound groups reported by Upham (see above) and noted that one appeared to represent marsh mounds (Lewis, 1881–1895b, letter dated October 7, 1884, from Waseca, Minnesota). Both sites occurred in flat upland settings.

Winchell (1884b; Thomas, 1891) appended a notice of a Waseca County site to his brief mound list from another county. This site was described as being "in the vicinity of Woodville [Township, Waseca County]. According to Mr. J.F. Murphy there are 21 mounds, from four to five feet in height, near the center of section 3 [T107N, R22W], between Watkins and Rice lakes, some being 30 feet in diameter" (Winchell, 1884b, p. 673). A ridge exists between these lakes near the boundary line separating Sections 3 and 4 and there is an extensive marsh to the north. Sites 21WE1 and 21WE2, which include anthropic mounds (Winchell, 1911), are recorded near this location and may include possible natural prairie mounds.

Washington County

Two examples of sites with possible natural prairie mounds exist along the Mississippi River floodplain in Washington County. The first consists of Mounds 5 and 10 at the Schilling Mound Group (21WA1). These possible natural prairie mounds are associated with a known anthropic mound group on Grey Cloud Island. According to Lewis, Mound 5 (Section 36) was 36 feet in diameter and 2.5 feet high. A 5-feet-wide trench excavated from one side to the other across the mound center by Wilford in 1947 yielded negative results for burials and artifacts (Wilford et al., 1969). Mound 10 (Section 32) was 38 feet in diameter and 4 feet tall. In 1947 Wilford measured a mound height of 1.75 feet. The result of digging a 30 × 5 feet trench across the mound was negative (Arzigian and Stevenson, 2003; Wilford et al., 1969).

A possible natural prairie mound is reported for an anthropic mound group on Grey Cloud Island. It is Mound 14 at the Michaud Mound Group (21WA2) with dimensions of 35 feet in diameter and 3 feet high. In 1947 Wilford excavated a 20 feet circle over the mound center but found no burials or prehistoric artifacts. The only materials listed in the site description were modern trash from a shallow looters pit (Wilford et al., 1969). Wilford noted in his mound excavations that previous diggers did not habitually remove all human bones if present (Arzigian and Stevenson, 2003; Wilford, 1970; Wilford et al., 1969). For that reason Mound 14 can be considered a possible natural prairie mound associated with a known anthropic mound group (21WA2).

Wright County

The list of aboriginal earthworks at the end of a county geology report includes both human burial and noncultural mounds (Upham, 1888f). Three probable Mima mounds were on a flat prairie at 21WRh on the William Hoar farm. This site position is approximately four miles south of the Mississippi River and near the south edge of T121N, R25W. On the 1901 county atlas the W.H. Hoar property is in the west half of Section 34 and the SW, SE of Section 27 (Northwest Publishing, 1901). Upham mentioned three mounds at 21WRh and ascribed a prehistoric origin without providing evidence (Winchell, 1911). In fact, he presented the data necessary for identifying a seasonal wet site setting in a low or slightly closed depression, and a second possible Mima mound site on the Smithson farm in Section 8, T120N, R25W is mentioned at the end. In 1901 the Smithson farm is in the SE¼ of Section 8 (Northwest Publishing, 1901).

Four miles south of Monticello, on William H. Hoar's farm, are several aboriginal mounds [21WRh], of the same round form [as a previously described anthropic site], but smaller, situated on as level prairie, within an area thirty or forty rods [495 or 660 feet] in extent. Two of them are each two rods [33 feet] across, and were originally four feet high; but their height has been reduced by plowing to two and a half feet. From center to center of these is about sixty feet in a southwest direction. Some six hundred feet farther east is another mound, two-thirds as large. No bones nor other relics have been struck in plowing over them. (Upham, 1888f, p. 262)

At Mr. Hoar's, the mounds [21WRh], situated on an area of modified drift, were surrounded by a shallow depression, perhaps six inches below the average surface; but at Mr. Smithson's [another previously described anthropic site] this is not perceptible. About a sixth of a mile farther south, near Mr. Smithson's house, are four or five other small mounds. (Upham, 1888f, p. 263)

The second possible Mima mound group at the Smithson farm is likely part of site 21WR29 (Winchell, 1911). A housing subdivision exists at 21WR29 and an undated ca. 1980 note in the Minnesota site file states that the Naglich family had recently removed 1.5 feet of soil from a mound for use in leveling their backyard, but nothing was found and the Naglich's have never heard of archaeological finds in the area.

Yellow Medicine County

The Mound Springs Prairie Scientific and Natural Area on the Prairie Coteau is a possible noncultural mound site. The low wetlands at Mound Springs have domed features associated with a fen environment.

Iowa

Reports can be found for the existence of scattered Mima mounds on the Des Moines Lobe and the Iowan Erosion Surface in north-central Iowa (Fig. 1B). This is remarkable because <0.1% of the original Iowa prairie is extant (Rosburg, 2001; Smith, 1998). In addition natural prairie mounds are believed to have existed in at least three localities along the Mississippi River where special environmental conditions, i.e., near surface wetness, existed.

Allamakee County

Harpers Ferry lies on a sandy terrace of the Mississippi River with an extensive prairie covering most of the ~2.6-square-mile landform. In 1892 Lewis recorded 900 mounds in the area as the Harpers Ferry Great Mound Group (13AM79). Most were small dome-shaped conical mounds, but several were effigy mounds. If Lewis' mound count was accurate, it would represent the single greatest concentration of human-built mounds in North America. Since 1892 other researchers have doubted the existence of this great mound group, at least one of such magnitude. A recent investigation and taphonomic assessment (Finney and Johnson, 2010) indicates that the upper surface of the terrace resulted from the sediment discharge of two creeks draining the adjacent uplands. In addition, these investigations enumerated evidence that in the past wetter environmental conditions existed in the terrace than today. For example, a sequential series of beaver dams along Cota Creek had raised the water table, the Government Land Office survey field notes noted the presence of a near surface gravel layer along the lower reaches of Cota Creek, and the U.S. Army Corps mapped wetlands along Cota Creek in the late nineteenth century. A recent paper concluded that in 1892 Lewis saw a landscape dominated by natural prairie mounds at Harpers Ferry—and that he suspected as much because nearly all were omitted from his mound totals for Iowa (Finney and Johnson, 2010).

A second Allamakee County site is the Rhomberg Street Mounds (13AM260) positioned within the village of Harpers Ferry. Two (Mounds 1 and 2) of the seven possible 13AM260 mounds were 5–6 m in diameter, 0.15–0.3 m in height, and conical in shape. Michael Fokken (1980) interpreted Mounds 1 and 2 as resulting from rodent activity. His argument was supported by roughly horizontal wavy stained lines interpreted as indications of a perched/high water table in the stratigraphic profiles, extensive evidence for bioturbation, and the absence of prehistoric cultural materials (Fokken, 1980).

Buena Vista County

The geologist Thomas MacBride (1902) described a series of hills and mounds, some of which were low in height, along with marshes, in the Des Moines Lobe portion of Buena Vista County. His depiction of these landforms could have included features of glacial origin as well as possible natural prairie mounds in association with the marshes. The MacBride description was not referring to features on the Altamont Moraine.

Butler County

Possible Mima mounds are reported for Butler County although the location is unclear, being either Wolters Prairie Preserve or Leeper Prairie Preserve in the uplands in the vicinity of Clarksville (Rosburg, 2001).

Carroll County

Possible Mima mounds were noted in a closed basin depression encompassing at least 160 acres positioned at the corner of Sections 5–8 in T82N, R33W, on the outer edge of the Des Moines Lobe. This location is southeast of Carrollton. As described by geologist Charles White:

In a shallow depression or plain, below Carrollton, on the east side of the Middle Raccoon, several interesting spring-mounds occur which have excited much attention. The plain is thirty to forty feet above the present level of the river, from which it is separated by a well-defined drift ridge which, in places, rises into considerable knob-like eminences from one hundred to one hundred and fifty feet above the stream. The plain, however, communicates with the valley both above and below, and was probably once the channel of the river. The spring mounds are situated along an irregular line more or less in the middle of the depression; they are four to six feet in height and as many yards in diameter, and are apparently entirely composed of vegetable matter, forming a peaty deposit which is largely mixed with the exuviate of shells and other animal remains. The crests of the mounds are covered by a rank growth of tall, flag or marsh grass, but upon the sides there are usually two well-marked bands of short herbage and moss encircling the mounds and separated by a narrow belt of taller grass. The disposition of the vegetation upon these places is exceedingly interesting, though the mounds themselves, doubtless, owe their origin to the existence of pools of water, indicating, more or less accurately, the course of a former water channel, and which, being fed from higher sources, the tendency is what we observe—a gradual building up of a peaty formation. The surface of the plain beyond the limits of the mounds is perfectly level, and the deposit consists of decayed vegetable matter mixed with sand, forming a sandy muck. (White, 1870, p. 140–141)

These features are not discussed in the subsequent 1899 report on Carroll County geology, suggesting that the natural depression had been drained.

Cerro Gordo County

In 1868 Orestes St. John, the assistant geologist who illustrated the geology reports of Charles White, depicted small mounds surrounding the west and north edges of a peat swamp located 2.5 miles north of Clear Lake. The location should be near the south edge of T97N, R22W. This drawing included a horse and wagon for giving an approximate size perspective (White, 1870). Another possible Mima mound location in Cerro Gordo County is the wet prairie and marshes at Hoffman Prairie located west of Clear Lake.

Clay County

Two purported Mima mound locations are in Clay County. The first is Dewey's Pasture that has a wet mesic prairie (Rosburg, 2001). It is part of a marsh and lake complex straddling the Clay and Palo Alto counties line.

The second Clay County location is Kirchner Prairie Wildlife Management Area that consists of 166 acres on the Des Moines Lobe. It is in the W½, SW¼, Section 19, T97N R35W, Lake Township and in the E½, Section 24, T97N R36W (appropriately named) Meadow Township. The east half is a former prairie hay field with mesic to wet prairie, sedge meadow, and pothole marshes; while the west half is a marsh formerly used for grazing (Rosburg, 2001).

Clayton County

At least one conical mound from the densely packed Sny Magill Mound Group (13CT18), consisting of 128 known mounds (120 conical, 3 linear, and 5 effigy) arranged in three or four rows, was noted as a possible Mima mound in a contract report for a ground penetrating radar survey (Whittaker and Storey, 2005). The site setting is on a low terrace adjacent to the Mississippi River. The published article version of this research noted that a total of 18 (conical, no distinct strata) 13CT18 mounds were interpreted as probable or possible noncultural in origin, and that the specific identification of Mima mounds would require ground truth excavations (Whittaker and Storey, 2008).

Dickinson County

Upham's (1884d) summary for the geology of Jackson County, Minnesota, mentioned the presence of mounds near the Minnesota border between Spirit Lake (in Iowa) and Little Spirit Lake (mostly in Minnesota). This site location would be near Bakers Point in the north half of Section 17, T100N, R36W, Dickinson County, Iowa. They should be considered probable natural prairie mounds.

Dubuque County

One site at the Dubuque harbor along the Mississippi River has pertinent information in the C.R. Keyes research file for Dubuque County that is based on information received from Richard Hermann about a village and possibly 10 mounds. This is site 13DB363 at Lake Peosta. H.T. Woodman (1873) had observed a much larger number of mounds and wrote additional details for 13DB363:

Large numbers of these mounds have been destroyed by the building of our city [Dubuque], how many I do not know, for it is true that a majority of persons do not recognize their true character even if living in daily contact with them. Indeed, so far as I know, no person had either publically or privately recognized the group I am about to describe, although located within the corporate limits of our city, until they were discovered by myself, and their existence made known to the public through the medium of one of our newspapers only a few months ago. This group [13DB363] is located within the northern limit of the city, adjacent to the narrow body of water known by the local name of Lake Peosta, and about fifty feet above its surface.

They are circular, or nearly so, except the three larger ones along the edge of the terrace facing the east, the Mississippi River and the lake before mentioned. They are almost invariably fifteen paces apart from center to center, the smaller ones being from two to two and a half feet high and about twenty feet in diameter. The material of which they are composed is the ordinary alluvial soil of the terrace. They are seventy in number, and are now shaded by a pleasant oak grove of comparatively recent growth.

What is most remarkable about this group of mounds is their number and the great regularity of their arrangement, being arranged in straight or slightly curved lines (some of them being parallel), and their nearly uniform distance apart, namely, about fifteen paces.

That the shape, size, and position of the larger mounds in relation to the others, together with the arrangement of nearly all in lines with almost uniform interspaces, formed part of the plan of their builders, cannot be doubted; but the knowledge as to what that plan had reference doubtless perished with those that constructed them.

No other traces of man or his works have as yet been discovered in connection with them, but only a few rods to the northward and eastward of the limits of the group I discovered fragments of the ancient pottery and several flint arrow-heads some years ago. (Woodman, 1873, p. 226–227)

Emmet County

In the botanical literature Mima mounds are reported at the 200 acre Anderson Prairie State Preserve on the west side of the upper Des Moines River in Section 33, T100N, R34W (Klaas et al., 2000; Wolfe-Bellin and Moloney, 2001).

Fayette County

A number of low mounds were described in the SW¼ of Section 26, T92N, R8W, by the assistant state geologist T.E. Savage (1905). This level upland prairie location in Smithfield Township is near the eastern boundary of the Iowan Erosion Surface. Since major waterways are absent, this site setting represents probable natural prairie mounds.

The natural prairie mound illustrated by McGee (1891) in his *Pleistocene History of Northeastern Iowa* was described as being in the uplands south of the Turkey River. The most likely placement for this site would be northern Fayette County (Fig. A1-2).

Floyd County

One of the best documented Iowa Mima mound sites is located approximately two miles northeast of Rockford on the nineteenth-century Webster farmstead. While the total site extent is unknown, a portion is illustrated in a dramatic drawing by Clement Webster (1900). This view shows at least 10 natural prairie mounds (Fig. A1-1). Curiously Webster depicted animal burrows in a cross-section view of a mound. According to the distance measurement and the 1895 Floyd County atlas the mounds were in the southeast quarter of Section 2, T95N, R18W, where I.C. Webster is listed as the landowner (Union Publishing Company, 1895). In other papers Webster provides informative discussions of natural prairie mounds in north-central Iowa but only the unpublished illustration gave a precise location (Webster, 1889, 1897).

Hancock County

This north-central Iowa locality appears to possess the largest documented sample of natural prairie mounds in the state along or near the Winnebago River. As noted by an archaeologist, "The Hovey 'mounds' [13HK10] also are only a small portion of a much larger number of similar features which occur throughout the northeast corner of Hancock County. This writer is not aware of another situation where individual man-made mounds are found scattered over such a large area" (Benn, 1976, p. 6–7). In the 1970s local resident Arlo Johnson (1974–1979) noted their presence extending for approximately seven miles from Hovey Marsh and site 13HK10 on the east end along a lengthy glacial drainage channel to the Winnebago River and along that waterway to Forest City at the Winnebago County line on the west end. This locality is south of Pilot Knob State Park. Unfortunately the majority of these natural prairie mounds have been cultivated and thus destroyed in the past three decades.

The best documented Hancock County site consists of 16 mounds referred to as "prairie blisters" at the Hovey Mound Group (13HK10) placed on low ridges above a wetland in the northeast part of the county. Site 13HK10 is located in the S½, NW ¼, Section 14, T97N, R23W, Ellington Township, at an elevation of ~1225 feet. The site setting in Hovey Marsh is part of a lengthy glacial-age channel to the Winnebago River. The channel starts near 13HK10, is ~150–200 feet wide, ~8000 feet in length, and a small quantity of water flows west from the wetland to the river. David Benn described the Hovey Mounds:

The sixteen mound-like features observed by the Luther College crew on the low ridges are very similar in all characteristics. They vary in size between ca. 8 and 13m at their maximum diameter and between ca. 0.25 and 0.50m in height. All of them have smooth contours, relatively distinct edges, and most appear to be slightly oval in shape. In many cases the longest diameter of these features is oriented roughly east-west. It cannot be concluded that the 16 mound-like features are grouped in distinct patterns, for they are scattered about on the limited areas of the narrow low ridges at varying distances from one another (e.g., distances vary from ~27m to more than 149m). Some narrow low ridges, which project 'finger-like' into the swamp, have as many as four features which may be construed as lying in a single line. (Benn, 1976, p. 3–4)

Unfortunately site 13HK10 and other nearby reported natural prairie mounds were put into cultivation after 1976. In that year the Iowa Conservation Commission attempted to purchase Hovey Marsh and the nearby wetlands but could not offer the price desired by the majority of the landowners. A suite of newspaper clippings, memorandums, and letters kept by local landowner, farmer, and Iowa Archeological Society member Arlo Johnson indicated that Roy Hovey wished to receive $2300 per acre for his portion of the marsh. When a potential deal with the state of Iowa fell through, the Hancock County Drainage District No. 174 was formed and drained Hovey Marsh for agricultural purposes (Johnson, 1974–1979). The archaeological survey and mound testing at 13HK10 was accomplished after a request by State Archaeologist Duane Anderson to determine whether the proposed drainage project might impact any human burial mounds protected by Iowa state law. Earlier in 1976, Iowa became the first state to enact modern legislation prohibiting the disturbance of unmarked as well as marked human burials and cemeteries (Benn, 1976; Johnson, 1974–1979; Myers, 1978).

Howard County

Small Mima mounds exist in the 240 acre Hayden Prairie State Preserve. Benn and Petersen (1976, p. 7) commented on their variable size and "rougher contours" instead of being round in shape. The size was recorded as several meters in diameter and 0.25–0.50 m in height (Benn and Petersen, 1976; Donald L. Johnson, 2008, personal commun.).

Kossuth County

Natural prairie mounds are reported in the Stinson Prairie, which is known for a mesic to wet prairie and moraine topography (Rosburg, 2001).

Muscatine County

Reportedly there are numerous small mounds from low topographic settings in the Mississippi River floodplain south of the city of Muscatine on Muscatine Island. This wide segment of the floodplain on the Iowa side is the Muscatine Bottom, a low and sandy landform with marshes and ponds that extends into the northeast corner of Louisa County. It includes a large sandy terrace known as Big Sand Mound. One early account that may have included possible natural prairie mounds was J.E. Stevenson's (1879) paper noting the presence of an estimated 2500 mounds between Muscatine and Toolesboro. The Stevenson sample should be predominated with anthropic mounds on the adjacent bluff line. However, Frederick Starr (1897, p. 106) summarized a Theron Thompson (1880) excavation site as: "Containing 20 mounds of which 10 were opened with no result except an occasional bit of charcoal or a fragment of a shell." These mounds are examples of possible natural prairie mounds. Thompson (1880, p. 359) stated: "The mounds are in size from 3 feet to 30 feet in diameter and 6 inches to 5 feet in height." Again, the smallest mounds are likely noncultural in origin.

Palo Alto County

In a geological overview MacBride (1905) made a general reference about the presence of low mounds at several upland locations in the central and western portions of Palo Alto County. He noted these examples as "hills of construction" but did not attribute their formation to glacial events nor Indians. Some of these mounds could have resulted from the Altamont Moraine stage of the Des Moines Lobe in western Palo Alto County but it is unlikely that MacBride would have missed that fact. MacBride (1905) had an additional comment that low erosional hills followed streams in the county.

Pocahontas County

A former prairie hayfield comprises the bulk of the 160 acre Kalsow Prairie State Preserve in the NE¼, Section 36, T90N, R32W, Pocahontas County, Iowa. Kalsow Prairie is located on the Des Moines Lobe. A waterway bisects the northwest corner of the preserve and 14 prairie potholes present. At least 128 Mima mounds are extant in the ~140 acre unplowed hayfield segment at Kalsow. The mounds range from 1.8 to 21.9 m (6–72 feet) in diameter with a maximum height of 0.9 m (3 feet) (Brotherson, 1982, 1983; Dornbush, 2004). An aerial view of this prairie provides a startling perspective in contrast to the surrounding row crop agricultural landscape. Brotherson (1982) mapped the 128 Mima mounds in plan view. Kalsow Prairie may have been among the mound sites observed by MacBride (1905).

Sac County

At Sac City on the Des Moines Lobe, Charles White (1868, 1870; Anonymous, 1868) saw a group of eight mounds in the village during his geological fieldwork. Local resident Carr Early provided White with details for these mounds. From northeast to southwest they were noted as having the following dimensions: (1) 96 × 36 feet oval and 2 feet high, (2) 60 × 30 feet oval and two feet high, (3) 66 feet in diameter and 5 feet high, (4) 80 feet in diameter and 6 feet high, (5), (6), and (8) 60 feet in diameter and 3 feet high, and (7) 50 feet in diameter and 2.5 feet high. It was reported that excavations had been made in Mounds 4, 5, and 6, however "nothing of human remains or works of art have been discovered" (White, 1868, p. 22).

Story County

Mounds were possibly on the Des Moines Lobe at Ames, Iowa, according to nineteenth-century resident Charles Taylor:

It was the ninth day of March, 1868 when I pulled into Ames, then a straggling slough town, which boasted of little else than a new railroad. The first night I put up at the old Sherwood, later the West House, which stood on the spot where the Masonic Temple [is now]. Well, sir, when I came to Ames at 24 years of age, the little town was fit for nothing but buffalo wallows. It was the wettest ground I ever saw in my life. Only a few prairie mounds offered a really dry place to stand. (Taylor, 1928)

In subsequent paragraphs Taylor (1928) repeated the theme that wetness was a common condition at Ames. A Story County history states that the frequently encountered prairie sloughs presented a seasonal problem prior to the installation of field tile drainage (Payne, 1911).

Worth County

Worth County, located adjacent to the Minnesota border, has a substantial number of natural prairie mound sites. These Mima mounds are found along the Shell Rock River, Deer Creek, and the open prairie.

Schmidt (1928) documented three Mima mound sites. They are present (1) on the banks of Shell Rock River at Northwood; (2) on the prairie three miles north of Kensett; and (3) four miles northeast of Hanlontown.

Several sites have received late-twentieth-century documentation. Benn recorded "prairie blisters" at sites 13WT4, 13WT6, 13WT8, and 13WT9 along Deer Creek in the northeast part of the county (Benn and Petersen, 1976).

At site 13WT4 there is a single natural prairie mound measuring 17.25 m (56.6 feet) in diameter and 0.46 m (1.5 feet) high. This site is in the NW, SE, SE, SE, Section 27, T100N, R19W. A 1 × 1 m square excavated in the mound center yielded one chert flake at a depth of 0.49 m. The mound profile consisted of a dark, homogeneous loam that is friable and structureless from 0 to 0.66 m in depth. There is a dark gray-brown loam from 0.66 to 0.92 m with a friable to weak blocky soil structure. The third soil profile unit is a dark gray-brown clay loam from 0.92 to 1.10 m with a moderately well-developed blocky structure. The excavators continued digging the natural soil strata into the C horizon and terminated at 1.62 m below the ground surface. The homogeneous and structureless sediments in the top 0.66 m are consistent with the expectations for a natural prairie mound (Benn and Petersen, 1976).

Another isolated mound was found at site 13WT6 in the NE, NE, NE, SW, Section 17, T100N, R19W. This feature is ~12 m (39.4 feet) in diameter and 0.30 m (0.98 feet) tall. The site setting is a low rise near Deer Creek. Reportedly this field had cultivated for only two years before the investigation. The investigators noted that plowing was beginning to stretch the mound shape out of a true round shape into an oval plan view. One test unit (1 × 1 m) was placed in the mound center. This unit had a 0.22-m-thick fine dark brown silty loam plowzone. From 0.22 to 0.60 m the sediments were darker than the plowzone and exhibited a friable texture and contained small pebbles. Light brown sandy loam was encountered between 0.60 and 1.14 m in depth. This layer exhibited weak blocky structure and contained more glacial pebbles. The excavation terminated at 1.85 m below ground surface. No artifacts were recovered from 13WT6 (Benn and Petersen, 1976).

At site 13WT8 the field crew found one of two informant reported mounds in an overgrown grass setting near Deer Creek. This mound in the NW, NW, SE, Section 35, T100N, R19W, is ~12 m (39.4 feet)

in diameter and 0.6 m (1.97 feet) in height. A core taken in the mound center revealed the presence of a dark topsoil extending 0.6 m in depth. No artifacts were present (Benn and Petersen, 1976).

Based on a local informant report, site 13WT9 consisted of several mounds located in the SW, NW, Section 28, T100N, R19W. The site setting was a pasture. At the time of the survey the field crew could not relocate the mounds (Benn and Petersen, 1976).

Wisconsin

While all previous research indicates that Mima mounds are rare east of the Mississippi River, a few definite occurrences are known (Fig. 1C). These sites are spatially dispersed.

Crawford County

For the Prairie du Chien terrace and lower bottomlands locality along the Mississippi River north of its junction with the Wisconsin River there is one statement for probable and four more noting possible natural prairie mounds among the known anthropic mounds on this landform. Lewis wrote from Prairie du Chien about the presence of numerous marsh mounds in the vicinity (Lewis, 1881–1895b, letter dated June 11, 1884, from Prairie du Chien, Wisconsin). Lyman C. Draper (1888) collected early settler accounts about the Prairie du Chien locality. Two of these commented about possible natural prairie mounds. O.B. Thomas stated, "There are many mounds on the prairie ..." (Draper, 1888, p. 350) while Horace Beach said, "Mound works are very common, and existed all over this prairie" (Draper, 1888, p. 351). Alfred Brunson (1850, p. 182) discussed the numerous mounds and specifically noted the presence of smaller round examples, "some being not over 10 feet on the base and two feet high." The archaeologist, surveyor, and scientist I.A. Lapham stated:

Traces of mounds were discovered by me (in 1852) along the whole extent of the prairie, apparently similar to others found in the vicinity; but from cultivation, and the light sandy nature of the materials, they are now almost entirely obliterated. (Lapham, 1855, p. 66)

While there are known anthropic mounds on the terrace at Prairie du Chien, five separate recollections about a greater quantity also suggest the reasonable possibility for natural prairie mounds at this locality.

Dodge County

Lapham (1855) noted more than 500 mounds around the extensive Horicon Marsh on the upper Rock River. Of these, ~200 small conical mounds were in the vicinity of Horicon, at the south end of the marsh. He excavated one and found nothing. For this large number of mounds, it is likely that both cultural and noncultural mounds are present. In addition historic Indian burials are known from the Horicon locality.

It is difficult to comprehend for what purpose the very numerous small tumuli were made, if not for burial; and yet it is hardly probable that all evidence of such use would have disappeared. They are here commonly made of the black vegetable mold, but slightly mixed with the subsoil, which has a lighter color. (Lapham, 1855, p. 56)

In his summary of Wisconsin mounds, Lapham (1855, p. 80) returned to this theme, "Great numbers of the smallest conical tumuli are also destitute of any remains."

Iowa County

A possible Mima mound location is the Shea Prairie, a remnant of the Military Ridge Prairie, in the W½, SW¼, Section 26, T6N, R5E, south of Barneveld. It includes wet prairie and there is a thin soil in the southeast corner (Henderson, 2006).

Juneau County

There are 200–300 mounds reported for the extensive Cranberry Creek Archaeological District (47JU2, 47JU5, and 47JU6) in Sections 30–31, T20N, R4E, north of Necedah (Boszhardt, 1987, 1988). Included in this group are a number of effigy mounds, several large Middle Woodland conical mounds, and hundreds of small conicals in a setting adjacent to an extensive wetland along Cranberry Creek. This area is part of the bed of glacial Lake Wisconsin. In this region T.H. Lewis never mapped a site of this great extent. This fact combined with the total mound count is indicative of possible Mima mounds occurring among the hundreds of small conical mounds.

Monroe County

A local resident initially noted the presence of the West Prairie Mound Group in 1863, and it was mentioned by Peet (1883c) as consisting of 19 mounds in 1883. By 1912 Charles E. Brown (1940) mapped 14 mounds at site 47Mo7, and this exercise was repeated the following year by Cole and Flint (1913). All of these researchers assumed that 47Mo7 represented anthropic mounds. The West Prairie Mound Group in Section 23, T17N, R3W, lies north of Cliff Court Road within the Fort McCoy Military Reservation approximately five miles east of Sparta. The setting is a former cultivated field on the south side of Coon Valley near Smith Creek, a tributary of Silver Creek, in a former prairie. Recent investigations include a Phase I survey (Holtz, 1995) and a Phase II site evaluation (Mier et al., 1996). Based on the archaeological and geomorphological investigations, the mounds were interpreted as natural in origin. Shovel testing between the mounds yielded only two chert flakes. Test excavations in four mounds found evidence of prior disturbances and no artifacts or features. Michael Kolb concluded:

> The mounds do not exhibit soil attributes typical of earthworks constructed by humans … The Mo 7 mounds lack buried soil horizons and the "basket loading" or other structural elements expected in human constructed earthworks. Pedogenic continuity exists between the mounds and the surrounding soils. Consequently, the mounds and the surrounding soils exhibit similar horizon sequences and horizon morphology. This morphological similarity has one notable exception: the presence of higher percentages of silt and clay in the mounds relative to the surrounding soils. The origin of this textural difference is crucial to determining the origin of the mounds. (Kolb in Mier et al., 1996, p. 128)

The higher mound profile clay content represented a soil pedogenic process and the silt quantity was related to the former presence of an eroded loess cap that is up to ~0.65 m thick on nearby hills in Monroe County (Mier et al., 1996). The site evaluation included two mounds being examined by ground-penetrating radar and agreement with the natural origin interpretation (Kloehn et al., 2000). For the above reasons site 47Mo7 was recommended as not eligible for listing on the National Register of Historic Places. The 47Mo7 site evaluation included a comparison of mound diameter size between the 1913 and the 1994 field seasons (Table 3).

The expected result would be an increase in the diameter correlated with the plowing down of the mound height. Surprisingly the mound diameters were smaller in 1994. Mier et al. (1996) also plotted the mound positions and approximate extent from the 1939 U.S. Department of Agriculture aerial photograph. The largest mound diameters are associated with this view. A 1913 recorded height was available only for Mound A. Between 1913 and 1994 it showed a 17% reduction in elevation (Table 3).

Trempealeau County

In his account of regional history, Judge George Gale (1867) placed the number of mounds in Trempealeau County between 1000 and 2000. Of these, the greater quantity was stated to be near Trempealeau. To reach this large number Gale must have included natural mounds in his totals, as his mound count is too high for a single county.

As summarized by George Squier (1905, p. 25): "This [Gale mound count] can scarcely be more than an estimate, and although I have never attempted an enumeration, I am confident that it is greatly in excess of the true number. I suspect that not a few natural elevations have been included." In an observation that supports this idea Squier (1905) discussed "tumuli" from 10 to 15 feet in diameter and from 2 to 4 feet high, some of which were culturally sterile, near the Mississippi River in the vicinity of Trempealeau. More pointedly Squier commented that the sterile mounds appeared never to have been previously disturbed. The accounts made by Gale and Squier both indicate the presence of probable natural prairie mounds in the Trempealeau vicinity.

Vernon County

A site with 13 conical mounds scattered across approximately three-quarters of a square mile was located on the Millard Prairie several miles southwest of Elroy. This locality in Hillsborough Township lies within the northeastern part of Vernon County. Mound size ranged from 30 × 30 × 1 feet up to ~80 × 60 × 4 feet. Cole and Flint (1913) found no artifacts and reportedly those who preceded them in opening the mounds found nothing of value. No burials were mentioned for the previous explorations (Cole and Flint, 1913).

Washington County

Possible human burial mounds located in a rural setting near West Bend were reported to the Wisconsin Historical Society in the mid-1990s. Upon investigation, the so-called "mounds" proved to be small, low, round (~4–5 m diameter × 0.3 m high), mound-like features composed of dense clay. No cultural materials or burials were present (Amy Rosebrough, 2010, personal commun.).

Illinois

Scattered examples of Mima mounds exist east of the Mississippi River, with at least two sites in Illinois and other possible cases reported in the literature (Fig. 1D).

Grundy County

Mima mounds are present in the unplowed portion of the huge Goose Lake Prairie State Natural Area (Donald L. Johnson, 2009, personal commun.). This wide and low floodplain location is immediately downstream from the confluence of the Des Plaines and Kankakee Rivers that form the Illinois River. Unfortunately, these natural prairie mounds have not been recognized in the published Goose Lake literature. They remain virtually invisible in the tall grasses and flowers

TABLE 3. MOUND DIAMETERS AT WEST PRAIRIE MOUND GROUP (47MO7)

47Mo7 Mound	1913 diameter (m)	1913 height (m)	1994 diameter (m)	1994 height (m)
A	16.2	1.5	16	1.25
B	14.6	–	12	0.87
C	14.6	–	8	0.32
D	13.7	–	10	0.36
E	13.4	–	12	0.78
G	14.6	–	9	0.63
H	14	–	12	0.38
I	13.7	–	12	0.69
J	12.8	–	14	0.69
K	19.8 × 9.1	–	19 × 14	0.97
L	14	–	14	0.62
M	11.6	–	10	0.59

Note: Mounds F, N, O, and P were not recorded (or not visible).
Source: Mier et al. (1996).

except when the prairie is burned. Based on field observations the estimated mound dimensions are 20–30 feet in diameter and 1–2 feet high. Recent light detection and ranging (LIDAR) mapping suggests a possible mound field totaling more than 300 Mima mounds exists in Section 34 of T34N, R8E and Sections 3 and 4 of T33N, R8E (Donald L. Johnson, 2010, personal commun.).

Jersey County

The nineteenth-century geologist and archaeologist William Mc-Adams (1880, 1881, 1887, 1919) mentioned numerous small mounds occurring in low settings in Jersey County and in the American Bottom located a short distance to the southeast in Madison County. Unfortunately he used a similar description for both localities. The earlier publication although dated 1919 was for Jersey County because these findings were cited by MacLean (1879). For the Jersey County uplands McAdams mentioned mounds along Otter Creek approximately three to four miles from Jerseyville around Sections 5 and 8, T7N, R11W. These mounds were about two miles east of the McAdams farm (Andreas and Lyter, 1872). Under the heading of "House Mounds" McAdams wrote:

There is a class of mounds in this county that long have been a puzzle to me. They are generally situated about the borders of the prairie lands, near some spring or water course. They occur in groups of two or three to thirty or more, and are from two to six feet in height, round or oval in shape, and fifteen to twenty feet in diameter; some are forty to fifty feet in length.
… Numbers of these ancient mounds can still be seen three or four miles from Jerseyville, on the banks of Otter Creek. (McAdams, 1919, p. 29)

Despite a general lack of artifacts, no human burials, no evidence for postmolds, and some fish remains along with scattered finds of ashes and charcoal, McAdams (1919) interpreted these features as house mounds and compared them to the Swiss lake dwellings discovered in the mid-nineteenth century. He believed that the postmolds had disintegrated and thus indicated a great antiquity for the house mounds. As discussed earlier, this house mound interpretation was not supported by the archaeological record. It represented the widespread nature of the persistent attempts to view all mounds as cultural constructs made by the American Indians.

Lake County

While surveying in Lake County, T.H. Lewis noted the presence of meadow mounds along the upper Des Plaines River. This location would be within T43–46N, R11E.

Yesterday afternoon I looked up some of the mounds on the west side of the Des Plaines river in Lake Co Ill, and found them to be meadow mounds. (Lewis, 1881–1895b, letter dated September 8, 1893, from Burlington, Wisconsin)

Later in the same letter Lewis announced that this discovery has pushed up his plans for the next scheduled survey location in southeast Wisconsin.

I found out the class of mounds on the Des P[laines] river, so that I telegraphed for the Wis[consin] notes. (Lewis, 1881–1895b, letter dated September 8, 1893, from Burlington, Wisconsin)

The earlier geological accounts, written before the expansion of the Chicago suburbs, indicate that the upper Des Plaines River drainage experienced seasonally wet conditions. As noted by Goldthwait (1909, p. 88): "The [upper] Des Plaines river is subject to floods of unusually long duration, as compared with other rivers of its size. Each spring, as the snow melts rapidly out of its basin, the river is swollen by the additional volume, and rises several feet above its normal level. A portion of the valley floor is thus flooded for days or even weeks." The upper Des Plaines River essentially has no tributaries as it is situated between two north-south trending glacial moraines that bound the narrow valley on the east and west margins. The uppermost reaches of the Des Plaines River in adjacent Kenosha County, Wisconsin, included marshes and wet prairie tracts (Overstreet, 1992).

La Salle County

An 1877 La Salle County history contains the following description:

Within a circuit of three miles of Ottawa there are three thousand mounds. The major part of these are unquestionably of Indian origin, the remainder may probably be attributed to the Mound Builders. (Kett, 1877, p. 175)

The Moundbuilder attribution may be safely ignored as impossible. A figure of 3000 mounds near Ottawa cannot be correct for Indian mounds unless Kett had included numerous natural prairie mounds along the upper Illinois River terraces and bottomlands. Lewis investigated this claim between Marseilles and Utica and stated, "Have seen but few mounds so far" (Lewis, 1881–1895b, letter to Hill dated April 25, 1888, from Ottawa, Illinois). In the past decade the writer has conducted archaeological surveys east and west of Ottawa in close proximity to the river. This work revealed the existence of an extensive near-surface sandstone bedrock bench terrace in this segment of the Illinois River. Its dimensions are greater than 11.3 km (7 miles) in length, from ~1.6 to 2.4 km (1–1.5 miles) in width, and typically found less than 0.6 m (2 feet) of the ground surface. Such a natural feature would present a barrier to burrowing animals, and a rationale for the existence of prairie mounds in this vicinity. If Kett's mound count estimate is accurate, then the occurrence of Mima mounds near Ottawa should be considered probable. The bedrock bench terrace is a legacy of the Kankakee Torrent when a catastrophic glacial lake drainage event in the terminal Pleistocene created the upper Illinois River Valley (Ferguson, 1995). Other researchers have noted the presence of widespread wet prairies along the lower Fox River in the vicinity of Ottawa (Schennum, 1992; White, 2000).

Madison and St. Clair Counties

The northern portion of the immense American Bottom floodplain of the Mississippi River in Madison and St. Clair counties contains three separate localities with probable natural mounds (Porter, 1974; Brackenridge, 1814; Flagg, 1838).

American Bottom. Writing about mounds in the American Bottom floodplain, McAdams (1881) made a statement similar to that previously issued for the Jersey County uplands.

There is in this region a peculiar class of mounds that was for a long time a puzzle to me. They are usually found in groups of from two or three to twenty or thirty, and even more, and are generally on some pleasant knoll, or rising ground, in the vicinity of a spring or watercourse, especially in the vicinity of our prairies or level areas of land. These mounds are from one to three, and, in a few instances, even four feet in height, and from twenty to fifty feet in diameter. One mound of the group is always larger than the rest, and always occupies the commanding position. (McAdams, 1881, p. 710)

For the interpretation McAdams (1881, p. 710–711) stated, "I am induced to believe that they are possibly the remains of ancient dwellings … in the same manner as many Indian tribes make their mud [earth] lodges, as, for instance, the Mandans and the Omahas."

East St. Louis site. Henry Brackenridge (1814) and Edmund Flagg (1838) indicated 45–50 mounds existed near Illinois Town, i.e., modern East St. Louis. These early description of the East St. Louis Mound Center (11S706) along the Madison and St. Clair counties line are unique for the mound total. All subsequent observers never saw a quantity approaching this number at East St. Louis. The next highest total is the 21 mounds that John Caspar Wild (1841) noted in 1841. The nineteenth-century urban and industrial developments undoubtedly resulted in mound destruction (Finney, 2000, 2007; Galloy and Kolb, 2008). When combining all reported East St. Louis tumuli, other than those of Brackenridge and Flagg, the count is no more than 25 documented mounds (Finney, 2007).

Mitchell site. The Mitchell site (11MS30) is a Mississippian temple mound center in Madison County. A massive report on highway salvage excavations at the site included a discussion of mounded soil features believed to have been surplus construction materials. The 1962 investigations revealed an absence of features, artifacts, and no basketloading in Mounds K and L. Mound K was described as consisting of gray clay sediments while Mound L was yellowish brown silts. Mound K was 2.5 feet in height, ~115 feet in diameter, and "described as a low rounded bump" (Porter, 1974, p. 336). The dimensions for Mound L were recorded as 2 feet in height, ~75 feet in diameter, and it was placed on the bank of Long Lake. The excavator interpreted these mounds as being soil stockpiles maintained by the prehistoric Mississippian inhabitants at the Mitchell site for future mound construction and that the gray clay had a second probable function for use in ceramic vessel production (Porter, 1974). Specifically, the spatial separation by soil color and texture could have served as stockpiles for making mound stage additions to the existing mounds placed around the plaza at the Mitchell site. Mississippian mound construction in the American Bottom used soil engineering including: (1) alternating soil colors in mound fill units, (2) lighter colored sediments for thin soil layers applied during annual world renewal ceremonies, and (3) clay sediments for increased soil stability and mound edge buttressing (Finney, 2007; Fortier and Finney, 2007; Moorehead, 1929b; Pauketat, 1993). However, based on the attribute descriptions from the salvage excavation it is equally possible that Mounds K and L at 11MS30 represent natural prairie mounds.

Mercer County

A nineteenth-century observer estimated that more than 1000 mounds existed in Mercer County (McWhorter, 1875). North-northeast of New Boston the aerial photographs and the county soil survey (Elmer, 1991) depict a series of low mounds on the Mississippi River floodplain in T14N, R5W. These mounds could be sand dunes, anthropic mounds, or possible natural prairie mounds (cf. McWhorter, 1875). They are positioned across the Mississippi River from the Muscatine Bottom in Iowa. McWhorter (1875, p. 351) reported another possible natural prairie mound site in the T15N, R5W uplands: "In this group may be found over two hundred mounds within the distance of a mile. This group seems to be a great antiquity, and is quite flattened down by the elements—only rising a few feet above the general level." A 1933 University of Chicago field survey crew recorded a few scattered mound groups in T15N, R5W. The site most likely to comprise a portion of the McWhorter site quoted above is the Hampton Mounds (11MC10) where an unspecified (probably fewer than 10) number of mounds were 30 feet in diameter and 2 feet in height (from Illinois Archaeological Survey site file).

REFERENCES CITED

Andreas, A.T., and Lyter, J., 1872, Atlas map of Jersey County, Illinois: Davenport, Iowa, Andreas, Lyter, and Company.

Anfinson, S.F., 1984, Cultural and natural aspects of mound distribution in Minnesota: The Minnesota Archaeologist, v. 43, no. 1, p. 3–30.

Anfinson, S.F., 1991, Recorded Archaeological Sites in Minnesota: Updated Version of 1988 Manuscript: Copy on file, State Historic Preservation Office, St. Paul, Minnesota Historical Society.

Anfinson, S.F., 1999, Giants in the earth: The archaeology of Northfield, *in* Clark, C., Zellie, C., Anfinson, S., Richardson, S., Rogness, D., and Swanson, S., eds., Northfield, the History and Architecture of a Community: Northfield, Minnesota, Northfield Heritage Preservation Commission, p. 38–42.

Anfinson, S.F., 2007, 21RC14—Dundas residential development, Rice County, *in* The State of Archaeology in Minnesota: Annual Report of the State Archaeologist Fiscal Year 2006: Fort Snelling, Minnesota, Office of the State Archaeologist, p. 27.

Anfinson, S.F., 2008, The State of Archaeology in Minnesota: Annual Report of the State Archaeologist Fiscal Year 2007: Fort Snelling, Minnesota, Office of the State Archaeologist, 45 p.

Anfinson, S.F., 2009, Archaeology in Minnesota: Annual Report of the State Archaeologist Fiscal Year 2008: Fort Snelling, Minnesota, Office of the State Archaeologist, 49 p.

Anfinson, S.F., 2010, Archaeology in Minnesota: Annual Report of the State Archaeologist Fiscal Year 2009: Fort Snelling, Minnesota, Office of the State Archaeologist, 43 p.

Anonymous, 1868, The geology of Iowa: American Naturalist, v. 2, no. 4, p. 207–208.

Anonymous, 1979, The 1979 resource inventory for Santee Prairie, Mahnomen county, Minnesota: Scientific and Natural Areas Section, Division of Parks and Recreation: St. Paul, Minnesota Department of Natural Resources, 6 p.

Anonymous, 2009, Schaefer Prairie: The Nature Conservancy: http://www.nature.org/wherewework/northamerica/states/minnesota/preserves/art7151.html (accessed 14 August 2009).

Arzigian, C.M., and Stevenson, K.P., 2003, Minnesota's Indian Mounds and Burial Sites: A Synthesis of Prehistoric and Early Historic Archaeological Data: St. Paul, Minnesota, Minnesota Office of the State Archaeologist, no. 1, 558 p.

Aten, L.E., and Bollich, C.N., 1981, Archeological evidence for pimple (prairie) mound genesis: Science, n.s., v. 213, no. 4514, p. 1375–1376.

Baisden, W.T., Amundson, R., Brenner, D.L., Cook, A.C., Kendall, C., and Harden, J.W., 2002, A multiisotope C and N modeling analysis of soil organic matter turnover and transport as a function of soil depth in a California annual grassland soil chronosequence: Global Biogeochemical Cycles, v. 16, p. 1135–1160, doi:10.1029/2001GB001823.

Barrett, S.A., and Hawkes, E.W., 1919, The Kratz Creek mound group: A study in Wisconsin Indian mounds: Bulletin of the Public Museum of the City of Milwaukee, v. 3, no. 1, p. 1–138.

Baxter, P.F., and Hole, F.D., 1967, Ant (*Formica cinerea*) pedoturbation in a prairie soil: Soil Science Society of America Journal, v. 31, p. 425–428, doi:10.2136/sssaj1967.03615995003100030036x.

Benn, D.W., 1976, Investigations at the so-called Hovey "mounds," Hancock County, Iowa: Iowa City, Iowa, Office of the State Archaeologist Research Papers: University of Iowa Research Reports, v. 1, no. 3, p. 1–13.

Benn, D.W., and Petersen, R.W., 1976, Archaeological Remains in the Deer Creek Watershed, Northwood, Iowa: Decorah, Iowa, Archaeological Research Center, Luther College, 31 p.

Bennet, F., Jr., Worthen, E.L., Willard, R.E., and Watson, E.B., 1911, Soil survey of Richland county, North Dakota: Bureau of Soils, Field Operations Bureau of Soils for 1908, 10th Report: Washington, D.C., U.S. Department of Agriculture, p. 1121–1151.

Berg, A.W., 1990a, Formation of mima mounds: A seismic hypothesis: Geology, v. 18, p. 281–284, doi:10.1130/0091-7613(1990)018<0281:FOMMAS>2.3.CO;2.

Berg, A.W., 1990b, Formation of mima mounds: A seismic hypothesis: Reply: Geology, v. 18, no. 12, p. 1259–1260, doi:10.1130/0091-7613(1990)018<1259:CAROFO>2.3.CO;2.

Berg, A.W., 1991, Formation of mima mounds: A seismic hypothesis: Reply: Geology, v. 19, no. 3, p. 284–285, doi:10.1130/0091-7613(1991)019<0284:CAROFO>2.3.CO;2.

Berry, B., Wrench, J.E., and Chapman, C., 1940, The archaeology of Wayne County: The Missouri Archaeologist, v. 6, no. 1, p. 1–40.

Bettis, E.A., 1988, Pedogenesis in late prehistoric Indian mounds, upper Mississippi valley: Physical Geography, v. 9, p. 263–278.

Bettis, E.A., Quade, D.J., and Kemmis, T.J., eds., 1996, Hogs, Bogs, and Logs: Quaternary Deposits and Environmental Geology of the Des Moines Lobe: Ames, Iowa State University Department of Natural Resources–Geological Survey Bureau, Guidebook Series No. 18, 170 p.

Bik, M.J.J., 1967, On the periglacial origin of prairie mounds, *in* Clayton, L., and Freers, T.F., eds., Glacial Geology of the Missouri Coteau and Adjacent Areas, Guidebook for the 18th Annual Field Conference of the Midwest Friends of the Pleistocene: Miscellaneous Series, no. 30: Grand Forks, North Dakota Geological Survey, p. 83–94.

Birmingham, R.A., and Eisenberg, L., 2000, Indian Mounds of Wisconsin: Madison, Wisconsin, University of Wisconsin Press, 245 p.

Blackmar, F., ed., 1912, Kansas: A Cyclopedia of State History: Chicago, Illinois, Standard Publishing Company, v. 2, 955 p.

Blair, C., Eldridge, J.L., Byers, O., McGovern, M., and Seal, U.S., eds., 2001, Sherburne National Wildlife Refuge Planning Workshop 2 Final Report: IUCN/SSC Conservation Breeding Specialist Group: Elk River, Minnesota, U.S. Fish and Wildlife Service, 102 p.

Blankinship, J.W., 1889, Peculiar earth-heaps in Missouri: The American Antiquarian, v. 11, p. 117.

Bluemle, J., 1983, Freckled land, *in* Bluemle, J., ed., NDGS Newsletter, December: Bismarck, North Dakota, North Dakota Geological Survey, Division of Mineral Resources, p. 35–38.

Blumer, A., 1883, Exploration of mounds in Louisa county, Iowa: Proceedings of the Davenport Academy of Natural Sciences, v. 3, p. 132–133.

Bocek, B., 1986, Rodent ecology and burrowing behavior: Predicted effects on archaeological site formation: American Antiquity, v. 51, no. 3, p. 589–603, doi:10.2307/281754.

Bogue, A.G., 1963, From Prairie to Corn Belt: Chicago, Illinois, University of Chicago Press, 310 p.

Bohannan, P., 1965, Introduction [to the reprint edition], *in* Morgan, L.H., Houses and House-Life of the American Aborigines: Chicago, Illinois, University of Chicago Press, p. v–xxi.

Borst, G., 1975, A cross section through a large Mima mound: Soil Survey Horizons, v. 16, no. 4, p. 20–23.

Boszhardt, R.F., 1987, Mapping mounds in the Cranberry Creek Archaeological District year 1: An example of cooperative archaeology in Wisconsin: Report of Investigation no. 58: La Crosse, Mississippi Valley Archaeological Center, University of Wisconsin–La Crosse.

Boszhardt, R.F., 1988, Mound mapping and related investigations at the Cranberry Archaeological District: Year 2: Report of Investigation no. 75: La Crosse, Mississippi Valley Archaeological Center, University of Wisconsin–La Crosse.

Boyle, B., 2010, Tamarac National Wildlife Refuge Comprehensive Conservation Plan: Rochert, Minnesota, U.S. Fish and Wildlife Service.

Brackenridge, H.M., 1814, Views of Louisiana Together with a Journal of a Voyage up the Missouri River in 1811: Pittsburgh, Pennsylvania, Cramer, Spear, and Eichbaum, 323 p.

Bragg, D.C., 2003, Natural presettlement features of the Ashley County, Arkansas: American Midland Naturalist, v. 149, no. 1, p. 1–20.

Bragg, D.C., and Weih, R.C., Jr., 2007, Notable environmental features in some historical aerial photographs from Ashley County, Arkansas: Journal of Arkansas Academy of Sciences, v. 61, p. 27–36.

Brotherson, J.D., 1982, Vegetation of the mima mounds of Kalsow Prairie, Iowa: The Great Basin Naturalist, v. 42, p. 246–261.

Brotherson, J.D., 1983, Species composition, distribution, and phytosociology of Kalsow Prairie, a mesic tall-grass prairie in Iowa: The Great Basin Naturalist, v. 43, no. 1, p. 137–167.

Brower, J.V., 1903, Minnesota: Discovery of Its Area, 1540–1665: Memoirs of Explorations in the Basin of the Mississippi, v. 6: St. Paul, Minnesota, H.L. Collins, 127 p.

Brower, J.V., 1904, Mandan: Memoirs of Explorations in the Basin of the Mississippi, v. 8: St. Paul, Minnesota, McGill-Warner, 158 p.

Brown, C.E., 1940, Charles E. Brown Archaeological Site Atlas: Copy on file, Archives: Madison, Wisconsin Historical Society.

Brunson, A., 1850, Ancient mounds or tumuli in Crawford county: Wisconsin Historical Collections, v. 3, p. 178–184.

Buikstra, J.E., Charles, D.K., and Rakita, G.F.M., 1998, Staging Ritual: Hopewell Ceremonialism at the Mound House Site, Greene County, Illinois: Kampsville Studies in Archeology and History, no. 1: Kampsville, Illinois, Center for American Archeology, 198 p.

Bullington, J., 1988, Middle Woodland mound structure: Social implications and regional context, *in* Charles, D.K., Leigh, S.R., and Buikstra, J.E., eds., The Archaic and Woodland Cemeteries at the Elizabeth Site in the Lower Illinois River Valley: Kampsville, Illinois, Center for American Archeology Research Series, v. 7, p. 218–241.

Buol, S.W., Southard, R.J., Graham, R.C., and McDaniel, P.A., 2003, Soil Genesis and Classification (fifth edition): Ames, Iowa State University Press, 510 p.

Bushnell, D.I., Jr., 1904, Archeology of the Ozark region of Missouri: American Anthropologist, v. 6, p. 294–298, doi:10.1525/aa.1904.6.2.02a00040.

Bushnell, D.I., Jr., 1905, The small mounds of the United States: Science, v. 22, no. 570, p. 712–714, doi:10.1126/science.22.570.712.

Byers, A.M., 2004, The Ohio Hopewell Episode: Paradigm Lost and Paradigm Gained: Akron, Ohio, University of Akron Press, 674 p.

Chappell, S., 2002, Cahokia: Mirror of the Cosmos: Chicago, Illinois, University of Chicago Press, 238 p.

Charles, D.K., and Buikstra, J.E., 2002, Siting, sighting, and citing the dead, *in* Silverman, H., and Small, D.B., eds., The Space and Place of Death: Archaeological Papers of the American Anthropological Association, v. 11, p. 13–25.

Charles, D.K., Van Nest, J.A., and Buikstra, J.E., 2004, From the earth: Minerals and meaning in the Hopewellian World, *in* Boivin, N., and Owoc, M.A., eds., Soils, Stones and Symbols: Cultural Perceptions of the Mineral World: London, UK, University College of London, p. 43–70.

Cole, H.E., and Flint, A.S., 1913, Archeological researches in the upper Baraboo Valley: The Wisconsin Archeologist, o.s., v. 12, no. 2, p. 41–52.

Collins, B., 1975, Range vegetation and Mima mounds in North Texas: Journal of Range Management, v. 28, no. 3, p. 209–211.

Conant, A.J., 1879, Footprints of Vanished Races in the Mississippi Valley: St. Louis, Missouri, Chancy R. Barns, 122 p.

Cox, G.W., 1984, Mounds of mystery: Natural History, v. 93, no. 6, p. 36–45.

Cox, G.W., 1990, Formation of Mima mounds: A seismic hypothesis: Comment: Geology, v. 18, no. 12, p. 1259–1260, doi:10.1130/0091-7613(1990)018 <1259:CAROFO>2.3.CO;2.

Cox, G.W., 2012, this volume, Alpine and montane Mima mounds of the western United States, *in* Horwath Burnham, J.L., and Johnson, D.L., eds., Mima Mounds: The Case for Polygenesis and Bioturbation: Geological Society of America Special Paper 490, doi:10.1130/2012.2490(03).

Cox, G.W., and Hunt, J., 1990, Nature and origin of stone stripes on the Columbia Plateau: Landscape Ecology, v. 5, p. 53–64, doi:10.1007 /BF00153803.

Cox, G.W., and Scheffer, V.B., 1991, Pocket gophers and mima terrain in North America: Natural Areas Journal, v. 11, p. 193–198.

Cramy, T.G., 1880, Untitled note on mounds in Meeker County, Minnesota, *in* Smithsonian Annual Report for 1879: Washington, D.C., Smithsonian Institution, p. 430.

Cronon, W., 1983, Changes in the Land: Indians, Colonists, and the Ecology of New England: New York, Hill and Wang, 241 p.

Curtis, E.S., 1924, The North American Indian: Volume 13—The Hupa. The Yurok. The Karok. The Wiyot. Tolowa and Tututni. The Shasta. The Achomawi. The Klamath: Seattle, Washington, E.S. Curtis, 316 p.

Curtis, J.T., 1959, The Vegetation of Wisconsin: Madison, University of Wisconsin Press, 640 p.

Curtiss-Wedge, F., ed., 1911, History of Mower County, Minnesota: Chicago, Illinois, H.C. Cooper, Jr., and Company, 1006 p.

Dalquest, W.W., and Scheffer, V.B., 1942, The origin of the Mima mounds of western Washington: The Journal of Geology, v. 50, no. 1, p. 68–84, doi:10.1086/625026.

Darwin, C., 1881, The Formation of Vegetable Mould through the Action of Worms with Observation of Their Habits: London, John Murray, 326 p.

Davids, R.C., 1968, The mystery of Mima mounds: Farm Journal, v. 92, no. 1, p. C-4.

de Nadaillac, M., 1884, Pre-Historic America (Translated by N. Bell d'Anvers): London, UK, G.P. Putnam's Sons, 566 p.

Deiss, R.W., 1992, A brief history and chronology of ceramic drainage and masonry tile produced in the United States: Journal of the International Brick Collectors Association, v. 10, no. 1, p. 6–18.

Dobbs, C.A., 1991, The Northwestern Archaeological Survey: An Appreciation and Guide to the Field Notebooks: Minneapolis, Minnesota, Institute for Minnesota Archaeology, Reports of Investigations, no. 135, 109 p.

Dornbush, M.E., 2004, Plant community change following fifty-years of management at Kalsow Prairie Preserve, Iowa, U.S.A.: American Midland Naturalist, v. 151, no. 2, p. 241–250, doi:10.1674/0003-0031(2004)151 [0241:PCCFFO]2.0.CO;2.

Dorney, J.R., 1983, Increase A. Lapham's pioneer observations and maps of land forms and natural disturbances: Wisconsin Academy of Arts: Sciences and Letters, v. 71, no. 2, p. 25–30.

Draper, L.C., 1888, Early French forts in western Wisconsin: Collections of the State Historical Society of Wisconsin, v. 10, p. 321–372.

Dudzik, M., 2005, Annual Report Fiscal Years 2004/05: Fort Snelling, Minnesota, Office of the State Archaeologist, 10 p.

Ellery, W.N., McCarthy, T.S., and Dangerfield, J.M., 1998, Biotic factors in Mima mound development: Evidence from the floodplains of the Okavango delta, Botswana: International Journal of Ecology and Environmental Sciences, v. 24, p. 293–313.

Elmer, S.L., 1991, Soil Survey of Mercer County, Illinois: Washington, D.C., U.S. Department of Agriculture, Soil Conservation Service, 205 p.

Embleton-Hamann, C., 2004, Processes responsible for the development of a pit and mound microrelief: Catena, v. 57, p. 175–188, doi:10.1016/j.catena .2003.10.017.

Engeman, R.M., and Witmer, G.W., 2000, Integrated management tactics for predicting and alleviating pocket gopher (*Thomomys* spp.) damage to conifer reforestation plantings: Integrated Pest Management Reviews, v. 5, p. 41–55.

Erlandson, J.M., 1984, A case study in faunalturbation: Delineating the effects of the burrowing pocket gopher on the distribution of archaeological materials: American Antiquity, v. 49, no. 4, p. 785–790, doi:10.2307/279743.

Estes, L.C., 1867, The antiquities on the banks of the Mississippi River and Lake Pepin, *in* Annual Report of the Smithsonian Institution for 1867: Washington, D.C., Smithsonian Institution, p. 366–367.

Eyles, N., Boyceb, J.I., and Barendregt, R.W., 1999, Hummocky moraine: Sedimentary record of stagnant Laurentide Ice Sheet lobes resting on soft beds: Sedimentary Geology, v. 123, p. 163–174, doi:10.1016/S0037 -0738(98)00129-8.

Farnsworth, K.B., 2004, Introduction, *in* Early Hopewell Mound Explorations: The First Fifty Years in the Illinois River Valley: Studies in Archaeology, no. 3, Illinois Transportation Archaeological Research Project: Urbana, University of Illinois, p. 1–93.

Ferguson, J.A., ed., 1995, Upper Illinois Valley Archaeology: The Cultural Resources of Starved Rock State Park: Technical Report 94-886-14, Quaternary Studies Program: Springfield, Illinois State Museum, 664 p.

Finney, F.A., 2000, Theodore H. Lewis and the Northwestern Archaeological Survey's 1891 fieldwork in the American Bottom: Illinois Archaeology, v. 12, p. 244–276.

Finney, F.A., 2001, An introduction to the Northwestern Archaeological Survey by Theodore H. Lewis: The Minnesota Archaeologist, v. 60, p. 13–29.

Finney, F.A., 2004, The 1878–1895 "Southern Archaeological Survey" of Theodore H. Lewis: Contract Completion Report, no. 40: St. Joseph, Illinois, Upper Midwest Archaeology, 56 p.

Finney, F.A., 2005, Theodore H. Lewis (1856–1930), an obituary: The Minnesota Archaeologist, v. 64, p. 11–20.

Finney, F.A., 2006, The archaeological legacy of Theodore H. Lewis: Letters, papers, and articles: The Wisconsin Archeologist, v. 87, p. 1–253.

Finney, F.A., 2007, History of site investigations, *in* Fortier, A.C., ed., The Archaeology of the East St. Louis Mound Center, Part 2: The Northside Excavations: Transportation Archaeological Research Reports, no. 22: Urbana, University of Illinois, p. 427–461.

Finney, F.A., 2008, A review of the Minnesota, Wisconsin, Iowa, and Illinois archaeological sites listed in William Pidgeon's *Traditions of De-Coo-Dah* that were relocated by T.H. Lewis: The Minnesota Archaeologist, v. 67, p. 89–105.

Finney, F.A., 2010, The Starved Rock Fort: Illinois Archaeology, v. 22, no. 1, p. 240–255.

Finney, F.A., and Johnson, D.L., 2010, Natural Prairie Mounds and a Taphonomic Assessment of Site 13AM79 at Harpers Ferry, Iowa: Contract Completion Report, no. 234: St. Joseph, Illinois, Upper Midwest Archaeology, 43 p.

Flagg, E., 1838, The Far West, or a Tour Beyond the Mountains, Vol. 1: New York, Harper and Brothers, 254 p.

Flemal, R.C., 1976, Pingos and pingo scars: Their characteristics, distribution, and utility in reconstructing former permafrost environments: Quaternary Research, v. 6, no. 1, p. 37–53, doi:10.1016/0033-5894(76)90039-9.

Fogelin, L., 2007, Inference to the best explanation: A common and effective form of archaeological reasoning: American Antiquity, v. 72, no. 4, p. 603–625, doi:10.2307/25470436.

Fokken, M., 1980, Phase II excavations at 13AM260 the Rhomberg St. Mounds, Harpers Ferry, *in* Fokken, M., and Cook, D., eds., A phase II report on four archaeological sites, 13AM26, 13AM152, 13AM251, 13AM260, in Allamakee County, Iowa: Iowa City, University of Iowa Highway Archaeology Program, Office of the State Archaeologist, p. 13–25.

Folgarait, P.J., 1998, Ant biodiversity and its relationship to ecosystem functioning: A review: Biodiversity and Conservation, v. 7, p. 1221–1244, doi:10.1023/A:1008891901953.

Fortier, A.C., and Finney, F.A., 2007, Features, *in* Fortier, A.C., ed., The Archaeology of the East St. Louis Mound Center, Part 2: The Northside Excavations: Transportation Archaeological Research Reports, no. 22: Urbana, University of Illinois, p. 513–661.

Fowke, G., 1910, Antiquities of Central and Southeastern Missouri Bulletin, no. 37, Bureau of American Ethnology: Washington, D.C., Smithsonian Institute, 116 p.

Fowke, G., 1922, Archeological Investigations: Washington, D.C., Smithsonian Institution, Bulletin, Bureau of American Ethnology, no. 76, 204 p.

Fowke, G., 1928, Archeological remains in Scott county, Ark., *in* Archaeological Investigations II: Forty-Fourth Annual Report of the Bureau of American Ethnology for 1926–1927: Washington, D.C., Smithsonian Institution, p. 464–466.

Fowke, G., 1929, Gerard Fowke: Ohio Archaeological and Historical Quarterly, v. 38, p. 201–218.

French, H.M., 1993, Cold-climate processes and landforms, *in* French, H.M., and Slaymaker, O., eds., Canada's Cold Environments: Montreal, Quebec, McGill-Queen's University Press, p. 143–167.

Fuller, M.L., 1912, The New Madrid Earthquake: Washington, D.C., U.S. Geological Survey Bulletin, no. 494, 119 p.

Gale, G., 1867, The Upper Mississippi: Historical Sketches of the Mound-Builders, the Indian Tribes, and the Progress of Civilization in the North-West: From A.D. 1600 to the Present Time: Chicago, Illinois, Clark, 460 p.

Gallaway, J.M., Martin, Y.M., and Johnson, E.A., 2009, Group sediment transport due to tree root throw: Integrating tree population dynamics, wildfire and geomorphic response: Earth Surface Processes and Landforms, v. 34, p. 1255–1269, doi:10.1002/esp.1813.

Galloy, J.M., and Kolb, M.F., 2008, Transposed mound fill in the CSX railyard, East St. Louis Mound Center (11S706): Illinois Archaeology, v. 20, p. 200–214.

Galm, J.R., and Keene, J.L., 2006, An archaeological perspective on the genesis of "mima" mounds in eastern Washington: Poster presented at the 71st Annual Meeting of the Society for American Archaeology, San Juan, Puerto Rico.

Gazin-Schwartz, A., and Holtorf, C., 1999, 'As long as I've ever known it …,' *in* Gazin-Schwartz, A., and Holtorf, C., eds., Archaeology and Folklore: London, UK, Routledge, p. 3–25.

Geertz, C., 1973, The Interpretation of Cultures: New York, Basic Books, 470 p.

Goldthwait, J.W., 1909, Physical Features of the Des Plaines Valley, Illinois State Geological Survey, Bulletin, no. 11: Urbana, University of Illinois, 103 p.

Goltz, G., 1993, Cultural resource reconnaissance survey of parcels affected by proposed wastewater collection project at Naytahwaush Community, White Earth Reservation: Cass Lake, Minnesota, All Nations Cultural Resource Protection.

Gravenor, C.P., 1955, The origin and significance of prairie mounds: American Journal of Science, v. 253, p. 475–481, doi:10.2475/ajs.253.8.475.

Greber, N.B., 1996, A commentary on the contexts and contents of large to small Ohio Hopewell deposits, *in* Pacheco, P.J., ed., A View from the Core: A Synthesis of Ohio Hopewell Archaeology: Columbus, Ohio Archaeological Council, p. 150–173.

Greene, G.S., 1975, Prehistoric utilization in the Channeled Scablands of eastern Washington [unpublished Ph.D. thesis]: Pullman, Washington State University, 149 p.

Gregg, R.E., 1944, The ants of the Chicago region: Annals of the Entomological Society of America, v. 37, p. 447–480.

Hall, R.L., 1979, In search of the ideology of the Adena-Hopewell climax, *in* Brose, D.S., and Greber, N., eds., Hopewell Archaeology: The Chillicothe Conference: Kent, Ohio, Kent State University Press, p. 258–265.

Hall, R.L., 1997, An Archaeology of the Soul: North American Indian Belief and Ritual: Urbana, University of Illinois Press, 240 p.

Hall, R.L., 2008, Inconstant companions: A review essay: Illinois Archaeology, v. 20, p. 228–234.

Harrington, M.W., 1884, The geology of Steele County, *in* The Geology of Minnesota: Minneapolis, Minnesota, Johnson, Smith and Harrison, State Printers, v. 1, p. 393–403.

Henderson, R., 2006, The sites we save: The Prairie Promoter, v. 21, no. 3, p. 5–6.

Henriksen, H.C., 1965, Utica Hopewell, a study of early Hopewellian occupation in the Illinois River Valley, *in* Herold, E.B., ed., Middle Woodland Sites in Illinois, Illinois Archaeological Survey Bulletin, no. 5: Urbana, Illinois, University of Illinois, p. 1–67.

Hinsley, C., 1981, Savages and Scientists: The Smithsonian Institution and the Development of American Anthropology: Washington, D.C., Smithsonian Institution Press, 336 p.

Hodder, I., 1999, The Archaeological Process: Oxford, Blackwell, 256 p.

Hole, F.D., 1981, Effects of animals on soil: Geoderma, v. 25, no. 1–2, p. 75–112, doi:10.1016/0016-7061(81)90008-2.

Holmes, W.H., 1893, Vestiges of early man in Minnesota: The American Geologist, v. 11, no. 4, p. 219–240.

Holtz, W., 1995, Phase II Investigations of the Sand Flats Site (47MO284) and the West Prairie Mound Group (47MO7) Monroe County, Wisconsin, Report of Investigation no. 211: La Crosse, Mississippi Valley Archaeological Center, University of Wisconsin–La Crosse.

Horwath, J.L., 2002, An Assessment of Mima Type Mounds, Their Soils, and Associated Vegetation, Newton County, Missouri [unpublished master's thesis]: Urbana, University of Illinois, 265 p.

Horwath, J.L., and Johnson, D.L., 2006, Mima-type mounds in southwest Missouri: Expressions of point-centered and locally thickened biomantles: Geomorphology, v. 77, p. 308–319, doi:10.1016/j.geomorph.2006.01.009.

Horwath, J.L., and Johnson, D.L., 2007, Mima-type mounds in southwest Missouri: Expressions of point-centered and locally thickened biomantles: Geomorphology, v. 83, p. 193–194, doi:10.1016/j.geomorph.2006.09.013.

Horwath, J.L., Johnson, D.L., and Stumpf, A.J., 2002, Evolution of a gravelly Mima-type moundfield in southwestern Missouri: Geological Society of America Abstracts with Programs, v. 34, no. 6, p. 369.

Horwath Burnham, J.L., Johnson, D.L., and Johnson, D.N., 2012, this volume, The biodynamic significance of double stone layers in Mima mounds, in Horwath Burnham, J.L., and Johnson, D.L., eds., Mima Mounds: The Case for Polygenesis and Bioturbation: Geological Society of America Special Paper 490, doi:10.1130/2012.2490(04).

Houck, L., 1908, A History of Missouri from the Earliest Explorations and Settlements until the Admission of the State into the Union: Chicago, Illinois, R.R. Donnelley and Sons, v. 1, 404 p.

Hoy, P.R., 1865, Journal of an exploration of western Missouri in 1854, under the auspices of the Smithsonian Institution: Annual Report of the Smithsonian Institution for 1864: Washington, D.C., Smithsonian Institution, v. 19, p. 431–438.

Hudak, G.J., and Ready, T.L., 1979, Cultural Resources Inventory of Lands Adjacent to Big Sandy Lake: St. Paul, Minnesota, The Science Museum of Minnesota: Copy on file, State Historic Preservation Office, Minnesota Historical Society, St. Paul.

Huntly, N., and Inouye, R.S., 1988, Pocket gophers in ecosystems: Patterns and mechanisms: Bioscience, v. 28, no. 11, p. 786–793.

Inouye, R.S., Huntly, N., and Wasley, G.A., 1997, Effects of pocket gophers on microtopographic variation: Journal of Mammalogy, v. 78, no. 4, p. 1144–1148.

Johnson, A.C., 1974–1979, Collected Information on Natural Prairie Mounds in Northeast Hancock County, Iowa: Manuscript on file, Office of the State Archaeologist: Iowa City, University of Iowa, 154 p.

Johnson, D.L., 1989, Subsurface stone lines, stone zones, artifact-manuport layers, and biomantles produced by bioturbation via pocket gophers (*Thomomys bottae*): American Antiquity, v. 54, no. 2, p. 370–389, doi:10.2307/281712.

Johnson, D.L., 1990, Biomantle evolution and the redistribution of earth materials and artifacts: Soil Science, v. 149, p. 84–102, doi:10.1097/00010694-199002000-00004.

Johnson, D.L., 2002, Darwin would be proud: Bioturbation, dynamic denudation, and the power of theory in science: Geoarchaeology, v. 17, p. 7–40, doi:10.1002/gea.10001.

Johnson, D.L., and Horwath Burnham, J.L., 2012, this volume, Introduction: Overview of concepts, definitions, and principles of soil mound studies, in Horwath Burnham, J.L., and Johnson, D.L., eds., Mima Mounds: The Case for Polygenesis and Bioturbation: Geological Society of America Special Paper 490, doi:10.1130/2012.2490(00).

Johnson, D.L., and Johnson, D.N., 2008, Some special case biomantle features and concepts: Stonelayers, Mima mounds, and hybrid mounds—with a focus on Texas and Louisiana soils, and some logical alternative "biomantle working hypotheses," Appendix 3 in Johnson, D.L., Mandel, R.D., and Frederick, C.D., eds., The Origin of the Sandy Mantle and Mima Mounds of the East Texas Gulf Coastal Plain: Geomorphological, Pedological, and Geoarchaeological Perspectives, GSA Field Trip Guidebook, Houston, Texas, Geological Society of America Annual Meeting, p. 113–122.

Johnson, D.L., Johnson, D.N., and West, R.C., 1999, Pocket gopher origins of some midcontinental Mima-type mounds: Regional and interregional genetic implications: Geological Society of America Abstracts with Programs, v. 31, no. 7, p. A232.

Johnson, D.L., Johnson, D.N., and Horwath, J.L., 2002, In praise of the coarse fraction and bioturbation: Gravelly Mima mounds as two-layered biomantles: Geological Society of America Abstracts with Programs, v. 34, no. 6, p. 369.

Johnson, D.L., Horwath, J.L., and Johnson, D.L., 2003, Mima and other animal mounds as point-centered biomantles: Geological Society of America Abstracts with Programs, v. 35, no. 6, p. 258.

Johnson, D.L., Domier, J.E.J., and Johnson, D.N., 2005, Reflections on the nature of soil and its biomantle: Annals of the Association of American Geographers, v. 95, p. 11–31, doi:10.1111/j.1467-8306.2005.00448.x.

Johnson, D.L., Johnson, D.N., Horwath, J.L., Wang, H., Hackley, K.C., and Cahill, R.A., 2006, Mima mounds as upper soil biomantles: What happens when the dominant bioturbators leave and invertebrates take over?: Philadelphia, Pennsylvania, 18th World Congress of Soil Science, abstract 122-6.

Johnson, D.L., Mandel, R., and Frederick, C., 2008, The Origin of the Sandy Mantle and Mima Mounds of the East Texas Gulf Coastal Plain: Geomorphological, Pedological, and Geoarchaeological Perspectives: GSA Field Trip Guidebook, Houston, Texas, Geological Society of America Annual Meeting, 122 p.

Johnson, R.R., Oslund, F.T., and Hertel, D.R., 2008, The past, present, and future of prairie potholes in the United States: Journal of Soil and Water Conservation, v. 63, no. 3, p. 84A–87A, doi:10.2489/jswc.63.3.84A.

Jones, S.R., and Cushman, R.C., 2004, A Field Guide to the North American Prairie: New York, Houghton Mifflin, 510 p.

Kaplan, J.O., Krumhardt, M., and Zimmermann, N., 2009, The prehistoric and preindustrial deforestation of Europe: Quaternary Science Reviews, v. 28, p. 3016–3034, doi:10.1016/j.quascirev.2009.09.028.

Kelley, J.H., and Hanen, M.P., 1988, Archaeology and the Methodology of Science: Albuquerque, University of New Mexico Press, 437 p.

Kessler, M.A., and Werner, B.T., 2003, Self-organization of sorted patterned ground: Science, v. 299, p. 380–383, doi:10.1126/science.1077309.

Kett, H.F., 1877, The Past and Present of La Salle County, Illinois: Chicago, Illinois, H.F. Kett, 655 p.

King, T.J., 1981, Ant-hills and grassland history: Journal of Biogeography, v. 8, no. 4, p. 329–334, doi:10.2307/2844766.

Klaas, B.A., Moloney, K.A., and Danielson, B.J., 2000, The tempo and mode of gopher mound production in a tall grass prairie remnant: Ecography, v. 23, no. 2, p. 246–256, doi:10.1111/j.1600-0587.2000.tb00280.x.

Kloehn, N.B., Junck, M.B., Jol, H.M., Running, G.L., Greek, D., and Caldwell, K., 2000, GPR investigation of the West Prairie Mound Group, central Wisconsin, U.S.A.: Are they burial mounds or natural landforms?: SPIE Proceedings series v. 4084, p. 590–595.

Kolb, M.F., 1998, Geoarchaeological investigations at the southern subgroup of 21-HE-19: Report of Investigations, no. 7: Sun Prairie, Wisconsin, Strata Morph Geoexploration.

Krinitzsky, E.L., 1949, Origin of pimple mounds: American Journal of Science, v. 247, p. 706–714, doi:10.2475/ajs.247.10.706.

Lane, D.R., and BassiriRad, H., 2005, Diminishing effects of ant mounds on soil heterogeneity across a chronosequence of prairie restoration sites: Pedobiologia, v. 49, no. 4, p. 359–366, doi:10.1016/j.pedobi.2005.04.003.

Lapham, I.A., 1855, The Antiquities of Wisconsin: Contributions to Knowledge, no. 7: Washington, D.C., Smithsonian Institution, 110 p.

Lass, B., 1981, A cultural resources survey of Sherburne County, Minnesota: Minneapolis, University of Minnesota: Copy on file, State Historic Preservation Office, Minnesota Historical Society, St. Paul.

Lee, B.D., and Carter, B.J., 2010, Soil morphological characteristics of prairie mounds in the forested region of south-central United States, in Soil Solutions for a Changing World: Brisbane, Australia, 19th World Congress of Soil Science, p. 37–40.

Lee, B.D., Carter, B.J., Ward, P.A., Taylor, R.S., and Lee, L.P., 2004, Morphology of mounded soils in the Ouachita physiographic province, eastern Oklahoma, in Agronomy Abstracts: Madison, Wisconsin, American Society of Agronomy, 200 p.

Lee, J.A., 1986, Origin of mounds under Creosote Bush (*Larrea tridentata*) on terraces of the Salt River, Arizona: Journal of the Arizona-Nevada Academy of Science, v. 21, no. 1, p. 23–28.

Lewis, T.H., 1881–1895a, Northwestern Archaeological Survey correspondence from T.H. Lewis to A.J. Hill: Manuscript on file, Archives: St. Paul, Minnesota Historical Society.

Lewis, T.H., 1881–1895b, Unpublished field notes, Notebook nos. 1–41. Manuscript on file, Archives: St. Paul, Minnesota, Minnesota Historical Society.

Lewis, T.H., 1883a, Swamp mounds: The American Antiquarian, v. 5, no. 4, p. 330–331.

Lewis, T.H., 1883b, In the Mandan mounds: T.H. Lewis, the archaeologist, reports the result of his recent delvings on the Missouri's banks: St. Paul and Minneapolis, Minnesota, Pioneer Press, v. 30, no. 329, p. 12, November 24.

Lewis, T.H., 1886, Mounds on the Red River of the north: The American Antiquarian, v. 8, no. 6, p. 369–371.

Lewis, T.H., 1898, The Northwestern Archaeological Survey: St. Paul, Minnesota, Pioneer Press, 16 p.

Lynott, M.J., 1982, An Evaluation of Three Archeological Sites in the Ozark National Scenic Riverways: Lincoln, Nebraska, Midwest Archeological Center, 115 p.

MacBride, T.H., 1902, The geology of Cherokee and Buena Vista counties, *in* Iowa Geological Survey Annual Report 1901: Iowa Geological Survey, v. 12, p. 305–354.

MacBride, T.H., 1905, The geology of Emmet, Palo Alto, and Pocahontas counties, *in* Iowa Geological Survey Annual Report 1904: Iowa Geological Survey, v. 15, p. 227–276.

Macgowan, E.S., 1948, Did the Dakota Indians build earth houses in Minnesota?: The Minnesota Archaeologist, v. 14, no. 3, p. 65–66.

MacLean, J.P., 1879, The Mound Builders, Archaeology of Butler County, O.: Cincinnati, Ohio, Robert Clarke, 233 p.

Madigan, T., and Bailey, T., 1998, Burial Site Authentication Investigations at 21-HE-0019 (south sub-group), Eden Prairie, Minnesota: Reports of Investigation, no. 534: Minneapolis, Minnesota, IMA Consulting.

Markley, M.C., 1948, Dakota County lowland mounds: The Minnesota Archaeologist, v. 14, no. 3, p. 63–64.

Martin, R.E., 1999, Taphonomy: A Process Approach: Cambridge, Massachusetts, Cambridge University Press, 508 p.

Mason, R.J., 2006, Inconstant Companions: Archaeology and North American Indian Oral Traditions: Tuscaloosa, Alabama, University of Alabama Press, 298 p.

McAdams, W., Jr., 1880, Mounds of Illinois: Science, v. 1, p. 138–139, doi: 10.1126/science.os-1.12.138-b.

McAdams, W., Jr., 1881, The ancient mounds of Illinois: Proceedings of the American Association for the Advancement of Science, v. 29, p. 710–721.

McAdams, W., Jr., 1887, Records of Ancient Races in the Mississippi Valley: St. Louis, Missouri, C.R. Barns Publishing Company, 120 p.

McAdams, W., Jr., 1919, Evidences of great antiquity, *in* Hamilton, O.B., ed., History of Jersey County, Illinois: Chicago, Illinois, Munsell Publishing Company, p. 26–40.

McGee, W.J., 1891, Pleistocene history of northeastern Iowa: Eleventh Annual Report of the U.S. Geological Survey for 1889–1890: Washington, D.C., U.S. Geological Survey, p. 189–577.

McKern, W.H., 1930, Kletzien and Nitschke mound groups: Bulletin of the Milwaukee Public Museum, v. 3, no. 4, p. 417–572.

McKusick, M., 1991, The Davenport Conspiracy Revisited: Ames, Iowa State University Press, 193 p.

McWhorter, T., 1875, Ancient mounds of Mercer County, Illinois: Annual Report of the Smithsonian Institution, v. 29, p. 351–353.

Melton, F.A., 1929, Natural mounds of northeastern Texas, southern Arkansas, and northern Louisiana: Proceedings of the Oklahoma Academy of Science, v. 9, p. 119–130.

Meltzer, D.J., 1985, North American archaeology and archaeologists, 1879–1934: American Antiquity, v. 50, no. 2, p. 249–260, doi:10.2307/280483.

Meltzer, D.J., 1998, Introduction, Ephraim Squier, Edwin Davis, and the making of an American archaeological classic, *in* Ancient Monuments of the Mississippi Valley (reprint edition): Washington, D.C., Smithsonian Institution Press, p. 1–95.

Meltzer, D.J., 2009, First Peoples in a New World: Colonizing Ice Age America: Berkeley, University of California Press, 446 p.

Meltzer, D.J., and Dunnell, R.C., 1992, Introduction, *in* Meltzer, D.J., and Dunnell, R.C., eds., The Archaeology of William Henry Holmes: Washington, D.C., Smithsonian Institution Press, p. vii–l.

Meysman, F.J.R., Middelburg, J.J., and Heip, C.H.R., 2006, Bioturbation: A fresh look at Darwin's last idea: Trends in Ecology & Evolution, v. 21, no. 12, p. 688–695.

MHS (Minnesota Historical Society), 1917, Minnesota Historical Society Nineteenth Biennial Report for the Years 1915 and 1916: St. Paul, Minnesota, Minnesota Historical Society.

Mielke, H.W., 1977, Mound building by pocket gophers (*Geomyidae*): Their impact on soils and vegetation in North America: Journal of Biogeography, v. 4, p. 171–180, doi:10.2307/3038161.

Mier, L., Kolb, M.F., and Richards, J.D., 1996, Archaeological Investigations at the West Prairie Mound Group (47Mo7): Reports of Investigations no. 378: Milwaukee, Wisconsin, Great Lakes Archaeological Research Center.

Montgomery, D.R., 2007, Dirt: The Erosion of Civilizations: Berkeley, California, University of California Press, 285 p.

Montgomery, H., 1906, Remains of prehistoric man in the Dakotas: American Anthropologist, v. 8, no. 4, p. 640–651, doi:10.1525/aa.1906.8.4.02a00040.

Moorehead, W.K., 1929a, The mound builder problem to date: American Anthropologist, v. 31, no. 3, p. 544–554, doi:10.1525/aa.1929.31.3.02a00280.

Moorehead, W.K., 1929b, The Cahokia Mounds: University of Illinois Bulletin, v. 26, no. 4, p. 9–273.

Morgan, L.H., 1881, Houses and house-life of the American aborigines: Contributions to North American Ethnology: Washington, D.C., Smithsonian Institution, v. 4, no. 1, 281 p.

Morse, D.F., and Morse, P.A., 1983, Archaeology of the Central Mississippi Valley: New York, Academic Press, 345 p.

Myers, T.P., ed., 1978, Current research: American Antiquity, v. 43, no. 1, p. 104–126.

Nabokov, P., 2002, A Forest of Time: American Indian Ways of History: New York, Cambridge University Press, 246 p.

Nelson, R.E., 1997, Implications of subfossil Coleoptera for the evolution of the mima mounds of southwestern Puget Lowland, Washington: Quaternary Research, v. 47, p. 356–358, doi:10.1006/qres.1997.1891.

Norman, S.A., Schaetzl, R.J., and Small, T.W., 1995, Effects of slope angle on mass movement by tree uprooting: Geomorphology, v. 14, p. 19–27, doi: 10.1016/0169-555X(95)00016-X.

Northwest Publishing, 1900, Plat Book of Rice County, Minnesota: Minneapolis, Minnesota, Northwest Publishing Company, 46 p.

Northwest Publishing, 1901, Plat Book of Wright County, Minnesota: Minneapolis, Minnesota, Northwest Publishing Company, 56 p.

Nystuen, D.W., 1973, Project 8285-53 in District 9, *in* The Minnesota Trunk Highway Archaeological Reconnaissance Survey Annual Report—1972: St. Paul, Minnesota, Minnesota Historical Society, Department of Archaeology.

O'Brien, M.J., 1996, Paradigms of the Past: The Story of Missouri Archaeology: Columbia, University of Missouri Press, 561 p.

O'Brien, M.J., and Wood, W.R., 1998, The Prehistory of Missouri: Columbia, University of Missouri Press, 417 p.

O'Brien, M.J., Lyman, R.L., and Holland, T.D., 1989, Geoarchaeological evidence for prairie-mound formation in the Mississippi alluvial valley, southeastern Missouri: Quaternary Research, v. 31, p. 83–93, doi:10.1016/0033-5894(89)90087-2.

Osborne, R., 2004, Hoards, votives, offerings, the archaeology of the dedicated object: World Archaeology, v. 36, no. 1, p. 1–10, doi:10.1080/00438240 42000192696.

Overstreet, D.F., 1992, Archaeological Studies on the Southeast Wisconsin Uplands: Case Studies in Great Lakes Archaeology No. 1: Milwaukee, Wisconsin, Great Lakes Archaeological Press, 131 p.

Owoc, M.A., 2002, Munselling the mound: The use of colour as metaphor in British Bronze Age funerary ritual, *in* Jones, A., and MacGregor, G., eds., Colouring the Past: The Significance of Colour in Archaeological Research: Oxford, UK, Berg, p. 127–140.

Parsons, R.B., 1962, Indian mounds of northeast Iowa as soil genesis benchmarks: Journal of the Iowa Archeological Society, v. 12, no. 2, p. 1–70.

Patterson, C.J., Hansel, A.K., Mickelson, D.M., Quade, D.J., Bettis, E.A., Kemmis, T.J., Colgan, P.M., McKay, E.D., and Stumpf, A.J., 2003, Contrasting glacial landscapes created by ice lobes of the southern Laurentide ice sheet, *in* Easterbrook, D.J., ed., Quaternary Geology of the United States: INQUA 2003 Field Guide Volume: Reno, Nevada, The Desert Research Institute, p. 135–153.

Pauketat, T.R., 1993, Temples for Cahokia Lords: Preston Holder's 1955–1956 Excavations of Kunnemann Mound: Memoirs, no. 26: Ann Arbor, University of Michigan, Museum of Anthropology, 166 p.

Pauketat, T.R., 2004, Ancient Cahokia and the Mississippians: Cambridge, Cambridge University Press, 218 p.

Pauketat, T.R., 2009, Cahokia: Ancient America's Great City on the Mississippi: New York, Viking-Penguin, 194 p.

Payne, W.O., 1911, History of Story County, Iowa, Vol. 1: Chicago, Illinois, S.J. Clarke, 546 p.

Peet, S.D., 1883a, Note: The American Antiquarian, v. 5, no. 4, p. 331.

Peet, S.D., 1883b, Swamp mounds: The American Antiquarian, v. 5, no. 4, p. 337.

Peet, S.D., 1883c, Mounds at Sparta and New Lisbon: The American Antiquarian, v. 5, no. 4, p. 337–338.

Peet, S.D., 1892, The Mound Builders: Their Works and Relics: Prehistoric America, v. 1: Chicago, Illinois, Office of the American Antiquarian, 370 p.

Perttula, T.K., 1996, Caddoan area archaeology since 1990: Journal of Archaeological Research, v. 4, no. 4, p. 295–348, doi:10.1007/BF02229090.

Petersen, R.W., 1984, A survey of the destruction of effigy mounds in Wisconsin and Iowa—A perspective: The Wisconsin Archeologist, v. 65, no. 1, p. 1–31.

Peterson, L.D., 1974, The Minnesota Trunk Highway Archaeological Reconnaissance Survey Annual Report 1973: St. Paul, Minnesota, Minnesota Historical Society, Department of Archaeology.

Pierce, C., 1992, Effects of pocket gopher burrowing on archaeological deposits: A simulation approach: Geoarchaeology, v. 7, no. 3, p. 185–208, doi:10.1002/gea.3340070302.

Pinkney, B.F., 1896, Plat Book of Dakota County, Minnesota: Philadelphia, Pennsylvania, Union Publishing Company, 55 p.

Pond, S.W., 1908, The Dakotas or Sioux in Minnesota as they were in 1834: Collections, v. 12: St. Paul, Minnesota: Historia y Sociedad (Rio Piedras, San Juan, P.R.), p. 319–501.

Porter, J.W., 1974, Cahokia archaeology as viewed from the Mitchell site: A satellite community at A.D. 1150–1200 [unpublished Ph.D. thesis]: Madison, University of Wisconsin, 1248 p.

Pratt, W.H., 1883, The President's annual address: Proceedings of the Davenport Academy of Natural Sciences, v. 3, p. 151–157.

Price, J.E., 1977, Anticipated impacts of the Little Black River Watershed on the finite cultural resource base, *in* Schiffer, M.B., and Gumerman, G.J., eds., Conservation Archaeology: New York, Academic Press, p. 303–308.

Price, J.E., Price, C.R., Cottier, J., Harris, S.E., and House, J., 1975, An Assessment of the Cultural Resources of the Little Black Watershed: Columbia, Missouri, University of Missouri, Division of American Archaeology.

Price, W.A., 1949, Pocket gophers as architects of mima (pimple) mounds of the western United States: The Texas Journal of Science, v. 1, p. 1–17.

Ramankutty, N., and Foley, J.A., 1999, Estimating historical changes in global land cover: Croplands from 1700 to 1992: Global Biogeochemical Cycles, v. 13, no. 4, p. 997–1027, doi:10.1029/1999GB900046.

Ready, T., 1977, Unpublished field notes for site 21AK28 dated August 26, 1977: Copy on file, Site 21AK28 in the Minnesota Archaeological Site File, Minnesota State Historic Preservation Office: St. Paul, Minnesota Historical Society.

Reed, S., and Amundson, R., 2012, this volume, Using LIDAR to model Mima mound evolution and regional energy balances in the Great Central Valley, California, *in* Horwath Burnham, J.L., and Johnson, D.L., eds., Mima Mounds: The Case for Polygenesis and Bioturbation: Geological Society of America Special Paper 490, doi:10.1130/2012.2490(01).

Rick, T.C., Erlandson, J.M., and Vellanoweth, R.L., 2006, Taphonomy and site formation on California's Channel Islands: Geoarchaeology, v. 21, no. 6, p. 567–589, doi:10.1002/gea.20124.

Riggs, R.E., 1989, Mounds and Mound Sites in Wisconsin: A Brief Study of Their Function: Burial Sites Preservation Program, Historic Preservation Division: Copy on file, State Historic Preservation Office: Madison, Wisconsin, State Historical Society of Wisconsin.

Rosburg, T.R., 2001, Iowa's Prairie Heritage: From the past, through the present, and into the future: Proceedings of the 17th North American Prairie Conference, p. 1–14.

Ross, B.A., Tester, J.R., and Breckenridge, W.J., 1968, Ecology of Mima-type mounds in northwestern Minnesota: Ecology, v. 49, p. 172–177, doi: 10.2307/1933579.

Salkin, P.H., 1976, Excavation of Earll Mound #1: Some hypotheses on the function of vacant mounds in the Effigy Mound tradition: The Wisconsin Archeologist, v. 57, no. 3, p. 152–164.

Salkin, P.H., and Hendrickson, C.F., 1983, A cultural resources reconnaissance survey and limited testing of the area of the permit application of Mr. Thomas Sullivan, Aitkin County, Minnesota: Madison, Wisconsin, Archaeological Consulting and Services.

Sasso, R.F., 2003, Vestiges of ancient cultivation: The antiquity of garden beds and corn hills in Wisconsin: Midcontinental Journal of Archaeology, v. 28, no. 2, p. 195–231.

Saucier, R., 1991, Formation of Mima mounds: A seismic hypothesis: Comment: Geology, v. 19, no. 3, p. 284, doi:10.1130/0091-7613(1991)019<0284:CAROFO>2.3.CO;2.

Saucier, R., 1994, Geomorphology and Quaternary Geologic History of the Lower Mississippi Valley, Vol. 1: Vicksburg, Mississippi, U.S. Army Engineer Waterways Experiment Station, 364 p.

Savage, T.E., 1905, The geology of Fayette County, *in* Iowa Geological Survey Annual Report 1904: Des Moines, Iowa, Iowa Geological Survey, v. 15, p. 433–546.

Schaetzl, R.J., and Anderson, S., 2005, Soils: Genesis and Geomorphology: New York, Cambridge University Press, 817 p.

Schaetzl, R.J., and Follmer, L.R., 1990, Longevity of treethrow microtopography: Implications for mass wasting: Geomorphology, v. 3, p. 113–123, doi:10.1016/0169-555X(90)90040-W.

Schaetzl, R.J., Johnson, D.L., Burns, S.F., and Small, T.W., 1989, Tree uprooting: Review of terminology, process, and environmental implications: Canadian Journal of Forest Research, v. 19, p. 1–11, doi:10.1139/x89-001.

Schennum, W.E., 1986, A comprehensive survey for prairie remnants in Iowa: Methods and preliminary results. Proceedings of the 9th North American Prairie Conference, p. 163–168.

Schennum, W.E., 1992, Wetlands: Reservoirs for prairie biota in "prairieless" landscapes: Proceedings of the 12th North American Prairie Conference, p. 95–100.

Schiffer, M.B., 1987, Formation Processes of the Archaeological Record: Albuquerque, University of New Mexico Press, 428 p.

Schmidt, E.W., 1908, Lowland mounds in Dakota, Rice and Goodhue Counties: Paper presented at the December meeting of the Minnesota Historical Society, St. Paul.

Schmidt, E.W., 1909, Evidence of the Mounds, *in* Curtiss-Wedge, F., ed., History of Goodhue County, Minnesota: Chicago, Illinois, H.C. Cooper, Jr., and Company, p. 18–30.

Schmidt, E.W., 1910, Prof. Schmidt's paper, *in* Curtiss-Wedge, F., ed., History of Rice and Steele Counties, Minnesota: Chicago, Illinois, H.C. Cooper, Jr., and Company, p. 9–17.

Schmidt, E.W., 1928, Earthen mounds in Cannon River Valley held to be work of prehistoric Indians: Northfield Independent, v. 41, no. 39, p. 27.

Schmidt, E.W., 1934a, Geology and lowland mounds history topics: Northfield Independent, v. 47, no. 16, p. 19.

Schmidt, E.W., 1934b, Remote antiquity of Rice County area pictured to history group: Dr. Gould discusses geological past, and Prof. Schmidt Indian remains: Northfield Independent, v. 47, no. 17, p. 26.

Schmidt, E.W., 1934c, Report on explorations of lowland mounds in Rice and Dakota counties: Northfield Independent, v. 47, no. 42, p. 18.

Schmidt, E.W., 1935a, A group of Minnesota lowland mounds: Minnesota History, v. 16, p. 307–312.

Schmidt, E.W., 1935b, Schmidt plans collection of Indian relics: Northfield News, v. 59, no. 17, p. 26.

Schmidt, E.W., 1935c, Beaver dams in Rice and adjoining parts of Dakota counties: Northfield Independent, v. 48, no. 4, p. 24.

Schmidt, E.W., 1937, Lowland Mounds in the Northfield Area: Northfield, Minnesota, Mohn Printing, 34 p.

Schmidt, E.W., 1941, A brief archaeological survey of the Red Wing area: The Minnesota Archaeologist, v. 7, p. 70–80.

Scott, D.D., Thiessen, T.D., Richner, J.J., and Stadler, S., 2006, An Archeological Inventory and Overview of Pipestone National Monument, Minnesota: Occasional Studies in Anthropology, no. 34: Lincoln, Nebraska, Midwest Archeological Center, 404 p.

Shermer, M., 1997, Why People Believe Weird Things: Pseudoscience, Superstition, and Other Confusions of Our Time: New York, W.H. Freeman, 306 p.

Sherwood, S.C., and Kidder, T.R., 2011, The DaVincis of dirt: Geoarchaeological perspectives on Native American mound building in the Mississippi River basin: Journal of Anthropological Archaeology, v. 30, p. 69–87.

Silverberg, R., 1968, Mound Builders of Ancient America: The Archaeology of a Myth: Greenwich, Connecticut, New York Graphic Society, 369 p.

Smith, A.C., 1877, A Random Historical Sketch of Meeker County, Minnesota: Litchfield, Minnesota, Belfoy and Joubert, 160 p.

Smith, B., 1985, Introduction, *in* Thomas, C., ed., Report on the Mound Explorations of the Bureau of Ethnology, reprinted: Washington, D.C., Smithsonian Institution Press, p. 5–18. (Originally published in 1894 in the Twelfth Annual Report of the Bureau of American Ethnology, 1890–1891: Washington, D.C., Smithsonian Institution, 742 p.)

Smith, D.D., 1998, Iowa prairie: Original extent and loss, preservation and recovery attempts: Journal of the Iowa Academy of Sciences, v. 105, no. 3, p. 94–108.

Smith, H.I., 1910, The archaeology of the Yakima valley: Anthropological Papers of the American Museum of Natural History, v. 6, no. 1, p. 1–170.

Smith, L.M., 1996, Fluvial geomorphic features of the Lower Mississippi alluvial valley: Engineering Geology, v. 45, p. 139–165, doi:10.1016/S0013-7952(96)00011-7.

Snyder, J.F., 1890, Were the Osages mound builders?: Smithsonian Annual Report for 1888: Washington, D.C., Smithsonian Institution, p. 587–596.

Squier, E.G., 1850, Aboriginal Monuments of the State of New York: Washington, D.C., Smithsonian Institution, Smithsonian Contributions to Knowledge, v. 2, no. 9, 228 p.

Squier, E.G., and Davis, E.H., 1848, Ancient monuments of the Mississippi valley: Washington, D.C., Smithsonian Institution, Smithsonian Contributions to Knowledge, v. 1, 306 p.

Squier, G.H., 1905, Certain archeological features of western Wisconsin: The Wisconsin Archeologist, o.s., v. 4, no. 2, p. 25–34.

Starr, F., 1897, Summary of the archaeology of Iowa: Proceeding of the Davenport Academy of Natural Sciences, v. 6, p. 53–124.

Stein, J.K., 2001, A review of archaeological site formation processes and their relevance to Geoarchaeology, in Goldberg, P., Holliday, V.T., and Ferring, C.R., eds., Earth Sciences and Archaeology: New York, Plenum, p. 37–51.

Stevenson, J.E., 1879, The mound builders: The American Antiquarian, v. 2, no. 2, p. 89–104.

Swanton, J.R., 1942, David Ives Bushnell, Jr: American Anthropologist, v. 44, no. 1, p. 104–110, doi:10.1525/aa.1942.44.1.02a00100.

Taylor, C., 1928, Uncle Charlie Taylor remembers coming to Ames in 1868: http://www.ameshistoricalsociety.org/stories/taylor.htm (accessed 5 January 2010).

Terrell, M., 2005, Trunk Highway 14—New Ulm to North Mankato Archaeological Survey Nicollet County, Minnesota: Phase I Archaeological and Geomorphological Survey and Phase II Archaeological Testing of 21NL58, 21NL59 and 21NL134: Shafer, Minnesota, Two Pines Resource Group.

Tester, J.R., 1995, Minnesota's Natural Heritage: An Ecological Perspective: Minneapolis, University of Minnesota Press, 332 p.

Tester, J.R., and Breckenridge, W.J., 1964, Population dynamics of the Manitoba Toad, Bufo Hemiophrys, in northwestern Minnesota: Ecology, v. 45, no. 3, p. 592–601, doi:10.2307/1936111.

Thoburn, J.B., 1929, The prehistoric cultures of Oklahoma: Chronicles of Oklahoma, v. 7, no. 3, p. 211–241.

Thoburn, J.B., 1937, The origin of the "natural" mounds of Oklahoma and adjacent states: Chronicles of Oklahoma, v. 15, no. 3, p. 322–343.

Thomas, C., 1884, The houses of the mound-builders: The Magazine of American History, v. 9, p. 110–116.

Thomas, C., 1891, Catalogue of Prehistoric Works East of the Rocky Mountains: Bulletin no. 12: Washington, D.C., Smithsonian Institution, Bureau of American Ethnology, 246 p.

Thomas, C., 1894, Report on the Mound Explorations of the Bureau of Ethnology: Twelfth Annual Report of the Bureau of American Ethnology for 1890–1891: Washington, D.C., Smithsonian Institution, p. 3–742.

Thompson, T., 1880, Mounds in Muscatine county, Iowa, and Rock Island county, Illinois: Annual Report of the Smithsonian Institution, v. 1879, p. 359–363.

Trigger, B., 2006, A History of Archaeological Thought (second edition): Cambridge, UK, Cambridge University Press, 710 p.

Union Publishing Company, 1895, Plat Book of Floyd County, Iowa: Philadelphia, Pennsylvania, Union Publishing, 47 p.

Upham, W., 1884a, The Geology of Fairbault county, in The Geology of Minnesota: Minneapolis, Minnesota, Johnson, Smith and Harrison, State Printers, v. 1, p. 454–471.

Upham, W., 1884b, The geology of Murray and Nobles counties, in The Geology of Minnesota: Minneapolis, Minnesota, Johnson, Smith and Harrison, State Printers, v. 1, p. 517–532.

Upham, W., 1884c, The geology of Waseca County, in The Geology of Minnesota: Minneapolis, Minnesota, Johnson, Smith and Harrison, State Printers, v. 1, p. 404–414.

Upham, W., 1884d, The geology of Cottonwood and Jackson counties, in The Geology of Minnesota: Minneapolis, Minnesota, Johnson, Smith and Harrison, State Printers, v. 1, p. 491–516.

Upham, W., 1887, The Upper Beaches and Deltas of the Glacial Lake Agassiz: Washington, D.C., U.S. Geological Survey Bulletin, no. 39, 84 p.

Upham, W., 1888a, The geology of Chisago, Isanti, and Anoka counties, in The Geology of Minnesota: Minneapolis, Minnesota, Johnson, Smith and Harrison, State Printers, v. 2, p. 399–425.

Upham, W., 1888b, The geology of Kandiyohi and Meeker counties, in The Geology of Minnesota: Minneapolis, Minnesota, Johnson, Smith and Harrison, State Printers, v. 2, p. 220–242.

Upham, W., 1888c, The geology of Crow Wing and Morrison counties, in The Geology of Minnesota: Minneapolis, Minnesota, Johnson, Smith and Harrison, State Printers, v. 2, p. 580–611.

Upham, W., 1888d, The geology of Otter Tail County, in The Geology of Minnesota: Minneapolis, Minnesota, Johnson, Smith and Harrison, State Printers, v. 2, p. 534–561.

Upham, W., 1888e, The geology of Benton and Sherburne counties, in The Geology of Minnesota: Minneapolis, Minnesota, Johnson, Smith and Harrison, State Printers, v. 2, p. 426–444.

Upham, W., 1888f, The geology of Wright County, in The Geology of Minnesota: Minneapolis, Minnesota, Johnson, Smith and Harrison, State Printers, v. 2, p. 243–263.

Upham, W., 1890, Report of exploration of the glacial Lake Agassiz in Manitoba: Annual report, n.s., no. 4, 1888–1889, part E. Geological and Natural History Survey of Canada: Montreal, Quebec, W.F. Brown, 156 p.

Upham, W., 1895, The Glacial Lake Agassiz: Monographs: Washington, D.C., U.S. Geological Survey, v. 25, 492 p.

Upham, W., 1916, The work of N. H. Winchell in glacial geology and archaeology: Economic Geology and the Bulletin of the Society of Economic Geologists, v. 11, no. 1, p. 63–72, doi:10.2113/gsecongeo.11.1.63.

Upham, W., 1920, Minnesota Geographic Names: Their Origin and Historic Significance: Collections, no. 17: St. Paul, Minnesota, Minnesota Historical Society, 735 p.

Urban, M.A., 2005, An uninhabited waste: Transforming the Grand Prairie in nineteenth century Illinois, USA: Journal of Historical Geography, v. 31, no. 4, p. 647–665, doi:10.1016/j.jhg.2004.10.001.

Van Nest, J., 2006, Rediscovering this earth: Some ethnogeological aspects of the Illinois valley Hopewell mounds, in Charles, D.K., and Buikstra, J.E., eds., Recreating Hopewell: Gainesville, University of Florida Press, p. 402–426.

Van Nest, J., Charles, D.K., Buikstra, J.E., and Asch, D.L., 2001, Sod blocks in Illinois Hopewell mounds: American Antiquity, v. 66, no. 4, p. 633–650, doi:10.2307/2694177.

Vogel, G., 2005, A view from the bottomlands: Physical and social landscapes and Late Prehistoric mound centers in the northern Caddo area [unpublished Ph.D. dissertation]: Fayetteville, University of Arkansas, 470 p.

Walker, D.A., Epstein, H.E., Romanovsky, V.E., Ping, C.L., Michaelson, G.J., Daanen, R.P., Shur, Y., Peterson, R.A., Krantz, W.B., Raynolds, M.K., Gould, W.A., Gonzalez, G., Nicolsky, D.J., Vonlanthen, C.M., Kade, A.N., Kuss, P., Kelley, A.M., Munger, C.A., Tarnocai, C.T., Matveyeva, N.V., and Daniels, F.J.A., 2008, Arctic patterned ground ecosystems: A synthesis of field studies and models along a North American transect: Journal of Geophysical Research, v. 113, p. G03S01, doi:10.1029/2007JG000504.

Washburn, A.L., 1988, Mima Mounds, an Evaluation of Proposed Origins with Special Reference to the Puget Lowlands. Report of Investigations, no. 29: Olympia, Washington, Division of Geology and Earth Resources, Department of Natural Resources, 53 p.

Waterman, T.T., 1924, North American Indian dwellings: Geographical Review, v. 14, no. 1, p. 1–25, doi:10.2307/208352.

Webb Publishing, 1915, Atlas and Farm Directory with Complete Survey in Township Plats, Rice County, Minnesota: St. Paul, Minnesota, Webb Publishing, 44 p.

Webb Publishing, 1916, Atlas and Farm Directory with Complete Survey in Township Plats, Dakota County, Minnesota: St. Paul, Minnesota, Webb Publishing, 50 p.

Webster, C.L., 1889, Mounds of the western prairies: Annual Report of the Smithsonian Institution for 1887: Washington, D.C., Smithsonian Institution, v. 42, p. 603–605.

Webster, C.L., 1897, History of Floyd County, Iowa: Charles City, Iowa, Intelligencer Print, 67 p.

Webster, C.L., 1900, Mounds of the prairies, in The Moundbuilders of Iowa: Unpublished manuscript on file, Office of the State Archaeologist: Iowa City, University of Iowa, v. 2, p. 39–40.

Wedding, K.C., 1985, Schmidt's Minnesota lowland mounds reconsidered: The Minnesota Archaeologist, v. 44, no. 1, p. 29–33.

White, C.A., 1868, Indian mounds: Annals of Iowa, v. 6, no. 1, p. 19–22.

White, C.A., 1869, A trip to the great red pipestone quarry: American Naturalist, v. 2, no. 12, p. 644–653, doi:10.1086/270332.

White, C.A., 1870, Report on the Geological Survey of the State of Iowa: Des Moines, Iowa, State of Iowa, v. 2, 487 p.

White, J., 2000, Early accounts of the ecology of the Fox River area: Volume 5, Fox River Area Assessment: Critical Trends Assessment Program, Illinois Department of Natural Resources, http://dnr.state.il.us/publications /pdf/00000400.pdf (accessed 28 April 2010).

White, J.A., and Glenn-Lewin, D.C., 1984, Regional and local variation in tallgrass prairie remnants of Iowa and Nebraska: Vegetation, v. 57, p. 65–78, doi:10.1007/BF00047300.

Whittaker, W.E., and Storey, G.R., 2005, Ground-Penetrating Radar Survey of the Effigy Mounds National Monument Sny Magill Mound Group (13CT18), Clayton County, Iowa: Contract Completion Report no. 1233, Office of the State Archaeologist: Iowa City, Iowa, University of Iowa, 97 p.

Whittaker, W.E., and Storey, G.R., 2008, Ground-penetrating radar survey of the Sny Magill Mound Group, Effigy Mounds National Monument, Iowa: Geoarchaeology, v. 23, no. 4, p. 474–499, doi:10.1002/gea.20229.

Wild, J.C., 1841, The Valley of the Mississippi Illustrated in a Series of Views: St. Louis, Missouri, Chambers and Knapp, 145 p.

Wilford, L.A., 1938, Memo on McLeod County, March 19, 1938: Manuscript report on file at the Wilford Archaeology Laboratory: Minneapolis, Minnesota, University of Minnesota.

Wilford, L.A., 1939a, Memorandum on Dakota County, May 16, 1939: Manuscript report on file at the Wilford Archaeology Laboratory: Minneapolis, Minnesota, University of Minnesota, 1 p.

Wilford, L.A., 1939b, Lake Villard Mounds: Unpublished manuscript on file: University of Minnesota Archaeology Collections: St. Paul, Minnesota, Minnesota Historical Society.

Wilford, L.A., 1939c, Memorandum on Rice County, May 16, 1939: Manuscript report on file at the Wilford Archaeology Laboratory: Minneapolis, Minnesota, University of Minnesota, 2 p.

Wilford, L.A., 1944, Memo on Renville County, August 13, 1944: Manuscript report on file at the Wilford Archaeology Laboratory: Minneapolis, Minnesota, University of Minnesota.

Wilford, L.A., 1945, Memorandum on Rice County, June 19–21, 1945: Manuscript report on file at the Wilford Archaeology Laboratory: Minneapolis, Minnesota, University of Minnesota.

Wilford, L.A., 1949, Memo on Washington County, May 16 to 19, 1949: Manuscript report on file at the Wilford Archaeology Laboratory: Minneapolis, Minnesota, University of Minnesota.

Wilford, L.A., 1951a, Memo on Kandiyohi County, June 6, 1951: Manuscript report on file at the Wilford Archaeology Laboratory: Minneapolis, Minnesota, University of Minnesota.

Wilford, L.A., 1951b, Memo on Renville County, June 6, 1951: Manuscript report on file at the Wilford Archaeology Laboratory: Minneapolis, Minnesota, University of Minnesota.

Wilford, L.A., 1953, Memo on Stevens County, August 6, 1953: Manuscript report on file at the Wilford Archaeology Laboratory: Minneapolis, Minnesota, University of Minnesota.

Wilford, L.A., 1954a, Memo on Pipestone County, May 21, 1954: Manuscript report on file at the Wilford Archaeology Laboratory: Minneapolis, Minnesota, University of Minnesota.

Wilford, L.A., 1954b, Memo on Stevens County, May 1954: Manuscript report on file at the Wilford Archaeology Laboratory: Minneapolis, Minnesota, University of Minnesota.

Wilford, L.A., 1956, Memorandum on Beltrami County, May, 1956: Manuscript report on file at the Wilford Archaeology Laboratory: Minneapolis, Minnesota, University of Minnesota, 4 p.

Wilford, L.A., 1970, Burial Mounds of the Red River Headwaters: Minnesota Prehistoric Archaeology Series, no. 4: St. Paul, Minnesota, Minnesota Historical Society, 36 p.

Wilford, L.A., Johnson, E., and Vicinus, J., 1969, Burial Mounds of Central Minnesota: Excavation Reports: Minnesota Prehistoric Archaeology Series, no. 4: St. Paul, Minnesota Historical Society, 72 p.

Williams, M., 2008, A new look at global forest histories of land clearing: Annual Review of Environment and Resources, v. 33, p. 345–367, doi:10.1146 /annurev.environ.33.040307.093859.

Wilson, G.L., 1917, Agriculture of the Hidatsa Indians: An Indian Interpretation: Studies in the Social Sciences, no. 9: Minneapolis, Minnesota, University of Minnesota, 129 p.

Winchell, N.H., 1873, First Annual Report of the Minnesota Geological Survey: Minneapolis, Minnesota, The Minnesota Geological and Natural History Survey, 129 p.

Winchell, N.H., 1877, Fifth Annual Report of the Minnesota Geological Survey: Minneapolis, Minnesota, The Minnesota Geological and Natural History Survey, 248 p.

Winchell, N.H., 1878, Sixth Annual Report of the Minnesota Geological Survey: Minneapolis, Minnesota, The Minnesota Geological and Natural History Survey, 226 p.

Winchell, N.H., 1884a, The geology of Mower County, in The Geology of Minnesota: Minneapolis, Minnesota, Johnson, Smith and Harrison, State Printers v. 1, p. 347–366.

Winchell, N.H., 1884b, The geology of Rice County, in The Geology of Minnesota: Minneapolis, Minnesota, Johnson, Smith and Harrison, State Printers, v. 1, p. 648–673.

Winchell, N.H., 1885, Thirteenth Annual Report of the Minnesota Geological Survey: Minneapolis, Minnesota, The Minnesota Geological and Natural History Survey, 196 p.

Winchell, N.H., 1888a, The geology of Dakota county, in The Geology of Minnesota: Minneapolis, Minnesota, Johnson, Smith and Harrison, State Printers, v. 2, p. 62–101.

Winchell, N.H., 1888b, The geology of Goodhue county, in The Geology of Minnesota: Minneapolis, Minnesota, Johnson, Smith and Harrison, State Printers, v. 2, p. 20–61.

Winchell, N.H., 1888c, The geology of Wabasha county, in The Geology of Minnesota: Minneapolis, Minnesota, Johnson, Smith and Harrison, State Printers, v. 2, p. 1–19.

Winchell, N.H., 1908, Habitations of the Sioux in Minnesota: The Wisconsin Archeologist, o.s., v. 7, no. 4, p. 155–164.

Winchell, N.H., 1911, The aborigines of Minnesota: A report based on the collections of Jacob V. Brower, and on the field surveys and notes of Alfred J. Hill and Theodore H. Lewis: St. Paul, Minnesota Historical Society, 761 p.

Winchell, N.H., and Upham, W., 1884, Geology of Minnesota: Minneapolis, Minnesota, Johnson, Smith and Harrison, State Printers, v. 1, 697 p.

Withrow, R., 2005, The enigmatic lowland mounds of Dakota and Rice counties: Over the Years, v. 46, no. 2: South Saint Paul, Minnesota, Dakota County Historical Society, p. 1–9.

Wolfe-Bellin, K.S., and Moloney, K.A., 2001, Successional vegetation dynamics on pocket gopher mounds in an Iowa tall grass prairie: Proceedings of the 17th North American Prairie Conference, p. 155–163.

Wood, W.R., and Johnson, D.L., 1978, A survey of disturbance processes in archaeological site formation, in Schiffer, M.B., ed., Advances in Archaeological Method and Theory: New York, Academic Press, v. 1, p. 315–381.

Woodbury, R.B., and Woodbury, N.F.S., 1999, The rise and fall of the Bureau of American Ethnology: Journal of the Southwest, v. 41, no. 3, p. 283–296.

Woodman, H.T., 1873, Ancient mounds of Dubuque and its vicinity: Proceedings of the American Association for the Advancement of Science, v. 21, p. 225–227.

Wright, G.B., 1873, Map of Hennepin County, Minnesota: St. Paul, Minnesota, G.B. Wright and G.J. Rice, 39 p.

MANUSCRIPT ACCEPTED BY THE SOCIETY 5 MARCH 2012

The Geological Society of America
Special Paper 490
2012

The polygenetic origin of prairie mounds in northeastern California

Donald L. Johnson*
Department of Geography, University of Illinois, Urbana, Illinois 61801, USA, and
Geosciences Consultants, 713 S. Lynn St., Champaign, Illinois 61820, USA

Diana N. Johnson†
Geosciences Consultants, 713 S. Lynn St., Champaign, Illinois 61820, USA

ABSTRACT

We studied low prairie (Mima) mounds and ridges with sorted stone borders separated by broad rubbly soil intermounds in areas near Mount Shasta, northern California. An earlier study ascribed a purely physical origin for these soil features based on a four-stage conceptual model. Mounds were interpreted as periglacially produced clay domes formed in polygonal ground and stone perimeters as loose gravity accumulations in unexplained shallow trenches at dome peripheries. The model was widely cited to account for similar stone-bordered prairie mounds and rubbly soil intermounds in the Pacific Northwest. Our observations and measurements indicate, however, that these mounded landscapes are more complex, and that a polygenetic origin best explains them. We suggest that combined bioturbation, seasonal frost action, and erosion processes, with occasional eolian inputs, best account for the mounds, their well sorted stone borders, and the poorly sorted rubbly soil intermound pavements. We propose a transitional, eight-stage conceptual model to explain this complex landscape. The model may generally explain the origin of other similar strongly bioturbated, cold winter-impacted, erosion-prone mounded tracts in the Pacific Northwest.

INTRODUCTION

Prairie mounds . . . are the result of surface erosion under peculiar conditions [because] a phenomenon so wide-spread must be due to a wide-spread agent. —J. LeConte (1875)

Prairie mounds . . . are the vestiges of hummocks thrown up by prairie dogs, or other burrowing animals. —G.K. Gilbert (1875)

The contrasting views of LeConte and Gilbert expressed in the epigraphs stem largely from their observations of mounds formed in thin soil over basalt—in Gilbert's case the White Mountains of east-central Arizona, and in LeConte's the Umatilla Plateau of north-central Oregon (cf. Table 3). Ignited partly by their writings, many geologists offered varying views on what became a particularly controversial topic—the origin of the prairie mounds of western North America. The debate temporarily climaxed in 1905–1906 when multiple hypotheses on the subject were proffered in the journal *Science*. Reflecting on the debate, geologist Marius R. Campbell (1906) suggested that a geographical comparison of soil mounds across the environments in which they occur might reveal whether a common denominator exists. Intrigued by LeConte's view that prairie mounds are due to a "widespread agent," that is, soil erosion operating under "peculiar conditions," and inspired by Campbell's suggestion, in the 1990s we began to field-assess mounded tracts in central and western North America that are particularly pertinent to the controversy about the origins of prairie (Mima) mounds (cf. Johnson, this volume, Appendix B).

While explaining the origin of soil mounds anywhere is not simple, the moundfields that primarily prompted LeConte's

**dljohns@uiuc.edu
†dn-jhnsn@uiuc.edu

Johnson, D.L., and Johnson, D.N., 2012, The polygenetic origin of prairie mounds in northeastern California, *in* Horwath Burnham, J.L., and Johnson, D.L., eds., Mima Mounds: The Case for Polygenesis and Bioturbation: Geological Society of America Special Paper 490, p. 135–159, doi:10.1130/2012.2490(06). For permission to copy, contact editing@geosociety.org. © 2012 The Geological Society of America. All rights reserved.

TABLE 1. TIMELINE OF SELECTED CONTRIBUTIONS TO MOUND STUDIES OF THE INTERIOR
PACIFIC NORTHWEST (PNW) VOLCANIC REGION OF NORTH AMERICA

Gibbs, 1854, Washington	Nelson, 1977, Oregon
Gibbs, 1855, *Washington*, Oregon	Wilson and Slupetsky, 1977, Idaho
Newberry, 1857, Oregon	Bork, 1978, Oregon
LeConte, 1874, 1875, 1877, California, *Oregon*, Washington	Winward and Youtie, 1978, Oregon
Gilbert, 1875, Arizona	Wilson, 1978, Idaho
Piper, 1905, Washington	Collett, 1980, Idaho
Van Duyne et al., 1919, Washington	Copeland, 1980, Oregon
Kocher and Strahorn, 1921, Washington	Sheldon, 1980, California
Watson and Cosby, 1924, California	Tallyn, 1980, Washington
Freeman, 1926, Washington	Green, 1981, 1982, Oregon
Waters and Flagler, 1929, *Oregon*, Washington, Idaho	Stockman, 1981, Washington
Freeman, 1932, Washington	Johnson, 1982, Oregon
Larrison, 1942, Washington	Newlun et al., 1983, California
Masson, 1949, California	Cox, 1984a, *California, Oregon,* Washington, etc.
Kaatz, 1959, Washington	Cahoon, 1985, Oregon
Mack, 1960, California	Almaraz, 1986, California, *Oregon*
Malde, 1961, Idaho	Cox and Allen, 1987a, Oregon
Peterson, 1961, Washington	Cox et al., 1987, Oregon
Brunnschweiler, 1962, California, *Oregon, Washington*	Washburn, 1988, California, Idaho, Oregon, *Washington*
Fosberg, 1963, Idaho	Cox, 1989, Oregon
Olmstead, 1963, Washington	Berg, 1989, 1990a, 1990b, Washington
Brunnschweiler, 1964, California, *Oregon, Washington*	Cox, 1990b, 1990c, Oregon
Fryxell, 1964, Washington	Cox and Hunt, 1990, Oregon
Fosberg, 1965, Idaho	Berg, 1991, Washington
Malde, 1964a, 1964b, Idaho	Gentry, 1991, Washington
Malde, 1965, Idaho	Noe, 1991, Idaho
Paeth, 1967, Oregon, Washington	Othberg, 1991, Idaho
Donaldson and Giese, 1968, Washington	Johnson, 1993, Oregon
Green, 1970, Oregon	Tullis, 1995, Idaho
Pyrch, 1973a, 1973b, Oregon	Geiger, 1998, Washington
Pyrch and Price, 1974, Oregon	MacDonald, 1998, Oregon
Gentry, 1974, Washington	MacDonald et al., 1999, Oregon
Green, 1975, Oregon	Gentry, 2006, Washington
Davis and Youtie, 1976, Oregon	Kienzle, 2007, Oregon
	Beieler and Kehne, 2008, Washington

Note: Italicized state(s) was (were) where most research was done or predominantly discussed.

proclamations—specifically those of the Pacific Northwest interior volcanic, cold-winter regions mantled by tephra-loess "eolian soils"—are an especially complex and polygenetic class, as illustrated by the studies in Table 1, and by Appendices A and B of this volume. The collective attributes of these mounds and the scabland-like tracts in which many have formed, are notably different than the much debated Olympia, Washington, moundfields, and most of those in California's Great Central Valley, both of which LeConte also discussed. They likewise differ from the Mississippi Valley and Gulf Coast mounds that were a primary focus of the 1905–1906 *Science* debates. The interior Pacific Northwest mounds, regionally termed "biscuits" in basaltic "biscuit scabland" terrains, are a special challenge, a point LeConte and many others understood—not to just explain mounds, but also to reasonably explain the apparent erosion, freeze-thaw, and bioturbation signatures genetically associated with them. As indicated in Table 1, erosion was emphasized by LeConte (1874, 1875, 1877; Johnson, this volume, Appendix B) and Waters and Flagler (1929), the freeze-thaw and periglacial aspects by Masson (1949), Kaatz (1959), Brunnschweiler (1962), and Fryxell (1964, this volume,

Appendix A), and the animal bioturbation aspects emphasized by Gilbert (1875; Johnson, this volume, Appendix B)[1], Peterson (1961), Cox and Allen (1987a), Cox et al. (1987), Cox and Hunt (1990), and Cox (1990a, 1990b, 1990c). The moundfields in the Mount Shasta–Modoc Plateau region of northeastern California and similar areas of Oregon, Washington, Idaho, and northern Nevada are all prime examples of such complex polygenetic landscapes. We include the Shasta-Modoc Plateau region as part of the greater Pacific Northwest area—hereafter PNW—because of regional similarities in eolian soil and volcanic landscapes, in having mounds with fringing stone borders (rings, nets), and having poorly sorted rubbly soil intermound pavements.

Mt. Shasta Region Mound Research

Shortly after World War II, geologist Peter Masson (1949) described and hypothesized an origin of ridges and soil mounds and their conspicuous stone borders—what he called "circular soil structures"—in the greater Mount Shasta volcanic region of northern California (see Table 2 for terms and definitions). He

---

[1]While not part of the Pacific Northwest, we nevertheless include Gilbert's (1875) White Mountains of east-central Arizona mound work in Table 1 for two reasons: (1) because of close parallels between the interior, cold-winter, ash-loess "eolian" mantled volcanic-basalt landscapes of both regions; and (2) LeConte (1877), by virtue of disagreeing with him, drew Gilbert's burrowing animal conclusions into the Pacific Northwest mound controversy.

TABLE 2. TERMS AND DEFINITIONS USED FOR PACIFIC NORTHWEST (PNW) MOUNDFIELDS

Activity centers (AC)	*Fixed* and/or *unfixed* living-reproduction-food cache loci of burrowing animals (rodents, ants, etc.). If location confers survival advantages, relatively fixed loci will form, and when occupied over many generations, soil mounds evolve; unfixed loci are survival neutral, and thus ephemeral.
Associated and unassociated sorted stone gutters	Gutters that border mounds and ridges are *associated* with them, but other gutters occur in intermound areas that are presently *unassociated* with either.
Biomantle	Bioturbated upper part of soil; usually includes A and E horizons, and, in gravelly soil, the stonelayer.
Biosorting	Particle sorting by organisms, normally burrowing animals, in mixed clast soils or sediments; clasts larger than burrow diameters settle downward to form a basal stonelayer. Promotes biostratification.
Biostratification	Layering in soil or sediment caused by organisms. Caused by biosorting.
Biotransfers	Movements of soil or sediment by organisms from one place to another.
Bioturbation	Biotransfers, biosorting, biomixing, and biocomminution of soil or sediment by animals, plants, fungi, proctocists, and microbes (refers to animals unless otherwise specified).
Biscuits of "biscuit scablands"	Regional PNW term for prairie (Mima) mounds formed on rubbly or scoured basalt or other volcanics.
Centripetal biotransfers	Movement of soil by burrowing animals, such as pocket gophers, from their foraging domains to their activity (nesting, food-storage) centers-loci.
Circular soil structures	Prairie mounds in Mount Shasta and PNW region bordered by sorted stone gutters and separated by rubbly soil intermounds (cf. Masson, 1949).
Coalesced spoil heaps	Two or more spoil heaps overlapped (merged) to form debris islands.
Daughter mound (DM)	Prairie mound derived from a larger soil parent body (SPB) or soil mantle.
Debris islands	See "coalesced spoil heaps."
Dominant bioturbator(s)	Organism(s) most responsible for particle size distributions in the biomantle; rodents, moles, ants, worms in mid-latitudes; ants, termites, worms in low latitudes; commonly trees (uprooting activity) in forested lands; etc.
DSC model	Dalquest-Scheffer-Cox model of *Mima-type* mound formation, where thin soil constitutes a barrier to vertical burrowing, often involving wetness, promotes centripetal soil movements to activity centers. This involves many generations of burrowing animals.
Duripan (DP)	A subsurface horizon cemented by illuvial silica into a hardpan that functions as an aquitard or aquiclude; example of a barrier to vertical burrowing.
Erosion removals (ER)	Erosion from spoil heaps, mounds, and mound saddles, especially where the latter is in conjunction with centripetal biotransfers to activity centers.
French drains (FD)	Stone-covered and stone-filled shallow gutter, net, bed, trench, or depression that collects and transmits water, usually with suspended sediment, often as interconnected flow channels.
Gopher eskers (gopher garlands, cf. snow eskers)	Gopher-dug snow tunnels backfilled with soil; appear after snowmelt as "soil ropes" on surface; common in cold winter PNW moundfields: a means for burrowing animals to access intermounds for wintertime food foraging below snowpack.
Gutters (sorted stone borders, perimeters, rims)	Shallow (~50 cm deep, ~1 m wide) trenches with frost sorted, fitted, and oriented stones that collect and transmit water, subject to strong seasonal frost action.
Heaps (spoil, or burrow heaps)	Surface spoil of burrowing animals; in PNW mounds-intermounds, usually composed of pebbly loam and ash-loess "eolian soils."
Interflow zone (IFZ)	Subsurface movement of soil water above an impermeable layer.
Krotovina (crotovina)	Infilled animal burrows, often of a different color than enclosing soil. Mima mounds are often composed predominantly of indistinguishable krotovina. Gopher eskers are krotovina created in snow.

(Continued)

mentioned mounds in three areas of Siskiyou County, and various mounded areas adjacent to Highway 299 and Pit River in Shasta, Lassen, and Modoc counties (Fig. 1; see Table 3 for key interior PNW and other select moundfield locations). His interest was piqued by then recent (1944) airphotos on which mounds were notably obvious (Fig. 2A), and which some authors attributed to humans (Westman, 1946; cf. Noble, 1951, 1985). What most captured Masson's attention were "circular structures" consisting (he wrote) of "stone-free" low mounds of "clay," some with bordering "stone rings" and intermound pavements on glacial outwash at Soulé Ranch in the Leaf-Bray area, and in Shasta Valley northwest of Weed (Fig. 1). Freeman (1997; cf. also COSL, 2011) later usefully reviewed aspects of the Shasta Valley mounds.

Masson presciently emphasized that the "Siskiyou structures" were of two kinds: mounds fringed by conspicuous shallow stone-filled trenches, as at Soulé Ranch, and mounds that lack them, as do most at Shasta Valley. Because most moundfields in North America lack them and thus are the more common kind, we designate them Type 1 mounds and moundfields. The Soulé Ranch mounds—those bordered by conspicuous, tightly fitted, well sorted with oriented stones that occupy shallow trenches, or gutters, and sometimes expressed as "sorted stone nets" or "beds," and "polygonal nets"—are typical of many interior PNW moundfields formed in volcanic soils. A key caveat is that soils must be thick enough and must contain sufficient mixed size clasts for frost sorting to produce such features. We designate them Type 2 mounds and moundfields. Both kinds may or may not have rubbly soil intermounds (most do). A particular goal of this chapter is to determine what processes and conditions are required, or absent, for both types of moundfields to form, especially Type 2.

Masson focused his effort on how Type 2 mounds form; namely, those with fringing, well sorted, and tightly fitted

TABLE 2. TERMS AND DEFINITIONS USED FOR PACIFIC NORTHWEST (PNW) MOUNDFIELDS (*Continued*)

Mima mounds (prairie, Mima-type, Mima-like, soil mounds)	Activity centers of burrowing animals; point centered, locally thickened biomantles; named for "Mound Prairie" (Mima Prairie), type locality for Mima-type mounds near Olympia, Washington; synonym for "prairie mounds."
Mima-like mounds	Mounded biomantles with unfixed (impermanent) activity centers.
Mima-type mounds	Mounded biomantles with fixed (relatively permanent) activity centers.
Mound disks	Flattish, circular, pebbly loam residual patches in intermounds left when soil mounds completely erode and waste away.
Mound saddles	Removal areas between activity centers in thin soil parent bodies acted upon by both erosion and centripetal biotransfers.
Paternoster mounds	Prairie (Mima) mounds that appear linked in a row, as beads on a necklace or chain. They occur on slight linear rises, low stream levees, meander scroll highs, rills and divides of drainage ways, along fracture traces in bedrock, and so on.
Patterned ground	General term in PNW for mounds and sorted stone borders, rings, nets, beds, and stripes; may or may not imply prolonged freeze-thaw cycles.
Pebbly debris islands (PDI)	Pebbly spoil piles of burrowing animals, often weather-wasted and/or somewhat coalesced; typically dot PNW rubbly soil intermounds and nets.
Pocket gopher (PG)	One of 40+ species of the superburrowing Geomyidae family of herbivorous rodents of Central and North America; major mound-makers.
Prairie mounds (Mima mounds)	Mounded biomantles; either fixed or unfixed activity centers; generic term.
Process polygenesis	Where multiple co-dominant processes operate concurrently to produce soil and landform features; different than climatic polygenesis and polycyclicity.
Proto-mounds	Incipient mounds so subtle as to be barely perceived as mounds.
Rubbly soil intermounds	Intermound areas impacted by process polygenesis (bioturbation, frost action, erosion); aggregated and coalesced former stone gutters, more or less equally bioturbated, frost sorted, and sheet eroded.
Snow eskers (snow rolls, soil eskers, soil ropes, winter cores)	Cf. gopher eskers.
Soil mounds	Generic term for Mima, prairie, pimple, or natural mounds.
Soil parent bodies (SPBs)	In PNW, eolian-derived (usually) flat-topped soil mantles from which flat-topped (initially) daughter mounds derive; usually edge-thinned to ±1 m above a subsoil restriction to burrowing.
Stone beds, nets, stripes	Sorted stone gutters, but wider; usually in low slope, seasonally wet, frost-prone PNW interior moundfields, often with wasted spoil heaps or debris islands; they become rubbly soil intermounds when bioturbation is greater than frost heaving.
Stone gutters, sorted stone gutters (cf. French drains, stone beds, stone nets)	Frost-sorted, fitted openwork stones in shallow, narrow (≤1 m) trenches that border many prairie mounds and SPBs in frost-prone areas; minimally bioturbated; function as French drains; called stone nets when >1 m wide.
Stone layer (stonelayer, stone line)	Basal layer of two-layered biomantles; most formed by biosorting in gravelly soils; some due to erosion.
Stone pavements (SP)	Surface pavements produced by erosion, frost heaving, tree uprooting and/or other biotransfers.
Stone rings	Same as sorted stone borders that encircle mounds; usually on low slopes of PNW moundfields.
Type 1 and Type 2 mounds and moundfields	Type 1 mounds and moundfields lack fringing sorted stone gutters and nets in intermound pavements; Type 2 mounds possess them.
Wasted heaps (burrow heaps)	Weather-eroded and or snowpack-flattened burrow spoil, or heaps.

stone-filled trenches. His explanatory model became both a template and a launch pad of ideas for many researchers for explaining mounds with sorted stone borders and nets in the interior PNW volcanic region, features that collectively came to be called "patterned ground." All Type 2 mounds are subject to some degree of bioturbation, erosion, and seasonal freezing and thawing. (While clear genetic similarities as indicated exist, we strongly emphasize that *every* PNW moundfield we have examined— mounds, their sorted stone borders, nets, and intermounds—are also somewhat different from one another, some notably.) Unfortunately, some researchers used the Masson model to explain Type 1 mounds, which has left a legacy of confusion.

For these reasons, and others explained below, we studied the soils and surface features in the Mount Shasta region, and focused on Soulé Ranch (hereafter Soulé Terrace, or Soulé). Our main field activities took place in mid-October 2010 near the end of the dry season, in late March 2011 during full spring snowmelt, and in mid-September 2011. We found that Masson's

purely physical "periglacial-clay dome-stone ring" conclusions lacked a biological component. We specifically found that (1) the prairie mounds in both Shasta Valley and Soulé Ranch primarily stem from bioturbation and erosion; (2) the well-formed and sorted stone gutters at elevationally higher and seasonally colder Soulé Terrace (1445 m elevation), but largely absent at elevationally lower and warmer Shasta Valley (915 m), are due to more intense and frequent freeze-thaw cycles at Soulé; and (3) the poorly sorted rubbly soil intermound pavements of both areas formed by a mix of all three processes: bioturbation, erosion, and frost sorting and orientation of stones. These processes are augmented by occasional ash-loess dustings. We believe these findings are relevant to explain the origin of the Siskiyou and other PNW moundfields.

Our Soulé Terrace research is supported by laboratory data, and by comparative field observations at Shasta Valley, Pit-Sacramento River terraces, Modoc Plateau, and many similar PNW moundfields. It draws substantially on Google Earth

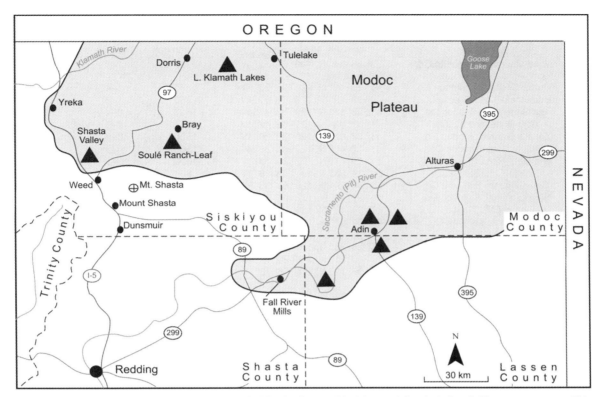

Figure 1. Map of mound localities in northeastern California discussed by Masson (triangles). Our field assessments are of his sites, plus similarly mounded sites across the region outlined by shading defined bold line, including the abundantly mounded Modoc Plateau. Courtesy of Jennifer Horwath Burnham and Illinois Archaeology Survey archaeologists Mike Lewis and Thomas C. Emerson.

satellite-map technologies. This chapter also adds to a working model that prairie mounds are microrelief variations on a polygenetic theme, one that involves intimately coupled biological and physical processes—polygenesis. It also supports the proposition that while prairie mounds are generally polygenetic, they are also invariably centers of bioturbation. That is, they are the nesting, denning, food storage, and overwintering loci of burrowing animals that, once formed, confer survival advantages to the animals (Cox, 1984a). Such raised activity centers—low bulk density areas of intense bioturbation—are the common denominator of prairie mounds. This applies to microtopographic highs now strongly bioturbated that were originally produced mainly by physical processes, termed "hybrid mounds"[2] (Johnson and Johnson, 2008).

Prairie mounds are best defined as point-centered, locally thickened biomantles (Johnson et al., 2003). These point centers are raised activity centers that become relatively *fixed* if they confer a survival advantage that fosters repeated occupation by generations of animals. Fixed activity centers usually form in thin soil with some barrier to vertical burrowing, which can include hard bedrock, dense gravels, heavy clay, hardpan, high water tables, or wherever periodic wetness and or flooding occur (Cox,

1984a; and many personal observations). Such occupation leads to Mima-type mound formation and maintenance. If on the other hand survival advantage is neutral, raised centers are ephemeral and *unfixed*. Unfixed centers promote Mima-like mounds. The fixed-unfixed continuum is controlled by local soil-geologic conditions that change with time.

Environmental Context and Site Descriptions of Shasta Valley and Soulé Ranch Moundfields

Shasta Valley and Soulé Ranch moundfields are part of a scenically grandiose southern Cascades volcanic region (Williams, 1949). Mount Shasta and nearby ancillary volcanoes dominate the local and regional Siskiyou landscape, and have collectively left a notable collage of process histories (Christiansen et al., 1977). Approximately 300–380 ka ago a large portion of ancestral Mount Shasta collapsed, generating an enormous ~49-km-long debris avalanche-lahar that flowed northward nearly to Klamath River, filling Shasta Valley with hill-size volcanic blocks embedded in flat areas underlain by an unsorted and unstratified mudflow-like deposit of sand, silt, clay, and rock fragments (Crandell, 1989; Crandell et al., 1984). The deposits cover ~675 km² with an estimated volume of ~45 km³. Glaciers, which later developed on the slopes of a reconstituted Mount Shasta, moved down along similar paths into Shasta Valley, depositing till and outwash in its

---

[2]The hybrid mound concept was preceded by the "biogenic maintenance hypothesis" formalized by Riefner and Pryor (1996).

TABLE 3. LOCATIONS OF REPRESENTATIVE PACIFIC NORTHWEST AND WESTERN MOUNDFIELDS, FROM GOOGLE EARTH; CF. HISTORICAL IMAGERY MENU, APPROXIMATED TO NEAREST SECOND; BEST OBSERVED AT ~1 KM ELEVATION

State	Location	Latitude and longitude	Elevation (m)	Slope (%)	Burrow-limiting barrier
Arizona	White Mountains, Apache Co.	33° 58′ 36″ N, 109° 24′ 53″ W	2793	1–3	Basalt
California	Alturas-Cedarville, Modoc Co.	41° 33′ 45″ N, 120° 23′ 28″ W	1535	1–4	Basalt
	Bella Vista T.M., Shasta Co.	40° 39′ 19″ N, 122° 09′ 57″ W	250	1–3	Basalt
	Big Sage Res., Modoc Co.	41° 33′ 58″ N, 120° 38′ 47″ W	1537	0–4	Basalt
	Carr Butte, Modoc Co.	41° 57′ 03″ N, 121° 13′ 42″ W	1425	2–5	Basalt
	Devil's Garden, Modoc Co.	41° 32′ 02″ N, 120° 39′ 20″ W	1520	0–2	Basalt
	Graven Res., Modoc Co.	41° 17′ 48″ N, 120° 41′ 33″ W	1580	0–3	Basalt
	Iron Gate Res., Siskiyou Co.	41° 59′ 41″ N, 122° 20′ 33″ W	1003	0–3	Andesite
	North T.M., Butte Co.	39° 34′ 55″ N, 121° 33′ 16″ W	354	0–2	Andesite
	Soulé Ranch, Siskiyou Co.	41° 38′ 09″ N, 122° 01′ 33″ W	1445	0–3	Duripan
	Shasta Valley, Siskiyou Co.	41° 27′ 27″ N, 122° 26′ 37″ W	915	0–2	Duripan
	Tuolumne T.M., Tuolum. Co.	37° 54′ 42″ N, 120° 28′ 41″ W	495	1–3	Basalt
	White Lake, Siskiyou Co.	42° 00′ 01″ N, 121° 37′ 47″ W	1247	0	Water table
Idaho	Bonneville Overlook, Ada Co.	43° 28′ 28″ N, 116° 04′ 32″ W	1056	0–12	Basalt
	Glenns Ferry 1, Elmore Co.	43° 05′ 31″ N, 115° 20′ 12″ W	1117	0–8	Basalt
	Glenns Ferry 2, Elmore Co.	43° 06′ 12″ N, 114° 56′ 12″ W	1404	4–14	Basalt
Oregon	Chenoweth, Columbia Gorge	45° 37′ 09″ N, 121° 14′ 10″ W	238	0–2	Basalt
	Falls Creek, Jackson Co.	42° 01′ 01″ N, 122° 19′ 36″ W	1028	0–3	Andesite
	Shaniko, Wasco Co.	44° 56′ 59″ N, 120° 47′ 23″ W	1055	0–8	Basalt
	Rowena, Columbia Gorge	45° 41′ 18″ N, 121° 18′ 15″ W	181	0–2	Basalt
	L. Table Rock, Medford	42° 27′ 23″ N, 122° 57′ 12″ W	614	0–3	Basalt
	U. Table Rock, Medford	42° 27′ 59″ N, 122° 53′ 42″ W	620	0–3	Basalt
	Zumwalt Road, Imnaha	45° 39′ 35″ N, 116° 57′ 24″ W	1433	0–12	Basalt
	Zumwalt, Alder Springs Rd.	45° 47′ 44″ N, 117° 00′ 32″ W	1420	0–4	Basalt
Washington	Badger Mountain, Doug. Co.	47° 31′ 54″ N, 120° 02′ 59″ W	1046	0–15	Basalt
	Cheney-Palouse Scabland	47° 18′ 34″ N, 117° 37′ 13″ W	695	1–2	Scoured basalt
	Manastash Ridge, Kitittas Co.	46° 54′ 47″ N, 120° 38′ 09″ W	857	0–12	Basalt
	Mima Prairie, Thurston Co.	46° 53′ 56″ N, 123° 02′ 40″ W	74	1–2	Coarse gravels
	Sprague Gravel Bar, Linc. Co.	47° 17′ 06″ N, 117° 49′ 52″ W	638	1–2	Coarse gravels
	Steptoe Butte, Whitman Co.	47° 01′ 40″ N, 117° 17′ 55″ W	881–1096	15–30	Quartzite
	Tallyn Site A, Spokane Co.	47° 18′ 33″ N, 117° 37′ 12″ W	695	1–3	Scoured basalt
	Waterville Plateau, Doug. Co.	47° 58′ 26″ N, 119° 04′ 56″ W	731	0–2	Basalt and till

Note: All sites have thin soil over a barrier to burrowing and/or wetness. T.M.—Table Mountain.

southern sector. The now relict glacial deposits, first described and mapped in any detail by Mack (1960), are being remapped and updated in U.S. Geological Survey work by R.L. Christiansen (November 2011, personal commun.). Shasta Valley soils, many bearing well-expressed profiles with strong duripans, commonly contain Mima-type mounds formed on glacial outwash terraces and on some sectors of the debris flow (Watson et al., 1923; Masson, 1949; Newlun et al., 1983; Freeman, 1997).

Glacial events in Butte Creek drainage, some 40 km to the west, the site of Soulé Terrace moundfield—and our major focus (Fig. 1)—may parallel those of Shasta Valley. Soulé mounds, like many in Shasta Valley, are formed on a dissected glacial outwash terrace that, in the case of Soulé terrace, lies ~11 m above Butte Creek, in front of morainal sediments. Butte Creek, which occupies a striking U-shaped valley glacially scoured into volcanic bedrock, heads in the Rainbow Mountain highlands to the south. R.L. Christiansen, who has studied the geohistory and age of the greater Mount Shasta area over much of his career, in response to our earlier query, had this to say (3 November 2011) about the Butte Creek glaciation:

I am, in fact, quite familiar with the glacial valley you ask about and have included it in the geologic mapping I have done (and continue to do) in the Mount Shasta segment of the Cascades. The glacier that carved this quite spectacular feature headed in the upper reaches of Butte Creek, within the volcanic terrain now marked by Haight Mountain, Dry Creek Peak, and Rainbow Mountain. The source area is no longer high enough to have sustained such a large glacier system as that which carved the valley and left conspicuous lateral and terminal moraines in the valley of Butte Creek. The source area, however, was once considerably higher and larger than it now would appear and ~0.8–1.2 Ma was a volcanic center roughly equivalent in size to present-day Mount Shasta. I have named it the Rainbow Mountain volcanic center. The left-lateral moraine is overlain by an extensive tholeiitic basalt that erupted in the saddle between The Whaleback and Deer Mountain. That flow field extends westward to Shasta Valley, where it overlies the great sector-collapse debris avalanche of ~300,000 years ago and eastward into the Butte Creek drainage. We have one 40Ar/39Ar age determination on that basalt of 160 ± 55 ka and hope to get another, better date. That at least puts a minimum age constraint on the age of the glaciation, but I would guess from the size and weathering profiles of the deposits, that it was considerably older. Incidentally, the patterned ground of the Butte Creek area described by Masson in the Howell Williams paper on the Macdoel 1:125K quad is developed on the outwash from that glacier.

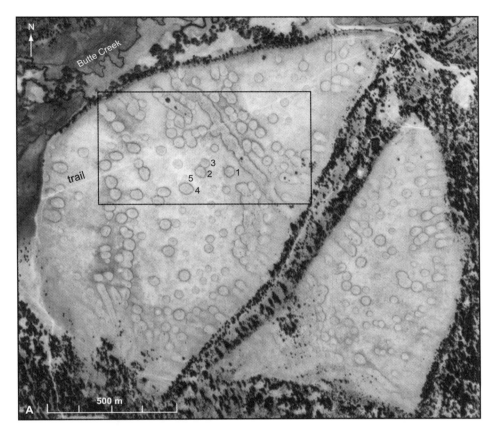

Figure 2 (*on this and following page*). (A) Soulé Terrace, cropped from larger 1944 airphoto, with NW-SE arrayed residual lobed and sinused linear ridges and circular mounds of pebbly loam, defined by well-sorted, frost-fitted stone gutters. Mounds and ridges are elongate at terrace edges, and open downslope where frost-fitted gutters become "stone stripes" as erosion trumps freeze-thaw processes.

The climate of the region is cold-winter Mediterranean, with warm usually dry summers, and cold snowy winters punctuated by many freeze-thaw cycles (Newlun et al., 1983). The soils on the terrace are classified as Haplic Durixeralfs, morphologically similar to Louie stony loam (Typic Durixeralfs) and Salisbury gravelly clay-loam (Palexerollic Durixerolls) of Shasta Valley, both of which are mounded (NRCS, 2011; USDA-USFS, 2011; Newlun et al., 1983; cf. Watson et al., 1923; Freeman, 1997). This hardpan-bearing Durixeralf is marked by many low, circular, buff-colored, soil mounds and linear ridges, each bordered by shallow (<50 cm deep) and narrow (±1 m wide) depressions (channels, gutters, and trenches) that contain dark-colored and lichen covered, well-sorted volcanic clasts—boulders, cobbles, and pebbles—sorted top to bottom in that order. Broad rubbly soil intermound pavements separate and lie between the mounds and ridges (Figs. 2 and 3). Pavements cover ~65% of the terrace surface, with mounds and ridges ~35%.

The grassy-shrubby mounds and ridges and grassy intermounds are ubiquitously dotted with old and new pebbly loam spoil heaps and some "snow eskers" of burrowing animals, most produced by the Northern Pocket Gopher (*Thomomys talpoides*), possibly a few by moles. Ground squirrels (Belding's, *Spermophilous beldingi*) and various other burrowers (badgers, ants, etc.) are also present. Because the mounds and ridges are elevationally slightly higher (~0.3–1.0 m) and drier than the surrounding seasonally wet and saturated intermounds, we assume they function as primary activity centers (havens) for denning, nesting, living, overwintering, and food storage for the animals that inhabit them (cf. Cox papers).

The conspicuous black sorted stone gutters genetically associated with the mounds and ridges are tightly fitted and openwork, with tabular clasts commonly vertically oriented. Rodent spoil heaps in them are rare. In fact, except for the rare heap, the mound- and ridge-bordering gutters so evident in Figure 2 are the *only* surface features on Soulé that are not thoroughly bioturbated by burrowing animals. Explaining these sorted stone features in conjunction with the mounds has been one of the more contentious and problematic issues of the soil mound literature in the volcanic interior lands of the PNW.

As indicated, the Soulé Terrace moundfield is somewhat similar to, though still notably different from, others in the interior volcanic region of the PNW mantled by ash-loess (eolian) loamy-silty soils (Tables 4 and 5). However, while intermound pavements of some PNW moundfields have few to common debris islands and patches of pebbly gopher spoil, those on Soulé, except for gutters, are essentially ubiquitous. The spoil is very similar in nature to the ashy loam spoil that fills "spaces between the blocks" of the "erosion furrows" in the intermound pavemented areas described by Waters and Flagler (1929, their figure 4 caption). We have personally observed this to be common in each of their three study areas—Maupin-Shaniko areas of Oregon, and Horse Heaven Hills and Channeled Scablands

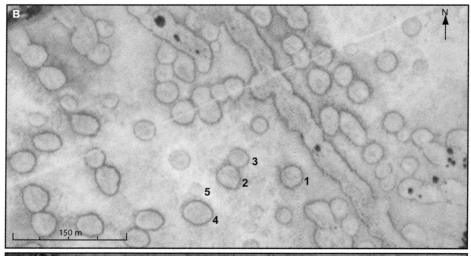

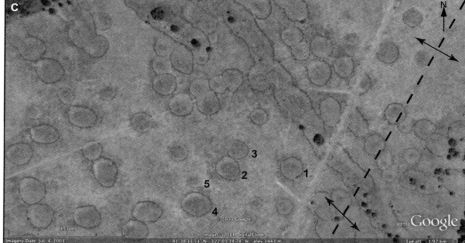

Figure 2 (*continued*). (B) Boxed area of A, enlarged. Thin light-colored line is an early trail. (C) Google Earth image of 4 July 2003; same area as B. Little change is evident in nearly 50 years, except 2.5-m-wide diagonal ranch road bulldozed in 1945 (or early 1946) that exposes the reddish subsoil Bt horizon and duripan. The nearly level terrace slopes gently (0%–1%) NW-SE indicated by double-tipped arrows, roughly parallel with lobed and sinused ridges, and away from nearly level apex at dashed line. Most low mounds and ridges measure <1 m high, with diameters of mounds 1–5 measured on ground at 29, 29, 31, 26, and 15 m. The mounds and ridges, remnants of a once continuous and thicker soil parent body, are primary living sites—*fixed activity centers*—of burrowing animals. Intermound pavements consist of poorly sorted rubbly clasts (boulders, cobbles, and pebbles) embedded in a thin pebbly loam soil derived from weather-wasted and coalesced former, now-eroded mounds, rodent spoil heaps, and "debris islands" (cf. Figs. 3A and 3C). Note that the smallest, most eroded mounds do not preserve gutters.

of Washington. Like Soulé, much of the gopher spoil in these areas is old and weather-wasted, and barely recognizable as such. In places the spoil has coalesced into a nearly continuous thin mantle between cobble-boulder pavement clasts (Fig. 3A). Fresh, unvegetated, erosion-prone gopher spoil also occurs on both mounds and intermounds of the three areas studied by Waters and Flagler, with every stage in-between represented, from fresh heaps to old, barely recognizable spoil (cf. Fig. 3C). Duric chips in the spoil indicate that burrowing intercepts the hardpan of this Durixeralf.

Openwork stone gutter extensions and cog-like protuberances that are presently unassociated with either mounds or ridges but connected to some perimeter gutters, also occupy the broad intermounds of Soulé Terrace. They are only faintly visible on airphotos, and appear to be residual remnants of gutters that were once associated with mounds that no longer exist. During our October 2010 work, we hypothesized that cold season frost sorting actively maintains the gutters, gutter segments, and protuberances. We further hypothesized that during spring snowmelt and freshets together with overland flow and subsurface interflow

that gutters function as part of a water-conducting, terrace-wide conduit network. The gutters on Soulé thus would be equivalent to the "channels" of the "minutely adjusted drainage system" of Waters and Flagler (1929) and the broader "stone beds" of Cox and Allen (1987a). Such a drainage net system reasonably explains how soil fines are seasonally flushed from mounds and intermounds: (1) as suspensions in concentrated flow in gutters; (2) as suspensions below the rubbly soil intermound pavements but above the hardpan in a bioturbationally porous interflow zone; and (3) as overland flow suspensions across the sloping pavements during freshets and snowmelt.

The hypotheses were visually confirmed on 29–30 March 2011, in transects across Soulé Terrace during peak snowmelt. Seepage at edges of mounds and ridges was observed to fill perimeter gutters where water freely flowed (Fig. 4). Many intermound pavements functioned as sluiceways for robust overland flow. Mounds and linear ridges were the only areas of Soulé Terrace not visibly saturated or impacted by snowmelt runoff. Because gutter trench bottoms are several decimeters lower than intermound pavements, and because gutters receive both surface

Figure 3. (A) Low mound on which men stand is mound 1 in Figures 2B and 2C. Note relatively unbioturbated stone gutter (arrows) that encircles the mound. Poorly sorted rubbly soil intermound is in foreground. Except for the gutter, all areas shown are profusely bioturbated. (B) Unbioturbated sorted stone gutter that encircles low mound at left where vegetation is shrubbier. Gutters are hypothesized to be part of a larger, terrace-wide French drain conduit-interflow system that operates above the subsoil Bt-duripan aquiclude. Runoff from mounds and ridges, subdrainage from beneath them, and runoff from intermounds into gutters are sources of seasonal free water for gutter frost sorting, maintenance, and sediment flushing. Where residual water collects, frost heaving and sorting trumps bioturbation. (C) Three weather-wasted and coalesced pebbly loam gopher heaps—now flattened spoil—that contain yellowish-brown chips of hardened duric material, at arrow, bioturbated up from underlying duripan-aquiclude.

runoff from mounds and slow residual under-drainage from beneath them, we hypothesize that water persists in gutters and broader stone nets in PNW moundfields into late spring. Runoff would render stone gutters and broad nets particularly susceptible to sediment flushing when above freezing, and to frost sorting and maintenance when below (cf. Johnson and Hansen, 1974).

Similar observations were made several days later in neighboring Modoc County moundfields, where snowmelt was essentially complete, as at Carr Butte (Table 3). Surface features there are similar to Soulé Terrace, with many scattered mounds and associated larger, erosion-thinned soil parent bodies, many fringed by sorted stone gutters. Also like Soulé, broad rubbly soil

TABLE 4. PERCENT FINE FRACTIONS (≤2 MM) FROM FIVE SAMPLED POCKET GOPHER HEAPS, SAMPLED 16 OCTOBER 2010

Gopher heap/mound	Sand: 2–0.05 mm	Total silt: 0.05–0.002 mm	Coarse silt: 0.05–0.02 mm	Fine silt: 0.02–0.002 mm	Clay: <0.002 mm
1.0	47.2	32.8	11.5	21.3	19.9
2.0	45.0	35.3	14.3	21.0	19.7
3.0	45.9	35.0	13.5	21.5	19.1
4.0	45.5	34.8	13.8	21.1	19.7
5.0	45.7	34.4	12.9	22.3	19.9
Averages	45.9	34.5	13.0	21.4	19.7

Note: One sample each from mounds 1–5 (Fig. 2); sampled using procedure of Franzmeier et al. (1977).

TABLE 5. WEIGHT PERCENTAGES OF FINES TO PEBBLES FROM FIVE SAMPLED
POCKET GOPHER HEAPS, SAMPLED 16 OCTOBER 2010

Gopher heap/mound	Sample wt (g)	wt (g) > 2 mm (pebbles)	wt (g) ≤ 2 mm (fines)	Percent fines (by weight)	Percent pebbles (by weight)
1	1126	162	964	86	14
2	1197	290	907	76	24
3	1128	202	926	82	18
4	1485	229	1256	85	15
5	979	276	703	72	28
Averages	1183	232	951	80	20

Note: One sample each from mounds 1–5 (Fig. 2); all pebbles <6 cm diameter.

intermound pavements dominate large sectors of the surface. In the greater Modoc Plateau area, however, the substrate commonly is hard basalt, not gravelly glaciovolcanic outwash with strong subsoil duripans. (Duripans are present, however, in some Modoc Plateau soils, as in the Devil's Garden area near Alturas [personal observations]). Nevertheless, the flow of residual snowmelt water was similar to that observed at Soulé several days earlier. Specifically, basal mound seepage was observed to be feeding sorted stone gutters in which water trickled, intermounds were soggy from waning overland flow, and dry areas were only in the uppermost mounds and soil parent bodies. Also like Soulé, essentially continuous debris islands and wasted rodent spoil typifies Modoc intermounds, with fresher spoil on drier mounds, all indicative of widespread bioturbation. Hence, the general surface and internal processes that sculpt these numerous northeastern California moundfields appear similar, even when substrates differ and moundfields look different.

The soil-geomorphologic processes that produce the soils and surface features on Soulé Terrace and neighboring Siskiyou-Modoc moundfields shed explanatory light on the evolution of moundfields generally throughout the volcanic interior PNW. It is noteworthy that, based on our observations, areally much larger and vegetated soil bodies, eolian mantles, and mounds with perimeter gutters very similar to the mounds and ridges of Soulé are found in many of the PNW moundfields listed in Table 3. Some of these ridges and soil bodies, like Soulé, likewise exhibit perimeters with escalloped and sinused mound-like lobes.

Masson's Explanatory Model

Drawing from Sharp's (1942) work in the Yukon, Antev's (1932) on Mount Washington, New Hampshire, Taber's (1930) frost heaving data, and Conrad's (1946) polygon net ideas, Masson hypothesized that the Siskiyou mounds formed at centers of frost polygons created under a Pleistocene periglacial climate. He cited none of the deep and extensive prairie (Mima) mound literature that then existed. Masson summarized his conclusions in a simple graphic model reproduced as Figure 5. He suggested that under repeated and intense freeze-thaw conditions clays in the glacial outwash migrated to polygon centers, became concentrated, and "domed" upwards forming clay mounds. Coincident with clay mounding, he surmised that under combined frost-displacement and gravity processes, surface rocks migrated off the rising clay mounds to accumulate as unsorted loose perimeter stone rings that occupied unexplained trenches. He viewed the mounds as increasing in size and coalescing, ultimately to form the ridges on Soulé Terrace.

Masson's conclusions, which significantly differ from ours, later appeared in many PNW documents that attempted to explain similar mounds and bordering rings. Among the more notable differences is our determination that the Soulé Terrace mounds and ridges are composed of pebbly loam, *not* pebble-free clays as Masson stated (cf. Tables 4 and 5). In fact, pebbly loam soils typically compose many of the PNW mounds we have investigated. Mounds become bioturbationally pebblier as they become erosionally lower and smaller with time. Other differences are discussed below.

Animal bioturbation is an unmistakably important and ubiquitous soil process on Soulé Terrace, as in soils generally. Nevertheless, Masson—understandable for his time—formulated a purely physical mechanism to explain the low mounds and ridges. We say "understandable" because in 1949 the idea of animal bioturbation and all the process nuances now associated with it, such as one- and two-layered biomantles, soil biosorting, biotransfers, biostratification, stonelayers (stone lines), etc., were not then part of a geologic-soils-geomorphic thought process, nor part of any explanatory paradigm. The term "bioturbation," which is foundational to all these concepts, was coined three

Figure 4. Photos taken during full snowmelt, Soulé Terrace, Siskiyou County, California, 29 March 2011. (A) Flooded intermound pavement on more level part of terrace (0%–1% slope); arrows denote low, brush-covered prairie mounds in background. Snow- and water-flattened pocket gopher spoil is visible above and below the water surface, typically present throughout the cobbly soil intermound pavement. (B) Man standing on water-filled gutter, with low brushy mound on left; arrow denotes slightly turbid water flow within gutter below snow toward man (0.5%–1% slope). (C) Water-charged gutter systems linked by brisk, anastomosing concentrated flow (arrow) combined with brisk sheet flow across inter-mound pavements (scale: cm left, inches right). (D) Close-up of flattened gopher spoil (arrow) and burrow in saturated intermound. (E) Close-up of flooded intermound showing turbid overland flow with suspended sediment. As the terrace system dewaters and slowly dries in spring, free water stands last in gutters, which renders them particularly susceptible to diurnal frost heaving, sorting, and maintenance.

years *after* Masson's article appeared (Richter, 1952; Schäfer, 1952). Hence, the impacts of burrowing animals on the soils and surface topography of Soulé Terrace went unmentioned—and presumably unnoticed—by Masson. The details of such key soil formative processes appeared much later (e.g., Soil Survey Staff, 1975, p. 21, 26, 36; Johnson and Watson-Stegner, 1987; John-son et al., 1987; Humphreys and Mitchell, 1988; Johnson, 1989, 1990, 2002; Johnson et al., 2002; Paton et al., 1995; Schaetzl and Anderson, 2005; Morrás et al., 2009; Wilkinson et al., 2009; Fey, 2010).

An Alternative Broader View

We propose alternative hypotheses to explain the origin of soils and surface features of Soulé Terrace and other PNW moundfields. The hypotheses flow from the dynamic denudation-biomantle-process vector explanatory approach (Horwath, 2002; Horwath et al., 2002; Horwath and Johnson, 2006, 2007; John-son et al., 2002, 2005a, 2005b). We also use updated Soil Taxo-nomic concepts (USDA-NRCS, 1999) and recently formulated hydropedological ideas (Lin, 2010), and integrate them with the

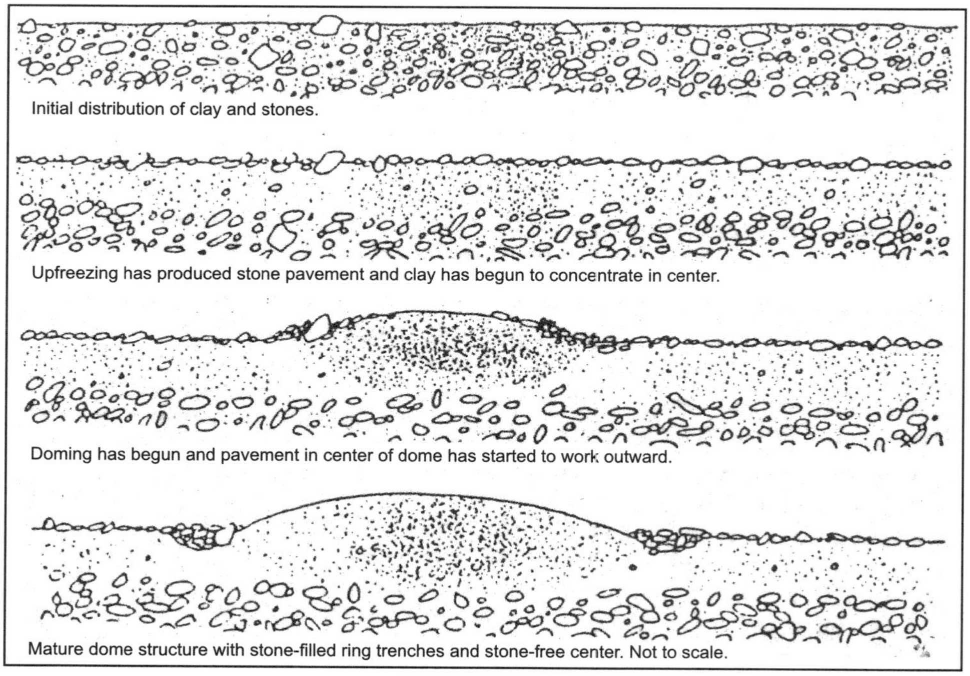

Figure 5. Masson's four-stage frost polygon-periglacial model for explaining mounds and bordering loose, gravity-accumulated "stone rings" (Masson, 1949, his figure 3, p. 65).

Dalquest-Scheffer-Cox burrowing animal/centripetal soil bio-transfers model (DSC model) of mound formation (Dalquest and Scheffer, 1942, 1944; Scheffer, 1947, 1958, 1969; Cox, 1984a, 1984b, 1990a, 1990b; Cox and Allen, 1987a, 1987b; Cox et al., 1987; Cox and Scheffer, 1991; Johnson and Horwath Burnham, this volume, Introduction). Drawing on this pool of concepts, we build on and augment several of Masson's early observations on Soulé Terrace, and provide some alternative conclusions.

Our integrative approach is based on observable and measurable processes and conditions rather than speculations about processes that may have occurred in the past. It is also based on the fact that most burrowing animals operate from activity centers—nesting, denning, overwintering, and food-storage loci—in meeting their survival imperatives. On Soulé Terrace these centers are in mounds and ridges, and are mainly those of rodents, most notably pocket gophers. Repeated radial "forage burrowing" by these animals in thin soil over a duripan barrier to vertical burrowing automatically results in centripetal soil movements back toward the activity centers. Centripetal burrowing involving multiple generations of pocket gophers leads to mound formation. In interior cold-winter areas like Soulé Ranch where multiple freeze-thaw cycles, bioturbation, and erosion are dominant processes, Type 2 prairie mounds—those with sorted stone gutters, rings, nets, beds, etc.—are formed (cf. Fryxell, 1964, this volume, Appendix A).

We present our soil-geomorphic observations and findings at Soulé Ranch within this broader framework, and follow with an eight-stage graphic model that links to them. We conclude that a polygenetic approach, where combined bioturbation, seasonal frost heaving, and erosion with occasional eolian inputs best accounts for this complex landscape. We further conclude that this approach and model best explains PNW moundfields in general.

SOIL-GEOMORPHIC OBSERVATIONS, FINDINGS, EXPLANATORY HYPOTHESES

Soulé Mounds as "Daughter Products" (Fixed Activity Centers) and Ridges as Remnants of a Once Thicker and Extensive "Soil Parent Body"

As noted by Masson, some of the circular mounds are linearly arrayed and parallel the lobed ridges. This geometry suggests that the mounds are daughter remnants of erosionally separated paternoster-like lobes of former parent ridges, not mounds coalescing into ridges as Masson proposed. The arrangement and array of mounds also suggests that the ridges themselves are erosional remnants of a once larger, and probably thicker and continuous, soil parent body that mantled Soulé Terrace. Observations and measurements at other PNW moundfields, as on Manastash Ridge and Badger Mountain in Washington, and the Shaniko-Maupin area of Oregon (Table 3), show that flattish and thick (>1 m) soil parent bodies (SPBs) of ash-loess dominated soil mantles become erosionally thin and lobed near their edges. When edge soils are ≤1 m thick above any barrier (bedrock, dense gravels, pan, etc.) that limits vertical burrowing, fixed activity centers, as noted earlier, begin to form. Low saddles—soil depletion zones—initiated by centripetal biotransfers to nutrient-rich activity centers become erosion-prone sinuses. This leads to even thinner, escalloped, and sinused SPB edges—polyp-like precursors to "daughter mound" (DM) formation that initiates and precedes erosional detachment. Stages of DM precursors are evident on Soulé ridges and other PNW moundfields, from lobate protrusions at sinused SPB edges, to "young" adjacent flat-topped mounds only recently detached (Fig. 2A). With time, intermounds expand at the expense of eroding loamy low bulk density mounds, which commensurately become more isolated, dome-like, and eventually smaller, thinner, more shield-shaped, and pebbly. At this stage low mounds are usually dotted with wasted and fresh pebbly gopher heaps, especially at their edges. Ultimately mounds erode away, leaving behind "mound disks."

We interpret sinused lobes on Soulé ridges as fixed activity centers, future daughter mounds (DM) that undergo slow erosional separation from ridges and one another, yet still joined at low mound saddles (MS). In essence, saddles are low midpoints between adjacent fixed centers (future mounds), produced jointly by centripetal biotransfers and erosion removals (ER). Centripetal bioturbation thus dictates where surface runoff collects, flows, and erodes. Such combined processes create Mima-type mounds.

The low circular soil mounds on Soulé Terrace are expressed singly and as mound clusters in couplet, triplet, quadruplet, and paternoster arrangements. Former, completely eroded mounds are now expressed as residual disks (Fig. 2A, area below box). Locally, two mounds form binary couplets, still conjoined but undergoing slow separation at mound saddles. Mound saddles—being early manifestations of mound separation—reflect both centripetal biotransfers to activity centers and erosion removals. Such processes promote intermound formation, as reasoned by Dalquest and Scheffer (1942, p. 72–73), Peterson (1961, p. 127–138), and Johnson (1982, p. 15). Mound saddles occur in proto-mound groupings near edges of soil parent bodies in many PNW moundfields. The discrete single mounds to which binaries, triplets, and quadruplets ultimately evolve temporarily share co-evolved stone gutters.

As indicated, the low mounds and ridges are bioturbation-ally modified and shaped erosional remnants of a thicker soil that mantled Soulé Terrace. Their bordering stone gutters are, we hypothesize, owed to combined frost sorting and runoff flushing. We further hypothesize that the rubbly soil intermound pavements are owed to combined bioturbation, frost action, and erosion. All of these processes operate above a duripan aquiclude to produce the soil features present.

Sorted Stone Borders and Rubbly Soil Pavements as Products of Process Polygenesis

Our model combines multiple processes to explain mound-intermound features that may be genetically associated in one region but absent in another. Examples are the dark, sorted, tightly fitted and openwork stone gutters and nets that are commonly associated with interior, cold-winter, Type 2 PNW volcanic moundfields, but largely absent in warmer winter volcanic moundfields elsewhere (Table 3). Runoff and subdrainage from mounds and soil parent bodies initially concentrates at their edges in gutters, nets, and remnant gutter segments, and then undergoes multiple freeze-thaw cycles before exiting the system as free water. These processes, we hypothesize, create and maintain the well-sorted, interlocking tight fit of stones (gutters, nets)

that border PNW mounds. Tightness of fit makes gutters resistant to burrowing rodents, which explains the absence of tan-colored spoil on visually striking black gutter surfaces (Fig. 2A). It is primarily the lack of spoil on the naturally black volcanic rock, made darker by lichens, that accounts for the stark contrast between black gutters, lighter-colored mounds, and spoil-dappled intermounds of Soulé Terrace and other PNW moundfields. Lack of spoil allows gutters, maintained by seasonal frost sorting, to function as openwork conduits for runoff in an integrated, French drain-like interflow-overland flow system.

Snowmelt and surface runoff from mounds, plus subdrainage at mound edges, are the sources of gutter water and the energy-force behind gutter freeze-thaw sorting, maintenance, and consequent low-slope conduit-interflow integration. If we are correct, mound-gutter systems are co-evolved, where gutters, "nets," and "stone beds" at other PNW Type 2 moundfields are collectors of runoff and effect fine fraction removals. Hence, as mounds erode via loss of fines and shrink in size, their genetically linked gutters centripetally contract to where gutters are no longer shared—although still integrated, we hypothesize, within the greater conduit-interflow system.

The concept of slow concurrent mound-gutter erosional contractions evokes complicated dynamic process-scenarios, with terms lacking to readily describe them. Gutter circumferences contract as associated mounds and soil bodies erode. Thus, outermost gutter-stones slowly coalesce with rubbly intermounds while innermost gutters concurrently incorporate new stones exposed via soil loss at mound edges. The gutters are functionally maintained by freeze-thaw cycles with fines flushed through the gutter-conduit system. As mound-gutter systems dissipate only rubbly disks remain. At this point gutters become unassociated and may either remain a segment of the interconnected gutter conduit-interflow frost-sorted system (Fig. 6A), or succumb to bioturbation and coalesce with rubbly soil intermounds. Where unassociated gutters continue functioning as openwork drains and remain free of gopher spoil, we presume freeze-thaw maintenance is responsible.

Such complex processes occur with apparently still active interconnected gutters and/or nets regardless of substrate. They are Type 2 mounds, fringed by interconnected frost-sorted stone nets, and beds, and typical of many cold-winter PNW moundfields (Table 3). Figure 6 aptly illustrates mounds with perimeter stone borders linked to unassociated "anastomosing" gutters and nets on broad intermounds on a gentle-to-steepening slope. The cross-sectioned mound is relatively flat-topped, and thus similar to many PNW mounds that originate from eolian-derived thicker parent mantles with accordant surfaces. Once detached from their larger parent mantle, mounds gradually become more rounded and dome-like under erosion and bioturbation. In the absence of eolian infall mounds also become more pebbly and smaller in size.

Accordingly, intermound pavements are polygenetic, partly composed of former gutters, old and new rodent spoil, mound disks, and spoil-coalesced debris islands. These elements are all impacted by seasonal frost action, bioturbation, and erosion. Explanatory permutations far more complex than those suggested here are of course possible. Indeed, because of the intertwined physical and biological complexities of prairie mounds, it is not surprising that the literature is controversial, particularly that dealing with seasonally cold PNW Type 2 mounds.

As at most moundfields, mounds on Soulé Ranch are obvious from the air and on airphotos. Large single mounds average 28 m in diameter and ±0.5 m high (Fig. 2C). Larger and smaller mounds are also present, the smallest, as indicated, approaching intermound debris island and mound disk sizes. Again, without inputs of eolian ash and loess, Soulé Terrace mounds will ero-

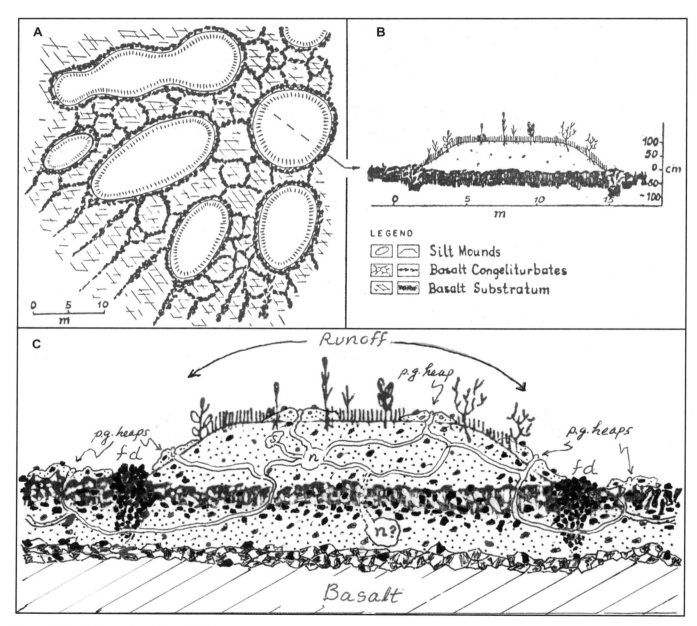

Figure 6. A and B constitute figure 3 of Brunnschweiler (1962), captioned: "Schematic ground plan of a group of stone-ringed 'Columbia' mounds and cross section of a typical individual." On Soulé, Brunnschweiler's cross-angled marks equate to rubbly soil intermounds, and his "basalt congeliturbates" to gutters. (A) Mounds and interconnected gutters—some associated with mounds, and others no longer associated with them. As slope steepens to SW, gutters become stone stripes. (C) Our modified enlargement of B, with added complexities based on our surveys of PNW moundfields that reflect combined bioturbation, frost action, and erosion. Frost-sorted and fitted stone gutters lack gopher spoil (are unbioturbated). Poorly frost-sorted intermound pavement contains much spoil and hence is strongly bioturbated. Pavement continues under mound as a stonelayer. (Symbols: n—nest-brood chamber; s—food storage chamber; fd—French drain; p.g.—pocket gopher; parallel lines represent active, open gopher tunnels.) Myriad krotovina that theoretically must exist cannot be shown due to lack of soil color contrasts.

sionally dissipate and evolve into pebbly loam disk-like debris islands and intermound pavements.

The stone borders of Soulé Ranch mounds and ridges are not "loose stones" as Masson stated, but vertically sorted, tightly frost-fitted, openwork gutters (Fig. 7). Tabular-shaped clasts are commonly vertically oriented and fitted in typical frost-sorted fashions. Uppermost is a zone of boulders and large cobbles, followed by large- to medium-sized cobbles very thinly coated with buff-colored silty-clay matrix material. The coatings are not apparent until the frost-fitted surface layer of boulders and large cobbles is removed. Below the coated cobbles, at 30–40 cm, is a zone of medium to small cobbles and large pebbles that bear thicker silt-clay coatings, and that contain some matrix silt and fine sand between stones. At 40–50 cm depth medium to small pebbles are present, with a matrix of silt and fine sand in spaces between pebbles. The pebbles here appear to "float" in the matrix.

An open, presumably active gopher burrow (black arrow, Fig. 7) was encountered at 42 cm that passed under the stone border. At other sections, several wasted, pebbly loam gopher eskers, which form only under snowpack, were observed draped across the stone borders. The deep 42 cm burrow and surface eskers indicate gopher movements between nesting mounds and intermounds both under and over the stone borders, presumably for food procurement (cf. Cox, 1989; Cox and Allen, 1987a).

Evidence that stone gutters function as French drains at Soulé and other PNW moundfields is based on (1) their open-work nature; (2) preserved signatures of seasonal wetness and free-standing water, such as dried moss, water rings, and stains—evident during dry seasons; and (3) our observations of water moving within them in late March 2011, in addition to similar observations by Fosberg (1963, p. 231) in Idaho gutters, and by Goldthwait (1976) in Arctic gutters.

The cross-sectional geometry of stone borders is urn-shaped; their widths narrowing with depth to ~40–50 cm. Gutter depths and geometries, however, likely vary between moundfields. Our gutter geometry and measurements fall within the range of the 21 determined by Pyrch (1973a) for stone stripe gutters on steep slopes in north-central Oregon. Stripes, like low slope gutters, are preferred snowmelt, ice-melt, and water-flushed channel-ways, apparently simply steeper versions of gutters (Fig. 6A). This con-clusion is supported by the presence of "ice waterfalls" and "ice cascades" that are observed to develop in some PNW stripes dur-ing winters.

Both the mound and intermound on either side of our ex-cavated gutter have abundant old and new pocket gopher spoil, as do all the mounds, ridges, and intermounds we surveyed. As indicated, active stone gutters are the only areas on Soulé Ter-race in which gopher spoil is largely absent. This fact reflects the known efficacy, over other processes, of frost sorting and soil heaving under both saturated and standing-water conditions (cf. Johnson and Hansen, 1974; Johnson et al., 1977). The pattern is consistent with our observations at other PNW moundfields, and with those of Cox and Allen (1987a) in the Shaniko Oregon area. In short, active frost heaving in gutters and nets trumps bioturba-tion, whereas in intermounds bioturbation trumps frost sorting.

We use terms "active" and "inactive" to describe gutters for several reasons. We observe that gopher spoil often spills onto gutter edges on both mound and intermound sides, a process that appears to be ongoing and regular. That spoil does not cover and clog Soulé gutters, and some stone nets and beds elsewhere, sug-gests that such openwork gutters, nets, and beds actively function as French drains and that they are seasonally flushed of fines. It also suggests that such stone gutter-net-bed systems are main-tained by frost sorting. "Inactive" gutters, nets, and beds, on the other hand, are those clogged by spoil, or are in the process of being clogged, and cease functioning as French drains. Clogged gutters then become part of rubbly soil intermound pavements. In this case, bioturbation largely trumps frost heaving and sorting.

Figure 7. Hand-excavated sorted stone gutter of mound 1 (Fig. 2). The stones in order of removal, right to left, are (1) large, blackish-colored, lichen-coated surface boulders; (2) small boul-ders, large cobbles; (3) medium cobbles, large pebbles; (4) small pebbles, gran-ules; and (5) fines with pebble-granule "floaters" at 40–50 cm. An open and ac-tive, 5-cm-diameter, gopher tunnel was encountered near gutter bottom (~42 cm) at black arrow (cf. text for discussion).

Mounds as Nesting Sites and Intermounds as Foraging Domains of Pocket Gophers (*T. talpoides*)

The mounds and ridges are gopher nesting and overwintering sites on Soulé Terrace. However, the presence in intermounds of abundant fresh, weather-wasted, and coalesced spoil suggests that gophers may spend as much time food foraging, if not more, in intermounds as in mounds. Presumably they do so during the main spring-summer-fall food-foraging seasons (cf. Cox, 1989; Cox and Allen, 1987a, and references therein). Toward the end of the dry season in mid-October 2010, we estimated that 80% of all fresh-appearing heaps were on intermounds, not mounds. Accordingly, these heaps had to have formed after the 2009–2010 winter season. Assuming the animals normally cannot burrow through the tightly fitted stone gutters, they must go over or under them—or both. The gopher tunnel exposed at 40 cm near the base of our excavated gutter (Fig. 7), and gopher eskers observed draped over gutters on Soulé—and observed by us elsewhere (Devil's Garden, Table 3)—prove they do both (cf. Marshall, 1941; Larrison, 1942).

Figure 8 illustrates a 0.28-ha mound-intermound area typical of some bedrock (basalt) hillcrests near Shaniko in north-central Oregon (Cox and Allen, 1987a, their figure 3). This figure, like Figure 6, usefully depicts similarities and differences with the Soulé Terrace moundfield and other moundfields scattered throughout the PNW (Table 3).

According to Cox and Allen (1987a, p. 179), "A distinctive feature of much of this landscape is the presence of sharply defined beds of bare, sorted stones that partially or completely encircle the mounds and occasionally form intermound stripes and polygonal networks. Well-developed stone circles and nets of this type have been reported from eastern Washington. . . ." Both Shaniko and Soulé Terrace intermound areas have abundant gopher heaps and debris islands of pebbly loam soil—what Cox and Allen call "islands of soil and rock." On Figure 8 the debris islands are shown as white patches within the dark sorted stone nets. Fresh heaps and wasted gopher spoil (black patches) are shown on debris islands and along mound edges, with several as small spoil patches within the stone nets proper. Cox and Allen (1987a) stated that debris islands range in size from 1 to 100 m^2 and that most bore fresh pocket gopher spoil likely less than two years old.

Other similarities and differences between moundfields of Shaniko and Soulé Terrace exist. From the air, mound-intermound resemblances and differences are broadly apparent, namely patterns of DMs derived from multiple sinused and lobed ridges and SPBs. But, unlike Soulé, many Shaniko mounds are on hilly terrain, as are many—though not all—PNW moundfields, often arrayed from summits to footslopes in visually striking starburst-like and paternoster-like patterns (Table 3; cf. Plate 2B in Fryxell, 1964, Appendix A, this volume). As is the case, however, when comparing PNW moundfields, differences exist due to wide variations in relief, geology, soils, erosion, aspect, and various other environmental factors.

Most Soulé mounds and ridges are low, <1 m, whereas at Shaniko they range from similarly low to commonly 1.5+ m. Soulé stone gutters are generally narrow, ±1 m wide, while those at Shaniko range from similarly narrow to 10+-m-wide stone nets (Fig. 8), often, as indicated, with scattered and embedded debris islands. At low slope sites at Shaniko, as represented in Figure 8, where snowmelt and freshet water stands longest and frost action is greatest, frost sorting trumps bioturbation and rodent spoil production. The balance of rates and interplay between

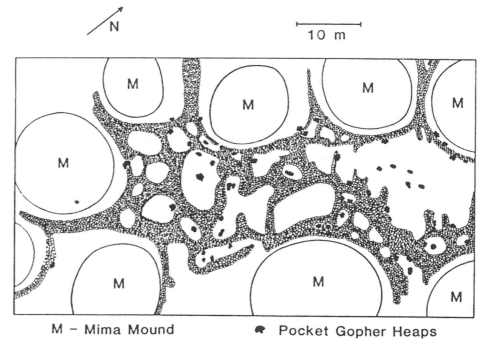

Figure 8. Mound-intermound and stone gutter-net complex in Shaniko area of north-central Oregon (figure 3 of Cox and Allen, 1987a). Their caption reads: "Pattern of Mima mounds, stone circles, intermound stone nets and islands of soil and rock on a 40 × 70 m plot centered on a large intermound flat on land adjacent to the Lawrence Memorial Grassland Preserve, southern Wasco County, Oregon. Pocket gopher heaps are shown only within and adjacent to the network of stone beds; heaps were also abundant on the mound surfaces."

M – Mima Mound   ● Pocket Gopher Heaps

these prime processes—frost heaving versus bioturbational spoil production—must, at any given site, depend on local biological-physical factors that doubtless change with time.

Cox and Allen (1987a) offer useful quantitative and observational details from Shaniko that generally apply to Soulé and other PNW moundfields. Their "soil-and-rock islands" equate to the debris islands and spoil patches at Soulé Terrace. They specify (p. 182–184):

Recent pocket gopher heaps were present throughout the intermound complex of stone nets and soil-and-rock islands. These heaps (n = 56) were concentrated at the edges of the stone beds (34), although many lay in the interior of the soil-and-rock islands (19). Three heaps, however, lay on the surface of stone beds. Numerous unmapped heaps were also present on the surfaces of the adjacent mounds. The pocket gopher heaps consisted of fine soil, gravel, and pebbles up to a maximum diameter of ~60 mm. . . . [and that] . . . The close proximity of sorted stone circles to edges of mounds and the significant relationship of stone circle development to mound size suggests that the mounds and sorted stone circles form concurrently . . . [and that] . . . Pocket gophers maintain their permanent nest and food storage chambers in the mounds because of frozen or waterlogged conditions of intermound areas during late winter and early spring. They tunnel extensively into the intermound zone during the dry seasons. This tunneling is evidenced by the deposition of heaps on the intermound margins of stone networks and on islands of soil and rock enclosed by stone nets. The distribution of heaps also demonstrates that the animals commonly tunnel beneath the stone beds. The low frequency of heaps in areas of otherwise soil-free stone beds suggests that the animals are largely unable to penetrate the stone beds themselves.

This part of their explanatory model firmly matches our observations and interpretations at Soulé. There is no question that gopher tunneling beneath Soulé intermounds and many PNW moundfields, including Shaniko, is a major biodynamic force in producing subsoil macroporosity. It is a porosity producing process that relaxes only during late winter and early spring when intermounds are saturated and/or flooded. The Cox-Allen study is commendable for calling attention to a previously unrecognized, though clearly key and important hydropedological process, one that invites future study and attempts at quantification.

We question only the authors' subsequent conclusion that gutter-net sorting is owed to differential settling of stones caused by gopher tunneling, rather than to frost sorting. We do so because of the sorted, oriented, and keyed character of stones in gutters of Soulé and nets of Shaniko and other interior cold-winter PNW volcanic tracts. Such stone orientation and sorting is absent in similarly mounded gopher-inhabited volcanic tracts under mild winters. Well bioturbated "eolian soil" in volcanic moundfields along California's Great Central Valley and elsewhere are examples (Table 3). Rubbly soil intermound pavements are invariably present at these elevationally lower and warmer moundfields, but the well-sorted and fitted gutter-net features in almost all cases are not. We agree that pocket gophers produce mounds by centripetal soil biotransfers extracted mainly from intermounds, which ultimately results in rubbly soil intermound pavements. We also conclude, as did LeConte (1874) and Waters and Flagler (1929), that erosion plays a key role in sculpting the character

of PNW mounded landscapes. These processes, we believe, are fundamental in initiating mound formation that begins at mound saddles when parent mantles thin. We likewise agree that gopher tunneling for food procurement significantly impacts intermounds in the other ways proposed by Cox and Allen (cf. Miller, 1948). We also believe, as indicated, that tunneling greatly increases porosity that augments subsurface interflow.

In sum, we suggest seasonal frost action produces and maintains the well-sorted, fitted, and openwork nature of stone gutters, beds, and nets. We hypothesize that seasonal flushing of the gutters and nets, together with subsurface interflow below the pavement and above the bedrock or hardpan, function as an interconnected sediment removal drainage system. We recognize that these hypotheses need testing at select PNW moundfields.

EIGHT-STAGE MODEL OF SOIL AND MOUNDFIELD EVOLUTION

We hypothesize in Figure 9 a transitional eight-stage landscape evolution model of the Soulé Ranch moundfield. The key process vectors involved at each stage, beginning with Stage 2, are approximated in each panel and keyed to the text. Glaciers that headed in the andesitic Rainbow Mountain uplands above and immediately south of Soulé Ranch carved the U-shaped valley occupied by Butte Creek, which now cuts Soulé Terrace. Age of the terrace gravels is unknown, but the strength of its Durixeralfic subsoil suggests an $0^{18}/0^{16}$ MIS (marine isotope stage) 4 or 6 age, or possibly even earlier (R.J. Shlemon, 2010, personal commun.). This age estimate is supported by the 160 ± 55 ka age on basalt, mentioned earlier, that overlies morainal material with which the mounded outwash is associated.

The elevation of Soulé Ranch lies in a zone of frost-prone snow-packed winters. Hence, seasonal frost sorting and heaving are assumed to always have occurred. Freeze-thaw processes are presumably reduced through later model Stages 3–8 compared to colder initial end-glacial and previous periglacial Pleistocene levels. The earlier model stages are obviously more uncertain than the present model Stage 8. Through time, soil formation processes are increasingly complex and polygenetic. The first two stages approximate the first two stages of Masson's model (Fig. 5).

Some terms defined in Table 2 and used in the model come from works in Table 1, especially from Copeland (1980), Johnson (1982), Winward and Youtie (1978), Cox and Allen (1987a), and Dalquest and Scheffer (1942, 1944), and to a lesser degree from Goldthwait (1976) and Washburn (1956, 1973, 1980, 1988). The concepts "soil parent bodies" and detached "daughter mounds" are drawn partly from Peterson (1961), Waters and Flagler (1929), and Kaatz (1959) and partly from our PNW mound work.

Stage 1

This very brief initial stage marks the cessation of outwash gravel deposition across the Soulé surface. The andesite gravels constitute the initial soil parent material. This begins (time zero) postglacial soil evolution. Physical processes begin (eluviation-

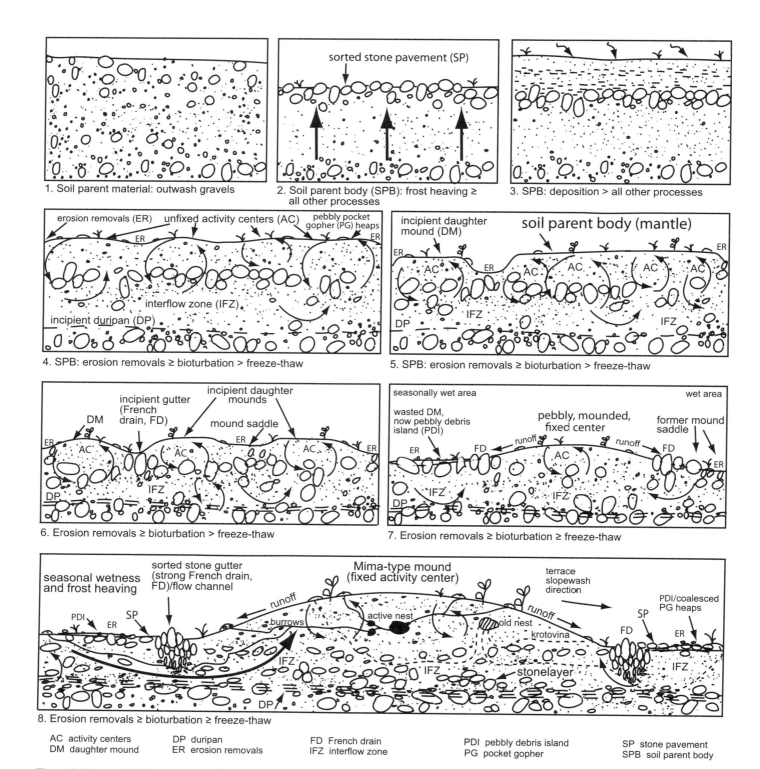

Figure 9. Proposed eight-stage generalized model for the evolution of soils and surface features on Soulé Ranch over variable time frames. Model Stages 1–3 represent short time frames (decades-centuries); 4–6 intermediate time frames (many millennia); and 7–8 long time frames (tens of millennia). Stage 8 is the landscape as presently observed. Horizontal-spatial scales change between Stages 6–7 and 7–8.

illuviation, weathering, solution-precipitation, oxidation-reduction, freeze-thaw, erosions, etc.), and then, with biogenic inputs, operate through this and all subsequent stages. Lower lifeforms begin colonizing, and bioturbating, this incipient soil.

Stage 2

Stage 2 is the formation of sorted stone pavement produced by frost heaving, as postulated by Masson (1949). Frost sorting and heaving processes occur under Pleistocene periglacial freeze-thaw conditions. Adjacent Butte Creek (Fig. 2A) begins entrenching its bed; terrace formation begins. Further entrenchment, water table lowering, and terrace processes continue through subsequent stages. Higher lifeforms have now colonized the soil, initiating accelerated bioturbation and biochemical inputs and incipient biomantle formation.

Stage 3

Stage 3 is marked by episodic overflow accretions (dashed lines) of sands and small gravels from Butte Creek. Pulses of silt-sized eolian ash and loess (dots, squiggly arrows) augment sand-gravel fluvial accretions. The new sediments bury the stone pavement that formed during Stage 2. This overflow and loess unit is new material for later soil evolution (deposition > bioturbation). Continued development of plant cover attracts more diverse soil biota. Near the end of this stage, initial destratification begins under bioturbation, and biomantle formation continues.

Stage 4

Soulé Terrace is now vegetated and undergoing plant-animal bioturbations. In-migration from surrounding unglaciated refugia of the more dominant small mammal burrowers and other bioturbating animals begins, including the ancestral northern pocket gopher *Thomomys*. These burrowers, and their burrowing predators (e.g., badgers, coyotes, wolves, and bears), are now the dominant process vectors in the evolution of the soil. Curved arrows represent dominant bioturbation in the biomantle, which thickens owing to surface spoil accumulations, occasional deep burrowings, and ash-loess dustings. The arrival of these dominant burrowers genetically links to certain processes and events:

1. Establishment of nesting-food storage sites at initially unfixed activity centers (AC), generally at higher and drier or other soil locations that confer slight survival advantages. Centers are spatially arranged by local conditions, food, biotic competition and carrying capacity, and territorial-dispersal imperatives.

2. Pocket gophers (PG) begin foraging and tunneling radially outward from centers, marking the onset of centripetal biotransfers of soil toward centers, giving rise to many spoil heaps.

3. PG heaps, unvegetated and easily eroded (Fig. 3C), dapple the Soulé surface and promote slow elutriation washing (thinning) of fines during this and all later stages. A subsoil Bt (clay-rich) horizon begins forming via eluviation-illuviation processes.

As volcanic ash weathers during this and all later stages, decomposition products are translocated downward, forming an incipient duripan (DP) of precipitated silica (SiO_2—shown as dashed line) below the Bt horizon. At this stage the duripan logically *must* form below the main zone of bioturbation—not within it (duripan formation > bioturbation). In later stages the duripan increases thickness, density, cemented strength, and upbuilds into the lower zone of bioturbation. Duric chips begin appearing in PG heaps (duripan thickening ≈ bioturbation). As the Bt horizon forms, its upper part is bioturbationally incorporated into the biomantle (Bt horizon formation = bioturbation). As the soil thins, activity centers become fixed, and terrace soil generally becomes more loamy and pebbly, with gradual loss of clay and silt via surface elutriation of eroding PG heaps. This process operates through all subsequent stages.

4. Soil movement to fixed centers during radial tunneling and foraging (*DSC model*) equates to soil depletions in distal foraging areas, leading to incipient mound saddles. Being low areas, saddles collect runoff and become zones of erosion removals (ER), and during later model stages ultimately evolve to become intermounds.

5. Deep burrowing by gophers results in fines and pebbly materials, smaller than burrow diameters (<6 cm), being biotransferred-biosorted up into the biomantle.

As Stage 4 ends, a porous interflow zone (IFZ) has bioturbationally developed above the Bt-DP subsoil layer; as DP thickens and strengthens through later stages, the IFZ plays a water transmission role in mound-intermound subdrainage.

Stage 5

Some activity centers have now evolved to incipient daughter mounds (DM) concomitant with erosion and centripetal removals from mound saddles. Bioturbation, namely sorting, lateral biotransfers, and occasional deep burrowing, becomes more pronounced. At this point and through the remaining stages (6–8), the now loamy biomantle consists of pebble-bearing organic and fine materials (pebbles ≤6 cm in diameter). Owing to gopher burrowing and biosorting, clasts larger (>6 cm) than gopher tunnel diameters settle downward. The duripan (Fig. 9, dashed line) strengthens. Elutriation-erosions of pebbly loam PG heaps that regularly erupt continue to promote soil thinning as fine fraction removals exit the system via overland flow and subdrainage. A mound saddle (at far right) indicated by ER portends a future daughter mound and intermound.

Stage 6

Pebbly loam daughter mounds (DM), now fully fixed activity centers (AC), and incipient mounds are now formed. Incipient sorted stone gutters begin developing, some connected to others via through-drainage systems. Surface runoff from mounds and larger residual soil parent bodies, together with submound drainage and leakage, concentrates water at mound perimeters, giving rise to intense seasonal frost sorting of gutter stones and flushing of clay-silt-fine-sand suspensions. The duripan (two dashed lines) continues to thicken and functions as an aquitard-aquiclude. The Soulé SPB slowly thins, loses fines, and becomes texturally coarser and pebblier, owing to all these processes.

Stage 7

The Soulé Terrace moundfield now consists of many, now domed-shaped mounds and several adjacent larger, now rounded soil parent bodies and ridges whose edges bear multiple mound saddles. Activity centers, prairie mounds (Mima-type mounds), are now *fixed* owing to survival-reproductive advantages for the burrowing animals. Some daughter mounds have eroded, leaving pebbly loam mound disks and large debris islands as residues. Other debris islands of coalesced pebbly gopher heaps dapple the intermounds and gradually expand with time. The duripan is now a fully developed aquiclude.

Stage 8

The present strongly bioturbated and erosionally impacted Soulé Terrace Haplic-Paleoxerollic Durixeralf (Louie stony loam-like and Salisbury loam-like) soil landscape has formed (Figs. 2 and 3). It consists of a co-evolved mound-gutter-intermound pavement system characteristic of many severe-winter PNW mound-fields. The duripan is now thick and well formed. Rubbly inter-mounds, too thin and seasonally wet for nesting or food storage, are key domains for food foraging during growing seasons. The thin soil, duripan, and wetness confront burrowing animals that need thicker and drier soil. Mima-type mounds evolve from fixed activity centers over multigenerational habitations. Frost sorting maintains bordering stone gutters, and seasonally competes with bioturbation in maintaining rubbly soil intermounds. Seasonal free and intergrain mound pore waters and runoff feed gutters, promoting strong perimeter frost sorting (more water equates to more frost sorting). The gutters function as openwork French drains (FD) for runoff, subdrainage, and removal of fines from the Soulé Terrace soil system. Mound saddles evolve first to sorted stone gutters, then to rubbly soil intermound pavements, which expand at the expense of the shrinking daughter mound-soil parent body system (Stages 6–8). As the coupled DM-SPB system gradually erodes and thins, gopher heaps increasingly exhibit duric chips bioexcavated from the duripan, indicating that gophers burrow down to it. The mound is now biostratified as a two-layered bio-mantle, with a pebbly loam mound above a stone line. Without eolian or other inputs to offset soil removals, mounds gradually become thinner, smaller, pebblier, and ultimately erode away.

Model Retrospections

In this conceptual model, the broad and gently sloping inter-mound pavements constitute innumerable former, coalesced gutters that evolved from mound saddles—soil removal zones caused by coupled centripetal biotransfers and surface erosion. As mound saddles deepened, they became gutters, and ultimately evolved to become part of the gently sloping intermounds, with fines removed through gutter conduits, by interflow and overland flow systems. Rubbly intermounds concomitantly were subject to frost heaving, extensive and intensive forage bioturbations, and sheet erosions. As the duripan (DP) gained density, thickness, and strength, it first inhibited, then restricted, and finally prevented vertical burrowing by *T. talpoides* and other burrowers. And because the DP functioned as an aquiclude, wetness conditions were created, which augmented fixed mound formation. All these processes and conditions operated in producing Mima-type mounds. Abrasion comminutions caused by animal ingestion, biomixing, and biotransfers, plus frost heaving and chemical weathering, produced smaller soil particles, all of which augmented erosion-elutriation processes on Soulé Terrace.

On-site observations show that many Soulé Terrace mounds are almost entirely eroded away; leaving flat circular disks where mounds once stood (Fig. 2). As indicated, all low mounds, ridges, and intermounds of Soulé Terrace are strongly bioturbated by burrowing rodents, principally by Northern Pocket Gophers *T. talpoides*. Belding ground squirrels are also important burrowers, as are badgers, canids, and other predators (R. Soulé, owner, based on many years of observations). Abundant spoil produced by *T. talpoides* ranges from old, flat, weather-wasted and tan-colored pebbly spoil barely recognizable as such, to moist, slightly reddish, fresh pebbly soil heaps and piles that seemingly were extruded hours or days before our observations of October 2010 and September 2011.

SUMMARY AND CONCLUSIONS

Joseph LeConte's "peculiar conditions" are simply G.K. Gilberts' burrowing animals that create fixed activity centers—survival advantage loci—in shallow soil, either above a pan, bedrock, dense gravels, and or in periodically wet soil. All of these conditions restrict vertical burrowing, and induce centripetal (radial) burrowing from activity-nesting centers. Other processes impact this bioturbational common denominator and establish that mounds and mounded landscapes are polygenetic.

The origin of prairie mounds and mounded landscapes can be viewed from several genetic perspectives. Our work in Mount Shasta region moundfields at Soulé Ranch and Modoc Plateau, and at other interior Pacific Northwest moundfields with sharp cold seasons, suggests that animal bioturbations, erosions, and multiple freeze-thaw cycles—coupled invariably with episodic eolian infalls—are dominant processes responsible for the surface features[3]. The eight-stage conceptual model illustrates these processes. The model appears applicable to most bioturbated,

---

[3]Evidence is strong (cf. Table 1 references) that many PNW volcanic lands have undergone multiple Miocene and post-Miocene erosion stripping and eolian deposition cycles, with stone pavements formed following erosion, and stone-layers and Mima-type mounds formed following eolian (ash-loess) burial and bioturbation.

erosion-prone, cold-winter interior PNW moundfields. Flat-topped daughter mounds, for example, like those at Manastash Ridge and Badger Mountain, Washington (Table 3), evolve first as activity center lobes along thinned and sinused edges of larger flat-topped, ashy-loess soil parent bodies. They detach from edges at mound saddles, produced by centripetal biotransfers of soil to activity centers; saddles then become susceptible to erosion via concentrated flow off and around mounds. Over time, detached mounds gradually become smaller, thinner, more dome-shaped and bioturbated, pebblier, and, in the absence of eolian inputs, they ultimately succumb to erosion. They constitute Type 2 mounds as defined, those bordered by sorted, tightly fitted, openwork stone gutters and/or broader stone nets. These gutters and nets are part of an interconnected French drain, seasonally flushed, conduit-interflow-overflow "minutely adjusted" drainage system (cf. Waters and Flagler, 1929).

However, for those PNW moundfields with milder or mild winters, typical of the more common Type 1 mounds, only two major processes—bioturbation and erosion—are co-dominant in producing mounds and intermounds. Such mild winter moundfields, specifically those *lacking* sorted stone gutters and nets, include Mima and other mounded prairies near Olympia, Washington, the Rowena-Dalles overlooks in Columbia Gorge of Oregon, Lower and Upper Table Rocks near Medford, Oregon, the Jenny-Fall Creeks area above Iron Gate-Copco Reservoirs on the California-Oregon border, and most mounds in Shasta Valley (Table 3). These, along with most Great Central Valley and Coast Range moundfields of California, and moundfields along the Louisiana-Texas Gulf Coast and Mississippi Valley regions, are Type 1 mounds and moundfields as defined. There may be no sharp defining climatic or altitudinal-latitudinal boundary that easily differentiates Type 1 from Type 2 mounds and moundfields, beyond the presence or absence of tightly fitted, sorted stone borders and/or nets. Such soils, however, must be sufficiently thick and contain a sufficient array of mixed size clasts for frost heaving and sorting to occur.

From another perspective, prairie mounds may be viewed as microrelief variations on a processual polygenetic theme, one that involves intimately coupled biological and physical processes. While mounds like those in the Pacific Northwest are polygenetic, they are in the first instance raised, low bulk density centers of bioturbational activity, specifically the nesting, denning, food-storage, and overwintering loci of burrowing animals. They commonly occupy preexistent high spots above either dense layers or periodically wet areas, or both, or create their own raised loci through in situ bioturbations. Such activity centers are the common denominator of point-centered, locally thickened biomantles, which characterize prairie mounds everywhere. If the activity centers form in thin soil over a barrier or restriction to vertical burrowing, like wetness, and thus afford a survival advantage to animals by virtue of being raised, they become fixed and give rise to *Mima-type* mounds. But if such centers form in deep soils without restrictions to burrowing, they are ephemeral, unfixed, and form *Mima-like* mounds for relatively short periods of time.

From yet a third perspective, a full understanding of the origin of North American prairie mounds requires understanding and appreciating how bioturbation processes and subprocesses give rise to generally nutrient-rich biomantles. Biomantles are texturally either one-layered when formed in nongravelly soils, or two-layered when they form in gravelly soils and thus possess a stonelayer (i.e., they are biosorted and biostratified). Many if not most "eolian soils" in the Pacific Northwest volcanic landscapes are initially loamy, and when they become erosionally thin (≤ 1 m) they incorporate gravels from below via bioturbating rodents and burrowing predators. Their associated, detached, relatively nutrient-rich mounds become gradually and increasingly biosorted and pebbly, and thus two-layered, consisting of pebbly loam mounds above a basal stonelayer (stone line) of coarse clasts. Without eolian renewals, erosion eventually claims them.

From a broader holistic and final perspective, all landscape elements, large and small, embed signatures of complex genetic processes—polygenetic processes—that vary with time and are difficult to interpret. Nevertheless, one process invariably plays a dominant role over others in producing any landscape attribute, and thus should impart readable signatures. Mounds of the kind discussed here manifest such polygenesis, where bioturbation is the process denominator, leaving readable signatures of life.

ACKNOWLEDGMENTS

We especially acknowledge Ray (Skip) Soulé for his friendship and repeated access to his ranch. We also acknowledge the thoughtful editing and evaluation of this chapter by co-editor Jennifer Horwath Burnham, Randy Schaetzl, and Roy Shlemon, each of whom significantly improved it. We thank Horwath Burnham for Figure 1. We also thank the University of California, Berkeley Libraries for 1944 airphotos, Google Inc. for use of their invaluable maps-images technologies (Fig. 2, Table 3), and Art Bettis for particle size analyses. We further thank the following people, some no longer with us, who assisted us directly or indirectly with our mound studies over several decades (for those not acknowledged, out of inexcusable oversight, we especially thank you, with sincere apologies): Douglas Alexander, John Alford, Russell Almaraz, Ron Amundson, Chuck Bancroft, Mary Barkworth, Tom Bass, Angie Bell, Albin Bills, James Bown, Shane Butler, Rich Cahill, Bill Collins, George Cox, Liz Clark, Jim Cotter, Bob Christiansen, Janis Dale, Kathy Danner, Bruce Delgado, Darlene Depner, Jane Domier, Leslie Durham, Lisa Ely, Thomas E. Emerson, Charles Frederick, Dennis Freeman, Linda Freeman, Roald Fryxell, Juan de la Fuente, Gerry Gates, Jamie Gentner, Patricia Healy, Paul Heinrich, Charles Higgins, Bill Hirt, Bob Holland, Randy Hughes, Lee-Ann Irvine, Rich Jaros, Tim Jensen, Jenny Marie Johnson, David Johnston, Lura Joseph, Martin Kaatz, Jo Kibbee, Dan Kirk, Michael Kobseff, Mike Lewis, Tony Lewis, Karl Lillquist, Don Luman, Susan Leslie, Hal Malde, Rolfe Mandel, Bob Massey, Sandi Matsumoto, Bruce McMillan, William (Bill) Mischo, Eric Morgan, Wesley Newman, Stevi and Dale Odem, Samuel H.

Ordway, III, Beth Painter, Kit Paris, Victor Parslow, Jonathan Phillips, Michael Piep, Judy Powell, Larry Price, Laura Prugh, Joe Pyrch, Jessie Quinn, Sarah Reed, Tom Reeves, Rick Riefner, Kurt Robins, Deborah Salvestrin, Brad Samuelson, Vic Scheffer, Mary C. Schlembach, Kathy Sharum, Mike Smith, Barbara Soulé, Joann Stenton, Andy Stumpf, Sister Teresita, James Thorp, Joel Tuhy, Pat Veesart, Larry M. Vredenburgh, Lee Walkling, Hong Wang, Ray Wheeler, Jack White, Deborah Wilson, Carol Witham, Ray Wood, Vern Yadon, and, notably, many unidentified and helpful librarians.

REFERENCES CITED

Almaraz, R.A., 1986, The characteristics of patterned ground in the southern Cascade Mountains [unpublished M.S. thesis]: Ashland, Southern Oregon State College, 68 p.

Antevs, E., 1932, Alpine Zone of Mt. Washington Range: Auburn, Maine, Merrill & Weber, 118 p.

Beieler, V.E., and Kehne, J., 2008, Soil Survey of Douglas County, Washington: U.S. Department of Agriculture, Soil Conservation Service, in cooperation with Washington State University Agricultural Research Center: Washington, D.C., Government Printing Office, 1854 p.

Berg, A.W., 1989, Formation of Mima mounds: A seismic hypothesis: Geological Society of America Abstracts with Programs, v. 21, no. 5, p. 56.

Berg, A.W., 1990a, Formation of Mima mounds: A seismic hypothesis: Geology, v. 18, p. 281–284, doi:10.1130/0091-7613(1990)018<0281:FOMMAS >2.3.CO;2.

Berg, A.W., 1990b, Formation of Mima mounds: A seismic hypothesis: Reply: Geology, v. 18, p. 1260–1261, doi:10.1130/0091-7613(1990)018 <0281:FOMMAS>2.3.CO;2.

Berg, A.W., 1991, Formation of Mima mounds: A seismic hypothesis: Reply: Geology, v. 19, p. 284–285, doi:10.1130/0091-7613(1991)019<0284: CAROFO>2.3.CO;2.

Bork, J.L., 1978, A survey of the vascular plants and vertebrates of Upper Table Rock [Medford area, Oregon] [unpublished M.S. thesis]: Ashland, Southern Oregon State College, 91 p.

Brunnschweiler, D., 1962, The periglacial realm in North America during the Wisconsin glaciation: Biuletyn Peryglacjalny (Lodzkie Towarzystwo Naukowe - Societas Scientiarium Lodziensis), Series 3, no. 11, p. 15–27.

Brunnschweiler, D., 1964, Der pleistozane Periblazialbereich in Nordamerika: Zeitschrift fur Geomorphologie, v. 8, p. 223–231.

Cahoon, J., 1985, Soil Survey of Klamath County, Oregon, Southern Part: U.S. Department of Agriculture, Soil Conservation Service, in cooperation with Oregon Agricultural Experiment Station: Washington, D.C., Government Printing Office, 165 p.

Campbell, M.R., 1906, Natural mounds: The Journal of Geology, v. 14, p. 708–717, doi:10.1086/621357.

Christiansen, R.L., Kleinhampl, F.J., Blakely, R.J., Tuchek, E.T., Johnson, F.L., and Conyak, M.D., 1977, Resource Appraisal of the Mount Shasta Wilderness Study Area, Siskiyou County, California: U.S. Geological Survey Open-File Report 77-250, 53 p.

Collett, R.A., 1980, Soil Survey of Ada County, Idaho: U.S. Department of Agriculture, Soil Conservation Service, in cooperation with U.S. Forest Service, U.S. Department of the Interior, Bureau of Land Management, University of Idaho College of Agriculture, Idaho Agriculture Experiment Station, and Idaho Soil Conservation Commission: Washington, D.C., Government Printing Office, 180 p.

Conrad, V., 1946, Polygon nets and their physical development: American Journal of Science, v. 244, no. 4, p. 277–296, doi:10.2475/ajs.244.4.277.

Copeland, W.N., 1980, The Lawrence Memorial Grassland Preserve: A biophysical inventory with management recommendations: Unpublished report (corrected 1983): Portland, Oregon Chapter, The Nature Conservancy, 168 p.

COSL (College of Siskiyous Library), 2011, Mount Shasta: An Annotated Bibliography, Chapter 3, Science: Weed, California, Geology and Climate, College of Siskiyous Library, http://www.siskiyous.edu/library/shasta/.

Cox, G.W., 1984a, Mounds of mystery: Natural History, v. 93, no. 6, p. 36–45.

Cox, G.W., 1984b, Soil transport by pocket gophers in Mima mound and vernal pool microterrain, *in* Jain, S., and Moyle, P., eds., Vernal Pools and Inter-

mittent Streams: Institute of Ecology Publication no. 28: Davis, University of California, p. 37–45.

Cox, G.W., 1989, Early summer diet and food preferences of Northern Pocket Gophers in North Central Oregon: Northwest Science, v. 63, no. 3, p. 77–82.

Cox, G.W., 1990a, Soil mining by pocket gophers along topographic gradients in a Mima moundfield: Ecology, v. 71, no. 3, p. 837–843, doi:10.2307 /1937355.

Cox, G.W., 1990b, Comment and reply on "Formation of Mima mounds: A seismic hypothesis": Geology, v. 18, no. 12, p. 1259–1260, doi:10.1130/0091 -7613(1990)018<1259:CAROFO>2.3.CO;2.

Cox, G.W., 1990c, Form and dispersion of Mima mounds in relation to slope steepness and aspect on the Columbia Plateau: The Great Basin Naturalist, v. 50, p. 21–31.

Cox, G.W., and Allen, D.W., 1987a, Sorted stone nets and circles of the Columbia Plateau: A hypothesis: Northwest Science, v. 61, p. 179–185.

Cox, G.W., and Allen, D.W., 1987b, Soil translocation by pocket gophers in a Mima moundfield: Oecologia, v. 72, p. 207–210, doi:10.1007 /BF00379269.

Cox, G.W., and Hunt, J., 1990, Nature and origin of stone stripes on the Columbia Plateau: Landscape Ecology, v. 5, no. 1, p. 53–64, doi:10.1007 /BF00153803.

Cox, G.W., and Scheffer, V.B., 1991, Pocket gophers and Mima terrain in North America: Natural Areas Journal, v. 11, no. 4, p. 193–198.

Cox, G.W., Gakahu, C.G., and Allen, D.W., 1987, The small stone content of Mima mounds of the Columbia Plateau and Rocky Mountain regions: Implications for mound origin: The Great Basin Naturalist, v. 47, p. 609–619.

Crandell, D.R., 1989, Gigantic Debris Avalanche of Pleistocene Age from Ancestral Mount Shasta Volcano, California, and Debris-Avalanche Hazard Zonation: U.S. Geological Survey Bulletin 1861, p. 32, http://vulcan.wr.usgs .gov/Volcanoes/Shasta/DebrisAval/description_shasta_debris_aval.html.

Crandell, D.R., Miller, C.D., Glicken, H.X., Christiansen, R.L., and Newhall, C.G., 1984, Catastrophic debris avalanche from ancestral Mount Shasta volcano, California: Geology, v. 12, p. 143–146, doi:10.1130/0091-7613 (1984)12<143:CDAFAM>2.0.CO;2.

Dalquest, W.W., and Scheffer, V.B., 1942, The origin of the mounds of western Washington: The Journal of Geology, v. 50, no. 1, p. 68–84, doi:10.1086 /625026.

Dalquest, W.W., and Scheffer, V.B., 1944, Distribution and variation in pocket gophers, *Thomomys talpoides*, in the state of Washington: American Naturalist, v. 78, p. 308–333, doi:10.1086/281203.

Davis, M., and Youtie, B.A., 1976, The biscuit-scabland range, Oregon's natural heritage rediscovered: Mazama, v. 58, no. 13, p. 30–34.

Donaldson, N.C., and Geise, L.D., 1968, Soil Survey of Spokane County, Washington: Washington, D.C., U.S. Department of Agriculture, Soil Conservation Service in cooperation with the Washington Agricultural Experiment Station, 143 p.

Fey, M., 2010, The Soils of South Africa: Cambridge, UK, Cambridge University Press, 287 p.

Fosberg, M.A., 1963, Genesis of some soils associated with low and big sagebrush complexes in the Brown, Chestnut and Chernozem-prairie zones in southcentral and southwestern Idaho [Ph.D. thesis]: Madison, University of Wisconsin, 308 p.

Fosberg, M.A., 1965, Characteristics and genesis of patterned ground in Wisconsin time in a chestnut soil zone of southern Idaho: Soil Science, v. 99, p. 30–37, doi:10.1097/00010694-196501000-00006.

Franzmeier, D.P., Steinhardt, G.C., Crum, J.R., and Norton, L.D., 1977, Soil characterization in Indiana: I. field and laboratory procedures: Research Bulletin, no. 943, p. 13–14.

Freeman, L., 1997, Parks Creek patterned ground: Manuscript MS2160, College of Siskiyous Library, Mount Shasta Collection: Weed, California, http://www.siskiyous.edu/library/shasta/

Freeman, O.W., 1926, Scabland mounds of eastern Washington: Science, v. 64, no. 1662, p. 450–451.

Freeman, O.W., 1932, Origin and economic value of the scabland mounds of eastern Washington: Northwest Science, v. 6, no. 2, p. 37–40.

Fryxell, R., 1964, The late Wisconsin age of mounds on the Columbia Plateau of eastern Washington: Unpublished report: Pullman, Washington, Laboratory of Anthropology, Washington State University, 16 p. (Appendix A, this volume.)

Geiger, B., 1998, Heaps of confusion: Earth (Waukesha, Wis.), v. 7, no. 4, p. 34–37.

Gentry, H.R., 1974, Geomorphology of some selected soil-landscapes in Whitman County, Washington [unpublished M.S. thesis]: Pullman, Washington State University, 130 p.

Gentry, H.R., 1991, Soil Survey of Asotin County Area, Washington, Parts of Asotin and Garfield Counties: Washington, D.C., U.S. Department of Agriculture, Soil Conservation Service in cooperation with Washington State Department of Natural Resources and Washington State Agricultural Research Center, 187 p.

Gentry, H.R., 2006, Soil Survey of Yakima Training Center, Parts of Kittitas and Yakima Counties, Washington: U.S. Department of Agriculture, Soil Conservation Service, in Cooperation with U.S. Department of Army, Yakima Training Center, U.S. Army Corps of Engineers, and Washington State University Agricultural Research Center: Washington, D.C., Government Printing Office, 865 p.

Gibbs, G., 1854, Report of George Gibbs upon the geology of the central portion of Washington Territory: U.S. Congress, 33rd, 1st Session, House Executive Document 129, v. 18, pt. 1, p. 494–512: Washington, D.C., Beverley Tucker, Printer.

Gibbs, G., 1855, Report of George Gibbs upon the geology of the central portion of Washington Territory: U.S. War Department: Reports of Explorations and Surveys, v. 1, p. 473–486.

Gilbert, G.K., 1875, Prairie mounds, *in* Report upon Geographical and Geological Explorations and Surveys West of the 100th Meridian, pt. 5, v. 3—Geology: Washington, D.C., Government Printing Office, p. 539–540.

Goldthwait, R.P., 1976, Frost sorted patterned ground: Quaternary Research, v. 6, p. 27–35, doi:10.1016/0033-5894(76)90038-7.

Green, G.L., 1970, Soil Survey of the Trout Creek-Shaniko Area, Oregon: U.S. Department of Agriculture, Soil Conservation Service, and U.S. Forest Service in cooperation with Oregon Agricultural Experimental Station: Washington, D.C., Government Printing Office, 83 p.

Green, G.L., 1975, Soil Survey of Trout Creek-Shaniko Area, Oregon (Parts of Jefferson, Wasco, and Crook Counties, OR): U.S. Department of Agriculture, Soil Conservation Service, and U.S. Forest Service in cooperation with Oregon Agricultural Experiment Station: Washington, D.C., Government Printing Office, 69 p.

Green, G.L., 1981, Soil Survey of Hood River Area, Oregon: Washington, D.C., U.S. Department of Agriculture, Soil Conservation Service, in cooperation with Oregon Agricultural Experiment Station, 94 p.

Green, G.L., 1982, Soil Survey of Wasco County, Oregon, Northern Part: Washington, D.C., U.S. Department of Agriculture, Soil Conservation Service, in cooperation with Oregon Agricultural Experiment Station, 125 p.

Henderson, J.A., 1974, Composition, distribution and succession of subalpine meadows in Mount Rainier National Park [unpublished Ph.D. thesis]: Corvallis, Oregon State University, 153 p.

Horwath, J.L., 2002, An assessment of Mima type mounds, their soils, and associated vegetation, Newton County, Missouri [unpublished M.S. thesis]: Urbana, University of Illinois, 265 p.

Horwath, J.L., and Johnson, D.L., 2006, Mima-type mounds in southwest Missouri: Expressions of point-centered and locally thickened biomantles: Geomorphology, v. 77, no. 3–4, p. 308–319, doi:10.1016/j.geomorph.2006.01.009.

Horwath, J.L., and Johnson, D.L., 2007, Erratum to "Mima-type mounds in southwest Missouri: Expressions of point-centered and locally thickened biomantles": Geomorphology, v. 83, p. 193–194, doi:10.1016/j.geomorph.2006.09.013.

Horwath, J.L., Johnson, D.L., and Stumpf, A.J., 2002, Evolution of a gravelly Mima-type moundfield in southwestern Missouri: Geological Society of America Abstracts with Programs, v. 34, no. 6, p. 369.

Humphreys, G.S., and Mitchell, P.B., 1988, Bioturbation: An important pedological and geomorphological process: Abstracts, v. 1, p. 265, 26th Congress, International Geographic Union, Sydney.

Johnson, C.B., 1982, Soil mounds and patterned ground of the Lawrence Memorial Grassland Preserve [unpublished M.S. thesis]: Corvallis, Oregon State University, 52 p.

Johnson, D.L., 1989, Subsurface stone lines, stone zones, artifact-manuport layers, and biomantles produced by bioturbation via pocket gophers (*Thomomys bottae*): American Antiquity, v. 54, p. 370–389, doi:10.2307/281712.

Johnson, D.L., 1990, Biomantle evolution and the redistribution of earth materials and artifacts: Soil Science, v. 149, p. 84–102, doi:10.1097/00010694-199002000-00004.

Johnson, D.L., 2002, Darwin would be proud: Bioturbation, dynamic denudation, and the power of theory in science: Geoarchaeology: An International Journal, v. 17, no. 1, p. 7–40, 631–632.

Johnson, D.L., and Hansen, K.L., 1974, The effects of frost heaving on objects in soils: Plains Anthropologist, v. 19, no. 64, p. 81–98.

Johnson, D.L., and Johnson, D.N., 2008, White paper: Stonelayers, mima mounds and hybrid mounds of Texas and Louisiana—Logical alternative biomantle working hypotheses: Appendix 3, *in* Johnson, D.L., Mandel, R.D., and Frederick, C.D., compilers and eds., Field trip guidebook, The Origin of the Sandy Mantle and Mima Mounds of the East Texas Gulf Coastal Plains: Geomorphological, Pedological, and Geoarchaeological Perspectives: Geological Society of America Annual Meeting, Houston, Texas, 4 October 2008.

Johnson, D.L., and Watson-Stegner, D., 1987, Evolution model of pedogenesis: Soil Science, v. 143, p. 349–366, doi:10.1097/00010694-198705000-00005.

Johnson, D.L., Muhs, D.R., and Barnhardt, M.L., 1977, The effects of frost heaving on objects in soils, II: Laboratory experiments: Plains Anthropologist, v. 22, no. 76, p. 133–147.

Johnson, D.L., Watson-Stegner, D., Johnson, D.N., and Schaetzl, R.J., 1987, Proisotropic and proanisotropic processes of pedoturbation: Soil Science, v. 143, p. 278–292, doi:10.1097/00010694-198704000-00005.

Johnson, D.L., Horwath, J., and Johnson, D.N., 2002, In praise of the coarse fraction and bioturbation: Gravelly Mima mounds as two-layered biomantles: Geological Society of America Abstracts with Programs, v. 34, no. 6, p. 369.

Johnson, D.L., Horwath, J., and Johnson, D.N., 2003, Mima and other animal mounds as point centered biomantles: Geological Society of America Abstracts with Programs, v. 34, no. 7, p. 258.

Johnson, D.L., Domier, J.E.J., and Johnson, D.N., 2005a, Animating the biodynamics of soil thickness using process vector analysis: A dynamic denudation approach to soil formation: Geomorphology, v. 67, no. 1–2, p. 23–46, doi:10.1016/j.geomorph.2004.08.014.

Johnson, D.L., Domier, J.E.J., and Johnson, D.N., 2005b, Reflections on the nature of soil and its biomantle: Annals of the Association of American Geographers, v. 95, no. 1, p. 11–31, doi:10.1111/j.1467-8306.2005.00448.x.

Johnson, D.R., 1993, Soil Survey of Jackson County Area, Oregon: U.S. Department of Agriculture, Soil Conservation Service, in cooperation with U.S. Forest Service, Bureau of Land Management, and Oregon Agricultural Experiment Station: Washington, D.C., Government Printing Office, 694 p.

Kaatz, M.R., 1959, Patterned ground in central Washington—A preliminary report: Northwest Science, v. 33, no. 4, p. 145–156.

Kienzle, J., 2007, Soil Survey of Wallowa County Area, Oregon: Washington, D.C., U.S. Department of Agriculture, Natural Resources Conservation Service, in Cooperation with U.S. Forest Service, U.S. Department of the Interior–Bureau of Land Management, Oregon State University Agricultural Experiment Station, and Wallowa Soil and Water Conservation District, 2307 p.

Kocher, A.E., and Strahorn, A.T., 1921, Soil Survey of Benton County, Washington: U.S. Department of Agriculture: Field Operations of the Bureau of Soils (for Year 1916), 18th Report, p. 2203–2270.

Larrison, E.J., 1942, Pocket gophers and ecological succession in the Wenas region of Washington: The Murrelet, v. 23, no. 2, p. 34–41, doi:10.2307/3535581.

LeConte, J., 1874, On the great lava-flood of the Northwest, and on the structure and age of the Cascade Mountains: American Journal of Science (series 3), v. 7, p. 167–180, 259–267.

LeConte, J., 1875, On the great lava-flood of the Northwest, and on the structure and age of the Cascade Mountains: Proceedings of the California Academy of Sciences, v. 5, p. 215–220.

LeConte, J., 1877, Hog wallows or prairie mounds?: Nature, v. 15, no. 390, p. 530–531, doi:10.1038/015530d0.

Lin, H., 2010, Earth's critical zone and hydropedology: Concepts, characteristics, and advances: Hydrology and Earth System Sciences, v. 14, p. 25–45, doi:10.5194/hess-14-25-2010.

Mack, S., 1960, U.S. Geological Survey: Geology and Ground Water Features of Shasta Valley, Siskiyou County, California: U.S. Geological Survey Water Supply Paper 1484, 115 p.

Malde, H.E., 1961, Patterned ground of possible solifluction origin at low altitude in the western Snake River Plain, Idaho: U.S. Geological Survey Professional Paper, v. 424-B, p. 170–173.

Malde, H.E., 1964a, Patterned ground in the western Snake River Plain, Idaho, and its possible cold-climate origin: Geological Society of America Bulletin, v. 75, p. 191–208, doi:10.1130/0016-7606(1964)75[191:PGITWS]2.0.CO;2.

Malde, H.E., 1964b, The ecologic significance of some unfamiliar geologic processes, *in* Hester, R.H., and Schoenwetter, J., eds., The Reconstruction of Past Environments: Proceedings of the Fort Burgwin Conference on Paleoecology no. 3: Fort Burgwin Research Center, SMU-Taos, New Mexico, p. 11–15.

Malde, H.E., 1965, Mountain Home to Malad Springs, INQUA Field Guidebook for Field Conference E, Northern and Middle Rocky Mountains: Denver, Colorado, INQUA 7th Congress, p. 94–95.

Marshall, W.H., 1941, *Thomomys* as burrowers in the snow: Journal of Mammalogy, v. 22, p. 196–197.

Masson, P.H., 1949, Circular soil structures in northeastern California: California Division of Mines Bulletin, no. 151, p. 61–71.

MacDonald, G.D., 1998, Soil Survey of Warm Springs Indian Reservation, Oregon: U.S. Department of Agriculture, Natural Resources Conservation Service in cooperation with Confederated Tribes of the Warm Springs Reservation, U.S. Department of the Interior–Bureau of Indian Affairs, and Oregon Agricultural Experimental Station: Washington, D.C., Government Printing Office, 266 p.

MacDonald, G.D., Lamkin, J.M., and Borine, R.H., 1999, Soil Survey of Sherman County, Oregon: U.S. Department of Agriculture, Natural Resources Conservation Service in cooperation with Oregon Agricultural Experiment Station: Washington, D.C., Government Printing Office, 82 p.

Miller, M.A., 1948, Seasonal trends in burrowing of pocket gophers (Thomomys): Journal of Mammalogy, v. 29, p. 38–44, doi:10.2307/1375279.

Morrás, H., Moretti, L., Píccolo, G., and Zech, W., 2009, Genesis of subtropical soils with stony horizons in NE Argentina: Autochthony and polygenesis: Quaternary International, v. 196, no. 1–2, p. 137–159, doi:10.1016/j.quaint .2008.07.001.

Nelson, C.A., 1977, The origin and characteristics of soil mounds and patterned ground in north central Oregon [unpublished M.S. thesis]: Corvallis, Oregon State University, 45 p.

Newberry, J.S., 1857, Mounds: Chapter 6, Geology of the Des Chutes basin, *in* Reports of Explorations and Surveys, v. 6, p. 52: House of Representatives, 33rd Congress, 2nd Session, Executive Document no. 91: Washington, D.C., A.O.P. Nicholson, Printer.

Newlun, J.J., Lindsey, W.C., Jahnke, J.J., and Day, L.A., 1983, Soil Survey of Siskiyou County, California, Central Part: U.S. Department of Agriculture, Soil Conservation Service, in cooperation with U.S. Forest Service, and University of California Agricultural Experiment Station: Washington, D.C., Government Printing Office, 291 p.

Noble, J., 1951, The mysterious circles of Shasta: Westways, v. 43, no. 1, p. 18–19.

Noble, J., 1985, The mysterious circles of Shasta, *in* O'Neill, T., ed. and compiler, Out of Time Out of Place: St. Paul, Minnesota, Llewellyn Publications, p. 55–59.

Noe, H.R., 1991, Soil Survey of Elmore County Area, Idaho, Parts of Elmore, Owyhee, and Ada Counties: U.S. Department of Agriculture, Soil Conservation Service, in cooperation with U.S. Forest Service, U.S. Department of the Interior, Bureau of Land Management, University of Idaho College of Agriculture, and Idaho Soil Conservation Commission: Washington, D.C., Government Printing Office, 270 p.

NRCS (Natural Resources Conservation Service), 2011, Klamath National Forest Area, Parts of Siskiyou County, California, and Jackson County, Oregon (Mapping level 4): Washington, D.C., U.S. Department of Agriculture: http://websoilsurvey.nrcs.usda.gov/app/WebSoilSurvey.aspx (accessed 15 April 2011).

Olmsted, R.K., 1963, Silt mounds of Missoula flood surfaces: Geological Society of America Bulletin, v. 74, no. 1, p. 47–53, doi:10.1130/0016 -7606(1963)74[47:SMOMFS]2.0.CO;2.

Othberg, K.L., 1991, Geology and geomorphology of the Boise Valley and adjoining areas, western Snake River Plain, Idaho [unpublished Ph.D. thesis]: Boise, Idaho, University of Idaho, 124 p.

Paeth, R.C., 1967, Depositional origin of Mima mounds [unpublished M.S. thesis]: Corvallis, Oregon State University, 61 p.

Paton, T.R., Humphreys, G.S., and Mitchell, P.B., 1995, Soils: A New Global View: New Haven, Connecticut, Yale University Press, 213 p.

Peterson, F.F., 1961, Solodized solonetz soils occurring in the uplands of the Palouse Loess [unpublished Ph.D. thesis]: Pullman, Washington State University, 280 p.

Piper, C.V., 1905, The basalt mounds of the Columbia lava: Science, v. 21, p. 824–825, doi:10.1126/science.21.543.824.

Pyrch, J.B., 1973a, The characteristics and genesis of stone stripes in North Central Oregon: [unpublished M.S. thesis]: Portland, Oregon, Portland State University, 134 p.

Pyrch, J.B., 1973b, North central Oregon stones stripes: Yearbook of the Association of Pacific Coast Geographers, v. 35, p. 191.

Pyrch, J.B., and Price, L., 1973, Rates of mass wasting on selected slopes in North Central Oregon—A preliminary report: Proceedings of the Oregon Academy of Science, v. 11, p. 34.

Richter, R., 1952, Fluidal-textur in sediment-gesteinen und über sedifluction überhaupt: Notizblatt des Hessischen Landesantes Bodenforschung zu Weisbaden, v. 6, p. 67–81.

Riefner, R.E., Jr., and Pryor, D., 1996, New locations and interpretation of vernal pools in southern California: Phytologia, v. 80, no. 4, p. 296–327.

Schaetzl, R.J., and Anderson, S., 2005, Soils—Genesis and Geomorphology: Cambridge, UK, Cambridge University Press, 817 p.

Schäfer, W., 1952, Biogene sedimentation im gefolge von bioturbation: Senckenbergiana, v. 33, p. 1–12.

Scheffer, V.B., 1947, The mystery of the mima mounds: The Scientific Monthly, v. 65, no. 5, p. 283–294.

Scheffer, V.B., 1958, Do fossorial rodents originate mima-type microrelief?: American Midland Naturalist, v. 59, p. 505–510, doi:10.2307/2422495.

Scheffer, V.B., 1969, Mima mounds: Their mysterious origin: Pacific Search, v. 3, no. 5, p. 3.

Sharp, R.P., 1942, Soil structures in the St. Elias Range: Yukon Territory: Journal of Geomorphology, v. 6, no. 4, p. 274–301.

Sheldon, W.B., 1980, Soil Survey of Modoc County, California, Alturas Area: U.S. Department of Agriculture, Soil Conservation Service, in cooperation with University of California Agricultural Experiment Station: Washington, D.C., Government Printing Office.

Soil Survey Staff, 1975, Soil Taxonomy: A Basic System of Soil Classification for Making and Interpreting Soil Surveys: Washington, D.C., U.S. Department of Agriculture Handbook 436, 754 p.

Stockman, D.D., 1981, Soil Survey of Lincoln County, Washington: Washington, D.C., United States Department of Agriculture, Soil Conservation Service in cooperation with the Washington State University Agricultural Research Center, 93 p.

Taber, S., 1930, The mechanics of frost heaving: The Journal of Geology, v. 38, p. 303–317, doi:10.1086/623720.

Tallyn, L.A.K., 1980, Scabland mounds of the Cheney quadrangle [unpublished M.S. thesis]: Cheney, Eastern Washington University, 94 p.

Tullis, J.A., 1995, Characteristics and origin of earth mounds on the eastern Snake River plain, Idaho [unpublished M.S. thesis]: Pocatello, Idaho State University, 164 p.

USDA-NRCS (U.S. Department of Agriculture, Natural Resources Conservation Service), 1999, Soil Taxonomy, A Basic System of Soil Classification for Making and Interpreting Soil Surveys (second edition): Agricultural Handbook No. 436: Washington, D.C., Government Printing Office: http://soils.usda.gov/technical/classification/taxonomy/

USDA-USFS (U.S. Department of Agriculture, U.S. Forest Service), 2011, Modoc National Forest Soil Survey, Area 702 (Order 4 mapping level, map unit 135): http://soils.usda.gov/technical/classification/osd/index .html (accessed 6 November 2011).

Van Duyne, C., Agee, J.H., and Ashton, F.W., 1919, Soil Survey of Franklin County, Washington: U.S. Department of Agriculture, Bureau of Soils: Field Operations of the Bureau of Soils for Year 1914, 16th Report, p. 2531–2627.

Washburn, A.L., 1956, Classification of patterned ground and review of suggested origins: Geological Society of America Bulletin, v. 67, p. 823–866, doi:10.1130/0016-7606(1956)67[823:COPGAR]2.0.CO;2.

Washburn, A.L., 1973, Periglacial Processes and Environments: London, Edward Arnold, 320 p.

Washburn, A.L., 1980, Geocryology—A Survey of Periglacial Processes and Environments: New York, John Wiley (Halsted Press Book), 406 p.

Washburn, A.L., 1988, Mima Mounds, an Evaluation of Proposed Origins with Special Reference to the Puget Lowlands: Report of Investigations no. 29: Olympia, State of Washington Department of Natural Resources, Division of Geology and Earth Resources, 53 p.

Waters, A.C., and Flagler, C.W., 1929, Origin of small mounds on the Columbia River Plateau: American Journal of Science (fifth series), v. 18, no. 105, p. 209–224.

Watson, E.B., and Cosby, S.W., 1924, Soil Survey of the Big Valley [Lassen-Modoc Cos.], California: U.S. Department of Agriculture, Bureau of Soils, Field Operations of the Bureau of Soils for Year 1920, 22nd Report, p. 1005–1031.

Watson, E.B., Wank, M.E., and Smith, A., 1923, Soil Survey of the Shasta Valley area, California: U.S. Department of Agriculture, Bureau of Soils, Field Operations of the Bureau of Soils for Year 1919, 22nd Report, p. 99–152.

Westman, B.J., 1946, Observations made of the unique Indian mounds near Bray, California: Unpublished manuscript (4 p.) on file: Yreka, California, Siskiyou County Historical Society.

Wilkinson, M.T., Richards, P.J., and Humphreys, G.S., 2009, Breaking ground: Pedological, geological, and ecological implications of soil bioturbation: Earth-Science Reviews, v. 97, p. 257–272, doi:10.1016/j.earscirev.2009.09.005.

Williams, H., 1949, The geology of the Macdoel quadrangle: San Francisco, California Division of Mines and Geology, Bulletin no. 151, p. 7–60.

Wilson, M.D., 1978, Patterned ground of erosional origin southwestern Idaho: Third International Conference on Permafrost Programme, Edmonton, Alberta: Ottawa, National Research Council of Canada, p. 60

Wilson, M.D., and Slupetzky, H., 1977, Origin of patterned ground near Boise, Idaho: Geological Society of America Abstracts with Programs, v. 9, no. 6, p. 775–776.

Winward, A.H., and Youtie, B.A., 1978, Ecological inventory of the Lawrence Memorial Grassland Preserve: Proceedings of the Oregon Academy of Science, v. 14, p. 50–65.

MANUSCRIPT ACCEPTED BY THE SOCIETY 5 MARCH 2012

The Geological Society of America
Special Paper 490
2012

Foreword to "Appendix A. The Late Wisconsin age of mounds on the Columbia Plateau of eastern Washington" (Third and final version of R.H. Fryxell's 1964 heretofore unpublished manuscript)

D.L. Johnson
Department of Geography, University of Illinois, Urbana, Illinois 61801, USA, and
Geosciences Consultants, 713 S. Lynn St., Champaign, Illinois 61820, USA

With assistance from D.N. Johnson

Roald Fryxell (1934–1974), earth scientist in the fields of geoarchaeology, geochronology, Quaternary geology, soils, and astropedology, was a professor of anthropology at Washington State University (WSU) at Pullman, Washington. In 1974 his life was tragically cut short by an automobile accident. Though I never met him, I strongly identify with him because of our close career interests, including parallel interests in Mima mounds. We also were the same age, both born in 1934. Further, in 1976, two years after his passing I interviewed for a position in anthropology at WSU, almost certainly Fryxell's, though I do not now recall being aware of that circumstance.

In 1964, Fryxell produced a draft of his mound manuscript that he circulated among several of his professional, mound-interested colleagues for feedback, namely M.A. Fosberg, H.E. Malde, D.R. Mullineaux, T.L. Péwé, and A.C. Waters (cf. his Acknowledgments). But the paper was never quite finished. The reference list was missing, as were plates 1 and 2. For some reason Fryxell apparently shelved the manuscript at its penultimate stage of completion.

When I first read the paper and recognized its worth, I wondered why he never published it. Was it an issue, or issues, raised by the scholars among whom the draft was circulated? Was it related to pressures associated with Fryxell's rather eclectic research involvements at the time which included co-producing a substantial field guide of the Channeled Scablands (Fryxell and Cook, 1964), writing-research involvements with the then-impending first INQUA (International Union for Quaternary Research) meetings (Richmond et al., 1965), added to various other commitments and involvements (e.g., Fryxell, 1964, 1965; Fryxell and Daugherty, 1963; Steen and Fryxell, 1965, etc.)—not

to mention his later astropedology involvements with NASA? Or was it—and this drawn from immemorable personal experience—perhaps the intense self- and administratively imposed pressures invariably placed on young pre-dissertation instructors, like Fryxell, to complete their Ph.D. programs (Fryxell, 1971)? Did all of these issues come into play?

Perhaps, but I am inclined to suspect that it was also at least partly, possibly mainly, because the problem with explaining Mima mounds anywhere, much less across the incredibly diverse and geologically complex Columbia Intermontane Province, with its tumultuous and catastrophically complicated geohistories and diversity of geomorphic surfaces, is fraught with conceptual-explanatory quicksand. I say this because Fryxell well knew, and early stated, that the hundreds of thousands, if not possibly millions, of mounds that dapple Washington's Channeled Scablands and Okanagan Plateau, and other bordering and regional glaciated areas, had to have formed after the last Missoula outburst megaflood, and after the last Pleistocene glaciation.[1] Of chronological necessity this would require that all the Mima mounds now present on such impacted surfaces be of postglacial age (and before mid-Holocene warming)—again as Fryxell knew, confirmed by the title of his paper and his carefully composed footnotes 1 and 2.

One strongly suspects that Fryxell ultimately came to question the wisdom of putting all his mound-explanatory "eggs" in

---

[1]The glaciated and mounded areas north of the Columbia River, including the Waterville Plateau and Colville Indian Reservation, and the ubiquitously mounded Channeled Scablands, plus many surrounding mounded basaltic terrains of the Columbia Basin, all fell within Fryxell's research domain (e.g., Fryxell, 1959, 1960, 1965, 1971; Fryxell and Daugherty, 1963; Fryxell and Cook, 1964; Richmond et al., 1965).

Johnson, D.L., 2012, Foreword to Appendix A: "The Late Wisconsin age of mounds on the Columbia Plateau of eastern Washington" (Third and final version of R.H. Fryxell's 1964 heretofore unpublished manuscript), *in* Horwath Burnham, J.L., and Johnson, D.L., eds., Mima Mounds: The Case for Polygenesis and Bioturbation: Geological Society of America Special Paper 490, p. 161–163, doi:10.1130/2012.2490(07). For permission to copy, contact editing@geosociety.org.

an exclusive freeze thaw-periglacial and regionally constrained genetic "basket." Regarding regional constraints, also reflected in his paper's title—and especially insofar as Fryxell was a seasoned and experienced scholar, he surely was aware of at least *some* of the extensive and deep literature that documents mound distributions in environments altogether different and much warmer than the Columbia Intermontane Province (e.g., San Diego–Baja California, Great Central Valley of California, Texas–Louisiana Gulf Coast, Mississippi Valley, etc.). These lower latitude regions were comparatively balmy even during full-glacial episodes.

I also cannot help but speculate that Fryxell eventually rethought his initial dismissal of the role of burrowing animals in mound formation (p. 168). I say this because burrowing activity by rodents and their predators is ubiquitous throughout the entire Columbia Plateau region—including archaeological sites that Fryxell himself worked on, and even pointedly noted (e.g., Fryxell, 1960, 1961, 1962; Fryxell and Daugherty, 1962, 1963). It is also worth consideration that Fred F. Peterson, pedologist colleague of Fryxell's at WSU whose thesis work he favorably cited for basic soil concepts, and mound locations in sodic-affected tracts, was an advocate of animal burrowing as a partial genetic factor in the origin of Columbia Plateau mounds (Peterson, 1961, p. 97, 115, 138, etc.). The sum of these points may have given Fryxell pause to genetically and chronologically rethink his conclusions, and to hold off finalizing and submitting his paper for publication.

Aside from these considerations, Fryxell produced an important and interesting paper on Mima mounds, one of such interest that, even though unfinished and unpublished, it was circulated among a select group while he was alive, and to others after his death, which includes this author. His paper is also historically important insofar as it provides insights not only to Fryxell's thinking about mounds and landscape evolution across the Columbia Intermontane Province in the 1960s, but likewise the thinking of several of his rather prominent contemporaries on such themes. A huge plus is that his figure 1 aptly summarizes and very creatively captures the range of specific landscape-landform contexts of mound distributions across the region.

I first heard about the manuscript in early 2010, and—after beating the bushes via innumerable emails—finally acquired a poor, somewhat illegible hand-edited (possibly by Fryxell) typescript, a poor xerocopy of his original first draft (edits were readable, but some text was not). That copy was kindly sent to me by Lee Walkling, geology librarian at Washington Department of Natural Resources (WDNR) in Olympia. Some months later, in August 2010, Martin Kaatz, retired professor at Central Washington University in Ellensburg—a colleague with long and abiding interests in Columbia Plateau mounds (Kaatz, 1959)—graciously sent me, from his files, another typescript. This copy was unedited but fully legible (presumably a copy of the original typescript). This copy allowed me to fill in the illegible sections of the edited version acquired from the WDNR library in Olympia. The passage of time between the first and edited second versions is not known, but could have been days, weeks, months, or years.

Later communications in the fall of 2010 involved WSU people, specifically Mary Collins of the Anthropology Department, and Betty Galbraith, WSU librarian. From them I learned that upon Fryxell's death, his papers had been donated to the American Heritage Center (AHC) in Laramie, Wyoming.

This Appendix is the "third version" and incorporates the edits. The References Cited section, which earlier versions lacked, was assembled from my professional files, from Internet sources, and with the help of our unfailingly proficient University of Illinois librarians. All the "in press" citations of the earlier manuscript versions were replaced with subsequently published ones.

The location of the moundfield on the Colville Indian Reservation just northeast of Grand Coulee Dam, where Fryxell states "solonetz patches occur between mounds in loess a short distance east of the Late Wisconsin ice margin," was slightly off. Fryxell's description and figure 1 indicate that he more likely meant the broad alluvial, episodically wet sodic area of 47° 59′ 51″ N, 118° 53′ 56″ W (confirmed in Peterson's 1961 thesis, with its location possibly gleaned by Fryxell). The location and presence of mounds there is confirmed both by Google Earth images, and by an October 2010 visit to that tract (by myself and my collaborator-wife D.N. Johnson). During that month we also visited the AHC in Laramie, which curates Fryxell's papers, for the hoped-for purpose of locating plates 1 and 2.

Fryxell's figures 1 and 2 were included in the circulated manuscripts, but plates 1 and 2 were missing from those provided by WDNR and by Kaatz. Fortunately, during our examination of Fryxell's papers at AHC, 10 unlabeled photographs of Columbia Plateau mounds and moundfields were found together in a file *with* figures 1 and 2. Assuming that the missing plates 1 and 2 were likely among the 10, we selected two that seemed to most closely convey points Fryxell made on his manuscript. We took the liberty of selecting one additional photo, plate 3, for the manuscript because it aptly captured several of Fryxell's main points. Notably, another readable copy of the first version of the manuscript was discovered at this time.

In sum, with the publication of this Appendix A nearly 50 years after Fryxell's manuscript was first circulated among a small group of Pacific Northwest researchers, his paper is now accessible to anybody interested in the multitude of processes that might lead to the formation of Mima mounds, and—in the Pacific Northwest—to such regionally associated mound features as sorted stone rings, nets, polygons, and beds. The manuscript is also now available to those interested in the history and development of soil-geomorphic-ecologic thought on this most interesting and historically contentious subject.

ACKNOWLEDGMENTS

Thanks to Lee Walkling of the Geology Library at Washington Department of Natural Resources (WDNR), Olympia, Washington, and Marty Kaatz of Central Washington University at Ellensburg, for making available to us their typescript copies of Fryxell's manuscript. Mary Collins, director of the

Museum of Anthropology, and Betty Galbraith, librarian, both at Washington State University, in Pullman, generously aided in aspects of this effort. The final completed manuscript was retyped by Barbara Bonnell, Department of Geography, University of Illinois, Urbana, Illinois, with subsequent minor final edits by myself. Figure 2 was redrafted for clarity, but otherwise unchanged, by graphics specialist Mike Lewis, courtesy of the Illinois State Archaeological Survey, University of Illinois, Urbana, Illinois. The American Heritage Center, University of Wyoming in Laramie, Wyoming, kindly assisted in making Fryxell's papers available for our inspection, granting permission for use of Fryxell's photos (plates 1, 2, and 3, from Box 20 of the Roald Fryxell Collection), and endorsing the completion and publication of Fryxell's 1964 paper in Geological Society of America Special Paper 490.

REFERENCES CITED

Fryxell, R., 1959, Soil and grazing resources inventory, Colville Indian Reservation, Washington: U.S. Department of the Interior, Bureau of Indian Affairs, Branch of Land Operations Survey, p. 200–216a.

Fryxell, R., 1960, Appendix 1, Stratigraphic description, *in* Sprague, R., Archaeology in the Sun Lakes area of Central Washington: Laboratory of Anthropology, Report of Investigations no. 6: Pullman, Washington State University, p. 42–45.

Fryxell, R., 1961, Geological examination of the Ford Island Archaeological Site (45-FR-47), Washington: Laboratory of Anthropology, Report of Investigations no. 18: Pullman, Washington State University, 16 p.

Fryxell, R., 1962, Geological field examination of the Park Lake Housepit Site (45-GR-90), Lower Grand Coulee, Washington: Laboratory of Anthropology, Report of Investigations no. 18: Pullman, Washington State University, 7 p.

Fryxell, R., 1964, Regional patterns of sedimentation recorded by cave and rock-shelter stratigraphy in the Columbia Plateau, Washington, *in* Abstracts for 1963: Geological Society of America Special Paper 76, p. 273.

Fryxell, R., 1965, Mazama and Glacier Peak volcanic ash layers: Relative ages: Science, v. 147, p. 1288–1290.

Fryxell, R., 1971, The contribution of interdisciplinary research to geologic investigation of prehistory, eastern Washington [Unpublished Ph.D. thesis]: Moscow, University of Idaho, 691 p.

Fryxell, R., and Cook, E.F., 1964, A field guide to the loess deposits and Channeled Scablands of the Palouse area, eastern Washington (in conjunction with the Geological Society of America, Rocky Mountain Section, May 3–6, 1964): Laboratory of Anthropology, Report of Investigations no. 27: Pullman, Washington State University.

Fryxell, R., and Daugherty, R.D., 1962, Interim report: Archaeological Salvage in the Lower Monumental Reservoir, Washington: Laboratory of Archaeology and Geochronology, Report of Investigations no. 21: Pullman, Washington State University, 39 p.

Fryxell, R., and Daugherty, R.D., 1963, Late glacial and post glacial geological and archaeological chronology of the Columbia Plateau, Washington: Laboratory of Anthropology, Division of Archaeology and Geochronology, Report of Investigations no. 23: Pullman, Washington State University, 21 p.

Kaatz, M.R., 1959, Patterned ground in central Washington—A preliminary report: Northwest Science, v. 33, no. 4, p. 145–156.

Peterson, F.F., 1961, Solodized solonetz soils occurring in the uplands of the Palouse loess [Unpublished Ph.D. thesis]: Pullman, Washington State University.

Richmond, G.M., Fryxell, R., Neff, G.E., and Weis, P.L., 1965, The Cordilleran ice sheet of the northern Rocky Mountains, and related Quaternary history of the Columbia Plateau, *in* Wright, H.E., Jr., and Frey, D.G., eds., The Quaternary of the United States: A Review Volume for the VII Congress of the International Association for Quaternary Research: Princeton, New Jersey, Princeton University Press, p. 231–242.

Steen, V.C., and Fryxell, R., 1965, Mazama and Glacier Peak pumice glass: Uniformity of refractive index after weathering: Science, v. 150, no. 3698, p. 878–890.

MANUSCRIPT ACCEPTED BY THE SOCIETY 5 MARCH 2012

The Geological Society of America
Special Paper 490
2012

Appendix A. The Late Wisconsin age of mounds on the Columbia Plateau of eastern Washington

Roald H. Fryxell*

Laboratory of Anthropology, Washington State University, Pullman, Washington, USA

March 1964

INTRODUCTION

Recent renewal of interest in patterned ground features of the Columbia Plateau has revived not only the question of their origin, but has raised also the question of their age.

Discussion of patterned ground features in eastern Washington has been devoted almost entirely to the ubiquitous circular or elliptical mounds which dot thousands of square miles within the area. Although these mounds have been the subject of much controversy in the past, as have been similar mounds in other areas, most recent papers dealing with such features consider them to be a product of intense frost activity.[1]

Critical consideration of a possible periglacial origin for these mounds requires consideration of their age, and of climatic conditions inferred for the period during which the development of mounds occurred. Thus, the temporal relationship of mound development in eastern Washington to Late Wisconsin glaciation should be noted, for this relationship is demonstrated through simple physiographic and stratigraphic relationships, by paleontologic and pedologic evidence, and by a few limiting radiocarbon dates.

DESCRIPTION OF THE MOUNDS

A transitional series of patterned ground features (Washburn, 1956) ranging from soil mottling through crudely polygonal mound-and-depression patterns, partially segregated mounds or mound strips, isolated but regularly spaced mounds, and mound remnants, as well as interrelated rock polygons, stone nets and stripes, rubble sheets, and solifluction deposits, are present as a continuum over virtually the entire Columbia Intermontane Province (Freeman et al., 1945). The most conspicuous of these forms are the mounds, which first were described in detail by Waters and Flagler (1929).

Most mounds typical of those on the Columbia Plateau (see Fig. 1) are circular to elliptical rises a few tens of feet across, with a maximum center height of less than ten, and generally less than five, feet. Elongation in the direction of slope increases with slope angle, frequently producing a pattern radiating outward from the crest of hills. Spacing usually is remarkably regular. Trenching of mounds shows them to be comprised most often of loess or of colluvium including large amounts of loess, but similar mounds of ablation moraine have been reported from the Okanogan Plateau (Fryxell, 1959). Generally these materials are found to overlie bedrock (including nonbasaltic rock types as well as basalt), although mounds occur also on unconsolidated materials such as gravel bars in scabland channels (Olmsted, 1963, p. 48), on glacial till (Kaatz, 1959, p. 147) and outwash, or on alluvium. As Kaatz has noted, the association of sorted stone features with mounds is not invariable; it is, however, the general rule.

Particularly at certain seasons of the year, mounds are strongly accentuated visually by coloration of the distinctive plant communities established on mound and intermound areas. Most common are the *Artemisia tridentata/Agropyron spicatum* and *Artemisia rigida/Poa secunda* associations of Daubenmire (1942), distribution of the more mesic *A. tridentata/Agropyron* association being restricted to the mounds because of the shallow soil and prolonged periods of drying characteristic of intermound areas. In areas where incipient mounds have not been well segregated, color differences in plowed fields accentuate the presence of mounds through differential drying.

---

*published posthumously

[1]"Intense," "severe," or "vigorous" frost activity are terms intended to imply: (1) freezing to significantly greater soil depths than at present; (2) a longer annual duration of existence of such conditions; and (3) a longer duration of total annual frost conditions involving alternate freezing and thawing due either to diurnal temperature fluctuations or to cyclic weather changes. Permafrost conditions are regarded as probably but not yet demonstrated by known field evidence.

Fryxell, R.H., 2012, Appendix A. The Late Wisconsin age of mounds on the Columbia Plateau of eastern Washington, *in* Horwath Burnham, J.L., and Johnson, D.L., eds., Mima Mounds: The Case for Polygenesis and Bioturbation: Geological Society of America Special Paper 490, p. 164–172, doi:10.1130/2012.2490(08).

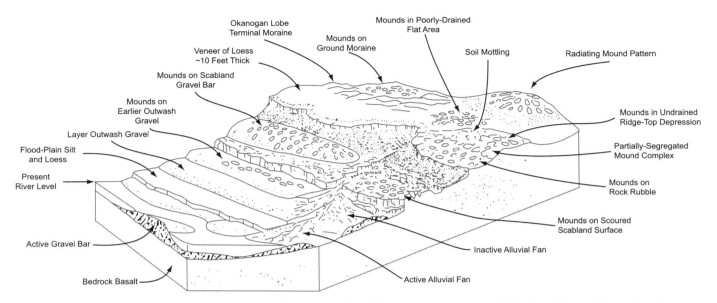

Figure 1. Composite illustration of major physiographic situations in which mounds occur on the Columbia Plateau. Selected landforms of known age indicate the physiographic basis for defining the age of mounds. Note that some mounds are younger than either the scabland channel or the maximum stand of the Late Wisconsin Okanogan ice lobe; yet no mounds are found on the later outwash terrace or on the postglacial flood-plain sediments. Thus time and conditions were favorable to the formation of mounds during the closing episode of glaciation, but have not permitted mounds to form since that time. The diagram is not to scale.

In most respects, mounds and sorted stone features associated with them on the Columbia Plateau are closely similar to those recently described in southern Idaho by Malde (1964) and Fosberg (1963).

ORIGIN OF THE MOUNDS

Mounds occurring on the Columbia Plateau have been regarded, successively, as produced by:

1. Rill dissection of a thin loess blanket (LeConte, 1874; Waters and Flagler, 1929).

2. Residual remnants of columnar basalt (Piper, 1905).

3. Gradual accumulation of eolian sediment trapped by patches of dense vegetation, which have been localized by moisture concentrated in bedrock depressions (Freeman, 1926; Olmsted, 1963).

4. Burrowing by pocket gophers (Larrison, 1942).

5. Control of surface run-off by subsurface jointing in the underlying basalt (Knechtel, 1952); and

6. Periglacial frost activity (Kaatz, 1959; Malde, 1961, 1964; Brunnschweiler, 1962, 1964).

A similar gamut of origins has been proposed for the controversial Mima mounds of southwestern Washington (Bretz, 1913; Dalquest and Scheffer, 1942; Scheffer, 1947; Krinitzsky, 1949; Péwé, 1948; Newcomb, 1952; Ritchie, 1953; Jackson, 1956). Although mounds in both areas share such common traits as regular size and spacing, and similar drainage characteristics, mounds on the Columbia Plateau generally are less closely spaced, less perfectly shaped, and are less abrupt in their relief. Sorted stone nets are intimately associated with mounds in both areas.

The processes of control by polygonal patterns of cracks and concurrent as well as subsequent stream dissection are demonstrated by field relationships, and the two are interrelated at many localities.

Two apparently contradictory features make clear the necessity for considering both erosional and frost activity in accounting for the development of these mounds:

1. Mounds exist which are bounded by undrained, linear, polygonal depressions (Plate 1a) which may follow stone nets without interstitial fill between fragments (Plate 1b), on level or concave slopes; and

2. Mounds also exist which are integrated with a normal dendritic drainage pattern (Plate 2a), and which show clearly transitional series from incipient but connected mound complexes (Plate 2b) to completely isolated mounds.

Both situations may be represented in a single mound field. Stone nets, breaking with increasing slope into stone stripes, may or may not be associated with mounds representing either case. Where stone nets are absent, intermound areas usually are veneered with non-sorted colluvium and stone rubble.

From these occurrences, it must be concluded that erosion of the intermound areas by water alone cannot account for development of mounds bounded by undrained depressions; even elutriation of fine fractions is impossible for mounds overlying bedrock (Plate 3). On the other hand, the frequent integration of mounds on sloping ground with well-developed drainage networks (often followed by stone nets and stripes) linking all intermound areas requires active development of the drainage system concurrent, or nearly so, with formation of the intervening

Plate 1. (A) Unlabeled photo, probably of the Channeled Scablands of east-central Washington, showing mounds in foreground, Missoula flood-stripped basalt layers in background, and large, flood-streamlined "loess islands" in right distance. (B) Unlabeled photo, possibly of Channeled Scablands, showing mounds surrounded by a poorly integrated intermound drainage system, with streamlined "loess islands" in background. The immediate foreground shows a segment of a sorted boulder-cobble "stone ring," with Jacob's staff for scale (inset is close-up photo—marks on staff are probably 1 foot intervals). Photos from Roald Fryxell Collection, courtesy American Heritage Center, Laramie, Wyoming.

mounds. Thus the interrelationship of control by development of polygons and dissection by erosion are recorded at many localities, and it may be concluded that solifluction and rill erosion have been channeled by the formation and spacing of cracks in a polygonal network,[2] probably initiated as large-scale thermal contraction cracks (Leffingwell, 1919). Erosional dissection without such control fails to account for the uniform size and spacing of the mounds. Thinning and truncation of soil A hori-

zons at the margin of some mounds reflect continuing dissection of the fine-grained mantle by present drainage along channels established earlier. Such erosion in progress has been observed in the field during the spring months, and at the present time is most active in the case of mounds overlying bedrock, accounting for the steep margins and scarplets often present on mounds over basalt, but usually absent from mounds resting on substrata, which now are pervious.

Poor subsurface drainage of material finer in texture than that underlying it is a common denominator of mounds observed in all cases but one. An exception is that of loess mounds overlying scabland gravel bars, which are both porous and permeable,

---

[2]An explanation with which Waters is in agreement (1959, personal commun.); patterned ground features in present-day temperate times were neither recognized nor understood as frost-related at the time of the 1929 study.

Plate 2. (A) Unlabeled photo of a loess-veneered, Mima mounded bedrock (basalt?) upland, with an integrated drainage net, surrounded by a thicker cultivated and less sloping loess mantle. Telephone poles in distance, and fence line provide scales. Location is unknown, but could be any bedrock upland area or steptoe of the Columbia Intermontane Province. (B) Airphoto labeled "Depner area"—though most likely meant to be Heppner area—of Oregon, in the Umatilla Plateau southern section of the Columbia Intermontane Province. The photo is typical of many thin-loess-over-bedrock areas in broadly dissected landscape segments of that part of Oregon, and typical of many areas elsewhere in the Columbia Intermontane Province (e.g., Manastash Ridge, Badger Mountain, etc.). Fryxell may have intended this photo to show how mounds become elongate with slope to produce a pattern of mounds radiating outward and downward on hillcrests. Photos from Roald Fryxell Collection, courtesy American Heritage Center, Laramie, Wyoming.

Plate 3. An unlabeled airphoto of an unknown area (possibly on the Colville Indian Reservation several km northeast of Grand Coulee Dam) that displays one of many poorly drained and sodic flattish areas of the Columbia Intermontane Province. The tract is obviously cultivated, but not so much as to blur and/or completely erase the mounds and intermounds that dot the swale. The plate is extraneous to Fryxell's typescript, but we have included it because the photo was in his files and because he mentioned such poorly drained mounded areas in his text, and gave an example in Figure 1. Photo from Roald Fryxell Collection, courtesy American Heritage Center, Laramie, Wyoming.

and often are the best-drained portions of the landscape. This exception can be accounted for through freezing of the gravel at a time of high water table, thus rendering it impervious, as would be the case under climatic conditions implied by considering these mounds to be frost-related features.

Accretional build-up of windblown material trapped by vegetation is of dubious importance in the formation of mounds, as the regularity of spacing, relationship of long axes to slope, the association with sorted stone nets and stripes, and the presence of large boulders and irregularly distributed rock fragments throughout many mounds cannot be accounted for by eolian sedimentation. Neither can the theory be considered for mounds of materials other than loess, though they appear to be morphologically and genetically similar. Most of the same objections apply to an origin ascribed solely to erosion along cracks bounding large-scale dissection polygons.

Other theories attributing these mounds to aboriginal inhabitants, prehistoric gophers, ants, or tree-throw disturbance are totally unacceptable for typical mounds encountered on the Columbia Plateau.

PHYSIOGRAPHIC EVIDENCE FOR THE AGE OF THE MOUNDS

Limiting dates for the period of development of these mounds are provided by the following physiographic positions in which they have been found (see Figs. 1 and 2)[3]:

1. Mounds are found on the bedrock floors of scabland channels and on scabland gravel deposits; thus they postdate discharge through these channels.

2. Mounds on the Waterville Plateau lie north of the Late Wisconsin terminal moraine; thus they were formed after the Late Wisconsin glacial maximum.

3. Mounds and weakly developed stone nets are found on the earlier of two Late Wisconsin outwash terraces along the middle Columbia River, at its confluence with the Spokane and at Vantage, but have not been observed on the younger; thus mounds either did not have an opportunity to develop because the younger terrace was actively worked by the Columbia, or because conditions favoring development of the mounds had ceased by the time this terrace was abandoned.

[3]In the absence of formal terminology for divisions of Pleistocene time in eastern Washington, the terms "Early" and "Late" Wisconsin are used to denote the two glaciations best known in this area. The former corresponds to the "Spokane glaciation" of Bretz (in Bretz et al., 1956), and the latter to the last major glacial episode to affect eastern Washington. Richmond et al. (1965) have correlated these glaciations with the Pinedale and Bull Lake Glaciations of the northern Rocky Mountains. For the purposes of this paper, the Late Wisconsin glacial maximum is represented best by the terminal moraine of the Okanogan ice lobe (Waters, 1933; Flint, 1935); Late Wisconsin glaciation is estimated to have begun about 25,000 years ago. This time span includes the climatic flood that swept the Channeled Scablands (Bretz et al., 1956) before the maximum Late Wisconsin glacial advance. All deposits and landforms referred to in this paper as Late Wisconsin time are younger than erosional features and deposits of this flood, as demonstrated through cut-and-fill relationships, stratigraphic superposition, or geographic position north of the terminal moraine of the Okanogan Lobe (see Fig. 1). Arbitrarily, Late Wisconsin time is terminated at 7500 years B.P., the time of major shift to warmer climatic conditions in the Columbia Plateau area (see Fig. 2). No direction correlation with details of present midwestern terminology is intended.

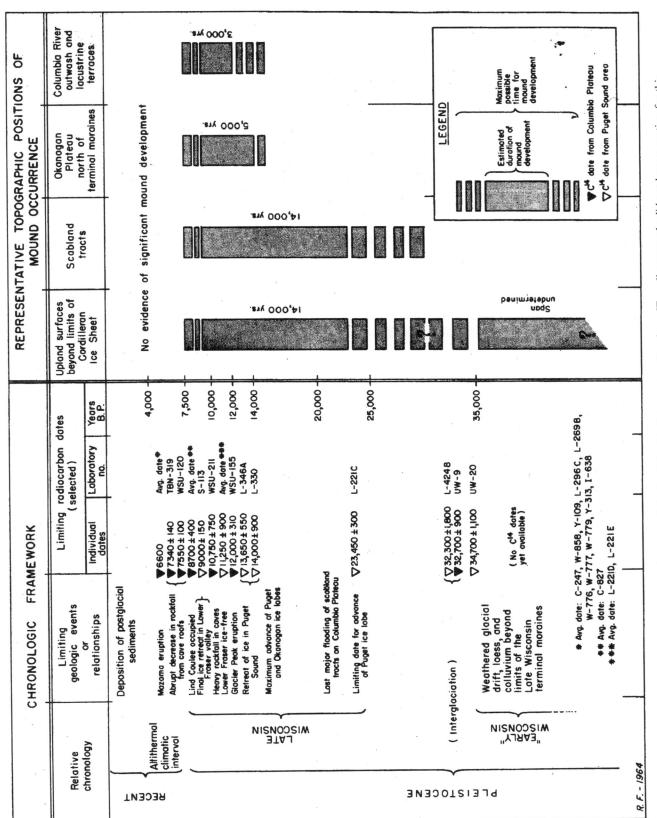

Figure 2. Chronologic framework for Mima mounds in the Columbia Intermontane Province. [Fryxell apparently did not have a caption for this self-explanatory figure, or at least we did not locate one. He apparently viewed this figure as fundamental to his age of mounds hypothesis, which, based on the title and gist of his text, was the primary theme of this manuscript. The caption is our best approximation.]

4. Mounds occur on most parts of the landscape surrounding tributaries to the Snake and Columbia Rivers, but have not yet been found on alluvial fans graded from these tributary canyons to Late Wisconsin terraces; thus the mounds had no opportunity to develop while these fans were actively growing, and mound formation had ceased by the time alluviation of the fans was completed.

5. Mounds are present on hills adjacent to Grand Coulee; individual mound fields extend onto flood-scoured bedrock and gravel bars of the Coulee floor, but are absent in both the upper and lower Coulee from those basins which remained as lakes throughout post-flood Late Wisconsin time; thus the lacustrine sediments were protected by water while mounds formed at the edge of the lakes, but no mounds were formed after the lakes dropped to their post-Wisconsin level.

6. Mounds of this type have not been found on any sediments or surfaces known to be post-Wisconsin in age; thus post-Wisconsin time has been insufficient, and/or post-Wisconsin conditions unsuitable, for the development of this type of patterned ground.

From a strictly temporal standpoint, these observations strongly support and are entirely in agreement with suggested origins requiring a stringent frost environment for the origin of mounds on the Columbia Plateau.

STRATIGRAPHIC, PALEONTOLOGIC, AND PEDOLOGIC EVIDENCE FOR AGE OF THE MOUNDS

Stratigraphic, paleontologic, and pedologic evidence support the Late Wisconsin age suggested for the most recent episode of mound development.

Mammoth and Bison antiquus bones have been collected from loess postdating the scabland flood at North Deadman Creek (NE ¼ S.35, T.13N, R.42E) south of the Snake River, in a deposit traceable (with interbedded colluvium) to loess in the mound fields on adjacent slopes. Soil profiles developed on loess containing extinct fauna are comparable to those developed on loess found on scabland channel floors; in many localities these profiles are overlain by volcanic ash of the postglacial Mazama eruption at Crater Lake, Oregon (Powers and Wilcox, 1964; Fryxell, 1965). The Lake Wisconsin loess and its associated colluvium and paleosols are overlain also by postglacial loess which lacks the extinct Wisconsin faunal remains found in the earlier loess, and instead contains human bones and artifacts recovered from archaeological sites along the Snake River.

Far greater lengths of time are available for the development of mounds found on older surfaces and materials and more strongly developed soil profiles are associated with such mound fields. This is reflected particularly by the development of solonetzic soil complexes (Peterson, 1961) in which solodized solonetz profiles may be associated with intermound areas. Such soil complexes are found in many places on the Columbia Plateau adjacent to scabland or normal stream channels, but are absent within them except on the finely textured sediments of

the Touchet Formation (Flint, 1935), which are initially high in soluble salt content (H.W. Smith, 1964, personal commun.). At Grand Coulee Dam, solonetz patches occur between mounds in loess a short distance east of the Late Wisconsin ice margin (SW ¼, S.27, T.29N, R.21E). Neither the loess nor the solonetz profiles have been found westward within the glaciated area, though soils there have been examined in some detail (Fryxell, 1959). The direct association of solonetz patches with patterned ground and intermound areas suggests that solonetz development in many localities was influenced by periglacial conditions and that these profiles also are of Wisconsin age. Similar relationships are strikingly evident south of Flathead Lake, Montana, at the source area for the scabland floods (Bretz et al., 1956). Solonetz networks are widespread on lacustrine sediments of "Early" (Bull Lake equivalent) Wisconsin age (Alden, 1953), but are absent on immediately adjacent sediments of younger age under the same or similar plant communities. Elsewhere on the Columbia Plateau, poor subsurface drainage of parent materials high in original soluble salt content may be of major importance.

Land surfaces older than Wisconsin undoubtedly have undergone additional cycles of intense frost activity, and traces of mounds having much greater age eventually may be found. The chances for their preservation, however, are slight, because of repeated stripping of these surfaces by solifluction, and by erosion during interglacial and interstadial times. It is much more likely that repeated episodes of mound development due to repeated Wisconsin glacial advances will be found, recorded by the developmental stratigraphy of associated soil profiles.

LIMITING RADIOCARBON DATES FOR MOUND DEVELOPMENT

Only four radiocarbon-limiting dates directly applicable to the chronology of mound development are available from eastern Washington. The earliest, a date of 32,700 ± 900 B.P., has been attributed to interstadial wood picked up and redeposited by waters of a scabland flood discharged into the Columbia River (Fryxell, 1962), and so serves as a limiting date for both scabland flooding and the maximum Late Wisconsin ice advance that followed the flood. Therefore, no mounds developed on Late Wisconsin landforms in eastern Washington can be older than about 30,000 years.

Disappearance of ice from the northern edge of the Columbia Plateau preceded deposition of valley deposits at Lind Coulee for which a weighted average of 8700 ± 400 B.P. has been made for two radiocarbon dates (Daugherty, 1956). Because mounds are absent from deposits younger than this age, and are not present on some older sediments, no significant development of mounds occurred after 9600 years ago. This conclusion is reinforced by radiocarbon dates of 7500 ± 200 (R.M. Chatters, 1963, personal commun.) and 7340 ± 140 B.P. (B. Robert Butler, 1963, personal commun.) which bracket the cessation of rockfall in cave deposits as a result of restricted frost activity during the Altithermal climatic phase (Fryxell and Daugherty, 1963). Numerous radiocarbon dates place the age of the Mazama eruption, which occurred still later, at about 6600 B.P.

Considerably closer estimates of limiting dates for the possible duration of mound development may be made by correlation of events in eastern Washington with those available for the detailed sequence of glaciation recognized in the Puget Sound area. Such tentative correlation is justified by the comparable weathering and topographic characteristics of drift in the two areas, and by the fact that the Okanogan and Puget ice lobes share a common catchment divide and so were subject to the same gross climatic controls (Fryxell, 1962, p. 117).

The radiocarbon chronology emerging from studies in Puget Sound and areas to the north (Armstrong, 1956; Crandell, 1963; Easterbrook, 1963; Waldron et al., 1957) places onset of the Vashon Glaciation less than 23,000 years ago, and its initial recession about 14,000 years ago; by 11,300 years B.P. the lower Fraser Valley was ice-free. Although Sumas ice readvanced after that time, it too had retreated beyond the Yale archaeological site in southern British Columbia by 9000 B.P. (Borden, 1965).

If the same relative sequence of events occurred in eastern Washington, as is suggested by the sketchy reconnaissance data now available, it is probably that mound-covered areas north of the Okanogan Lobe terminal moraine were first exposed about 14,000 years ago, allowing not more than 7500 years total for development of the mounds, and perhaps as little as 5000; mounds on the Columbia River outwash terraces may have had a period of little more than 3000 years in which to form.[4]

The relative spans of time available in eastern Washington for mound formation on surfaces of various ages, with limiting geologic events and radiocarbon dates, are shown in Figure 2. These limiting dates do not necessarily apply to other patterned ground features. Stone stripes, for example, have been reactivated at least twice in the past 4000 years, and freshly over-turned rock fragments and partially buried shrubs show some stripes to be active now.

SUMMARY AND CONCLUSIONS

Patterned ground features, including a transitional series of stages in the development and location of mounds, as well as interrelated rock polygons, stone nets and stripes, rubble sheets, and solifluction deposits, are present as a continuum over virtually the entire Columbia Intermontane Province. Only in those areas where slopes are too precipitous to permit their development and preservation, or where deposits or surfaces are too young to have been affected by severe Wisconsin frost activity, are these features consistently absent.

The intimate association of these mounds with other features generally accepted as products of intense frost activity, as well as the external and internal morphology of the mounds, suggest that development of the mounds took place under Pleistocene "periglacial" climatic conditions. This conclusion is confirmed, and the origin of mounds of this type as frost-related features themselves is strongly supported, by physiographic, stratigraphic, paleontologic, and pedologic evidence placing the age of the last episode of mound development within Late Wisconsin time.

Moreover, the maximum span of time possible for development of the youngest mounds may be restricted in terms of relative chronology to the period between retreat of the Late Wisconsin glacial maximum and the close of waning frost activity which came to a halt with the Altithermal climatic phase. On the basis of the few radiocarbon dates available locally, the correlation of Late Wisconsin events east and west of the Columbia Plateau, the maximum time span available for development of the youngest mounds appears to have been between 14,000 and 7500 years ago. The lack of radiocarbon dates available makes closer limiting dates hazardous guesses, but the actual duration of intense frost activity available for development of some mounds may have been less than half this amount; about 3000 years.

The estimated limit of 3000 years, and not more than 7500 years, for the rate of development of mounds in patterned ground network applies only to those mounds developed on surfaces dating from the recession of the Late Wisconsin glacial maximum; mounds on surfaces in existence throughout the Wisconsin undoubtedly were subjected to longer, and possibly to several, episodes of development.

Although some patterned ground features, such as stone stripes, have been weakly rejuvenated on at least two occasions since the end of the Altithermal, there is no evidence that the mounds have undergone any significant constructional change in the last 7500 years. Further segregation of mound patterns already established, however, continues through erosion of intermound channels where drainage conditions permit.

ACKNOWLEDGMENTS

The writer is indebted for discussion and for critical review of the manuscript to Maynard Fosberg, Harold E. Malde, Donal E. Mullineaux, Troy L. Péwé, and Aaron C. Waters. Field relationships on which these comments are based were recorded in part during investigations supported by the National Science Foundation through grant NSF G-24959.

REFERENCES CITED

Alden, W.C., 1953, Physiography and Glacial Geology of Western Montana: U.S. Geological Survey Professional Paper, v. 231, 190 p.

Armstrong, J.E., 1956, Mankato drift in the lower Fraser valley of British Columbia, Canada: Geological Society of America Bulletin, v. 67, no. 12, p. 1666–1667.

Borden, C.E., 1965, Radiocarbon and geological dating of the Lower Fraser Canyon archaeological sequence: Proceedings of the Sixth International Congress on Radiocarbon and Tritium Dating: Pullman, Washington State University, p. 165–178.

Bretz, J.H., 1913, Glaciation of the Puget Sound region: Washington Geological Society Bulletin No. 8, 244 p.

---

[4]On the basis of annual development of thermal-contraction cracks he had observed in Alaska, Leffingwell (1919, p. 211) concluded that ice wedges found between large, convex polygons could have formed within less than a thousand years.

Bretz, J.H., Smith, H.T.U., and Neff, G.E., 1956, Channeled Scabland of Washington: New data and interpretations: Geological Society of America Bulletin, v. 67, p. 957–1049, doi:10.1130/0016-7606(1956)67[957:CSOWND]2.0.CO;2.

Brunnschweiler, D., 1962, The periglacial realm in North America during the Wisconsin glaciation: Biuletyn Peryglacjalny (Lodzkie Towarzystwo Naukowe - Societas Scientiarium Lodziensis), Series 3, no. 11, p. 15–27.

Brunnschweiler, D., 1964, Der pleistozane Periblazialbereich in Nordamerika: Zeitschrift für Geomorphologie, v. 8, p. 223–231.

Crandell, D.R., 1963, Surficial Geology and Geomorphology of the Lake Tapps Quadrangle, Washington: U.S. Geological Survey Professional Paper 388A, 84 p.

Dalquest, W.W., and Scheffer, V.B., 1942, The origin of the mounds of western Washington: The Journal of Geology, v. 50, no. 1, p. 68–84, doi:10.1086/625026.

Daubenmire, R.F., 1942, An ecological study of the vegetation of southeastern Washington and adjacent Idaho: Ecological Monographs, v. 12, no. 1, p. 53–79, doi:10.2307/1948422.

Daugherty, R.D., 1956, Early Man in the Columbia Intermontane Province: Anthropological Papers no. 24: Salt Lake City, Department of Anthropology, University of Utah, 132 p.

Easterbrook, D.J., 1963, Late Pleistocene glacial events and relative sea-level changes in the Northern Puget Lowland, Washington: Geological Society of America Bulletin, v. 74, p. 1465–1484, doi:10.1130/0016-7606(1963)74[1465:LPGEAR]2.0.CO;2.

Flint, R.F., 1935, Glacial features of the southern Okanagan region: Geological Society of America Bulletin, v. 46, p. 169–194.

Fosberg, M.A., 1963, Genesis of some soils associated with low and big sagebrush complexes in the Brown, Chestnut and Chernozem-prairie zones in southcentral and southwestern Idaho: [Ph.D. thesis]: Madison, University of Wisconsin, 308 p.

Freeman, O.W., 1926, Scabland mounds of eastern Washington: Science, v. 64, no. 1662, p. 450–451.

Freeman, O.W., Forrester, J.D., and Lupher, R.L., 1945, Physiographic divisions of the Columbia Intermontane Province: Annals of the Association of American Geographers, v. 35, no. 2, p. 53–76.

Fryxell, R., 1959, Soil and grazing resources inventory, Colville Indian Reservation, Washington: U.S. Department of the Interior, Bureau of Indian Affairs, Branch of Land Operations Survey, p. 200–216a.

Fryxell, R., 1962, A radiocarbon limiting date for Scabland flooding: Northwest Science, v. 36, no. 4, p. 113–119.

Fryxell, R., 1965, Mazama and Glacier Peak volcanic ash layers: Relative ages: Science, v. 147, no. 3663, p. 1288–1290, doi:10.1126/science.147.3663.1288.

Fryxell, R., and Daugherty, R.D., 1963, Late glacial and postglacial geological and archaeological chronology of the Columbia Plateau, Washington: Laboratory of Anthropology, Reports of Investigations no. 23: Pullman, Washington State University, 22 p.

Jackson, H.E., 1956, The mystery of the Mima mounds: Natural History, v. 65, no. 3, p. 136–139.

Kaatz, M.R., 1959, Patterned ground in central Washington—A preliminary report: Northwest Science, v. 33, no. 4, p. 145–156.

Knechtel, M.M., 1952, Pimpled plains of eastern Oklahoma: Geological Society of America Bulletin, v. 63, no. 7, p. 689–699, doi:10.1130/0016-7606(1952)63[689:PPOEO]2.0.CO;2.

Krinitzsky, E.L., 1949, Origin of pimple mounds: American Journal of Science, v. 247, p. 706–714, doi:10.2475/ajs.247.10.706.

Larrison, E.J., 1942, Pocket gophers and ecological succession in the Wenas region of Washington: The Murrelet, v. 23, no. 2, p. 34–41, doi:10.2307/3535581.

LeConte, J., 1874, On the great lava-flood of the Northwest, and on the structure and age of the Cascade Mountains: American Journal Science, series 3, v. 7, p. 167–180, 259–267.

Leffingwell, E. de K., 1919, The Canning River Region, North Alaska: U.S. Geological Survey Professional Paper 109, 251 p.

Malde, H.E., 1961, Patterned ground of possible solifluction origin at low altitude in the western Snake River Plain, Idaho: U.S. Geological Survey Professional Paper 424-B, p. B170–B173.

Malde, H.E., 1964, Patterned ground in the western Snake River Plain, Idaho, and its possible cold-climate origin: Geological Society of America Bulletin, v. 75, p. 191–208, doi:10.1130/0016-7606(1964)75[191:PGITWS]2.0.CO;2.

Newcomb, R.C., 1952, On the origin of the Mima mounds, Thurston County region, Washington: The Journal of Geology, v. 60, no. 5, p. 461–472, doi:10.1086/625998.

Olmsted, R.K., 1963, Silt mounds of Missoula flood surfaces: Geological Society of America Bulletin, v. 74, no. 1, p. 47–53, doi:10.1130/0016-7606(1963)74[47:SMOMFS]2.0.CO;2.

Peterson, F.F., 1961, Solodized solonetz soils occurring in the uplands of the Palouse loess [Ph.D. thesis]: Pullman, Washington State University.

Péwé, T.L., 1948, Origin of the Mima mounds: The Scientific Monthly, v. 66, no. 4, p. 293–296.

Piper, C.V., 1905, The basalt mounds of the Columbia lava: Science, v. 21, p. 824–825, doi:10.1126/science.21.543.824.

Powers, H.A., and Wilcox, R.E., 1964, The volcanic ash deposits from Mazama (Crater Lake) and from Glacier Peak in northwestern United States and southwestern Canada: Science, v. 144, no. 3624, p. 1334–1336, doi:10.1126/science.144.3624.1334.

Richmond, G.M., Fryxell, R., Neff, G.E., and Weis, P.L., 1965, The Cordilleran ice sheet of the northern Rocky Mountains, and related Quaternary history of the Columbia Plateau, *in* Wright, H.E., Jr., and Frey, D.G., eds., The Quaternary of the United States: A Review Volume for the VII Congress of the International Association for Quaternary Research: Princeton, New Jersey, Princeton University Press, p. 231–242.

Ritchie, A.M., 1953, The erosional origin of the Mima mounds of southwest Washington: The Journal of Geology, v. 61, no. 1, p. 41–50, doi:10.1086/626035.

Scheffer, V.B., 1947, The mystery of the mima mounds: The Scientific Monthly, v. 65, no. 5, p. 283–294.

Waldron, H.H., Mullineaux, D.R., and Crandell, D.R., 1957, Age of the Vashon glaciation in the southern and central parts of the Puget Sound basin: Geological Society of America Bulletin, v. 68, no. 12, pt. 2, p. 1840–1850.

Washburn, A.L., 1956, Classification of patterned ground and review of suggested origins: Geological Society of America Bulletin, v. 67, p. 823–866, doi:10.1130/0016-7606(1956)67[823:COPGAR]2.0.CO;2.

Waters, A.C., 1933, Terraces and coulees along the Columbia River near Lake Chelan, Washington: Geological Society of America Bulletin, v. 44, p. 783–820.

Waters, A.C., and Flagler, C.W., 1929, Origin of small mounds on the Columbia River Plateau: American Journal of Science, 5th series, v. 18, no. 105, p. 209–224.

Printed in the USA

The Geological Society of America
Special Paper 490
2012

Appendix B. Early prairie mound observations by two celebrated geologists: Joseph LeConte and Grove Karl Gilbert

Donald L. Johnson
Department of Geography, University of Illinois, Urbana, Illinois 61801, USA, and
Geosciences Consultants, 713 S. Lynn St., Champaign, Illinois 61820, USA

In the 1870s, North American geologists Joseph LeConte and Grove Karl Gilbert—like many others before and after—made observations, but drew different conclusions, on the origin of prairie mounds in their areas of study, LeConte's in California and the Pacific Northwest, and Gilbert's in east-central Arizona. Because both were viewed as keen observers and their statements came early in the mound origins issue, and are among the more interesting, we have quoted verbatim their observations and statements. They represent contrasting views that, among others, have steeped the issue of prairie mound genesis in controversy for nearly 200 years. Their statements are windows to, and good early examples

of, this controversy. (Both also establish early 1870s priorities for the term "prairie mound," preceding by nearly 40 years the term "Mima mound" introduced by Bretz in 1913 for mounds formed in the Olympia area prairies of Washington.)

Brief sketches of both LeConte and Gilbert (Fig. 1) are provided to establish background contexts. Their mound statements then follow. Our comments, where appropriate within the texts of each, are in brackets. We follow each statement with a commentary that links to the overall theme of this volume. We end with an epilogue of how their views impacted the legacy of mound discussions.

Figure 1. Joseph LeConte, left, and Grove Karl Gilbert, right. (LeConte photo courtesy Bancroft Library, University of California, Berkeley; Gilbert photo courtesy U.S. Geological Survey.)

PROFESSIONAL SKETCH OF JOSEPH LeCONTE

Joseph LeConte was born 26 February 1823, in Liberty County, Georgia to Louis LeConte, patriarch of the noted LeConte family. He was educated at Franklin College in Athens, Georgia (now Franklin College of Arts and Sciences, University of Georgia). After graduation in 1841, he studied medicine and received his degree at the New York College of Physicians and Surgeons in 1845, and practiced medicine for several years in Macon, Georgia. He then entered Harvard University and studied under Louis Agassiz, where he developed a keen interest in natural history. After graduating and teaching at several institutions in Georgia and South Carolina, LeConte moved west from Carolina College (now University of South Carolina) in 1869 to join the faculty of the newly established (1868) University of California at Berkeley. He was appointed the first professor of geology, natural history, and botany, a position he held until his death in 1901. In 1874, he was nominated to the National Academy of Sciences; in 1892 he became president of the American Association for the Advancement of Science; and in 1896 was elected president of the Geological Society of America. He was also co-founder, with John Muir, of the Sierra Club in 1892, and its first Director from 1892 to 1898. He died of heart failure in 1901, in Yosemite, California (cf. Chisholm, 1911).

JOSEPH LeCONTE'S 1874–1877 MOUND STATEMENTS

LeConte's (1874) main mound statement is embedded in a paper titled "On the great lava-flood of the Northwest, and on the structure and age of the Cascade Mountains," in the *American Journal of Science (and Arts):*

Prairie mounds.—The irregularly ramifying grassy glades or prairies already described as existing at the southern extremity of Puget Sound, are studded over as thickly as possible with *mounds* about three to four feet high and thirty to forty feet in diameter at base. For this reason these glades are usually known as *"mound-prairies."* There are millions of these mounds, and they stand so thickly that they touch each other at their bases, leaving no level space between. Although accurate measurement would doubtless show some variety in shape, size, and position, yet the general impression upon the observer is that of perfect regularity. The mounds themselves consist wholly of a drift soil of earth, gravel and small pebbles: the intervals between are thickly strewn with larger pebbles and small boulders. The vegetation of the mounds is mostly ferns: the intervals are covered with fine grass only.

There has been much speculation as to the origin of these remarkable mounds. Some have supposed that they are Indian burial mounds, veritable cities of the dead. Others have thought that they are artificial mounds, upon which were built huts of Indian villages. Still others have thought that they were made by certain, perhaps unknown, species of fish, at a time when these prairies were still the bottoms of shallow inlets; that they were in fact large fish-nests. No one who has examined them can for a moment accept any of these views. They have been many of them opened, but nothing indicating artificial origin has ever been found*. Dr. Newberry has frankly acknowledged his inability to account for them (cf. Newberry, 1857).

*There have been many rumors of the opening of those mounds and the finding in them of human relics. Only very recently these rumors have been revived and extensively circulated in the daily prints. It was positively stated that the officers of the N.P.R.R. had opened several, and found undoubted evidences of human origin. I immediately wrote to the intelligent Superintendent, Mr. I.W. Sprague, asking for information. Only a few weeks ago I received [an] answer from him, stating that the road does not run through Mound Prairie, that they had never found any relics of any kind, and as far as he could learn, no relics had ever been found in the mounds of Mound Prairie.

When I first examined these mounds in 1871, they were to me perfectly inexplicable. But upon subsequent reflection, and from what I heard of similar phenomena in portions of California, I long ago arrived at the conclusion that they are the *result of surface-erosion under peculiar conditions.* On conferring with Mr. Condon [Oregon geologist] during my last visit, I found that he too had come to this conclusion. In my journey with him to the John Day Valley, I saw evidence which was perfectly convincing. The whole rolling country between the Dalles and the upper bridge of Des Chutes River [probably Sherar's Bridge, cf. Pyrch, 1973, p. 33; their route approximated present-day Hwy 197], a distance of about thirty miles [48 km], is literally covered with these mounds. From every eminence the whole face of the country, as far as the eye could reach, presented a singular appearance, as if thickly broken out with a popular eruption. But the true key to their mode of formation is given here, as it was not at Mound Prairie, by the great variety of forms, sizes, and degrees of regularity which they assume. They varied in *size* from scarcely detectable pimples to mounds five feet high, and forty feet diameter at the base, and in *form* from circular through elliptic and long-elliptic to ordinary hill-side erosion-furrows and ridges. In regularity of size and position, there was equal diversity; in some places being as complete as at Mound Prairie, i.e., they were all apparently of the same size, and touched each other at base; in other places they were of different sizes, and often separated by *wide pebble-covered spaces,* as if they were but the remnants of a general erosion of the surface-soil. Thus on the one hand, portions of many square miles in extent were covered with mounds as large and as regular as any portion of Mound Prairie, and *evidence produced by the same cause*: on the other hand, other portions were marked only by long hill-side furrows and ridges, *evidently produced by surface erosion*; and between these extremes every stage of gradation could be traced.

No one, I think, can ride over those thirty miles [48 km] and observe closely, without being convinced that these mounds are wholly the result of surface-erosion acting under peculiar conditions. These conditions are a *treeless country* and a *drift soil,* consisting of two layers, a finer and more movable one above and less movable one below*. Surface erosion cuts through the finer superficial layer into the pebble layer beneath, leaving, however, portions of the superficial layer as mounds. The *size* of the mounds depends on the *thickness* of the superficial layer; the *shape* of the mounds depends much upon the *slope* of the surface. The process once started, small shrubs and weeds take possession of the mounds as the *better soil,* and hold them by their roots, and thus increase their size by preventing or retarding erosion in these spots. The treelessness of the country in eastern Oregon has been produced gradually, since Post-Tertiary times, by the increasing dryness of the climate. We may imagine the mounds, therefore, as having been *held by the struggling remnants of a departing vegetation.* At Mound Prairie, however, the treelessness is probably produced by a contrary condition, viz: the extreme *wetness* of these lower level spots in winter. Here, therefore, the weeds and ferns hold and preserve the mounds, not only as the *better soil,* but also as the *drier spots.*

*The necessary condition, I believe, is the greater movableness of the surface soil, as compared with the subsoil, whatever may be the cause of the greater movableness. In Oregon and Washington, the cause is a *pebble-subsoil*; in other places mentioned below, the cause may be different.

When once attention is turned to the subject, the same phenomena in a less degree is observed in nearly all the treeless regions of California and Oregon, which have not yet been touched by the plough. On returning from Oregon, I observed it in the upper part of the Sacramento Valley, where, however, the mounds are so small and inconspicuous as to escape observation unless attention has been previously awakened—only a light dappling of the surface of the country. Similar mounds, more conspicuous than the last mentioned, but far less so than those of Oregon and Washington, under the name of 'hog wallow' are well known to exist over wide areas in middle and southern California. They dapple the whole surface of the San Joaquin and Tulare plains, and are found also in the country about Los Angeles. The smaller, sometimes scarcely noticeable mounds of California, graduate completely into the larger and more perfect ones of Mound Prairie. If the mounds of Mound Prairie were an unique phenomenon, we might resort to exceptional modes of origin; but a phenomenon so wide spread must be attributed to the action of a wide-spread agent.

Oakland, Cal., Dec. 15th, 1873. (p. 265–267, asterisked footnotes from original)

Commentary

For those unfamiliar with the prairie mound literature, which is vast, confusing, and challenging, the expression "hog wallow" in most regions refers to mound-and-depression topography produced by shrink-swell processes that operate in clayey soils—that is, in Vertisols. Unfortunately in mid-nineteenth century Texas and California, nonvertisol-produced prairie mounds early became visually and conceptually confused with Vertisol microrelief. The latter, as indicated, was referred to as "hog wallow" terrain, which technically meant the depressions of that microrelief, not the high spots. That prairie mounds and hog wallows were conflated is not surprising insofar as evidence is strong that before the plow leveled most landscapes—and contrary to conventional wisdom—many loamy-sandy prairie mounds *were*, in fact, developed on Vertisols (cf. Retzer, 1949; Watson et al., 1925; Carter et al., 1916; Carter and Patrick, 1919; Foster and Moran, 1935; Smith and Marshall, 1938, p. 2; Crout, 1976; Aronow, 2006, p. 380–384; Howard and Freeman, 1983, p. 10; Heinrich, 1986, p. 168; Ensor, 1987; and personal observations). Insofar as both argilliturbation and bioturbation operated simultaneously on such vertic landscapes, the former more generally than the latter—and before the plow blurred the relationships—it is easy to see why the term "hog wallow" came to be used in both Texas and California for both vertisolic and nonvertisolic mounded terrain, and confusion ensued.

We (myself and my colleague-wife Diana) have personally and repeatedly examined the mounded prairies south of Olympia, Washington, and likewise the Dalles-Des Chutes River mound transect (Hwy. 197) of east-central Oregon, and nearby regions, that so impressed LeConte. Mounds in the area appear commonly independent of slope and differences in local soil, as evident on the ground and on Google Earth images (e.g., 45° 22′ 52″ N, 121° 06′ 38″ W; 45° 00′ 52″ N, 120° 45′ 25″ W). In hindsight LeConte's powers of observation in linking erosion, slope, mound form, and vegetation anchoring—with emphasis on erosion—are very reasonable in light of knowledge and conventional wisdom at the time.

A condensed version of his 1874 article appeared a year later in which he reiterated that the mounds were a "result of surface erosion under peculiar conditions" (LeConte, 1875). He concluded with:

The hog-wallows of California may be traced by insensible gradations into the larger mounds of Eastern Oregon, and these, in their turn, into the more perfect mounds of Mound Prairie [Mima Prairie, WA]; they are all evidently due to the same cause. If the mounds of Mound Prairie were a unique phenomenon, we might resort to exceptional causes; but a phenomenon so wide-spread must be due to a wide-spread agent. (LeConte, 1875, p. 220)

While LeConte's powers of observation served him well in most geological-landscape regards, they fell short in one slope-soil process that was, understandably, missing from his explanation, and that is strikingly evident throughout every mounded region he discussed—bioturbation. The ubiquitous spoil heaps of burrowing animals, most notably those of pocket gophers (*Thomomys talpoides and T. mazama*)—ranging from fresh to extremely weather-wasted and barely perceptible spoil to every stage in between—are in evidence essentially everywhere throughout these moundfields. I say "understandably" because the process term "bioturbation" and various associated concepts was not part of any explanatory tool kit at that time.

In 1877 LeConte sent a letter to the journal *Nature* in response to noted evolutionist Alfred R. Wallace's (1877a, 1877b) statements that glaciation in California might be linked to prairie mound formation, but also responded to, among other things, G.K. Gilbert's (1875) suggestion that rodents might have made prairie mounds in Arizona. In his letter, part of which is reproduced below, LeConte (1877) pointedly dismissed burrowing animals, among other possible processes, as factors in producing mounds in Arizona, in California, and in the Pacific Northwest, arguing instead for the overriding effects of "surface erosion under peculiar conditions":

In Nature (vol. xv, p. 274), Mr. Wallace quotes a letter from his brother in regard to the so-called Hog-wallows of California, in which their origin is ascribed to *débris* left at the broad foot of a retiring glacier modified by the erosion of innumerable issuing rills, and asks if this structure is known to occur elsewhere. As I have observed the same formation in many parts of the Pacific slope and have tried to explain it, I hope I may be allowed to say a few words on the subject.

The peculiar configuration of [mounded] surface so well described by Mr. Wallace, is very widely diffused in America, and has been described under different names. In California the mounds are called *Hog-wallows*, but elsewhere they are known as *Prairie mounds*. This latter is the better name since they are found only in grassy, treeless, or nearly treeless regions. They occur over much of the Prairie region or *"Plains"* east of the Rocky Mountains; also over portions of the *basin* region, *e.g.*, in Arizona; also over much of the bare grassy portions of California, *e.g.*, along the lower foothills of the Sierra and adjacent portion of the San Joaquin plains; also over enormous areas in Middle Oregon, on the eastern slope of the Cascade mountains, an undulating grassy region; also on the level grassy Prairies about the southern end of Pugit [sic] Sound, Washington territory.

They have been ascribed to the most diverse causes. In Texas, where they are very small, Prof. Hilgard thinks they are *ant-hills*. In Arizona, where they are also imperfectly developed, Mr. Gilbert thinks they are the ruined habitations of departed *Prairie dogs*. In some portions of California, also, where they are small they been popularly ascribed to *burrowing squirrels*. In the Prairies, about Pugit Sound, where they are splendidly developed, their great size and extreme regularity has suggested that they are *burial mounds*, and that the Prairies are veritable cities of the dead. It is possible that the cause may be different in different places, but I am sure that no one who has examined them in California, and especially in Oregon and Washington, can for a moment entertain any of these theories for the Pacific slope. (p. 530)

LeConte then restates the main points made in his earlier 1874 and 1875 papers, attributing mounds to "surface erosion under peculiar conditions," refers readers to that paper, and expresses uncertainty and frustration about the exact nature and origin of the seemingly inexplicable (to him) "drift" in that part of the world from which the mounds had formed. (The 1870s, it should be noted, was an early period when concepts on the extent of glaciation, and the nature of loess, soil, pedogenesis, soil parent material, weathering, slope processes, and other now familiar soil-geomorphic concepts and issues were still in a nascent stage. For example, "drift" was a term commonly used to describe many genetically uncertain surface accumulations.) Toward the end of his 1877 letter, he expressly and colorfully describes the character of the mounded regions as follows:

. . . In the bare hilly regions of Middle Oregon, on the east side of the Cascade Mountains, every stage of gradation may be traced from circular mounds, through elliptic, long elliptic, to ordinary erosion furrows. [and] All the high, bare, grassy, hilly, slopes of the Cascades Mountains are covered evenly with a pebble and boulder drift, graduating upwards into a finer top soil. From this surface-soil are carved the mounds, which cover hill and dale so thickly that, viewed from an eminence the whole face of the country seems broken out with measles. (p. 531)

Le Conte concluded both his 1874 and 1875 papers with the statement that a phenomenon—like mounds—so widespread must be attributed to the action of a widespread agent. In addition to the widespread agent of erosion which he viewed as predominantly responsible for the mounds, another widespread agent—in Arizona, California, the Pacific Northwest, and elsewhere—is bioturbation, especially as expressed in the nest-centered centripetal style of burrowing by pocket gophers (cf. Dalquest and Scheffer, 1942, 1944; Cox, 1984).

PROFESSIONAL SKETCH OF GROVE KARL GILBERT

Grove Karl Gilbert was born 6 May 1843, in Rochester, New York. In 1862 he received a degree at age 19 from the University of Rochester after having specialized in Mathematics and Greek, taking but one course in geology. After teaching in Michigan for a few years, and working for a period in a scientific supply house, he joined George M. Wheeler's geographical survey as assistant geologist in 1871. This began his career as a member of the U.S. Geological Survey that covered three decades. Dur-

ing this time he produced, among other reports, two monumental monographs that focused on Utah, "Report of the Geology of the Henry Mountains" in 1877, and "Lake Bonneville" in 1890. He was elected President of the Geological Society of America in 1892, and again in 1909—the only geologist ever to be so honored twice. In 1899 Gilbert joined the Harriman Alaska Expedition, which resulted in a full volume on "Glaciers and Glaciation." In 1900 the Geological Society of London awarded him the Wollaston Medal, and in 1910 the American Geographical Society presented him with the Charles P. Daly Medal. Gilbert died 1 May 1918 in Jackson, Michigan.

In 1980 the Geological Society of America (GSA) published Special Paper 183 that honored and summarized Gilbert's scientific research. To further honor his accomplishments, in 1983 the GSA established the G.K. Gilbert Award for planetary geology (cf. Barrell, 1919; Davis, 1926; Mendenhall, 1920; Yochelson, 1980).

G.K. GILBERT'S BRIEF 1875 MOUND STATEMENT

Gilbert's (1875) mound statement is in Chapter 19, "The volcanic region," part of a larger document titled "Report upon Geographical and Geological Explorations and Surveys West of the 100th Meridian," and reads:

Prairie mounds.—The grassy plains that diversify the pine forests at the eastern base of Sierra Blanca [basalt-composed White Mountains, elevationally above and south of Springerville, AZ] are dotted with a system of low mounds, in a manner independent of the nature or slope of the soil. They are usually one or two rods broad [~5–10 m] and less than a foot high [~31 cm], and separated by interspaces several times as broad as themselves. There is frequently a notable difference of soil texture between the level ground and the mound, the mound being, in some instances, the more gravelly, and in others, the less so; and there is some difference of quality, that the eye detects only in its effect upon vegetation. The grass upon the mounds is distinguished by a ranker growth, and, as we saw it in August, by a deeper green. Viewed from a commanding position, the effect is peculiarly beautiful, the green spots dappling the plain like the figure of a carpet. These are not considered geological features, and they are mentioned in this place only to distinguish them from the prairie mounds of California, which, according to Professor Le Conte, are phenomena of erosion.* There is little question that they are the vestiges of hummocks thrown up by prairie dogs, or other burrowing animals. The manner of their distribution first suggested this explanation, and it appears to accord with all the observed facts. The subsoil, brought to the surface in the excavation of the burrows produces the superficial difference of texture; and the looser aggregation of the earth, together, perhaps, with the presence of animal manure, gives the grass a stronger growth. None of the mounds are inhabited, and no burrows, or other sign of recent occupation, were seen. I am acquainted with no colonies of prairie dogs at such an altitude—7,000 feet [2134 m]—and, if the mounds are the work of that species, they may point to a climate, in very recent time, of even greater warmth and aridity than the present. Some of the mounds were seen upon, and on the slope below, the oldest of the Reservoir Lake beaches, showing that the race which built them has given way since the diminution of the lake. (p. 539–540)

---

*American Journal of Science and Arts, April, 1874, p. 365. (footnote from original text)

Commentary

During our October 2005 and April 2007 surveys of the White Mountains basaltic moundfields briefly described by Gilbert, we confirmed that low prairie mounds, as he noted, do indeed dapple most grassland tracts and meadows in that part of the Apache-Sitgreaves National Forest of Arizona. They are also common on the lower desert grassland and open conifer-bearing basalt tracts between Eagar and McNary on Highway 60. Their occurrence is, as Gilbert observed, indeed independent of slope and local differences in soil, as is evident both in ground surveys and on Google Earth images (e.g., 33° 52′ 18″ N, 109° 25′ 20″ W; 33° 54′ 25″ N, 109° 20′ 25″ W).

Our surveys also confirm the presence of innumerable spoil heaps ranging from fresh, old, and weather-wasted, to those entirely flattened and barely recognizable as such produced by the dominant soil bioturbator of this region, the Botta pocket gopher (*Thomomys bottae*). In fact, in some tracts nearly every mound and intermound was observed to be dotted with the spoil heaps of this animal, yet in other tracts, sometimes over large areas, few or no recognizable spoil heaps—fresh or old—were evident. So, while these legendary burrowing animals are extremely common and ubiquitous in some mounded grassy basaltic tracts of the White Mountains, they are rare or temporarily absent from others, even over short distances, at least during the several times we observed them. Such is the nature of burrowing animal distributions, and their fluctuating amoeba-like changes with time (cf. Howell, 1923).

Gilbert's observations and conclusions were made early in his geological career. One wonders if, at the end, he would have departed from his earlier pronouncement that prairie mounds, so produced, were not both geologic and soil phenomena.

EPILOGUE

The 1874–1877 mound statements of LeConte and Gilbert played roles in reactivating the discussion of natural prairie mounds versus those of human origin that had been on-going for decades. The reactivation underscored the intense public and professional fascination with "Mound Builder" issues in pre-European North America, and mounds in general (cf. Silverberg, 1970/1986). After the LeConte-Gilbert statements, the debate on mound origins continued in various venues during the next few decades, reaching a peak in 1905–1906 when a dozen letters-to-the-editor on the subject appeared in *Science Magazine*. Origins were attributed to humans, ants, termites, pocket gophers, prairie dogs, basalt rock caps (in Channeled Scablands of Washington), tree-uprooting, spring-gas vent-artesian accumulations, differential sediment settling, climate change, wind (dunes), surface erosion, glaciation, fish nests exposed by elevation, concretionary accumulations owed to chemical precipitations, flint [chert] aggregations, and/or combinations of the above—polygenesis. (In order of appearance, the letters were by Morris, 1905 (reporting on Veatch); Branner, 1905; Hilgard, 1905; Spillman, 1905;

Purdue, 1905; Piper, 1905; Bushnell, 1905; Veatch, 1906; Farnsworth, 1906; Hill, 1906; Wentworth, 1906, and Udden, 1906.)

The debate, which continues (e.g., Seifert et al., 2009), led to the present volume.

REFERENCES CITED

Aronow, S., 2006, Surface geology, *in* Brown, S.E., Jr., ed., Soil Survey of Colorado County, Texas, U.S. Department of Agriculture, Soil Conservation Service, in cooperation with Texas Agricultural Experiment Station: Washington, D.C., U.S. Government Printing Office, p. 380–384.

Barrell, J., 1919, Grove Karl Gilbert, an appreciation: Sierra Club Bulletin, v. 10, no. 4, p. 397–399.

Branner, J.C., 1905, Natural mounds or 'hog-wallows': Science, v. 21, no. 535, p. 514–516, doi:10.1126/science.21.535.514-b.

Bushnell, D.I., Jr., 1905, The small mounds of the United States: Science, v. 22, no. 570, p. 712–714, doi:10.1126/science.22.570.712.

Carter, W.T., and Patrick, A.L., 1919, Soil Survey of Bryan County, Oklahoma: U.S. Department of Agriculture Bureau of Soils, Field Operations of Bureau of Soils for Year 1914, 16th Report, p. 2165–2212.

Carter, W.T., Jr., Schoenmann, L.R., Bushnell, T.M., and Maxon, E.T., 1916, Soil Survey of Jefferson County, Texas: U.S. Department of Agriculture Bureau of Soils, Field Operations of Bureau of Soils for Year 1913, 15th Report, p. 1001–1043.

Chisholm, H., ed., 1911, Joseph LeConte: Encyclopedia Britannica (eleventh edition): New York, Cambridge University Press. [Article incorporates text from an old publication now in the public domain.]

Cox, G.W., 1984, Mounds of mystery: Natural History, v. 93, no. 6, p. 36–45.

Crout, J.D., 1976, Soil Survey of Chambers County, Texas: U.S. Department of Agriculture, Soil Conservation Service, in Cooperation with Texas Agricultural Experiment Station, v. 52: Washington, D.C., U.S. Government Printing Office, 53 p.

Dalquest, W.W., and Scheffer, V.B., 1942, The origin of the mounds of western Washington: The Journal of Geology, v. 50, no. 1, p. 68–84, doi: 10.1086/625026.

Dalquest, W.W., and Scheffer, V.B., 1944, Distribution and variation in pocket gophers, *Thomomys talpoides*, in the state of Washington: American Naturalist, v. 78, no. 777, p. 308–333, 423–450, doi:10.1086/281203.

Davis, M., 1926, Biographical memoir, Grove Karl Gilbert, 1843–1918: Biographical Memoirs, National Academy of Sciences, v. 21, fifth memoir, presented 1922.

Ensor, H.B., 1987, The Cinco Ranch Sites, Barker Reservoir, Fort Bend County, Texas: Archaeological Research Laboratory, Report of Investigations 3: College Station, Texas, A&M University.

Farnsworth, P.J., 1906, On the origin of the small mounds of the lower Mississippi Valley and Texas: Science, v. 23, no. 589, p. 583–584, doi:10.1126/science.23.589.583-a.

Foster, Z.C., and Moran, W.J., 1935, Soil Survey of Galveston County, Texas: U.S. Department of Agriculture, Bureau of Chemistry and Soils, in cooperation with Texas Agricultural Experiment Station, Series 1930, no. 31, p. 1–18.

Gilbert, G.K., 1875, Prairie mounds: Report upon Geographical and Geological Explorations and Surveys West of the 100th Meridian, pt. 5, v. 3—Chapter 19, The volcanic region: Washington, D.C., Government Printing Office; available at http://books.google.com/books?id=2UI-AQAAIAAJ.

Heinrich, P.V., 1986, Geomorphology of the Barker Reservoir area: Appendix D, *in* Fields, R.C., Godwin, M.F., Freeman, M.D., and Lisk, S.V., Inventory and Assessment of Cultural Resources at Barker Reservoir, Fort Bend and Harris Counties, Texas, Reports of Investigations, no. 40: Austin, Texas, Prewitt & Associates, Consulting Archeologists p. 147–176.

Hilgard, E.W., 1905, The prairie mounds of Louisiana: Science, v. 21, no. 536, p. 551–552, doi:10.1126/science.21.536.551-a.

Hill, R.T., 1906, On the origin of the small mounds of the lower Mississippi Valley and Texas: Science, v. 23, no. 592, p. 704–706, doi:10.1126/science.23.592.704.

Howard, M.A., and Freeman, M.D., 1983, Inventory and Assessment of Cultural Resources at Bear Creek Park, Addicks Reservoir, Harris County, Texas, Reports of Investigations, no. 24: Austin, Texas, Prewitt & Associates, Consulting Archeologists, 92 p.

Howell, A.B., 1923, Periodic fluctuations in the numbers of small mammals: Journal of Mammalogy, v. 4, no. 3, p. 149–155, doi:10.2307/1373563.

LeConte, J., 1874, On the great lava-flood of the Northwest, and on the structure and age of the Cascade Mountains: American Journal of Science (series 3), v. 7, p. 167–180, 259–267: http://ajs.library.cmu.edu/books/pages.cgi?call =AJS_1874_007_1874&layout=vol0/part0/copy0&file=00000174 (continued at http://ajs.library.cmu.edu/books/pages.cgi?call=AJS_1874_007 _1874&layout=vol0/part0/copy0&file=00000266) (accessed 10 May 2011).

LeConte, J., 1875, On the great lava-flood of the northwest; and on the structure and age of the Cascade Mountains: Proceedings of the California Academy of Science, v. 5, p. 214–220.

LeConte, J., 1877, Hog wallows or prairie mounds: Nature, v. 15, no. 390, p. 530–531, doi:10.1038/015530d0.

LeConte family, New Georgia Encyclopedia, 2003: http://www.georgiaencyclopedia .org/nge/Article.jsp?path=/ScienceMedicine/Individuals-1&id=h-791 (accessed 10 May 2011).

Mendenhall, W.C., 1920, Memorial to Grove Karl Gilbert: Geological Society of America Bulletin, v. 31, p. 26–45.

Morris, E.L., 1905, The Biological Society of Washington [report on 14 January talk by A.C. Veatch]: Science, v. 21, no. 530, p. 310–311, doi:10.1126/science .21.530.310.

Newberry, J.S., 1857, Routes in California and Oregon explored by Lieut. R.S. Williamson, Corps of Topographical Engineers, and Lieut. Henry L. Abbot, Corps of Topographical Engineers, in 1855. Report upon the geology of the route, v. 6, part 2, *in* Reports of Explorations and Surveys, to Ascertain the Most Practicable and Economical Route for a Railroad from the Mississippi River to the Pacific Ocean: http://name.umdl.umich.edu /AFK4383.0006.002 (accessed 10 May 2011).

Piper, C.V., 1905, The basalt mounds of the Columbia lava: Science, v. 21, no. 543, p. 824–825, doi:10.1126/science.21.543.824.

Purdue, A.H., 1905, Concerning the natural mounds: Science, v. 21, no. 543, p. 823–824, doi:10.1126/science.21.543.823.

Pyrch, J.B., 1973, The characteristics and genesis of stone stripes in North Central Oregon [unpublished M.S. thesis]: Portland, Oregon, Portland State University, 134 p.

Retzer, J.L., 1945, Morphology and origin of some California mounds, *in* Proceedings of the Soil Science Society of America, v. 10, p. 360–367, doi:10.2136/sssaj1946.03615995001000C00062x.

Seifert, C.L., Cox, R.T., Foreman, S.L., Foti, T.L., Waskiewicz, T.A., and McColgan, A.T., 2009, Relict nebkhas (pimple mounds) record prolonged late Holocene drought in the forested region of south-central United States: Quaternary Research, v. 71, p. 329–339, doi:10.1016/j.yqres.2009 .01.006.

Silverberg, R., 1970/1986, The Mound Builders: Athens, Ohio University Press, 276 p. [Abridged edition of 1968, Mound Builders of Ancient America: The Archaeology of a Myth, N.Y. Graphic Society.]

Smith, H.M., and Marshall, R.M., 1938, Soil Survey of Bee County, Texas: U.S. Department of Agriculture, Bureau of Chemistry and Soils, in Cooperation with Texas Agricultural Experiment Station, Series 1932, no. 30, 34 p.

Spillman, W.J., 1905, Natural mounds: Science, v. 21, no. 538, p. 632, doi: 10.1126/science.21.538.632-a.

Udden, J.A., 1906, The origin of the small sand mounds in the Gulf Coast country: Science, v. 23, no. 596, p. 849–851, doi:10.1126/science.23.596.849.

Veatch, A.C., 1906, On the human origin of the small mounds of the lower Mississippi Valley and Texas: Science, v. 23, no. 575, p. 34–36, doi:10.1126 /science.23.575.34-a.

Wallace, A.R., 1877a, Glacial drift in California: Nature, v. 15, p. 274–275, doi:10.1038/015274c0.

Wallace, A.R., 1877b, The "Hog Wallows" of California: Nature, v. 15, p. 431–432, doi:10.1038/015431c0.

Watson, E.B., Wank, M.E., and Smith, A., 1925, Soil Survey of the Shasta Valley, Area, California: U.S. Department of Agriculture, Bureau of Soils, Field Operations of the Bureau of Soils for Year 1919, 22nd Report, p. 99–152.

Wentworth, I.H., 1906, A few notes on 'Indian mounds' in Texas: Science, v. 23, no. 595, p. 818–819, doi:10.1126/science.23.595.818-a.

Yochelson, E.L., ed., 1980, The Scientific Ideas of G.K. Gilbert: An Assessment on the Occasion of the Centennial of the United States Geological Survey (1879–1979): Geological Society of America Special Paper 183, 148 p.

MANUSCRIPT ACCEPTED BY THE SOCIETY 5 MARCH 2012

Appendix C. Literature-culled names for mounds and moundfields

APPENDIX C. LITERATURE-CULLED NAMES FOR MOUNDS AND MOUNDFIELDS

Active mounds	Wallace, 1991
Ancient mounds/remains	Webster, 1889a, 1889b, 1889c, 1889d; Taylor, 1843
Ant villages	Forshey, 1854
Baffling bumps	Jackson, 1956, 1973; Ith, 2004; Kirk, 1970
Basalt mounds	Piper, 1905
Beaded stripes	DeGraff, 1975, 1976
Beads	Southard and Williams, 1970
Beaver mounds	Kelly, 1948
Biscuit and swale topography	Daubenmire, 1970
Biscuit mounds	Topinka, 2009
Biscuits	Davis and Youtie, 1976; Winward and Youtie, 1978; Green, 1975, 1982; Copeland, 1980; Anderson et al., 1998; Donaldson and Geise, 1968; Dyksterhuis and High, 1985; Macdonald, 1998; Macdonald et al., 1999; Dyksterhuis et al., 1969; Green et al., 1969; Norgren et al., 1969
Biscuit scabland/topography	Green, 1975, 1982; Nelson, 1977; Johnson, D.R., 1993; Daubenmire, 1970; Davis and Youtie, 1976; Winward and Youtie, 1978; Copeland, 1980; Kienzle, 2007; Macdonald, 1998; Macdonald et al., 1999; Dyksterhuis and High, 1985; Pyrch, 1973; Anderson et al., 1998; Cox, G.W., 1989
Blister-like mounds	Easterbrook and Rahm, 1970
Bumps	Whitney, 1921a; Treasher and Reynolds, 1937; Kirk, 1970; Kohr, 1976; Malde, 1964a
Bunch of bumps	Treasher and Reynolds, 1937
Burial mounds	O'Brien et al., 1989
Cellular soils	Eakin, 1932
Channeled microrelief	Arroues et al., 2007
Characteristic knolls	Bennett et al., 1911
Circular elevations/heaps/lumps/rises	Mead, 1878; Malde, 1964a; Fenneman, 1906; Fryxell, 1964, Appendix A, this volume
Circular knolls/mounds	Lewis, 1883/1891a; Harris, 1902; Johnson, D.R., 1993; Slusher, 1967; Saucier, 1994; Green, 1975, 1982; Fields et al., 1986; Stockman, 1981; Cahoon, 1985; Smies et al., 1925; Crout, 1976; Golden et al., 1994; Aronow, 1976; Capps, 1994; Foster and Moran, 1935; McCune, 1999; Taylor and Cobb, 1921; Larance, 1961; Lounsbury and Deeter, 1919; Deeter et al., 1925; Green et al., 1969; Alt and Hyndman, 2000; Bragg, 2003; Kruckeberg, 1991, p. 290; Crout, 1976; Fryxell, 1964, Appendix A, this volume
Circular light spots	Melton, 1928; Riefner and Pryor, 1996
Circular sand(y) mounds/ridges	Sobecki, 1980; Carter and Patrick, 1919; Foster and Moran, 1935; Featherman, 1872; Fryxell, 1964, Appendix A, this volume
Circular soil structures	Masson, 1949; Malde, 1964a; Johnson and Johnson, this volume
Clay knolls/patches/domes	Wheat, 1953; Masson, 1949
Cleared circles	McDonald et al., 2000
Clustered knolls	Wheat, 1953
Coastal plain prairie mounds	O'Brien et al., 1989
Colossal gopher mounds	Scheffer, 1948
Columbia mounds	Brunnschweiler, 1962
Cone-shaped mounds	Turner, 1982
Conical ant hills	Dumble, 1918
Conical mounds	McGee, 1878; Lewis, 1883/1891a; Anonymous, 1866; Nikiforoff, 1941; Kloehn et al., 2000; Wilkes, 1845
Conspicuous microrelief	Nikiforoff, 1941
Coppice dunes	Gile, 1966; Holliday, 1987; Seifert et al., 2009; Mandel, 1987; Langford, 2000; Hall et al., 2010; Johnson, D.L.,1997
Corn hills/hillocks	Wilkes, 1841, *in* Meany, 1926; Clarke, 1878
Domes	Jenny, 1976
Domes of volcanic ash	Smith, 1910

(Continued)

179

APPENDIX C. LITERATURE-CULLED NAMES FOR MOUNDS AND MOUNDFIELDS (*Continued*)

Dome-shaped knolls/mounds/structures	Smith, 1910; Smies et al., 1925; Carter and Patrick, 1919; Knobel et al., 1926; Tillman et al., 1923; Deeter et al., 1925, 1928; Lounsbury and Rogers, 1924; Smies et al., 1925; Foster and Moran, 1935; Taylor and Cobb, 1921; Deeter and Cohn, 1923; Beck et al.,1923; Tillman et al., 1919; Deeter et al., 1925, 1928; Melton, 1929b; Shaw, E.W., 1957
Domiciliary mounds	Fowke, 1910
Drift mounds	Rogers, 1893; Bryson, 1893
Dwelling mounds	Lewis, 1883/1891a
Earthen formations/haycocks/mounds/pimples	Kloehn et al., 2000; Pearson, 1967; Bragg, 2003; Kruckeberg, 1991, p. 291; Williams, 1990
Earth hummocks	Sasaki, 1979*
Earthlodge ruins	Bushnell, *in* O'Brien, 1996
Earth mounds	Scheffer, 1947; Cox, G.W., 1990a, 1990b, 1990c; Cox, G.W., and Allen, 1987a,1987b; Cox, G.W., et al., 1987; Vitek, 1973, 1978; Gold, 2001; Tester and Breckenridge,1964; Tullis, 1995; Smith, 1949*; Bryan, 1940; Storm, 2004; Aldritt, 1986; Frederking, 1973
Earth polygons	Jackson, 1956
Earthworks	Lewis, 1883/1891a
Elongate(d)-elliptical mounds/compound mounds	Dumble, 1918; Green, 1982; Johnson, C.B., 1982; Crout, 1976; Stout, 1965; Wallace, 1991; Fryxell, 1964, Appendix A, this volume; LeConte, 1874, Appendix B, this volume
Embankments	de Nadaillac, 1885, and *in* Veatch, 1906c
Enigmatic landscapes/mounds	McKee, 1972; Withrow, 2005; Kruckeberg, 1991, p. 290; Newcomb, 1952; Saucier, 1994
Erosional remnants	Melton, 1929b; Vitek, 1973
Erosional soil hillocks	Melton, 1934
False mounds	Whittaker and Storey, 2005
Fish mounds/heaps	Agassiz, *in* Gibbs, 1855, 1873a, 1873b; Crossman, 1921; Kelly, 1948; Kelly and Dachille, 1953
Flat-topped/flattened mounds	N. de Lafora, 1767, *in* Kinnaird, 1967; Fowke, 1922
Floodplain mounds	Heinrich, 1986; Ensor et al., 1990; Mandel, 1987; Jones and Shuman, 1988; White and Wiegand, 1989; Fields et al., 1983, 1986; Howard and Freeman, 1983; Aronow, 1990
Floodplain sand mounds	White and Wiegand, 1989; Wiegand and White, 1989
Flower mounds	Branson, 1966
Fracture trace mounds	Shlemon et al., 1973; Johnson and Horwath Burnham, this volume
Freckled areas/hills/land	Deal, 1971; Vitek, 1973; Bluemle, 1983
Freckles	Bluemle, 1983; Malde, 1964a
Garden beds	de Nadaillac, 1885
Garden mounds	Fowke, 1910, 1922; Conant, 1979, *in* Thoburn, 1937; O'Brien, 1996
Gas blowout mounds	Koons, 1948
Gas mounds	Fenneman, 1906; Kennedy, 1917; Bailey, 1923
Giant ripple marks/mounds	Kelly, 1948; Kelly and Dachille, 1953
Gilgai	Abbott, 1982, 1983a, 1984; Aten and Bollich, 1981; Fields et al., 1983, 1986; Howard and Freeman, 1983; Heinrich, 1986
Gopher mounds/hills/knolls	Bailey, 1923; Hoy, 1865; Kennicott, 1958; Hall, 1838; Schmidt, 1937; Winchell and Upham, 1881
Granular soil mounds	Capps, 1994
Grass-covered "haycocks"	Kruckeberg, 1991, p. 291
Grassy humps	Seven, 2008
Gravel mounds	Murray, 1967
Hammocks/hammock land	Martin, 1902
Haycocks	Wilkes, 1845; Kruckeberg, 1991, p. 291
Heaps of confusion	Geiger, 1998
Hemispherical mounds	Lee and Carter, 2010
Hemis-spheroidal tumuli	Forshey, 1845
Hillocks	Darby, 1816; Barnes, 1879; Jenny, 1980; Krinitzsky, 1949; Abbott, 1982; Jackson, 1956, 1973; Caine and Kocher, 1904; Cleveland, 1893; Hilgard, 1873, 1884b; Stuart, 1837; Hertlein and Grant, 1944; Bjornstad and Kiver, 2012; Niiler, 1998; Kruckeberg, 1991, p. 293
Hog-wallow land/prairies/surface/areas	Riddell, 1839; Hilgard, 1884b; Holmes and Nelson, 1919; Means and Holmes, 1901; Nelson et al., 1919a; Holmes et al., 1919; Holmes and Pendleton, 1919
Hog-wallow mounds/microrelief/relief	Turner, 1896; Strahorn et al., 1914; Means and Holmes, 1901; Whitney, 1921b; Nikiforoff, 1941; Storie et al., 1944; Herbert and Begg, 1969; Nelson et al., 1919a; Holmes et al., 1919; Holmes and Pendleton, 1919; Arroues et al., 2007; Watson et al., 1923; Arkley, 1963; Hobson and Dahlgren, 1998a
Hog wallows/hogwallows/"Red" hog-wallows	Arkley and Brown, 1954; Crossman, 1921; Gabb, 1877; LeConte, 1877; Marbut et al., 1913; Branner, 1905; Surr, 1920; Whitney, 1921a, 1921b; Holland et al., 1952; Strahorn et al., 1912, 1914; Jenny, 1976; Watson and party, 1919; Holmes and Nelson, 1919; Hilgard, 1984b; Prokopovich, 1969; Nikiforoff, 1941; Nelson et al., 1919a, 1923; Lapham et al., 1905; Watson and Smith, 1921; Watson et al., 1923; Shlemon, 1967; Shaw, C.F., 1928; Stout, 1965; Wallace, 1877b; Hobson and Dahlgren, 2001
Hogwallow topography/country	Jenny, 1976; Nelson et al., 1923; Holmes et al., 1919; Kruckeberg, 1991, p. 290; Hobson and Dahlgren, 1998a, 1998b; Scheffer, 1983
House mounds	Lewis, 1883/1891b; O'Brien et al., 1989

(*Continued*)

APPENDIX C. LITERATURE-CULLED NAMES FOR MOUNDS AND MOUNDFIELDS (*Continued*)

Hovey mounds	Benn, 1976
Hummocks	Jenny, 1976; Arman and Thornton, 1972; Ellis and Lee, 1919; Grinnell and Linsdale, 1936; Van Duyne et al., 1919; Watson and party, 1919; Cahoon, 1985; Anonymous, 1921; Gilbert, 1875; Carter et al., 1916; Holland and Jain, 1977; Taylor and Cobb, 1921; Lounsbury and Deeter, 1919; Tillman et al., 1919; Hall et al., 1919; Herbert and Begg, 1969; Alt and Hyndman, 1994; Bennett et al., 1911; Nelson et al., 1919a; Holmes et al., 1919; Seven, 2008; Warrick, 1984; Simpson, 1841, *in* Williams, 1973; McGee, 1891, p. 209; Gilbert, 1875, Appendix B, this volume
Hummocky areas/earth mounds	Rockie, 1942*; Krantz et al., 1988*; Nelson et al., 1923
Hummocky plains/soils/microrelief/surface	Knechtel, 1949; Prokopovich, 1969; Herbert and Begg, 1969; Nelson et al., 1919a, 1923; Holmes et al., 1919; Arroues et al., 2007; Watson et al., 1923; Watson and Cosby, 1924
Humps	Seven, 2008
Hybrid/bio-maintained mounds	Johnson and Johnson, 2008; Riefner and Pryor, 1996
Indian corn hills/hillocks	Lubbock, 1863; Clarke, 1878
Indian garden beds/mounds	Harris and Veatch, 1899; Veatch, 1906a, 1906b; de Naidallac, *in* Veatch,1906c
Indian house mounds	Fowke, 1922
Indian mounds [or 'mounts']	Dunbar, 1804; Webster 1889e, 1897; Lewis, 1883/1891a,1883/1891b; Jeter et al., 1989; Wentworth, 1906; Westman, 1946; Davids, 1967; Gatchet, 1891
Inexplicable mounds	Forshey, *in* Foster, 1873, and *in* Veatch, 1906c; LeConte, 1874, Appendix B, this volume
Innumerable hillocks/small mounds	Orcutt, 1885; Loughridge, 1884b; Strahorn et al., 1912
Irregular hummocks/mounds	Taylor and Cobb, 1921; Hall et al., 1919
Islands	Southard and Williams, 1970
Isolated cherty heaps/mounds	Owen, 1858; Loughridge, 1884a
Kangaroo rat mounds	Ayarbe and Kieft, 2000; Moorhead et al., 1988; Mun and Whitford, 1990; Vorhies, 1922; Schroder and Geluso, 1975
Knoblike elevations	Lapham et al., 1905
Knolls	Bennett et al., 1911; Wheat, 1953; Smith, 1910; Anonymous, 1921
Knoll-sink topography	Jenny, 1976
Large earth mounds	Péwé, 1948*; Best et al., 1988
Large low mounds	Crenwelge et al., 1988; Watson and Cosby, 1924
Large mounds	Tillman et al., 1923; Crout, 1976
Large mounds of dirt	Schroder and Geluso, 1975
Light(er) colored patches/soil/spots	Melton, 1929b; Vogel, 2005
Linear/lineated mounds/patterns	Branson, 1966; Wallace, 1991
Little hills/knolls/mounds	Herrick, 1892; Hilgard, 1884a; Wilder and Shaw, 1908
Locally thickened biomantles	Johnson, D.L., et al., 2003, 2005b; Horwath and Johnson, D.L., 2006, 2007; Finney, this volume; Horwath Burnham et al., this volume; Johnson and Horwath Burnham, this volume; Johnson, D.L., and Johnson, D.N., this volume
Loess biscuits	Busacca et al., 2002
Loess mounds	Quinn, 1961; Topinka, 2009; Busacca et al., 2002; Peterson, 1961
Loessal soil mounds	Winward and Youtie, 1978
Low broad hummocks	Nelson et al., 1923
Low circular mounds/eminences	Hill 1906; Crout, 1976; Lounsbury and Deeter, 1919; Stephenson and Crider, 1916
Low, dome-shaped mounds	Deeter and Cohn, 1923; Deeter et al., 1925
Low, dome-shaped sandy mounds	Taylor and Cobb, 1921; Tillman et al., 1919
Low hummocky microrelief	Herbert and Begg, 1969
Low knolls	Bennett et al., 1911
Lowland mounds	Schmidt, 1935; Upham, *in* Schmidt, 1937; Markley, 1948; Withrow, 2005; Wedding, 1985; Finney, this volume
Low mounds/hummocks	Gilbert, 1875; Watson and Allen, 1912; Veatch, 1906a; Brezina, 2007; Owen, 1858; Loughridge, 1884a; Lounsbury and Deeter, 1919, 1925; Deeter et al., 1928; Alt and Hyndman, 1994; Smith et al., 1945; Nelson et al., 1919a, 1923; Watson and Smith, 1921; Watson et al., 1923; Watson and Cosby, 1924; Shlemon et al., 1973; Shaw, C.F., 1928; Masson, 1949; McGee, 1891, p. 208
Low mounds of earth	Thoburn, 1937
Low round(ed) hummocks/mounds	Nelson et al., 1919a, 1919b, 1923; Martin et al., 1990; Kerr et al., 1926; Lounsbury and Deeter, 1919; Holmes et al., 1919
Low sandy mounds	Lounsbury and Deeter, 1919
Low symmetrical mounds	Holmes et al., 1919
Magnificent microrelief	Arkley, 1948
Mammae/mammillae	Lerch, 1896; Mead, 1878
Mammal mounds	Ayarbe and Kieft, 2000
Manastash mounds	Kaatz, 1959
Marsh gas mounds	Jones and Shuman, 1988
Marsh mounds	Lewis, 1883/1891a; Winchell, 1911
Meadow mounds	Lewis, 1883/1891a; Schmidt, 1937
Measles-like bumps	Bjornstad and Kiver, 2012
Megamounds	Higgins, C.G., *in* Williams,1990
Micro-highs	Sobecki, 1980
Microknolls	Goebel, 1971

(*Continued*)

APPENDIX C. LITERATURE-CULLED NAMES FOR MOUNDS AND MOUNDFIELDS (*Continued*)

Microknoll topography	Goebel, 1971
Micro-landforms	Vitek, 1978
Microrelief	Scheffer, 1958; Vitek, 1978; Nikiforoff, 1937, 1941; Foster and Moran, 1935; Arkley and Brown, 1954; Knechtel, 1952; Storie et al., 1944; Herbert and Begg, 1969; Kruckeberg, 1991, p. 290; Schaetzl and Anderson, 2005, p. 503–508; Arkley, 1963; Johnson, D.L., and Johnson, D.N., this volume
Microterrain	Cox, G.W., 1984a, 1984b
Microtopography	Abbott, 1984; Schaetzl and Anderson, 2005, p. 503–508; Reed and Amundson, this volume
Mima mound microtopography	Cox, G.W., and Gakahu, 1985
Mima (Mima-type, Mima-like) mounds	Arkley, 1948; Arkley and Brown, 1954; Berg, 1989, 1990a,1990b, 1991; Bretz, 1913; Brotherson,1969, 1982; Cox, G.W., 1989, 1990a,1990b, 1990c; Cox, G.W., and Allen, 1987b; Cox, G.W., et al., 1987; Hansen and Morris, 1968; Hombs, 2001; McFaul, 1977; Price, 1949, 1950; Ritchie, 1953; Ross et al., 1968; Dalquest and Scheffer, 1942, 1944; Scheffer, 1947, 1948, 1954, 1956a, 1956b, 1958, 1983; Stallings, 1948; Collins, 1975; Corliss, 1988; Grant, 1948; Vitek, 1973; Hansen, 1962; Jackson, 1956; Kelly, 1948; Péwé, 1948*; Pearson, 1967; Noble and Molenaar, 1965; Kienzle, 2007; Amundson, 1998, 2000a, 2000b; Vogel, 2005; Lee and Carter, 2010; Riefner and Pryor, 1996; Riefner et al., 2007; Shlemon, 1967; Ricks et al., 1997; Johnson, D.L., et al., 2005b, 2008b; Stout, 1965; Arkley, 1963; Kruckeberg, 1991, p. 290; Washburn, 1988; Butler, 1979; Giles, 1970; Hill, 2001; Horwath, 2002a, 2002b; Huss, 1994; Klein, 1998; Miller, 1994; Paeth, 1967; Spackman, 1982; Tapler, 1996; Wallace, 1991; Washburn, 1980; Schaetzl and Anderson, 2005; Reed, n.d.; Reed et al., 2008; Johnson, D.L., and Horwath Burnham, this volume; Reed and Amundson, 2007; Irvine and Dale, this volume; Cox, G.W., this volume; Horwath Burnham et al., this volume; Finney, this volume; Johnson, D.L., and Johnson, D.N., this volume
Minor undulations	Nelson et al., 1923
Monuments	Wilkes, 1845
Mound and depression	Abbott, 1984
Mound-and-trough-configuration	Jenny, 1976
"Mound builder" mounds	Webster, 1889e
Mound-depression microrelief/terrain/topography	Riefner and Pryor, 1996; Holdredge and Wood, 1947
Mound-dotted landscapes	Rich, 1934
Mounded soils/landscapes	Allgood, 1972; Allgood and Gray, 1973, 1974; Lee and Carter, 2010; Riefner and Pryor, 1996
Mound formations	Barnes, 1879
Mound-intermound microrelief	Arroues et al., 2007
Mound microrelief	Johnson, D.L., 1988; Nikiforoff, 1941; Arkley, 1963
Mound microsites	McGinnies et al., 1976
Mound micro-topography/topography	Fenneman, 1931; Pyrch, 1973; Shlemon et al., 1973; Riefner and Pryor, 1996; Stout, 1965; Wallace, 1991
Mounds	Lewis, 1883/1891a, 1883/1891b; Brunnschweiler, 1964; Gangmark and Sanford, 1963; Owen, 1844a, 1844b; Myrum and Ferry, 1999; Rich, 1934; Green, 1975; Kienzle, 2007; WDFW, 1995; Hannemann, 1928; Grinnell and Linsdale, 1936; Veatch, 1900; Taylor, 1843; Melton, 1928, 1929a, 1929b, 1934, 1935, 1954; Hobbs, 1907; Vitek, 1973; Fenneman, 1931, 1938; Hilgard, 1906; Loughridge, 1884b; McDonald, 1965; Macdonald, 1998; Macdonald et al., 1999; Johnson, D.R.,1993; Easterbrook and Rahm, 1970; Wilder and Shaw, 1908; Larance, 1961; Burgess and Ely, 1909; Tillman et al., 1923; Beck et al., 1923; Stephenson and Crider, 1916; Sweet and McBeth, 1911; Herbert and Begg, 1969; Carter, 1999; Holmes and Pendleton, 1919; Watson et al., 1923; Bragg, 2003; Shlemon et al., 1997; Riefner and Pryor, 1996; DeGraff, 1975, 1976; Stout, 1965; Wilkes, 1845; Simpson, 1841, *in* Williams, 1973; Giles, 1970; Cross, 1964; Wallace, 1991; Fryxell, 1959, 1964, Appendix A, this volume
Mounts of earth	Darby, 1816
Mud lumps/volcanoes	Thomassy, 1860; Hopkins, 1870; Clendenin, 1896; Reagan, 1907; Hobbs, 1907
Muskrat mounds	Webster, 1897; McGee, 1891, p. 209
Mysterious circles/hummocks/mounds	Noble, 1951, 1985; Doughton, 2009; McDonald, 1965; Shaw, 1957; Pearson, 1967; Featherman, 1872; Cox, G.W.,1984a; Jackson, 1973; Kohr, 1976; Noble, 1951, 1985; Reddy, 1971; Scheffer, 1969, 1981; Bjornstad and Kiver, 2012; El Hult, 1955; Shaw, E.W., 1957; Seven, 2008; Williams, 1990
Natural mounds	Bernard, 1950; Branner, 1905; Campbell, 1906; Holland et al., 1952; Melton, 1929a, 1929b, 1934, 1954; Spillman, 1905; Thoburn, 1937; Veatch, 1905, 1906a, 1906b, 1906c; Howard and Freeman, 1983; Branson, 1966; Purdue, 1905; Treasher and Reynolds, 1937; Shaw, C.F., 1928, 1937; Bragg, 2003; Mead, 1878; Knechtel, 1949; Webb, 1957; Vitek, 1973; Finney, this volume
Natural sandy rises/clay knolls/knolls	Wheat, 1953
Nebkhas	Seifert and Cox, R.T., 2007; Seifert et al., 2009
Numerous hummocks/mounds/slight elevations	Watson and party, 1919; Nelson et al., 1919a; Holmes et al., 1919; Watson and Cosby, 1924; Burgess and Ely, 1909
Occasional low hummocks/mounds	Deeter et al., 1928; Bragg, 2003

(*Continued*)

APPENDIX C. LITERATURE-CULLED NAMES FOR MOUNDS AND MOUNDFIELDS (*Continued*)

Ordinary mounds	Lewis, 1883/1891a
Ordinary tumuli	Brower, 1893
Paleoseismic sand blows	Cox, R.T., et al., 2010
Patterned areas/ground/land/morphology	Almaraz, 1986; Johnson, C.B., 1982; Johnson, D.R., 1993; Kaatz, 1959; Fosberg, 1963, 1965; Malde, 1961, 1964a, 1964b; Vitek, 1973, 1978; Nelson, 1977; White and Agnew, 1968; Green, 1975; Kienzle, 2007; Macdonald, 1998; Macdonald et al., 1999; Collett, 1980; Gentry, 1991; Stockman, 1981; Dyksterhuis and High, 1985; Brunnschweiler, 1962; Green et al., 1969; Tugel, 1993; Southard and Williams, 1970; DeGraff, 1975, 1976; Wallace, 1991; Fryxell, 1959, 1964, Appendix A, this volume
Pebble mounds	Branson et al., 1965
Peculiar earth heaps/structures/tracts	Webster, 1897; Blankinship, 1889; Burgess et al., 1908
Peculiar microrelief/mounds/topography	Bennett et al., 1911; Burgess et al., 1908; Means and Holmes, 1901; Rice and Griswold, 1904; McAdams, 1881; Nikiforoff, 1937
Periglacial patterned ground	DeGraff, 1975, 1976; Parsons and Herriman, 1976; Fryxell, 1964, Appendix A, this volume
Pimpled ground/hummocks/mounds	Thornbury, 1965, p. 67; Jenny, 1976, 1980; Bik, 1967*; Price, 1949; Krinitzsky, 1949
Pimpled plains/topography	Fenneman, 1906, 1931, 1938; Knechtel, 1952; Koons, 1948, 1926; Nelson, 1977; Malde, 1964a, 1964b; Vitek, 1973; Pyrch, 1973; Lee and Carter, 2010; Scheffer, 1960, 1983
Pimple dunes	Cross, 1964; McMillan, 2007; McMillan and Day, 2010
Pimple mound microrelief	Irvine and Dale, this volume
Pimple mounds/plains/prairies	Aronow, 1963, 1976, 1978, 1990; Bernard, 1950; Krinitzsky, 1949, 1950; O'Brien et al., 1989; Heinrich, 1986; Slusher, 1967; Murray, 1948; Hansen and Morris, 1968; Holland et al., 1952; Fields et al., 1986; Price, 1949; Aten, 1979; Aten and Bollich, 1981; Kaczorowski and Aronow, 1978; Branson, 1966; Welch, 1942; Voellinger et al., 1987; Cain, 1974; McGuff and Cox, W.N., 1973; Arman and Thornton, 1972; Lee and Carter, 2010; Webb, 1957; Whitesides, 1957; Carty, 1980; Abbott, 1984; Dietz, 1945; Zedler, 1987; Veatch, 1906a; Schaetzl and Anderson, 2005, p. 258; Sobecki, 1980; Scheffer, 1960; Saucier, 1991; Irvine, 2005; Irvine et al., 2002; Irvine and Dale, 2007, 2008, this volume
Pimples	Fisk, 1938, 1940, 1948; Krinitzsky, 1949; Hannemann, 1928; Koons, 1948; Pearson, 1967; LeConte, 1874, Appendix B, this volume
Point-centered biomantles	Johnson, D.L., et al., 2003, 2005a, 2005b; Horwath and Johnson, D.L., 2007; Finney, this volume; Horwath Burnham et al., this volume; Johnson and Horwath Burnham, this volume; Johnson, D.L., and Johnson, D.N., this volume
Polygonboden	Knechtel, 1952
Prairie blisters/bumps	Local name, Siloam Springs area, Arkansas; Vogel, 2005; Morse and Morse, 1983; Benn, 1976; O'Brien et al., 1989; Saucier, 1978, 1994
Prairie mounds	Bernard, 1950; Bik, 1967*, 1968*, 1969*; Gravenor, 1955*; Gilbert, 1875; Hilgard, 1905; Holland et al., 1952; O'Brien et al., 1989; Quinn, 1961, 1968; LeConte,1877; Aten and Bollich, 1981; Hertlein and Grant, 1944; Aronow, 1963; Branson, 1966; Brotherson, 1982; Guccione et al., 1991; Kinahan, 1877a, 1877b; Page et al., 1977; Seifert, 2007; Vogel, 2004, 2005; Williams, 1877; Vitek, 1973; McCune, 1999; Lee and Carter, 2010; Fuller, 1912: Seifert and Cox, 2007; Arnold, 1960; Schaetzl and Anderson, 2005, p. 258–259; Johnson, D.L., and Johnson, D.N., this volume; Finney, this volume
Prairie pimples/pulpits	Irvine and Dale, 2007; Scheffer, 1984; Vogel, 2005; Lee and Carter, 2010; Washburn, 1988; local name, Siloam Springs area, Arkansas
Pre-historic mounds	Lewis, 1883/1891a
Pronounced/circular mounds/hog-wallow surfaces	Nelson et al., 1919a; Bragg, 2003
Radially aligned mounds	Branson, 1966
Raised garden plots	O'Brien et al., 1989
Residual soil hillocks	Bernard, 1950; Holland et al., 1952; Melton, 1929a, 1929b, 1954; Vitek, 1973
Reticulated soils	Eakin, 1932
Root mounds	Melton, 1929b
Rounded hillocks/hummocks/knolls	Heinrich, 1986; Hilgard, 1873; Nelson, 1919a, 1919b, 1923; Holmes et al., 1919
Rounded mounds/low mounds/soil mounds	Bennett et al., 1911; Vodrazka et al., 1971; Nelson et al., 1919b; Lounsbury and Deeter, 1919; Rich, 1934
Rounded mounds of earth	Gangmark and Sanford, 1963
Round elevations	Kane, 1847, *in* Kruckeberg, 1991, p. 295
Ruined habitations	Veatch, 1906c; LeConte, 1877, Appendix B, this volume
Sand blows/spots	Cox, R.T., et al., 2004, 2007, 2010; Hall et al., 1919
Sand(y) hillocks/hummocks	Caine and Kocher, 1904; Burgess et al., 1908; Carter et al., 1916; Deeter and Davis, 1921; Hall et al., 1919
Sandy islands/knolls	Veatch, 1906b; local name, Columbia County, Texas; Wheat, 1953; Carter et al., 1916
Sandy loam mounds	McGuff and Cox, W.N., 1973
Sand(y) mounds	Koons, 1926, 1948; Van Duyne and Byers, 1915; O'Brien et al., 1989; White and Wiegand, 1989; Aten and Bollich, 1981; Smies et al., 1925; Smith and Marshall, 1938; Crout, 1976; Carter et al., 1916; Foster and Moran, 1935; Featherman, 1872; Taylor and Cobb, 1921; Deeter and Davis, 1921; Tillman et al., 1919; Hall et al., 1919; Rice and Griswold, 1904

APPENDIX C. LITERATURE-CULLED NAMES FOR MOUNDS AND MOUNDFIELDS (*Continued*)

Scabland/scabland mounds	Freeman, 1926; Tallyn, 1980; Watson et al., 1923
Scattered dome-shaped mounds	Tillman et al., 1919
Scattered mounds	Tillman et al., 1923; Watson and party, 1919; Wallace, 1991
Schmidt mounds	Finney, this volume
Shrub mounds	McDonald et al., 2000
Silt mounds	Brunnschweiler, 1962; Olmsted, 1963; Washburn, 1980, p. 169
Singular mounds/rounded hillocks	Webster, 1889e; Hilgard, 1873, 1884b
Siskiyou structures	Masson, 1949
Sleeping circles	McDonald et al., 2000
Slight elevations	Beck et al., 1923; Deeter and Davis, 1921; Nelson et al., 1919a
Slight hog-wallow mounds	Strahorn et al., 1912
Slightly hummocky/undulating surfaces	Nelson et al., 1923; Holmes et al., 1919
Small circular/oval/elliptical mounds	Fisk, 1938, 1940; Fenneman, 1931; Taylor and Cobb, 1921; Fryxell, 1959
Small earth hummocks/mounds	Breckenridge and Tester, 1961; Fisk, 1948; Knechtel, 1949
Small hillocks/hills/hummocks/mounds	Bushnell, 1905; Van Duyne and Byers, 1915; Farnsworth, 1906; Dumble, 1918; Veatch, 1906c; Allgood and Gray, 1973, 1974; Hill, 1906; Watson and party, 1919; Cahoon, 1985; Stuart, 1837; Caine and Kocher, 1904; Corliss, 1988; Waters and Flagler, 1929; Dietz, 1945; Knechtel, 1952; Loughridge, 1884b; Pyrch, 1973; Deeter and Davis, 1921; Beck et al., 1923; Knobel et al., 1926; Hall et al., 1919; Stephenson and Crider, 1916; Dyksterhuis et al., 1969; Green et al., 1969; Norgren et al., 1969; Storie et al., 1944; Carter, 1999; Anderson et al., 1998; Amundson, 1998, 2000a; Lapham et al., 1905; Holmes and Nelson, 1919; Nelson et al., 1919a; Holmes et al., 1919; Melton, 1929b; Arkley, 1963; Branson et al., 1965
Small isolated/elevations/hummocks	Bushnell, 1905; Knechtel, 1952; Arman and Thornton, 1972; Watson and party, 1919; Lockett, S.H., *in* Veatch, 1906c
Small hog-wallow hummocks	Watson and party, 1919
Small mound-like elevations/formations	Loughridge, 1884a; Orcutt, 1885
Small natural mounds	Olmsted, 1963; Loughridge, 1884b
Small round(ed) hillocks/hummocks	Foster and Moran, 1935; Means and Holmes, 1901
Small sand mounds	Udden, 1906; Van Duyne and Byers, 1915
Small scattered mounds	Cahoon, 1985; Sweet and McBeth, 1911
Small, slightly domed, oval-shaped mounds	Allgood and Gray, 1974
Soil blisters/hillocks/mottlings	Nikiforoff, 1928*; Melton, 1934; Davis and Youtie, 1976; Rich, 1934; Thornbury, 1965, p. 67
Soil mounds	Cox, G.W., 1990; Melton, 1929, 1935; Nelson, 1977; Johnson, C.B., 1982; Davis and Youtie, 1976; Branson, 1966; Cahoon, 1985; Vitek, 1973; Rich, 1934; Goodarzi, 1978; Johnson, D.L., and Johnson, D.N., this volume; Irvine and Dale, this volume
Solifluction lobes	Williams, 1958
Spots/white spots	Bluemle, 1983
Spring mounds	White, 1870; Melton, 1929b
Stony mounds	Quinn, 1961
Strings of pearls/beads/islands	Busacca et al., 2002; Southard and Williams, 1970
Sun cups	Logan and Walsh, 2009; Doughton, 2009
Super mounds	Busacca et al., 2002
Swamp mounds	Lewis, 1883/1891b; Finney, this volume
Symmetrical mounds	Lewis, 1883/1891a
Tenino mounds	Horner, 1930; Newcomb, 1940; Jackson, 1973; Melton, 1929b
"Those mounds"	Hilgard, 1873
True pimple mounds	Ensor, 1987
Tumors	Roberts, R.C., *in* Retzer, 1945
Tumuli	Wilkes, 1845; McGee, 1878; Lewis, 1883/1891a; Robertson, 1867; Scheffer, 1948; Conant, A.J., *in* Thoburn, 1937; Lockett, 1870; Bretz, 1911; Lockett, S.H., *in* Veatch, 1906c
Tumuli colossi	Scheffer, 1948
Undulating areas/hillocks/microrelief/soils/topography	Niiler, 1998; Nelson et al., 1923; Holmes et al., 1919; Shlemon, 1967; Crout, 1976
Unique Indian mounds	Westman, 1946
Unique mounds/small "verma" mounds	Allgood and Gray, 1973
Unusual landscapes	Kruckeberg, 1991, p. 290
Usual hog-wallow mounds	Strahorn et al., 1912
Vernal pool landscapes/topography	Abbott, 1982, 1983a, 1983b, 1984, 1986; Holland and Jain, 1977; Holland, 1978; Riefner and Pryor, 1996
Very hummocky surfaces	Nelson et al., 1919a; Holmes et al., 1919
Village mounds	Beyer, 1898
Wallows (of buffaloes, elephants)	Kelly, 1948
Well-developed hummocks	Watson and party, 1919; Nelson et al., 1919a
Wigwam sites	Veatch, 1906c

*Sasaki (1979), Rockie (1942), Nikiforoff (1928), Péwé (1948), Smith (1949), Bik (1967, 1968, 1969), Gravenor (1955), and Krantz et al. (1988) describe cold region "Mima mounds," "hummocks," "hummocky earth mounds," "pitting," "palsen," "tuer," "giant doughnuts," and/or "soil blisters," which appear to be permafrost ice-melt and/or ground ice features, or versions thereof. While these lie well outside the domain of what are usually considered to be Mima (prairie, pimple) mounds, as do the bog ice-lens micromounds described by Kuntz (1974), they are included here because they have been cited in various Mima mound–focused articles.

Appendix D. Timeline of authors and theories offered to explain Mima mounds

APPENDIX D. TIMELINE OF AUTHORS AND THEORIES OFFERED TO EXPLAIN MIMA MOUNDS

Years	Authors	Hypotheses
1804	Dunbar, W.	Produced by Indians
1816	Darby, W.	Animal burrowing, 'a kind of mole'
1837	Stuart, J.	Animal burrowing, pocket gophers
1840	Douglas, Sir D. (in Jackson, 1956)	Volcanic eruptions
1845	Wilkes, C.	Marks of "Savage" labor
1854	Gibbs, G.	Vegetation anchoring-coppicing
1855	Gibbs, G.	Flowing water
1855	Agassiz, L. (in Gibbs, 1855)	Sucker fish nests
1860	Cooper, J.G.	River eddies and whirlpools
1860	Owen, D.D.	Differential weathering, erosion
1860	Thomassy, R.	Aqueous volcanoes, mud lumps
1865	Hoy, P.R.	Burrowing rodents, gophers
1865	Whitney, J.D.	Unclear origins, perhaps treethrow
1867	Robertson, J.B.	Upwelling and erupting gas
1870	Hopkins, F.V.	Water, gas, oil or mud spring deposits
1872	Featherman, A	Whirlwinds, mountain-like waves
1873a, 1873b	Gibbs, G.	Giant root, fish nests (Agassiz and others)
1873	Hilgard, E.W.	Animal burrowing, ant-hills
1874, 1875, 1877	LeConte, J.	Erosion and vegetation anchoring
1875	Gilbert, G.K.	Animal burrowing, rodents
1877	Williams, W.M.	Morainal deposits
1877a, 1877b	Kinahan, G.H.	Hummocky glacial drift, soil shrinkage
1877a, 1877b	Wallace, A.R.	Glacio-fluvial deposition
1878	Mead, S.H.	Glacio-fluvial deposition
1879	Barnes, G.W.	Wind coppicing, water erosion
1884a, 1884b	Hilgard, E.W.	Runoff erosion
1884a, 1884b	Loughridge, R.H.	Soil shrink-swell (gilgai)
1889a, 1889b, 1889c, 1889d, 1889e	Webster, C.L.	Gophers, badgers, coyotes . . .
1892	Herrick, C.L.	Burrowing animals, pocket gophers
1893	Rogers, G.O.	Drift-filled ice-melt depressions
1893	Bryson, J.	Esker-like fluvial deposition
1893	Cleveland, D.	Coppicing, then erosional leveling
1895	de Nadaillac, M.	Indian home sites, or garden beds
1896	Turner, H.W.	Uncertain cause, possibly rodents
1896	Clendenin, W.W.	Earthquake-erupted gas forms mud lumps
1897	Webster, C.L.	Burrowing by pocket gophers
1900	Veatch, A.C.	Gas-sediment eruptions
1901	Means, T.H., and Holmes, J.G.	Animal burrowing, ground squirrels
1904	Upham, W.	Morainal topography of peculiar type
1905	Shepard, E.M.	Seismic-induced artesian eruptions
1905	Veatch, A.C.	Possibly dunes and anthills
1905	Branner, J.C.	Ants, concretionary nucleation
1905	Hilgard, E.W.	Ants, eolian coppicing, water erosion
1905	Spillman, W.J.	Disintegrating flint concretions
1905	Purdue, A.H.	Chemical segregation by groundwater
1905	Piper, C.V.	Water erosion and basalt weathering
1905	Bushnell, D.I., Jr.	Polygenesis, and Indian house sites
1906c	Veatch, A.C.	Possibly dunes, anthills, but not anthropic
1906	Farnsworth, P.J.	Boles of uprooted trees
1906	Hill, R.T.	Polygenesis, differential soil settling
1906	Wentworth, I.H.	Stony Indian mounds

(Continued)

185

Appendix D

Years	Authors	Hypotheses
1906	Udden, J.A.	Uncertain origins, anthills plausible
1906	Hilgard, E.W.	Burrowing and mounding by ants
1906	Campbell, M.R.	Burrowing animals
1906	Fenneman, N.M.	Challenges oil-gas related origins
1908	Reagan, A.B.	Eruptions of semi-fluid clay, mudlumps
1907	Hobbs, W.H.	Eruptions of mud
1913	Bretz, J.H.	Debris-filled holes and pits in melting glacial ice
1914	Thoroddsen, T.*	Frost action periglacial processes
1917	Kennedy, W.	Gas and water eruptions
1919	Ellis, A.J., and Lee, C.H.	Wind erosion of shrub-anchored soil
1920	Surr, G.	Wind coppicing
1921	Crossman, L.C.	Sturgeons wallowing in mud
1921	Anonymous	Boles of uprooted oaks
1921	Boncquet, P.A., *in* Whitney, 1921a	Water coppicing, then erosion
1923	Bailey, T.L.	Pocket gopher burrowing
1924	Hawkes, L.*	Frost action processes
1926, 1932	Freeman, O.W.	Wind deposits in basalt depressions
1926	Koons, F.C. (Ph.D.)	Animal burrowing, pocket gophers
1928	Nikiforoff, C.C.*	Freeze-thaw artesian pressure eruptions
1928, 1937	Shaw, C.F.	Coppicing, wind deposition and erosion
1929a, 1929b, 1934, 1935, 1954	Melton, F.A.	Water erosion about vegetation clumps
1929	Waters, A.C., and Flagler, C.W.	Water erosion of ash-rich soils
1932	Eakin, H.M.	Periglacial frost action
1933	von Lozinski, W.	Periglacial frost action
1934	Rich, J.L.	Accretions around vegetation clumps
1936	Grinnell, J., and Linsdale, J.M.	Gophers may have modified mounds
1937	Shaw, C.F.	Coppice accumulations of soil
1938	Fisk, H.N.	Uncertain origins
1938	Porsild, A.E.*	Freeze-thaw hydraulic pressures
1940	Newcomb, R.C.	Fissure polygon-ice wedge melts
1941	Nikiforoff, C.C.	Polygenesis, artesian eruptions dominant
1942, 1944	Dalquest, W.W., and Scheffer, V.B.	Animal burrowing, pocket gophers
1942	Larrison, E.J.	Animal burrowing, pocket gophers
1942	Rockie, W.A.*	Permafrost processes
1942	Welch, R.N.	Fluvial erosion
1943	Holland, W.C.	Fluvial erosion
1944	Hertlein, L.G., and Grant, U.S., IV	Wind coppicing and erosion
1945	Dietz, R.S.	Water deposition
1945	Retzer, J.L.	Oozing mud by hydrostatic pressures
1946	Westman, B.J.	Indian mounds
1947	Holdredge, C.P., and Wood, H.B.	Water erosion
1947, 1948, 1958, 1969, 1984	Scheffer, V.B.	Animal burrowing, pocket gophers
1948	Arkley, R.J.	Animal burrowing, pocket gophers
1948	Grant, C.	Rejection of animal burrowing ideas
1948	Kelly, A.O.	Ripples of Noachian floods
1948	Koons, F.C.	Pocket gopher burrowing
1948	Péwé, T.*	Periglacial forms of ice wedges
1948	Stallings, J.H.	Nest-centered gopher burrowing
1948	Whitney, D.J.	Origins may always remain a mystery
1949, 1950	Krinitzsky, E.L.	Polygenetic dunes and hillocks
1949	Masson, P.H.	Periglacial freeze-thaw processes
1949, 1950	Price, W.A.	Pocket gopher burrowing
1949	Thorp, J.	Pocket gopher burrowing
1949	Smith, H.T.U.*	Periglacial origin plausible for some
1950	Bernard, H.A.	Slope erosion
1951	Horberg, L.	Tectonic adjustments
1952	Henderson, J.G. (Ph.D.).*	Glacial-periglacial processes
1952	Holland, W.C., et al.	Vegetation and erosion
1952	Knechtel, M.M.	Fissure polygon networks
1952	Newcomb, R.C.	Polygonal ice expansion and melting
1953	Ritchie, A.M.	Erosion of polygonal-fissure ice
1954	Arkley, R.J., and Brown, H.C.	Animal burrowing, pocket gophers
1955	Gravenor, C.P.*	Periglacial phenomena*
1956, 1973	Jackson, H.E.	Erosion of partly thawed polygons
1957	Hubbs, C.L.	Permafrost, ice-melt frost polygons
1959	Kaatz, M.R.	Periglacial ice wedge polygons
1960	Arnold, J.J. (M.A.)	Paleoclimatic desert coppicing
1960	McGinnies, W.J.	Animal burrowing, pocket gophers
1961	Breckenridge, W.J., and Tester, J.R.	Polygenesis, bioturbation, pocket gophers

APPENDIX D. TIMELINE OF AUTHORS AND THEORIES OFFERED TO EXPLAIN MIMA MOUNDS (*Continued*)

Years	Authors	Hypotheses
1961	Quinn, J.H.	Paleoclimatic desert coppicing
1961	Peterson, F.F. (Ph.D.)	Animal burrowing, pocket gophers
1961, 1964a, 1964b	Malde, H.E.	Freeze-thaw, uncertain processes
1962	Hansen, R.M.	Animal burrowing, pocket gophers
1962	Brunnschweiler, D.	Periglacial frost action
1963	Fosberg, M.A. (Ph.D.)	Periglacial frost action
1963	Gangmark, H.A., and Sanford, F.B.	Plant anchoring, fluvial deposition
1963	Olmsted, R.K.	Eolian coppicing, differential erosion
1963	Frye, J.C., and Leonard, A.B.	Eolian scour and coppicing under aridity
1964	Fryxell, R.H.	Periglacial processes
1965	Bernard, H.A., and LeBlanc, R.J.	On Gulf Coast, most due to erosion
1965	Stout, M.L.,1965	Shallow water marine currents
1965	Fosberg, M.A.	Periglacial frost action
1966	Scheffer, V.B., and Kruckeberg, A.	Animal burrowing, pocket gophers
1967	Murray, D.F.	Animal burrowing, pocket gophers
1967	Paeth, R.C. (M.S.)	Iceberg-rafted depositions
1967	Slusher, D.F.	Eolian coppicing under aridity
1967, 1968, 1969	Bik, M.J.J.*	Periglacial, pingo-like origin*
1968	Denny, C.G., and Goodlett, J.C.	Boles of uprooted trees
1968	Ross, B.A., et al.	Animal burrowing, gophers and others
1968	Hansen, R.M., and Morris, M.S.	Animal burrowing, pocket gophers
1969	Hallsworth, E.G., and Beckman, G.G.	Gilgai (vertisol) shrink-swell
1969	Brotherson, J.D. (Ph.D.)	Gophers produce and maintain mounds
1969	Prokopovich, N.P.	Possible permafrost processes
1969, 1970	Southard, A.R., and Williams, J.S.	Unstable soils, landscapes
1970	Easterbrook, D.J., and Rahm, D.A.	Periglacial-solifluction processes
1971	Goebel, J.E. (M.S.)	Random patterned erosional origin
1972	Allgood, F.P. (M.S.)	Organisms via wet soil avoidance
1972	Collins, O.B. (M.S.)	Erosional origin supported
1973	Vitek, J.D. (Ph.D.)	Periglacial frost action
1973	Frederking, R.L. (Ph. D.)	Periglacial frost action
1973	McGuff, P.R.	Alluvial origin shaped by winds
1973	Shlemon, R., et al.	Animal burrowing, pocket gophers
1974	Saucier, R.T.	Biological origin
1974	Allgood, F.P., and Gray, F.	Biological activity, water erosion
1974	Cain, R.H.	Tree-root anchoring and erosion
1974	Harksen, J.C., and Christensen, C.M.	Permafrost conditions
1974	Henderson, J.A. (Ph.D.)	Animal burrowing, pocket gophers
1975, 1976	DeGraff, J.V.	Relict frost action, heaving
1975	Borst, G.	Gilgai (Vertisol) shrink-swell
1975	Collins, B.	Eolian deflation and coppicing
1976	Davis, M., and Youtie, B.A.	Freeze-thaw probably dominant
1976	del Moral, R., and Deardorff, D.C.	Large, mound forming gophers
1976	Parsons, R.B., and Herriman, R.C.	Periglacial patterned ground
1976	Aronow, S.	Landscape deterioration
1976	Aronow, S.	Residual patches left by erosion
1977	Mielke, H.W.	Animal burrowing, pocket gophers
1977	Nelson, C.A. (M.S.)	Freeze-thaw and polygenesis
1977	Wilson, M.D., and Slupetzky, H.	Multigelation, with wash and creep
1977	Page, W.D. et al.	Animal burrowing, pocket gophers
1978	Bork, J.L. (M.S.)	Origin unstated, much burrowing evident
1978	Goodarzi, N.K. (M.S.)	Surface erosion
1978	Vitek, J.D.	Undetermined process
1978	Wilson, M.D.	Water erosion
1979	McFaul, M.	Ice wedge polygons in Histosols
1979	Sasaki, S.*	Glacial and periglacial frost action
1979	Herriman, R.C., and Parsons, R.B.	Periglacial patterned ground
1979	Zedler, P.H., and Ebert, T.A.	Differential settling and erosion
1980	Greenwood, N.H., and Abbott, P.C.	Expandable clays and wind erosion
1980	Archeuleta, T.E. (M.S.)	Stream deposition by eddy vortices
1980	Carty, D.J. (M.S)	Water-wind reworking, coppicing
1980	Tallyn, L.A.K. (M.S.)	Origin unclear, probably periglaciation
1980	Washburn, A.L.	Water erosion of dessication fissures
1980	Sobecki, T.M. (M.S.)	Uncertain origins, possibly polygenesis
1981	Aten, L.E., and Bollich, C.N.	Aggradation and intermound erosion
1982, 1983a, 1983b, 1984, 1986	Abbott, P.L.	Shrink-swell, erosion, coppicing events
1982, 1983	Sobecki, T.M., and Wilding, L.P.	Uncertain origins, possibly polygenesis
1982	Brotherson, J.D.	Primarily pocket gopher burrowing
1982	Johnson, C.B. (M.S.)	Polygenesis, frost action dominant

(*Continued*)

188 *Appendix D*

APPENDIX D. TIMELINE OF AUTHORS AND THEORIES OFFERED TO EXPLAIN MIMA MOUNDS (*Continued*)

Years	Authors	Hypotheses
1982	Spackman, L.K. (M.S.)	Cryostatic eruptions into sand wedges
1983	Howard, M.A., and Freeman, M.D.	Fluvial aggradation features
1983, 1984, 1985, 1986, 1987	Cox, G.W., and Gakahu, C.G.	Burrowing rodents
1984a, 1984b, 1990a, 1990b	Cox, G.W.	Animal burrowing, pocket gophers
1984	Gakahu, C.G., and Cox, G.W.	Animal burrowing rodents
1984	Spackman, L.K., and Munn, L.C.	Cryostatic eruptions into sand wedges
1986	Aldritt, J.E.	Polygenesis—rodents, duripan, plants
1986	Heinrich, P.V.	Flood scour and or eolian coppicing
1986	Cox, G.W., and Roig, V.G.	Animal burrowing, rodents
1986	Almaraz, R.A. (M.S.)	Freeze-thaw periglacial processes
1987	Holliday, V.T.	Slow eolian floodplain aggradation
1987a, 1987b	Cox, G.W., and Allen, D.W.	Animal burrowing, pocket gophers
1987	Cox, G.W., et al.	Animal burrowing, pocket gophers
1987	Zedler, P.H.	Favors Dalquest-Scheffer model
1988	Aronow, S.	Multiple explanations reviewed
1988	Johnson, D.L.	Probably pocket gopher burrowing
1988	Krantz, W.B., et al.*	Frost action processes
1988	Washburn, A.L.	Multiple explanations plausible
1989, 1990a, 1990b, 1991	Berg, A.W.	Seismic shaking over rigid substrate
1989	White, K.L., and Wiegand, K.C.	Remnants of fluvial linguloid bars
1990	Cox, G.W., and Hunt, J.	Animal burrowing, pocket gophers
1991	Kruckeberg, A.R.	Animal burrowing, rodents
1991	Cox, G.W., and Scheffer, V.B.	Animal burrowing, pocket gophers
1991	Othberg, K.L. (Ph.D.)	Possibly periglacial processes
1991	Wallace, R.E.	Ground squirrels, wind, other processes
1991	Guccione, M.J., et al.	Selective erosion of floodplain soils
1992	Reider, R.G.	Ground-water vortices
1994	Hallet, B., and Sletten, R.S.	Rodents and seismic shaking
1994	Sletten, R.S., et al.	Rodents and seismic shaking
1994	Saucier, R.T.	Animal burrowing, ants or termites
1994	Miller, T.W. (M.A.)	Hydrostatic (ground-water) processes
1994	Huss, J.M. (M.A.)	Ground-water vortices
1994	Cox, R.T., and Shingleur, R.A.	Gullying in prairies during drought
1995	Tullis, J.A. (M.S.)	Polygenesis during cold climates
1996	Issacson, J., and Johnson, D.L.	Polygenetic and complex processes
1996	Riefner, R.E., Jr., and Pryor, D.R.	Paleoliquefaction, and biogenic maintenance
1996, 1999	Reider, R.G., et al.	Ground-water vortices
1996	Tapler, J.F., Jr. (M.A.)	Ground-water vortices
1997	Ricks, D.K., et al.	Pocket gopher burrowing
1997	Shlemon, R.J., et al.	Complex processes, and bioturbation
1998	Anderson, E.W., et al.	Dust-filled pits in "sympathetic" glaciers
1998	Klein, J.L. (M.S.)	Probably polygenetic processes
1998, 2000a, 2000b	Amundson, R.	Pocket gopher burrowing
1999	Johnson, D.L., et al.	Pocket gopher burrowing, bioturbation
1999	Willson, C.J.Z. (M.S.)	Animal burrowing, bioturbation
2000	McDonald, E.V., et al.	Animal and plant bioturbation
2001	Hill, J.P. (M.S.)	Animal burrowing, bioturbation
2002b	Horwath, J.L (M.S.)	Animal burrowing, polygenesis
2002	Horwath, J.L., et al.	Animal burrowing, polygenesis
2002	Johnson, D.L., et al.	Mounds as two-layered biomantles
2003	Johnson, D.L., et al.	Mounds as point-centered biomantles
2004, 2007	Cox, R.T., et al.	Liquefaction-seismic sand blows
2005	Vogel, G. (Ph.D.)	Likely fluvial-eolian erosion processes
2005	Irvine, L.L-A. (M.S.)	Polygenesis, bioturbation dominant
2005	Kuhn, G.G.	Polygenesis, and seismic sand blows
2005a, 2005b	Johnson, D.L., et al.	Point-centered, locally thickened biomantles
2006	Horwath, J.L., and Johnson, D.L.	Mounds as locally thickened biomantles
2007, 2008, 2011	Irvine, L.L-A., and Dale, J.E.	Burrowing animals, polygenesis
2007, 2008	Reed, S.E., and Amundson, R.	Pocket gopher burrowing, bioturbation
2007	Seifert, C.L. (M.S.)	Coppicing under semiarid conditions
2007	Seifert, C.L., and Cox, R.T.	Semiarid nebkhas, eolian coppices
2008	Reed, S.E., et al.	Pocket gopher burrowing, bioturbation
2008	Horwath, J.L., and Johnson, D.L.	Pocket gopher burrowing, bioturbation
2008	Johnson, D.L., and Johnson, D.N.	Pocket gopher burrowing, bioturbation
2008	Fenton, M.M., and Paulen, R.C.*	Glacial and periglacial processes
2008	Cox, R.T., et al.	Liquefaction-seismic sand blows
2008	Cox, R.T., et al.	Coppicing under drought conditions
2008	Cox, G.W.	Burrowing animals, pocket gophers
2008	Finney, F.A.	Burrowing animals, pocket gophers

(*Continued*)

APPENDIX D. TIMELINE OF AUTHORS AND THEORIES OFFERED TO EXPLAIN MIMA MOUNDS (*Continued*)

Years	Authors	Hypotheses
2008a	Johnson, D.L., et al.	Burrowing animals, pocket gophers
2008	Thackray, G.D.	Coppicing, bioturbation, soil diffusion
2008	Cox, R.T., and Gordon, J.	Liquefaction-seismic sand blows
2009	Seifert, C.L., et al.	Coppicing under semiarid conditions
2009	Logan, R.L., and Welsh, T.J.	Sediment-filled sun cups on melting ice
2010	Fey, M.	Burrowing animals, vertebrates, invertebrates
2010	Johnson, D.L., and Johnson, D.N.	Rodent burrowing, bioturbation
2010	Cox, R.T., et al.	Paleoseismic sand blows
2011	Johnson, D.L., et al.	Point-centered biomantles

*The "prairie mounds" of Bik (1967, 1968, 1969), Gravenor (1955), Fenton and Paulen (2008), Sasaki (1979), Rockie (1942), Nikiforoff (1928), Porsild (1938), Péwé (1948), Thoroddsen (1914), Hawkes (1924), Smith (1949), and Krantz et al. (1988) are presumably glacial and periglacial phenomena, and lie outside the domain of what are usually considered to be Mima (prairie, pimple) mounds.

Appendix E. References cited in Appendices C and D

Abbott, P.L., 1982, Genesis of vernal pool topography in San Diego, *in* Abbott, P.L., ed., Geologic Studies in San Diego: San Diego, California, San Diego Association of Geologists Field Trips, San Diego State University, p. 118–126.

Abbott, P.L., 1983a, Vernal pool topography: Environment Southwest, v. 31, p. 9–11.

Abbott, P.L., 1983b, Genesis of vernal pool topography in San Diego, California: Geological Society of America Abstracts with Programs, v. 15, no. 5, p. 386.

Abbott, P.L., 1984, On the origin of vernal pool topography, San Diego County, California, *in* Jain, S., and Moyle, P., eds., Vernal Pools and Intermittent Streams, Institute of Ecology Publication no. 28: Davis, University of California, p. 18–29.

Abbott, P.L., 1986, Expandable clays and the origin of vernal pool topography, San Diego, California, *in* Proceedings of the Association of Engineering Geologists Annual Meeting, v. 29, p. 41.

Aldritt, J.E., 1986, Earth-mound characteristics and possible origins, Birch Creek Valley, East-Central Idaho [unpublished M.A. thesis]: Lawrence, University of Kansas, 114 p.

Allgood, F.P., 1972, Genesis and morphology of mounded soils [unpublished M.S. thesis]: Stillwater, Oklahoma State University, 98 p.

Allgood, F.P., and Gray, F., 1973, Genesis, morphology, and classification of mounded soils in eastern Oklahoma: Soil Science Society of America Proceedings, v. 37, p. 746–753, doi:10.2136/sssaj1973 .03615995003700050033x.

Allgood, F.P., and Gray, F., 1974, An ecological interpretation for the small mounds in landscapes of eastern Oklahoma: Journal of Environmental Quality, v. 3, p. 37–41, doi:10.2134/jeq1974.00472425000300010012x.

Almaraz, R.A., 1986, The characteristics of patterned ground in the southern Cascade Mountains [unpublished M.S. thesis]: Ashland, Southern Oregon State College, 68 p.

Alt, D.D., and Hyndman, D.W., 1994, Roadside Geology of Washington: Missoula, Mountain Press Publishing Co., 288 p.

Alt, D.D., and Hyndman, D.W., 2000, Roadside Geology of Northern and Central California: Missoula, Mountain Press Publishing Co., 369 p.

Amundson, R., 1998, Do soils need our protection?: Geotimes, v. 43, no. 3, p. 16–20.

Amundson, R., 2000a, Are soils endangered?: California Geology, v. 53, no. 3, p. 4–13.

Amundson, R., 2000b, Are soils endangered? *in* Schneiderman, J., ed., The Earth around Us: New York, W.H. Freeman, p. 4–13.

Anderson, E.W., Bormann, M.M., and Krueger, W.C., 1998, The ecological provinces of Oregon: A treatise on the basic ecological geography of the state: Corvallis, Oregon Agricultural Experiment Station, SR 990, 138 p.

Anonymous, 1866, From Vancouver Island to the mound prairies: Temple Bar, v. 18, p. 343–356.

Anonymous, 1921, Hog wallows: California Cultivator, v. 56, p. 127.

Archeuleta, T.E., 1980, Analysis of mima mounds in northeastern Arkansas [unpublished M.S. thesis]: New Orleans, Louisiana, University of New Orleans, 59 p.

Arkley, R.J., 1948, The Mima mounds: The Scientific Monthly, v. 66, no. 2, p. 175–176.

Arkley, R.J., 1963, Soils of the Santa Cruz Campus: Santa Cruz, University of California, 42 p. (Reprinted 1980; on file, Science and Engineering Library, University of California, Santa Cruz.)

Arkley, R.J., and Brown, H.C., 1954, The origin of Mima mound (hog-wallow) microrelief in the far western states: Proceedings of the Soil Science Society of America, v. 18, no. 2, p. 195–199, doi:10.2136/sssaj1954 .03615995001800020021x.

Arman, A., and Thornton, S.I., 1972, Collapsible soils in Louisiana: Engineering Research Bulletin, no. 111: Baton Rouge, Louisiana State University, 131 p.

Arnold, J.J., 1960, Prairie mounds and their climatic implications [unpublished M.S. thesis]: Fayetteville, University of Arkansas, 36 p.

Aronow, S., 1963, Internal characteristics of pimple (prairie) mounds in southeast Texas and southwest Louisiana, *in* Abstracts for 1962: Geological Society of America Special Paper 73, p. 106.

Aronow, S., 1976, Geology, *in* Mills, J.F., and Wilding, L.P., eds., Morphology, classification, and use of selected soils in Harris County, Texas: Department of Soils and Crop Sciences, Departmental Technical Report 76-48: College Station, Texas, Texas A&M University, p. 6–17.

Aronow, S., 1978, Day two stop descriptions, *in* Kaczorowski, R.T., and Aronow, S., eds., The Chenier Plain and modern coastal environments, southwestern Louisiana and geomorphology of the Pleistocene Beaumont Trinity River delta plain: Field trip guidebook: Houston, Texas, Houston Geological Society, 87 p.

Aronow, S., 1988, Hackberry salt dome and pimple mounds, *in* Birdseye, R.U., and Aronow, S., eds., Late Quaternary geology of southwestern Louisiana and southeastern Texas: Guidebook for south-central Friends of the Pleistocene Sixth Field Conference, March 25–27, p. 13–25: part 1, 63 p.; part 2, 19 p.; participants list, 15 p.

Aronow, S., 1990, Geomorphology of the project area, *in* Ensor, H.B., Aronow, S., Freeman, M.D., and Sanchez, J.M., An archeological survey of the proposed Greens Bayou regional stormwater detention facility, Greens Bayou, Harris County, Texas: Archaeological Surveys no. 7: College Station, Texas, Archaeological Research Laboratory, Texas A&M University, p. 21–44.

Arroues, K.D., Russell, E., and Regal, J., 2007, Soil Survey of Kern County, Northeastern Part, and Southeastern Part of Tulare County: U.S. Department of Agriculture–Natural Resources Conservation Service in cooperation with Bureau of Land Management, California Department of Conservation, and University of California Agricultural Experiment Station: Washington, D.C., Government Printing Office, 1449 p.

Aten, L.E., 1979, Indians of the upper Texas coast: Ethnohistoric and archeological framework [unpublished Ph.D. thesis]: Austin, University of Texas, 560 p.

Aten, L.E., and Bollich, C.N., 1981, Archaeological evidence for pimple (prairie) mound genesis: Science, v. 213, p. 1375–1376, doi:10.1126/science.213 .4514.1375.

Ayarbe, J.P., and Kieft, T.L., 2000, Mammal mounds stimulate microbial activities in a semiarid shrubland: Ecology, v. 81, p. 1150–1154, doi:10.1890 /0012-9658(2000)081[1150:MMSMAI]2.0.CO;2.

Bailey, T.L., 1923, The Geology and Natural Resources of Colorado County: University of Texas Bulletin, no. 2333, 163 p.

Barnes, G.W., 1879, The hillocks or mound-formations of San Diego, California: American Naturalist, v. 13, p. 565–571, doi:10.1086/272405.

Beck, M.W., Longacre, M.Y., Hayes, F.A., and Carter, W.T., Jr., 1923, Soil Survey of Howard County, Arkansas: U.S. Department of Agriculture Bureau of Soils, Field Operations of the Bureau of Soils for 1917, 19th Report, p. 1355–1398.

Benn, D.W., 1976, Investigations at the so-called Hovey "mounds," Hancock County, Iowa: Report of survey and testing for the State Archaeologist of Iowa: Des Moines, Iowa, Department of Natural Resources, 13 p.

Bennett, F., Worthen, E.L., Willard, R.E., and Watson, E.B., 1911, Soil survey of Richland County, North Dakota: Field Operations of the Bureau of Soils for 1908, 10th Report, p. 1121–1154.

Berg, A.W., 1989, Formation of Mima mounds: A seismic hypothesis: Geological Society of America Abstracts with Programs, v. 21, no. 5, p. 56.

Berg, A.W., 1990a, Formation of Mima mounds: A seismic hypothesis: Geology, v. 18, p. 281–284, doi:10.1130/0091-7613(1990)018<0281: FOMMAS>2.3.CO;2.

Berg, A.W., 1990b, Formation of Mima mounds: A seismic hypothesis: Reply: Geology, v. 18, p. 1260–1261, doi:10.1130/0091-7613(1990)018 <0281:FOMMAS>2.3.CO;2.

Berg, A.W., 1991, Formation of Mima mounds: A seismic hypothesis: Reply: Geology, v. 19, p. 284–285, doi:10.1130/0091-7613(1991)019<0284: CAROFO>2.3.CO;2.

Bernard, H.A., 1950, Quaternary geology of southeast Texas [unpublished Ph.D. dissertation]: Baton Rouge, Louisiana State University, 165 p.

Bernard, H.A., and LeBlanc, R.J., 1965, Résumé of Quaternary geology of the northwestern Gulf of Mexico province, *in* Wright, H.E., Jr., and Frey, D.G., eds., The Quaternary of the United States: Princeton, New Jersey, Princeton University Press, p. 137–186.

Best, T.L., Intress, C., and Shull, K.D., 1988, Mound structure in three taxa of Mexican kangaroo rats (*Dipodomys spectabilis cratodon, D. s. zygomaticus* and *D. nelsoni*): American Midland Naturalist, v. 119, no. 1, p. 216–220, doi:10.2307/2426071.

Beyer, G.E., 1898, The mounds of Louisiana: New Orleans: Publications of the Louisiana Historical Society, v. 2, no. 1, p. 7–27.

Bik, M.J.J., 1967, On the periglacial origin of prairie mounds: North Dakota Geological Survey Miscellaneous Series, no. 30, p. 83–94.

Bik, M.J.J., 1968, Morphoclimatic observations on prairie mounds: Zeitschrift fur Geomorphologie, v. 12, no. 4, p. 409–469.

Bik, M.J.J., 1969, The origin and age of the prairie mounds of southern Alberta, Canada: Biuletyn Peryglacjaln (Lodzkie Towarzstwo Naukowe-Societas Scientiarium Lodziensis), no. 19, p. 85–130.

Bjornstad, B., and Kiver, E., 2012, On the Trail of the Ice Age Floods: The Northern Reaches—A Geological Field Guide to Northern Idaho and the Channeled Scabland: Sand Point, Idaho, Keokee Company Publishing, 480 p.

Blankinship, J.W., 1889, Peculiar earth-heaps in Missouri: American Antiquarian and Oriental Journal (Chicago), v. 11, p. 117.

Bluemle, J.P., 1983, Freckled land: North Dakota Geological Survey Newsletter, December, p. 35–38.

Bork, J.L., 1978, A survey of the vascular plants and vertebrates of Upper Table Rock [unpublished M.S. thesis]: Ashland, Southern Oregon State College (now Southern Oregon University), 91 p.

Borst, G., 1975, A cross section through a large mima mound: Soil Survey Horizons, v. 16, no. 4, p. 20–24.

Bragg, D.C., 2003, Natural presettlement features of the Ashley County, Arkansas area: American Midland Naturalist, v. 149, p. 1–20, doi:10.1674/0003-0031 (2003)149[0001:NPFOTA]2.0.CO;2.

Branner, J.C., 1905, Natural mounds or 'hog-wallows': Science, v. 21, no. 535, p. 514–516, doi:10.1126/science.21.535.514-b.

Branson, C.C., 1966, Patterns of Oklahoma prairie mounds: Oklahoma Geology Notes, v. 26, no. 11, p. 263–273.

Branson, F.A., Miller, R.F., and McQueen, I.S., 1965, Plant communities and soil moisture relationships near Denver, Colorado: Ecology, v. 46, p. 311–319, doi:10.2307/1936334.

Breckenridge, W.J., and Tester, J.R., 1961, Growth, local movements and hibernation of the Manitoba toad, *Bufo hemiophrys*: Ecology, v. 42, no. 4, p. 637–646, doi:10.2307/1933495.

Bretz, J.H., 1911, The terminal moraine of the Puget Sound glacier: The Journal of Geology, v. 19, no. 2, p. 161–174, doi:10.1086/621826.

Bretz, J.H., 1913, The mounds of the Vashon Outwas, *in* Glaciation of the Puget Sound region: Washington Geological Society Bulletin No. 8: Olympia, Washington, Frank M. Lamborn, Printer, p. 81–108.

Brezina, D.N., 2007, Soil Survey of Padre Island National Seashore, Texas, Special Report: U.S. Department of Agriculture–Soil Conservation Service in cooperation with U.S. Department of Interior–National Park Service and Texas Agricultural Experiment Station: Washington, D.C., U.S. Government Printing Office, 275 p.

Brotherson, J.D., 1969, Species composition, distribution and phytosociology of Kalsow Prairie, a mesic tall-grass prairie in Iowa [unpublished Ph.D. thesis]: Ames, Iowa State University.

Brotherson, J.D., 1982, Vegetation of the Mima mounds of Kalsow Prairie, Iowa: The Great Basin Naturalist, v. 42, no. 2, p. 246–261.

Brower, J.V., 1893, The Mississippi River and its source: Minneapolis, Minnesota Historical Society Collections, v. 7, 361 p.

Brunnschweiler, D., 1962, The periglacial realm in North America during the Wisconsin glaciation: Biuletyn Peryglacjalny (Lodzkie Towarzystwo Naukowe - Societas Scientiarium Lodziensis), Series 3, no. 11, p. 15–27.

Brunnschweiler, D., 1964, Der pleistozane Periblazialbereich in Nordamerika: Zeitschrift fur Geomorphologie, v. 8, p. 223–231.

Bryan, K., 1940, Soils and periglacial phenomena in the Carolinas: Science, v. 91, p. 523–524, doi:10.1126/science.91.2370.523.

Bryson, J., 1893, The drift mounds of Olympia and of Long Island: The American Geologist, v. 12, p. 127–129.

Burgess, J.L., and Ely, C.W., 1909, Soil Survey of Conway County, Arkansas: U.S. Department of Agriculture Bureau of Soils, Field Operations of the Bureau of Soils for 1907, 9th Report, p. 753–771.

Burgess, J.L., Hurst, L.A., Wilder, H.J., and Shaw, C.F., 1908, Soil Survey of Caddo Parish, Louisiana: U.S. Department of Agriculture, Bureau of Soils, Field Operations of the Bureau of Soils for 1906, 8th Report, p. 427–458.

Busacca, A.J., Gaylord, D.R., Sweeney, M.R., and McDonald, E.V., 2002, Paired eolian deposits and megaflood features, Columbia Plateau, Washington: Friends of the Pleistocene 10th Annual Pacific Northwest Cell Field Trip, 16–18 August, 2002: Pullman, Washington, Department of Crop and Soil Sciences, Washington State University, 80 p.

Bushnell, D.I., Jr., 1905, The small mounds of the United States: Science, v. 22, p. 712–714, doi:10.1126/science.22.570.712.

Butler, A.C., 1979, Mima mounds grasslands of the upper coastal prairie of Texas [unpublished M.S. thesis]: College Station, Texas A&M University, 94 p.

Cahoon, J., 1985, Soil Survey of Klamath County, Oregon, Southern Part: U.S. Department of Agriculture–Soil Conservation Service in cooperation with Oregon Agricultural Experiment Station: Washington, D.C., Government Printing Office, 165 p.

Cain, R.H., 1974, Pimple mounds: A new viewpoint: Ecology, v. 55, p. 178–182, doi:10.2307/1934633.

Caine, T.A., and Kocher, A.E., 1904, Soil Survey of the Paris [Lamar Co.] area, Texas: U.S. Department of Agriculture Field Operations of Bureau of Soils for 1903, 5th Report, p. 533–562.

Campbell, M.R., 1906, Natural mounds: The Journal of Geology, v. 14, p. 708–717, doi:10.1086/621357.

Capps, G., 1994, Table Rock geologic field guide, *in* Reyes, C., compiler-editor, The Table Rocks of Jackson County [Oregon]: Islands in the Sky: Ashland, Oregon, Last Minute Publications, p. 35–49.

Carter, B.J., 1999, Brief overview: Mounded landscapes in Oklahoma, Stop 3, *in* Guidebook, South-Central Friends of the Pleistocene and DiGQuEST (Discussion Group for Quaternary Earth Scientists): Aspects of Soil Geomorphology in Eastern Oklahoma—A Pleistocene Medley, Eufaula, Oklahoma, 16–17 April 1999: Stillwater, Oklahoma, Department of Plant and Soil Sciences, Oklahoma State University, p. 1–2.

Carter, W.T., and Patrick, A.L., 1919, Soil Survey of Bryan County, Oklahoma: U.S. Department of Agriculture, Bureau of Soils, Field Operations of the Bureau of Soils for 1914, 16th Report, p. 2165–2212.

Carter, W.T., Jr., Schoenmann, L.R., Bushnell, T.M., and Maxon, E.T., 1916, Soil Survey of Jefferson County, Texas: U.S. Department of Soils, Bureau of Soils, Field Operations of the Bureau of Soils for Year 1913, 15th Report, p. 1001–1043.

Carty, D.J., 1980, Characteristics of pimple mounds associated with the Morey soil of SE Texas [unpublished M.S. thesis]: College Station, Texas A&M University, 211 p.

Clarke, J.W., 1878, On the origin and antiquity of the mounds of Arkansas, U.S.: Report of the 47th Meeting of the British Association for the Advancement of Science, held at Plymouth, August 1877: London, John Murray, 67 p.

Clendenin, W.W., 1896, A preliminary report upon the Florida Parishes of East Louisiana and the bluff, prairie and hill lands of southwest Louisiana: Part 3, Geology and Agriculture: Baton Rouge, Louisiana Geological Survey, p. 156–256.

Cleveland, D., 1893, Hillock and mound formations of Southern California: Science, v. 22, p. 4, doi:10.1126/science.ns-22.544.4.

Collett, R.A., 1980, Soil Survey of Ada County, Idaho: U.S. Department of Agriculture, Soil Conservation Service, in cooperation with U.S. Forest Service, U.S. Department of the Interior, Bureau of Land Management, University of Idaho College of Agriculture, Idaho Agriculture Experiment Station, and Idaho Soil Conservation Commission: Washington, D.C., Government Printing Office, 327 p.

Collins, B., 1975, Range vegetation and mima mounds in north Texas: Journal of Range Management, v. 28, no. 3, p. 209–211, doi:10.2307/3897527.

Collins, O.B., 1972, Climax vegetation and soils of the Blackland Prairie of Texas [unpublished M.S. thesis]: College Station, Texas A&M University, 102 p.

Cooper, J.G., 1860, Report on the botany of the route, in Route near the Forty-Seventh and Forty-Ninth Parallels, Explored by I.I. Stevens, Governor of Washington Territory, in 1853–55: Botanical report no. 1: U.S. War Department, Reports of Exploration and Surveys, v. 12, book 2, p. 19–24.

Copeland, W.N., 1980, The Lawrence Memorial Grassland Preserve: A biophysical inventory with management recommendations: Unpublished report (corrected 1983): Portland, Oregon, Oregon Chapter, The Nature Conservancy, 168 p.

Corliss, W.R., 1988, Carolina Bays, Mima Mounds, Submarine Canyons and Other Topographical Phenomena: A Catalog of Geological Anomalies: Glen Arm, Maryland, The Sourcebook Project, 245 p.

Cox, G.W., 1984a, Mounds of mystery: Natural History, v. 93, no. 6, p. 36–45.

Cox, G.W., 1984b, Soil transport by pocket gophers in Mima mound and vernal pool microterrain, *in* Jain, S., and Moyle, P., eds., Vernal Pools and Intermittent Streams, Institute of Ecology Publication No. 28: Davis, California, University of California, p. 37–45.

Cox, G.W., 1989, Early summer diet and food preferences of Northern Pocket Gophers in North Central Oregon: Northwest Science, v. 63, no. 3, p. 77–82.

Cox, G.W., 1990a, Soil mining by pocket gophers along topographic gradients in a Mima moundfield: Ecology, v. 71, no. 3, p. 837–843, doi:10.2307/1937355.

Cox, G.W., 1990b, Comment and reply on "Formation of Mima mounds: A seismic hypothesis": Geology, v. 18, no. 12, p. 1259–1260, doi:10.1130/0091-7613(1990)018<1259:CAROFO>2.3.CO;2.

Cox, G.W., 1990c, Form and dispersion of Mima mounds in relation to slope, steepness and aspect on the Columbia Plateau: The Great Basin Naturalist, v. 50, p. 21–31.

Cox, G.W., 2008, Mima mounds of the alpine west: Geological Society of America Abstracts with Programs, v. 40, no. 6, p. 208.

Cox, G.W., 2012, this volume, Alpine and montane Mima mounds of the western United States, *in* Horwath Burnham, J.L., and Johnson, D.L., eds., Mima Mounds: The Case for Polygenesis and Bioturbation: Geological Society of America Special Paper 490, doi:10.1130/2012.2490(03).

Cox, G.W., and Allen, D.W., 1987a, Sorted stone nets and circles of the Columbia Plateau; a hypothesis: Northwest Science, v. 61, p. 79–185.

Cox, G.W., and Allen, D.W., 1987b, Soil translocation by pocket gophers in a Mima moundfield: Oecologia, v. 72, p. 207–210, doi:10.1007/BF00379269.

Cox, G.W., and Gakahu, C.G., 1983, Mima mounds in the Kenya highlands: Significance for the Dalquest-Scheffer hypothesis: Oecologia, v. 57, p. 170–174, doi:10.1007/BF00379577.

Cox, G.W., and Gakahu, C.G., 1984, The formation of mima mounds in the Kenya highlands: A test of the Dalquest-Scheffer hypothesis: Journal of Mammalogy, v. 65, no. 1, p. 149–152, doi:10.2307/1381218.

Cox, G.W., and Gakahu, C.G., 1985, Mima mound microtopography and vegetation pattern in Kenyan savannas: Journal of Tropical Ecology, v. 1, p. 23–36, doi:10.1017/S0266467400000055.

Cox, G.W., and Gakahu, C.G., 1986, A latitudinal test of the fossorial rodent hypothesis of mima mound origin: Zeitschrift fur Geomorphologie, v. 30, no. 4, p. 485–501.

Cox, G.W., and Gakahu, C.G., 1987, Biogeographical relationships of rhizomyid mole rats with Mima mound terrain in the Kenya highlands: Pedobiologia, v. 30, p. 263–275.

Cox, G.W., and Hunt, J., 1990, Nature and origin of stone stripes on the Columbia Plateau: Landscape Ecology, v. 5, no. 1, p. 53–64, doi:10.1007/BF00153803.

Cox, G.W., and Roig, V.G., 1986, Argentinian mima mounds occupied by Ctenomyid rodents: Journal of Mammalogy, v. 67, p. 428–432, doi:10.2307/1380907.

Cox, G.W., and Scheffer, V.B., 1991, Pocket gophers and Mima terrain in North America: Natural Areas Journal, v. 11, no. 4, p. 193–198.

Cox, G.W., Gakahu, C.G., and Allen, D.W., 1987, Small-stone content of Mima mounds of the Columbia Plateau and Rocky Mountain regions: Implications for mound origin: The Great Basin Naturalist, v. 47, p. 609–619.

Cox, R.T., and Gordon, J., 2008, Sand blows on late Quaternary surfaces in northeast Louisiana: Geological Society of America Abstracts with Programs, v. 40, no. 6, p. 151.

Cox, R.T., and Shingleur, R.A., 1994, A new line of evidence for the erosional origin of pimple mounds: Geological Society of America Abstracts with Programs, v. 26, p. 4.

Cox, R.T., Larsen, D., Forman, S.L., Woods, J., Morat, J., and Galluzzi, J., 2004, Preliminary assessment of sand blows in the southern Mississippi Embayment: Bulletin of the Seismological Society of America, v. 94, no. 3, p. 1125–1142, doi:10.1785/0120030130.

Cox, R.T., Hill, A.A., Larsen, D., Holzer, T., Forman, S.L., Noce, T., Gardner, C., and Morat, J., 2007, Seismotectonic implications of sand blows in the southern Mississippi Embayment: Engineering Geology, v. 89, p. 278–299, doi:10.1016/j.enggeo.2006.11.002.

Cox, R.T., Forman, S.L., Siefert, C.S., and Foti, T.L., 2008, Pimple mound OSL ages: Late Holocene Great Plains drought chronology extended to the south central United States: Geological Society of America Abstracts with Programs, v. 40, no. 6, p. 208.

Cox, R.T., Gordon, J., Forman, S., Brezina, T., Negrau, M., Hill, A., Gardner, C., and Machin, S., 2010, Paleoseismic sand blows in north Louisiana and south Arkansas: Seismological Research Letters, v. 81, no. 6, p. 1032–1047, doi:10.1785/gssrl.81.6.1032.

Crenwelge, W.E., Griffin, E.L., and Baker, J.K., 1988, Soil Survey of Galveston County, Texas: U.S. Department of Agriculture–Natural Resources Conservation Service, in cooperation with Texas Agricultural Experiment Station and Texas State Soil and Water Conservation Board: Washington, D.C., U.S. Government Printing Office, 182 p.

Cross, C.I., 1964, The Parramore Island mounds of Virginia: Geographical Review, v. 54, p. 502–515, doi:10.2307/212979.

Crossman, L.C., 1921, Hogwallows again: California Cultivator, v. 56, p. 451.

Crout, J.D., 1976, Soil Survey of Chambers County, Texas: U.S. Department of Agriculture–Soil Conservation Service in cooperation with Texas Agricultural Experiment Station: Washington, D.C., U.S. Government Printing Office, 53 p.

Dalquest, W.W., and Scheffer, V.B., 1942, The origin of the mounds of western Washington: The Journal of Geology, v. 50, no. 1, p. 68–84, doi:10.1086/625026.

Dalquest, W.W., and Scheffer, V.B., 1944, Distribution and variation in pocket gophers, *Thomomys talpoides*, in the state of Washington: American Naturalist, v. 78, p. 308–333, doi:10.1086/281203.

Darby, W., 1816, A Geographical Description of the State of Louisiana: Philadelphia, John Melish, 270 p.

Daubenmire, R., 1970, Steppe vegetation of Washington: Washington Agriculture Experiment Station Technical Bulletin 62, 131 p.

Davids, R.C., 1967, The mystery of mima mounds: Farm Journal, v. 91, p. 17–18.

Davis, M., and Youtie, B.A., 1976, The biscuit-scabland range, Oregon's natural heritage rediscovered: Mazama, v. 58, no. 13, p. 30–34.

Deal, D., 1971, Areas with a freckled appearance on aerial photographs: North Dakota Geological Survey Bulletin, v. 58, no. part 1, p. 19–20.

Deeter, E.B., and Cohn, H.I., 1923, Soil Survey of Faulkner County, Arkansas: U.S. Department of Agriculture Bureau of Soils, Field Operations of the Bureau of Soils for 1917, 19th Report, p. 1323–1353.

Deeter, E.B., and Davis, L.V., 1921, Soil Survey of Craighead County, Arkansas: U.S. Department of Agriculture Bureau of Soils, Field Operations of the Bureau of Soils for 1916, 18th Report, p. 1161–1188.

Deeter, E.B., Bushnell, T.M., and Wolfanger, L.A., 1925, Soil Survey of Perry County, Arkansas: U.S. Department of Agriculture Bureau of Soils, Field Operations of Bureau of Soils for 1920, 22nd Report, p. 493–536.

Deeter, E.B., Meyer, A.H., and Benton, T.H., 1928, Soil Survey of Pulaski County, Arkansas: U.S. Department of Agriculture Bureau of Soils, Field Operations of Bureau of Soils for 1922, 24th Report, p. 721–785.

DeGraff, J.V., 1975, Inactive patterned ground, Bear River Range, north-central Utah: Geological Society of America Abstracts with Programs, v. 7, p. 1048–1049.

DeGraff, J.V., 1976, Relict patterned ground, Bear River Range, north-central Utah: Utah Geology, v. 3, p. 111–116.

del Moral, R., and Deardorf, D.C., 1976, Vegetation of the Mima mounds, Washington State: Ecology, v. 57, p. 520–530, doi:10.2307/1936436.

de Nadaillac, M., 1885, Pre-Historic America (translated by N. D'Anvers): London, Murray, 566 p.

Denny, C.G., and Goodlett, J.C., 1968, Microrelief resulting from fallen trees, *in* Denny, C.S., Surficial geology and geomorphology of Potter County, Pennsylvania: U.S. Geological Survey Professional Paper 288, p. 59–66.

Dietz, R.S., 1945, The small mounds of the Gulf Coastal Plain: Science, v. 102, p. 596–597, doi:10.1126/science.102.2658.596.

Donaldson, N.C., and Geise, L.D., 1968, Soil Survey of Spokane County, Washington: U.S. Department of Agriculture, Soil Conservation Service in cooperation with the Washington Agricultural Experiment Station: Washington, D.C., U.S. Government Printing Office, 143 p.

Doughton, S., 2009, Mima mounds: Mystery solved? The Seattle Times, 3 April: Seattle, Washington, p. A1, A16.

Douglas, S.D., 1840, cited *in* Jackson, H.E., 1973, Washington's mysterious mounds: National Parks and Conservation Magazine, v. 47, no. 4, p. 24–26.

Dumble, E.T., 1918, The geology of east Texas: University of Texas Bulletin no. 1869, p. 272–274.

Dunbar, W., 1804, Journal of an Expedition up the Ouachita River: Mississippi Department of Archives and History, p. 185.

Dyksterhuis, E.L., and High, C.T., 1985, Soil Survey of the Union County area, Oregon: U.S. Department of Agriculture, Soil Conservation Service, in cooperation with the Oregon Agricultural Experiment Station: Washington, D.C., Government Printing Office, 194 p.

Dyksterhuis, E.L., Simonson, G.H., Norgren, J.A., and Hosler, R.E., 1969, John Day drainage basin general soil map report with irrigable areas, *in* Oregon's Long-Range Requirements for Water: U.S. Department of Agriculture, Soil Conservation Service, in cooperation with the Oregon State Water Resources Board: Appendix I-6, State Water Resources Board: Salem, Oregon, 101 p.

Eakin, H.M., 1932, Periglacial phenomena in the Puget Sound region: Science, v. 75, no. 1951, p. 536.

Easterbrook, D.J., and Rahm, D.A., 1970, Landforms of Washington: The Geologic Environment: Bellingham, Washington, Union Printing Co., 155 p.

El Hult, R., 1955, A million mysterious mounds: American Mercury, v. 80, p. 55–56.

Ellis, A.J., and Lee, C.H., 1919, Geology and ground waters of the western part of San Diego County, California: U.S. Geological Survey Water-Supply Paper 446, 321 p.

Ensor, H.B., 1987, The Cinco Ranch sites, Barker Reservoir, Fort Bend County, Texas: Archaeological Research Laboratory Report of Investigations no. 3: College Station, Texas A&M University, 306 p.

Ensor, H.B., Aronow, S., Freeman, M.D., and Sanchez, J.M., 1990, An archeological survey of the proposed Greens Bayou regional storm water detention facility, Greens Bayou, Harris County, Texas: Archaeological Surveys no. 7: College Station, Texas, Archaeological Research Laboratory, Texas A&M University, 137 p.

Farnsworth, P.J., 1906, On the origin of the small mounds of the lower Mississippi Valley and Texas: Science, v. 23, no. 589, p. 583–584, doi:10.1126/science.23.589.583-a.

Featherman, A., 1872, Third annual report of botanical survey of southwest and northwest Louisiana: Annual Report of D.F. Boyd, Superintendent Louisiana State University for Year 1871: New Orleans, Republican Office Printers, p. 101–161.

Fenneman, N.M., 1906, Oil Fields of the Texas-Louisiana Gulf Coastal Plain: U.S. Geological Survey Bulletin 282, 146 p.

Fenneman, N.M., 1931, Physiography of Western United States: New York, McGraw-Hill, 534 p.

Fenneman, N.M., 1938, Physiography of Eastern United States: New York, McGraw-Hill, 714 p.

Fenton, M.M., and Paulen, R.C., 2008, Doughnuts: "Flavors" and genesis of hummocky glacial terrain, Alberta: Geological Society of America Abstracts with Programs, v. 40, no. 6, p. 208.

Fey, M., 2010, The Soils of South Africa: Cambridge, UK, Cambridge University Press, 287 p.

Fields, R.C., Freeman, M.D., and Kotter, S.M., 1983, Inventory and assessment of cultural resources at Addicks Reservoir, Harris County, Texas: Reports of Investigations 22: Austin, Texas, Prewitt and Associates, Consulting Archeologists, 127 p.

Fields, R.C., Godwin, M.F., Freeman, M.D., and Lisk, S.V., 1986, Inventory and assessment of cultural resources at Barker Reservoir, Fort Bend and Harris Counties, Texas: Reports of Investigations 40: Austin, Texas, Prewitt and Associates, Consulting Archeologists, 176 p.

Finney, F.A., 2008, Natural prairie mounds of the upper Midwest: Their abundance, distribution, origin, and archaeological implications: Geological Society of America Abstracts with Programs, v. 40, no. 6, p. 209.

Finney, F.A., 2012, this volume, The forgotten natural prairie mounds of the Upper Midwest: Their abundance, distribution, origin, and archaeological implications, *in* Horwath Burnham, J.L., and Johnson, D.L., eds., Mima Mounds: The Case for Polygenesis and Bioturbation: Geological Society of America Special Paper 490, doi:10.1130/2012.2490(05).

Fisk, H.N., 1938, Geology of Grant and LaSalle Parishes: Geological Bulletin no. 10: New Orleans Department of Conservation, and Louisiana Geological Survey, 246 p.

Fisk, H.N., 1940, Geology of Avoyelles and Rapides Parishes: Geological Bulletin no. 18: New Orleans Department of Conservation, and Louisiana Geological Survey, 240 p.

Fisk, H.N., 1948, Definite project report Mermentau River, Louisiana, Appendix II: Geology of Lower Mermentau River Basin, *in* Mermentau River Basin, Mermentau River, Louisiana, flood control, irrigation and navigation, definite project report: New Orleans, Louisiana, New Orleans District, Department of the Army, Corps of Engineers, p. 1–40.

Forshey, C.G., 1845, Description of some artificial mounds on Prairie Jefferson, Louisiana: American Journal of Science, v. 49, p. 38–42.

Forshey, C.G., 1851–1852, cited in Foster, 1873, Pre-Historic Races of the United States of America (second edition): Chicago, S.C. Griggs and Co. (cf. also cited in Veatch, 1906), 415 p.

Forshey, C.G., 1854, Communications on mounds: Proceedings of the New Orleans Academy of Science, v. 1, p. 18–19.

Fosberg, M.A., 1963, Genesis of some soils associated with low and big sagebrush complexes in the Brown, Chestnut and Chernozem-prairie zones in southcentral and southwestern Idaho [unpublished Ph.D. thesis]: Madison, University of Wisconsin, 308 p.

Fosberg, M.A., 1965, Characteristics and genesis of patterned ground in Wisconsin time in a chestnut soil zone of southern Idaho: Soil Science, v. 99, p. 30–37, doi:10.1097/00010694-196501000-00006.

Foster, J.W., 1873, Pre-Historic Races of the United States of America (second edition): Chicago, S.C. Griggs and Co., 415 p.

Foster, Z.C., and Moran, W.J., 1935, Soil Survey of Galveston County, Texas: U.S. Department of Agriculture Bureau of Chemistry and Soils, in cooperation with Texas Agricultural Experiment Station, Series 1930, no. 31, p. 1–18.

Fowke, G., 1910, Antiquities of central and south-eastern Missouri: Bureau of American Ethnology (Smithsonian Institution) Bulletin 37: Washington, D.C., Government Printing Office, 116 p.

Fowke, G., 1922, Archaeological investigations: Bureau of American Ethnology (Smithsonian Institution) Bulletin 76: Washington, D.C., Government Printing Office, 204 p.

Frederking, R.L., 1973, Spatial variation of the presence and form of earth mounds on a selected Alp surface: Sangre de Cristo Mountains, Colorado [unpublished Ph.D. dissertation]: Iowa City, University of Iowa, 201 p.

Freeman, O.W., 1926, Scabland mounds of eastern Washington: Science, v. 64, no. 1662, p. 450–451.

Freeman, O.W., 1932, Origin and economic value of the scabland mounds of eastern Washington: Northwest Science, v. 6, no. 2, p. 37–40.

Frye, J.C., and Leonard, A.B., 1963, Pleistocene geology of Red River Basin in Texas: Report of Investigations, Texas University: Bureau of Economic Geology, Report 49, 48 p.

Fryxell, R., 1959, Soil and grazing resources inventory, Colville Indian Reservation, Washington: U.S. Department of the Interior, Bureau of Indian Affairs, Branch of Land Operations Survey, p. 200–216a.

Fryxell, R., 1964, The late Wisconsin age of mounds on the Columbia Plateau of eastern Washington: unpublished report: Pullman, Washington, Laboratory of Anthropology, Washington State University, 16 p.

Fryxell, R.H., 2012, this volume, Appendix A. The Late Wisconsin age of mounds on the Columbia Plateau of eastern Washington, *in* Horwath Burnham, J.L., and Johnson, D.L., eds., Mima Mounds: The Case for Polygenesis and Bioturbation: Geological Society of America Special Paper 490, doi:10.1130/2012.2490(08).

Fuller, M.L., 1912, The New Madrid Earthquake: U.S. Geological Survey Bulletin 494: Washington, D.C., U.S. Department of the Interior, Government Printing Office, 119 p.

Gabb, W.M., 1877, Hog wallows: Nature, v. 16, p. 183–184, doi:10.1038/016183c0.

Gakahu, C.G., and Cox, G.W., 1984, The occurrence and origin of Mima mound terrain in Kenya: African Journal of Ecology, v. 22, p. 31–42.

Gangmark, H.A., and Sanford, F.B., 1963, Theory on development of mounds near Red Bluff, Calif. U.S.: Fish and Wildlife Service: Fish Bulletin, v. 63, p. 213–220.

Gatchet, A.S., 1891, The Karankara Indians, *in* The Karankawa Indians: The Coast People of Texas: Archaeological and Ethnological Papers of the Peabody Museum, v. 1, no. 2, p. 21–103.

Geiger, B., 1998, Heaps of confusion: Earth (Waukesha, Wis.), v. 7, no. 4, p. 34–37.

Gentry, H.R., 1991, Soil Survey of Asotin County Area, Washington, Parts of Asotin and Garfield Counties: U.S. Department of Agriculture, Soil Conservation Service in cooperation with Washington State Department of Natural Resources and Washington State Agricultural Research Center: Washington, D.C., Government Printing Office, 376 p.

Gibbs, G., 1854, Report of George Gibbs upon the geology of the central portion of Washington Territory: U.S. Congress, 33rd, 1st Session, House Executive Document 129, v. 18, pt. 1: Washington, D.C., Beverley Tucker, Printer, p. 494–512.

Gibbs, G., 1855, Report of George Gibbs upon the geology of the central portion of Washington Territory: U.S. War Department Reports of Explorations and Surveys, v. 1, p. 473–486.

Gibbs, G., 1873a, Physical geography of the north-western boundary of the United States: Journal of the American Geographical Society of New York, v. 3, p. 134–157, doi:10.2307/196415.

Gibbs, G., 1873b, Physical geography of the north-western boundary of the United States: Journal of the American Geographical Society of New York, v. 4, p. 298–392, doi:10.2307/196401.

Gilbert, G.K., 1875, Prairie mounds, *in* Report upon Geographical and Geological Explorations and Surveys West of the 100th Meridian, pt. 5, v. 3— Geology: Washington, D.C., Government Printing Office, p. 539–540.

Gile, L.H., 1966, Coppice dunes and the Rotura soil: Proceedings of the Soil Science Society of America, v. 30, p. 657–676, doi:10.2136/sssaj1966.03615995003000050035x.

Giles, L.J., 1970, The ecology of the mounds on Mima Prairie with special reference to Douglas-fir invasion [unpublished M.S. thesis]: Seattle, University of Washington, 99 p.

Goebel, J.E., 1971, Genesis of microknoll topography in northeast Texas [unpublished M.S. thesis]: Commerce, Texas, East Texas State University (now Texas A&M University at Commerce), 207 p.

Gold, T., 2001, The Deep Hot Biosphere: New York, Copernicus Books (Springer-Verlag), 243, p.

Golden, M.L., Peer, A.C., and Brown, S.E., Jr., 1994, Soil Survey of Harrison County, Texas: U.S. Department of Agriculture–Natural Resources Conservation Services, in cooperation with Texas Agricultural Experiment Station and Texas State Soil and Water Conservation Board: Washington, D.C., U.S. Government Printing Office, 252 p.

Goodarzi, N.K., 1978, Geomorphological and soil analysis of soil mounds in southwest Louisiana [unpublished M.S. thesis]: Baton Rouge, Louisiana State University, 70 p.

Grant, C., 1948, Mima mounds: The Journal of Geology, v. 56, p. 229–231, doi:10.1086/625505.

Gravenor, C.P., 1955, The origin and significance of prairie mounds: American Journal of Science, v. 253, p. 475–481, doi:10.2475/ajs.253.8.475.

Green, G.L., 1975, Soil Survey of Trout Creek-Shaniko Area, Oregon (Parts of Jefferson, Wasco, and Crook Counties, OR): U.S. Department of Agriculture, Soil Conservation Service, and U.S. Forest Service in cooperation with Oregon Agricultural Experiment Station: Washington, D.C., U.S. Government Printing Office, 69 p.

Green, G.L., 1982, Soil Survey of Wasco County, Oregon, Northern Part: U.S. Department of Agriculture, Soil Conservation Service, in cooperation with Oregon Agricultural Experiment Station: Washington, D.C., U.S. Government Printing Office, 83 p.

Green, G.L., Simonson, G.H., and Norgren, J.A., 1969, Hood drainage basin general soil map report with irrigable areas, *in* Oregon's Long-Range Requirements for Water: U.S. Department of Agriculture, Soil Conservation Service, in cooperation with the Oregon State Water Resources Board: Appendix I-4, State Water Resources Board: Salem, Oregon, 58 p.

Greenwood, N.H., and Abbott, P.C., 1980, The physical environment of H series vernal pools, Del Mar Mesa, San Diego County: San Diego State University: California Department of Transportation, Project R-11 988, 57 p.

Grinnell, J., and Linsdale, J.M., 1936, Vertebrate animals of Point Lobos Reserve, 1934–35: Carnegie Institution of Washington Publication 481, 159 p.

Guccione, M.J., Shingleur, R.A., van Arsdale, R.B., and Pratt, M., 1991, Origin of some prairie mounds in northwestern Arkansas: Geological Society of America Abstracts with Programs, v. 23, no. 4, p. 28.

Hall, E.C., Bushnell, T.M., Davis, L.V., Carter, W.T., Jr., and Patrick, A.L., 1919, Soil Survey of Mississippi County, Arkansas: U.S. Department of Agriculture Bureau of Soils, Field Operations of the Bureau of Soils for 1914, 16th Report, p. 1325–1362.

Hall, J., 1838, Notes on the Western States: Philadelphia, Harrison Hall Printers, 424 p.

Hall, S.A., Miller, M.R., and Goble, R.J., 2010, Geochronology of the Bolson sand sheet, New Mexico and Texas, and its archaeological significance: Geological Society of America Bulletin, v. 112, no. 11–12, p. 1950–1967.

Hallet, B., and Sletten, R.S., 1994, Mima mounds: Constraints on the timing of formation: Geological Society of America Abstracts with Programs, v. 26, no. 7, p. A-300.

Hallsworth, E.G., and Beckman, G.G., 1969, Gilgai in the Quaternary: Soil Science, v. 107, p. 409–420, doi:10.1097/00010694-196906000-00005.

Hannemann, M., 1928, "Mounds" und "pimples" in der küstenebene von Texas und Louisiana: Petermann's Mitteilungen, v. 74, p. 218–224.

Hansen, R.M., 1962, Movements and survival of Thomomys talpoides in a mima-mound habitat: Ecology, v. 43, p. 151–154, doi:10.2307/1932058.

Hansen, R.M., and Morris, M.S., 1968, Movement of rocks by northern pocket gophers: Journal of Mammalogy, v. 49, no. 3, p. 391–399, doi:10.2307/1378197.

Harksen, J.C., and Christensen, C.M., 1974, Proglacial frost action in west-central South Dakota: Geological Society of America Abstracts with Programs, v. 6, no. 6, p. 512.

Harris, G.D., 1902, The geology of the Mississippi Embayment, with special reference to the State of Louisiana: Special report no. 1, *in* A Report on the Geology of Louisiana: Part 6, Geology and Agriculture: Baton Rouge, Louisiana, State Experiment Station, p. 1–39.

Harris, G.D., and Veatch, A.C., 1899, A Preliminary Report on the Geology of Louisiana: Baton Rouge, Louisiana, State Experiment Station, Part 5, 354 p.

Hawkes, L., 1924, Frost action in superficial deposits, Iceland: Geological Magazine, v. 61, p. 509–513, doi:10.1017/S0016756800102018.

Heinrich, P.V., 1986, Geomorphology of the Barker Reservoir area, *in* Fields, R.C., Godwin, M.F., Freeman, M.D., and Lisk, S.V., Inventory and Assessment of Cultural Resources at Barker Reservoir, Fort Bend and Harris Counties, Texas, Reports of Investigations, no. 40: Austin, Texas, Prewitt and Associates, Consulting Archeologists, Appendix D, p. 148–176.

Henderson, J.A., 1974, Composition, distribution and succession of subalpine meadows in Mount Rainier National Park [unpublished Ph.D. dissertation]: Corvallis, Oregon State University, 153 p.

Henderson, J.G., 1952, Pleistocene geology of the Watino Quadrangle, Alberta [unpublished Ph.D. thesis]: Bloomington, Indiana University, 92 p.

Herbert, F.W., Jr., and Begg, E.L., 1969, Soils of the Yuba Area, California: Cooperative Project between the Department of Soils and Plant Nutrition, University of California, Davis, and County of Yuba, California, 170 p.

Herrick, C.L., 1892, The Mammals of Minnesota: Geological and Natural History Survey of Minnesota, Bulletin no. 7: Minneapolis, Harrison & Smith, State Printers, 299 p.

Herriman, R.C., and Parsons, R.B., 1979, Land use and management of soils with relict duripans, southwestern Oregon (U.S.A.): Geoderma, v. 22, p. 99–103, doi:10.1016/0016-7061(79)90010-7.

Hertlein, L.G., and Grant, U.S., IV, 1944, The geology and paleontology of the marine Pliocene of San Diego, California: Memoirs of the San Diego Society of Natural History, v. 2, no. part 1, p. 1–72.

Hilgard, E.W., 1873, Supplementary and final report of a geological reconnaissance of the State of Louisiana, under auspices of New Orleans Academy of Sciences and The Bureau of Immigration of the State of Louisiana, May–June 1869, 44 p. (cf. p. 11).

Hilgard, E.W., 1884a, Report on the cotton production of the state of Louisiana, with a discussion of the general agricultural features of the state: U.S. Census Office, 10th Census, v. 5, p. 95–195.

Hilgard, E.W., 1884b, Report on the physico-geographical and agricultural features of the state of California: U.S. Census Office: 10th Census, v. 6, p. 649–783.

Hilgard, E.W., 1905, The prairie mounds of Louisiana: Science, v. 21, p. 551–552, doi:10.1126/science.21.536.551-a.

Hilgard, E.W., 1906, Soils—Their Formation, Properties, Composition, and Relations to Climate and Plant Growth in the Humid and Arid Regions: New York, MacMillan Co., 593 p.

Hill, J.P., 2001, Ecophysiological basis of phenotypic differentiation in *Artemisia tridentata* ssp. *wyomingensis* growing on and off Mima-like mounds in southeast Wyoming [unpublished M.S. thesis]: Winston-Salem, North Carolina, Wake Forest University, 73 p.

Hill, R.T., 1906, On the origin of the small mounds of the lower Mississippi Valley and Texas: Science, v. 23, no. 592, p. 704–706, doi:10.1126/science.23.592.704.

Hobbs, W.H., 1907, Some topographic features formed at the time of earthquakes and the origin of mounds in the Gulf Plain: American Journal of Science (4th series), v. 23, p. 245–256.

Hobson, W.A., and Dahlgren, R.A., 1998a, A quantitative study of pedogenesis in California vernal pool wetlands, *in* Rabenhorst, M.C., Bell, J.C., and McDaniel, P.A., eds., Quantifying Soil Hydromorphology: Soil Science Society of America Special Publication no. 54, p. 107–127.

Hobson, W.A., and Dahlgren, R.A., 1998b, Soil forming processes in vernal pools of Northern California, Chico area, *in* Witham, C.W., Bauder, E.T., Belk, D., Ferren, W.R., Jr., and Ornduff, R., eds., Ecology, Conservation, and Management of Vernal Pool Ecosystems—Proceedings from a 1996 Conference: Sacramento, California Native Plant Society, p. 24–37.

Hobson, W.A., and Dahlgren, R.A., 2001, Wetland soils of basins and depressions: Case studies of vernal pools, *in* Richardson, J.L., and Vepraskas, M.J., eds., Wetland Soils: Genesis, Hydrology, Landscapes, and Classification: Boca Raton, Lewis Publishers (CRL Press), p. 267–281.

Holdredge, C.P., and Wood, H.B., 1947, Mound and depression topography of the Central Valley of California and other areas: Geological Society of America Bulletin, v. 58, no. 12, p. 1253–1254.

Holland, R.F., 1978, Biogeography and ecology of vernal pools in California [published Ph.D. dissertation]: Davis, University of California, 121 p.

Holland, R.F., and Jain, S.K., 1977, Vernal pools, *in* Barbour, M.G., and Major, J., eds., Terrestrial Vegetation of California: New York, John Wiley & Sons, p. 515–533.

Holland, W.C., 1943, Physiography of Beauregard and Allen Parishes, Louisiana [unpublished Ph.D. dissertation]: Baton Rouge, Louisiana State University, 178 p.

Holland, W.C., Hough, L.W., and Murray, G.E., 1952, Geology of Beauregard and Allen Parishes: Louisiana Geological Survey Geological Bulletin, v. 27, p. 11–68.

Holliday, V.T., 1987, Observations on the stratigraphy and origin of the Cinco Ranch mounds, *in* Ensor, H.B., The Cinco Ranch sites, Barker Reservoir, Fort Bend County, Texas, Archeological Research Laboratory Report of Investigations no. 3: College Station, Texas A&M University, Appendix 4, p. 275–280.

Holmes, L.C., and Nelson, J.W., 1919, Reconnaissance Soil Survey of the San Francisco Bay, California: U.S. Department of Agriculture Bureau of Soils, Field Operations of the Bureau of Soils for 1914, 16th Report, p. 2679–2784.

Holmes, L.C., and Pendleton, R.L., 1919, Reconnaissance Soil Survey of the San Diego Region, California: U.S. Department of Agriculture Bureau of Soils, Field Operations of the Bureau of Soils for 1915, 17th Report, p. 2509–2581.

Holmes, L.C., Eckmann, E.C., Nelson, J.W., and Guernsey, J.E., 1919, Reconnaissance Soil Survey of the Middle San Joaquin Valley, California: U.S. Department of Agriculture Bureau of Soils, Field Operations of the Bureau of Soils for 1916, 18th Report, p. 2421–2529.

Hombs, C., 2001, A mound of mystery: Joplin Globe, v. 1, no. June, p. 10A.

Hopkins, F.V., 1870, First Annual Report of the Louisiana State Geological Survey, *in* Annual Report of the Board of Supervisors of the Louisiana State Seminary of Learning and Military Academy, for Year ending Dec. 31, 1869: New Orleans, Louisiana, A.L. Lee, State Printer, p. 77–109.

Horberg, L., 1951, Intersecting minor ridges and periglacial features in the Lake Agassiz basin, North Dakota: The Journal of Geology, v. 59, p. 1–18, doi:10.1086/625798.

Horner, J.B., 1930, The million mystery mounds of Tenino: Privately published by Professor J.B. Horner, Director of Historical Research, Oregon State College, Corvallis, Oregon, 13 p.

Horwath, J.L., 2002a, The Mima mound story: Missouri Prairie Journal, v. 23, no. 3, p. 6–7.

Horwath, J.L., 2002b, An assessment of Mima type mounds, their soils, and associated vegetation, Newton County, Missouri [unpublished M.S. thesis]: Urbana, University of Illinois, 265 p.

Horwath, J.L., and Johnson, D.L., 2006, Mima-type mounds in southwest Missouri: Expressions of point centered and locally thickened biomantles: Geomorphology, v. 77, no. 3–4, p. 308–319.

Horwath, J.L., and Johnson, D.L., 2007, Erratum to "Mima-type mounds in southwest Missouri: Expressions of point centered and locally thickened biomantles": Geomorphology, v. 83, p. 193–194, doi:10.1016/j.geomorph.2006.09.013.

Horwath, J.L., and Johnson, D.L., 2008, The biodynamic significance of double stone-layers at Diamond Grove mima moundfield, southwest Missouri: Geological Society of America Abstracts with Programs, v. 40, no. 6, p. 208.

Horwath, J.L., Johnson, D.L., and Stumpf, A.J., 2002, Evolution of a gravelly Mima-type moundfield in southwestern Missouri: Geological Society of America Abstracts with Programs, v. 34, no. 6, p. 369.

Horwath Burnham, J.L., Johnson, D.L., and Johnson, D.N., 2012, this volume, The biodynamic significance of double stone layers in Mima mounds, *in* Horwath Burnham, J.L., and Johnson, D.L., eds., Mima Mounds: The Case for Polygenesis and Bioturbation: Geological Society of America Special Paper 490, doi:10.1130/2012.2490(04).

Howard, M.A., and Freeman, M.D., 1983, Inventory and assessment of cultural resources at Bear Creek Park, Addicks Reservoir, Harris County, Texas, Reports of Investigations, no. 24: Austin, Texas, Prewitt and Associates, Consulting Archeologists, 92 p.

Hoy, P.R., 1865, Journal of an exploration of western Missouri in 1854, under the auspices of the Smithsonian Institution: Smithsonian Institution Annual Report, v. 19, p. 431–438.

Hubbs, C.L., 1957, Recent climatic history in California and adjacent areas, *in* Craig, H., ed., Proceedings of Conference on Recent Research in Climatology: California University Water Resources Center Contribution no. 8, p. 10–22.

Huss, J.M., 1994, Stratigraphic and sodium analyses of Laramie Basin mima-like mounds, Wyoming [unpublished M.A. thesis]: Laramie, University of Wyoming, 176 p.

Irvine, L.L.-A., 2005, A study of pimple mounds in southern Saskatchewan [unpublished Master's thesis]: Regina, Saskatchewan, Canada, University of Regina, 188 p.

Irvine, L.L.-A., and Dale, J.E., 2007, "Pimple mound" topography in southern Saskatchewan: The Canadian Association of Geographers, Poster 16: Saskatoon, Saskatchewan, Canada.

Irvine, L.L.-A., and Dale, J.E., 2008, "Pimple" mound micro-relief in southern Saskatchewan, Canada: Geological Society of America Abstracts with Programs, v. 40, no. 6, p. 208.

Irvine, L.L.-A., and Dale, J.E., 2011, "Pimple" mound micro-relief in southern Saskatchewan, Canada: Geological Society of America Abstracts with Programs, v. 40, no. 6, p. 208.

Irvine, L.L.-A., and Dale, J.E., 2012, this volume, "Pimple" mound micro-relief in southern Saskatchewan, Canada, *in* Horwath Burnham, J.L., and Johnson, D.L., eds., Mima Mounds: The Case for Polygenesis and Bioturbation: Geological Society of America Special Paper 490, doi:10.1130/2012.2490(02).

Irvine, L.L.-A., Dale, J.E., and Sauchyn, D.J., 2002, Description and analysis of pimple mounds in the little Manitou Floodway Channel near Watrous, Saskatchewan: Geological Association of Canada and Mineralogical Association of Canada Joint Annual Meeting: Saskatoon, University of Saskatchewan, Abstract 27, p. 52.

Isaacson, J., and Johnson, D.L., 1996, The polygenetic and complex origins of "mima mounds" in the California coastal area; perspectives on a scientific controversy: Geological Society of America Abstracts with Programs, v. 28, no. 7, p. A111.

Ith, I., 2004, Ecologist shares findings on mystery of Mima mounds: The Seattle Times, 17 February: Seattle, Washington, p. A1.

Jackson, H.E., 1956, The mystery of the Mima Mounds: Natural History, v. 65, no. 3, p. 136–139.

Jackson, H.E., 1973, Washington's mysterious mounds: National Parks and Conservation Magazine, v. 47, no. 4, p. 24–26.

Jenny, H., 1976, The origin of Mima mounds and hogwallows: Fremontia, v. 4, no. 3, p. 27–28.

Jenny, H., 1980, The Soil Resource, Origin and Behaviour: New York, Springer-Verlag, 377 p.

Jeter, M.D., Rose, J.C., Williams, I.W., Jr., and Harmon, A.M., 1989, Archeology and bioarcheology of the lower Mississippi Valley and Trans-Mississippi south in Arkansas and Louisiana: Arkansas Archeological Survey Research Series no. 37, 419 p.

Johnson, C.B., 1982, Soil mounds and patterned ground of the Lawrence Memorial Grassland Preserve [unpublished M.S. thesis]: Corvallis, Oregon State University, 52 p.

Johnson, D.L., 1988, Soil geomorphology, paleopedology, and pedogenesis in the Vandenberg-Lompoc area, Santa Barbara County, California: A fieldguide for Soils Geomorphology Tour sponsored by the Soil Science Society of America, Annual Meetings of the American Society of Agronomy, Anaheim, California, 27 November–2 December, 79 p.

Johnson, D.L., 1997, Geomorphological, geoecological, geoarchaeological, and surficial mapping study of McGregor Guided Missile Range, Fort Bliss, New Mexico, v. 1–2: U.S. Army Corps of Engineers, Fort Worth District and Fort Bliss Military Reservation Miscellaneous Report of Investigations no. 157: Plano, Texas, Geomarine, Inc., 260 p.

Johnson, D.L., 2012, this volume, Appendix B. Early prairie mound observations by two celebrated geologists: Joseph LeConte and Grove Karl Gilbert, *in* Horwath Burnham, J.L., and Johnson, D.L., eds., Mima Mounds: The Case for Polygenesis and Bioturbation: Geological Society of America Special Paper 490, doi:10.1130/2012.2490(a2).

Johnson, D.L., and Horwath Burnham, J.L., 2012, this volume, Introduction: Overview of concepts, definitions, and principles of soil mound studies, *in* Horwath Burnham, J.L., and Johnson, D.L., eds., Mima Mounds: The Case for Polygenesis and Bioturbation: Geological Society of America Special Paper 490, doi:10.1130/2012.2490(00).

Johnson, D.L., and Johnson, D.N., 2008, White paper: Stonelayers, mima mounds and hybrid mounds of Texas and Louisiana—Logical alternative biomantle working hypotheses: Appendix 3, *in* Johnson, D.L., Mandel, R.D., and Frederick, C.D., compiler-editors, The origin of the sandy mantle and mima mounds of the east Texas Gulf Coastal Plains: Geomorphological, pedological, and geoarchaeological perspectives: Field trip guidebook, Geological Society of America Annual Meeting, Houston, Texas, 4 October 2008, p. 114–122.

Johnson, D.L., and Johnson, D.N., 2010, Why are mima mounds and pedogenic stone lines (stonelayers) absent on the California Channel Islands?: Geological Society of America Abstracts with Programs, v. 42, no. 5, p. 365.

Johnson, D.L., and Johnson, D.N., 2012, this volume, The polygenetic origin of prairie mounds in northeastern California, *in* Horwath Burnham, J.L., and Johnson, D.L., eds., Mima Mounds: The Case for Polygenesis and Bioturbation: Geological Society of America Special Paper 490, doi:10.1130/2012.2490(06).

Johnson, D.L., Johnson, D.N., and West, R.C., 1999, Pocket gopher origins of some midcontinental Mima-type mounds: Regional and interregional genetic implications: Geological Society of America Abstracts with Programs, v. 31, no. 7, p. A232.

Johnson, D.L., Horwath, J., and Johnson, D.N., 2002, In praise of the coarse fraction and bioturbation: Gravelly Mima mounds as two-layered biomantles: Geological Society of America Abstracts with Programs, v. 34, no. 6, p. 369.

Johnson, D.L., Horwath, J., and Johnson, D.N., 2003, Mima and other animal mounds as point centered biomantles: Geological Society of America Abstracts with Programs, v. 34, no. 7, p. 258.

Johnson, D.L., Domier, J.E.J., and Johnson, D.N., 2005a, Animating the biodynamics of soil thickness using process vector analysis: A dynamic denudation approach to soil formation: Geomorphology, v. 67, no. 1–2, p. 23–46, doi:10.1016/j.geomorph.2004.08.014.

Johnson, D.L., Domier, J.E.J., and Johnson, D.N., 2005b, Reflections on the nature of soil and its biomantle: Annals of the Association of American Geographers. v. 95, no. 1, p. 11–31, doi:10.1111/j.1467-8306.2005.00448.x.

Johnson, D.L., Johnson, D.N., Horwath, J.L., Wang, H., Hackley, K.C., and Cahill, R.A., 2008a, Predictive biodynamic principles resolve two longstanding topographic-landform-soil issues: Mima mounds and soil stonelayers: Geological Society of America Abstracts with Programs, v. 40, no. 6, p. 209.

Johnson, D.L., Mandel, R.D., and Frederick, C.D., 2008b, The Origin of the Sandy Mantle and Mima Mounds of the East Texas Gulf Coastal Plains: Geomorphological, Pedological, and Geoarchaeological Perspectives: Field trip guidebook, Geological Society of America Annual Meeting, Houston, Texas, 4 October 2008, 57 p., 30 plates, 3 appendices.

Johnson, D.L., Horwath Burnham, J., and Johnson, D.N., 2011, Historic formation and re-formation of mima mounds: Geological Society of America Abstracts with Programs, v. 43, no. 5, p. 509–510.

Johnson, D.R., 1993, Soil Survey of Jackson County Area, Oregon: U.S. Department of Agriculture, Soil Conservation Service, in cooperation with U.S. Forest Service, Bureau of Land Management, and Oregon Agricultural Experiment Station: Washington, D.C., U.S. Government Printing Office, 694 p.

Jones, D., and Shuman, M., 1988, Investigations into prehistoric occupation of pimple mounds in Vermillion Parish, Louisiana: Baton Rouge, Museum of Geoscience, Louisiana State University, 65 p.

Kaatz, M.R., 1959, Patterned ground in central Washington—A preliminary report: Northwest Science, v. 33, no. 4, p. 145–156.

Kaczorowski, R.T., and Aronow, S., 1978, The Chenier Plain and modern coastal environments, southwestern Louisiana and geomorphology of the Pleistocene Beaumont Trinity River delta plain: Field trip guidebook: Houston, Texas, Houston Geological Society, 87 p.

Kane, P., 1847, Paul Kane's diary (cited *in* Kruckeberg, A.R., 1991, The Natural History of Puget Sound Country: Seattle, University of Washington Press, p. 294–295).

Kelly, A.O., 1948, The mima mounds: The Scientific Monthly, v. 66, no. 2, p. 174–175.

Kelly, A.O., and Dachille, F., 1953, The Waters Mould the Earth: The Role of Large Meteors in Earth Science: Carlsbad, California, Target Earth, 263 p.

Kennedy, W., 1917, Coastal salt domes: Bulletin of the Southwest Association of Petroleum Geologists, v. 1, p. 34–59.

Kennicott, R., 1858, The quadrupeds of Illinois injurious and beneficial to the farmer: Annual Report of the U.S.: Department of Agriculture, v. 1857, p. 72–121.

Kerr, J.A., Hendrickson, B.H., Phillips, S.W., Elwell, J.A., Wolfanger, L.A., and Devereux, R.E., 1926, Soil Survey of Natchitoches Parish, Louisiana: U.S. Department of Agriculture Bureau of Soils, Field Operations of Bureau of Soils for 1921, 23rd Report, p. 1395–1441.

Kienzle, J., 2007, Soil Survey of Wallowa County Area, Oregon: U.S. Department of Agriculture, Natural Resources Conservation Service, in Cooperation with U.S. Forest Service, U.S. Department of the Interior–Bureau of Land Management, Oregon State University Agricultural Experiment Station, and Wallowa Soil and Water Conservation District: Washington, D.C., Government Printing Office, 2307 p.

Kinahan, G.H., 1877a, Hummocky moraine drift: Nature, v. 15, p. 379, doi:10.1038/015379a0.

Kinahan, G.H., 1877b, Hog-wallows and prairie mounds: Nature, v. 16, p. 7, doi:10.1038/016007a0.

Kinnaird, L., 1967, The frontiers of New Spain: Nicolas de Lafora's description 1766–1768: The Quivira Society, Berkeley, California, v. 13: New York, Arno Press, 243 p.

Kirk, R., 1970, The baffling bumps: Nature and Science, v. 7, no. 14, p. 10.

Klein, J.L., 1998, Vegetation patterning relative to Mima-like mounds and its implications for a tallgrass prairie restoration in southwestern Missouri [unpublished M.S. thesis]: Madison, University of Wisconsin, 167 p.

Kloehn, N.B., Junck, M.B., Jol, H.M., Running, G.L., IV, Greek, D., and Caldwell, K., 2000, GPR investigation of the west Prairie Mound Group, central Wisconsin, U.S.A.: Are they burial mounds or natural mounds?: Eighth International Conference on Ground Penetrating Radar, *in* Noon, D.A., Stickley, G.F., and Longstaff, D., eds., Proceedings of the International Society for Optical Engineering (SPIE), v. 4084, p. 590–595.

Knechtel, M.M., 1949, Geology and coal and natural gas resources of northern Le Flore County, Oklahoma: Oklahoma Geological Survey Bulletin, v. 68, p. 1–76.

Knechtel, M.M., 1952, Pimpled plains of eastern Oklahoma: Geological Society of America Bulletin, v. 63, no. 7, p. 689–699, doi:10.1130/0016-7606 (1952)63[689:PPOEO]2.0.CO;2.

Knobel, E.W., Lounsbury, C., Davis, L.V., Fowler, E.D., and Goke, A.W., 1926, Soil Survey of Lonoke County, Arkansas: U.S. Department of Agriculture Bureau of Soils, Field Operations of Bureau of Soils for 1921, 23rd Report, p. 1279–1327.

Kohr, L.H., 1976, Another theory on mysterious mounds: The Sunday Olympian, Feb. 29, 1976, p. 3.

Koons, F.C., 1926, Origin of the sand mounds of the pimpled plains of Louisiana and Texas [unpublished Master's thesis]: Chicago, University of Chicago, 36 p.

Koons, F.C., 1948, The sand mounds of Louisiana and Texas: The Scientific Monthly, v. 66, no. 4, p. 297–300.

Krantz, W.B., Gleason, K.J., and Caine, N., 1988, Patterned ground: Scientific American, v. 259, no. 6, p. 68–76, doi:10.1038/scientificamerican1288-68.

Krinitzsky, E.L., 1949, Origin of pimple mounds: American Journal of Science, v. 247, p. 706–714, doi:10.2475/ajs.247.10.706.

Krinitzsky, E.L., 1950, Origin of pimple mounds: American Journal of Science, v. 248, no. 5, p. 360.

Kruckeberg, A.R., 1991, The Natural History of Puget Sound Country: Seattle, University of Washington Press, 468 p.

Kuhn, G.G., 2005, Paleoseismic features as indicators of earthquake hazards in North Coastal, San Diego County, California, USA: Engineering Geology, v. 80, p. 115–150, doi:10.1016/j.enggeo.2005.04.006.

Kuntz, C.S., 1974, Ice lens mounds, Cedar Bog, Champaign County, Ohio: The Ohio Journal of Science, v. 74, no. 2, p. 133–139.

Langford, R.P., 2000, Nabkha (coppice dune) fields of south-central New Mexico, U.S.: Journal of Arid Environments, v. 46, p. 25–41, doi:10.1006/jare.2000.0650.

Lapham, M.H., Root, A.S., and Mackie, W.W., 1905, Soil Survey of the Sacramento Area: U.S. Department of Agriculture Field Operations of Bureau Soils for 1904, 5th Report, p. 1049–1087.

Larance, F.C., 1961, Soil Survey of Bradley County, Arkansas: U.S. Department of Agriculture–Soil Conservation Service in cooperation with Arkansas Agricultural Experiment Station, Series 1958, no. 1: Washington, D.C., Government Printing Office, 66 p.

Larrison, E.J., 1942, Pocket gophers and ecological succession in the Wenas region of Washington: The Murrelet, v. 23, no. 2, p. 34–41, doi:10.2307/3535581.

LeConte, J., 1874, On the great lava-flood of the Northwest, and on the structure and age of the Cascade Mountains: American Journal of Science (Series 3), v. 7, p. 167–180, 259–267.

LeConte, J., 1875, On the great lava-flood of the Northwest, and on the structure and age of the Cascade Mountains: Proceedings of the California Academy of Sciences, v. 5, p. 215–220.

LeConte, J., 1877, Hog wallows or prairie mounds?: Nature, v. 15, no. 390, p. 530–531, doi:10.1038/015530d0.

Lee, B., and Carter, B., 2010, Soil morphological characteristics of prairie mounds in the forested region of south-central United States: 19th World Congress of Soil Science, Soil Solutions for a Changing World, 1–6 August 2010, Brisbane, Australia: International Union of Soil Scientists, p. 37–40. Available on DVD from http://www.iuss.org.

Lerch, O., 1896, On the geology of North Louisiana, in A Preliminary Report upon The Hills of Louisiana, South of the Vicksburg, Shreveport and Pacific Railroad, to Alexandria, Louisiana: Part 2, Geology and Agriculture: Baton Rouge, Louisiana, State Experimental Station, p. 56–158.

Lewis, T.H., 1883/1891a, Ancient mounds of Minnesota: Originally published in "At Home" in St Paul, May 12, 1883; republished in Tracts for Archaeologists, 1891, St. Paul, First Series, 1880–1891.

Lewis, T.H., 1883/1891b, Swamp mounds: The American Antiquarian, v. 5, no. 4, p. 330–331.

Lewis, T.H., 1891, Tracts for Archaeologists, being reprints from various periodicals: First Series, 1880–1891: St. Paul, Minnesota, privately printed.

Lockett, S.H., 1870, First Annual Report of the Topographical Survey of Louisiana for 1869: Annual Report of the Board of Supervisors of Louisiana State University: New Orleans, Office of Republicans, p. 66–67.

Logan, R.L., and Walsh, T.J., 2009, Mima mounds formation and their implications for climate change: Program and Abstracts, Northwest Scientific Association 81st Annual Meeting: Seattle, University of Washington, 25–28 March 2009, p. 38–39.

Loughridge, R.H., 1884a, Report on the cotton production of the State of Arkansas: U.S. Census Office, 10th Census, v. 5, p. 531–652.

Loughridge, R.H., 1884b, Report on the cotton production of the State of Texas: U.S. Census Office, 10th Census, v. 5, p. 653–831.

Lounsbury, C., and Deeter, E.B., 1919, Soil Survey of Columbia County, Arkansas: U.S. Department of Agriculture Bureau of Soils, Field Operations of the Bureau of Soils for 1914, 16th Report, p. 1363–1396.

Lounsbury, C., and Rogers, R.F., 1924, Soil Survey of La Salle Parish, Louisiana: U.S. Department of Agriculture Bureau of Soils, Field Operations of Bureau of Soils for 1916, 18th Report, p. 677–714.

Lubbock, J., 1863, North American Archaeology: Smithsonian Annual Report, v. 17, p. 318–336.

Macdonald, G.D., 1998, Soil Survey of Warm Springs Indian Reservation, Oregon: U.S. Department of Agriculture–Natural Resources Conservation Service in cooperation with Confederated Tribes of the Warm Springs Reservation, U.S. Department of the Interior, Bureau of Indian Affairs, and Oregon Agricultural Experiment Station: Washington, D.C., Government Printing Office, 276 p.

Macdonald, G.D., Lamkin, J.M., and Borine, R.H., 1999, Soil Survey of Sherman County, Oregon: U.S. Department of Agriculture–Natural Resources Conservation Service in cooperation with Oregon Agricultural Experimental Station: Washington, D.C., Government Printing Office, 122 p.

Malde, H.E., 1961, Patterned ground of possible solifluction origin at low altitude in the western Snake River Plain, Idaho: U.S. Geological Survey Professional Paper 424-B, p. 170–173.

Malde, H.E., 1964a, Patterned ground in the western Snake River Plain, Idaho, and its possible cold climate origin: Geological Society of America Bulletin, v. 75, p. 191–208, doi:10.1130/0016-7606(1964)75[191:PGITWS]2.0.CO;2.

Malde, H.E., 1964b, The ecologic significance of some unfamiliar geologic processes, in Hester, R.H., and Schoenwetter, J., eds., The Reconstruction of Past Environments: Proceedings of the Fort Burgwin Conference on Paleoecology no. 3: Taos, New Mexico, Fort Burgwin Research Center, SMU, p. 11–15.

Mandel, R.D., 1987, Geomorphological investigations, in Bement, L.C., Mandel, R.D., de la Teja, J.F., Utley, D.K., and Turpin, S.A., Buried in the bottoms: The archeology of Lake Creek Reservoir, Montgomery County, Texas: Texas Archaeological Survey Research Report 97: Austin, University of Texas, p. 4.1–4.41.

Marbut, C.F., Bennett, H.H., Lapham, J.E., and Lapham, M.H., 1913, Soils of the United States (1913 edition): U.S. Department of Agriculture, Bureau of Soils Bulletin no. 96: Washington, D.C., Government Printing Office, 791 p.

Markley, M.C., 1948, Dakota County lowland mounds: The Minnesota Archaeologist, v. 14, no. 3, p. 63–64.

Martin, J.O., 1902, Soil Survey of the Willis Area [Montgomery County], Texas: U.S. Department of Agriculture, Bureau of Soils, Field Operations of the Bureau of Soils, 3rd Report, p. 607–619.

Martin, P.G., Jr., Butler, C.L., Scott, E., Lyles, J.E., Mariano, M., Ragus, J., and Schoelerman, L., 1990, Soil Survey of Natchitoches Parish, Louisiana: U.S. Department of Agriculture–Soil Conservation Service in cooperation with U.S. Forest Service, Louisiana Agriculture Experiment Station, and Louisiana Soil and Water Conservation Commission, v. 34, 133 p.

Masson, P.H., 1949, Circular soil structures in northeastern California: California Division of Mines Bulletin, no. 151, p. 61–71.

McAdams, W., 1881, Ancient mounds of Illinois: Proceedings of the American Association for the Advancement of Science, v. 29, Sec. B, p. 710–718.

McCune, D., 1999, Arkansas prairie becomes a natural legacy: Nature Conservancy, January–February, p. 24.

McDonald, E.V., Bruier, F., and McAuliffe, J., 2000, Interaction of soil and biologic processes in the formation of cleared circles in the desert southwest U.S.: Active rodents or weary humans?: Geological Society of America Abstracts with Programs, v. 32, no. 7, p. 328.

McDonald, L., 1965, Whodunit . . . Gophers? Glaciers? Indians? Buffalo? Washington to preserve portion of big prairie: The Sunday Oregonian [newspaper], 14 November 1965: Portland, Oregon, p. HG7.

McFaul, M., 1977, Preliminary results of a field study of the mima mounds at their type locality in Washington State: Great Plains–Rocky Mountain: The Geographical Journal, v. 6, no. 2, p. 405.

McFaul, M., 1979, A geomorphic and pedological interpretation of the mima-mounded prairies, South Puget Lowland, Washington State [unpublished M.A. thesis]: Laramie, University of Wyoming, 77 p.

McGee, W.J., 1878, On the artificial mounds of northeastern Iowa, and the evidence of the employment of a unit of measurement in their erection: American Journal of Science (Series 3), v. 16, p. 272–278.

McGee, W.J., 1891, The Pleistocene history of northeastern Iowa: Eleventh Annual Report of the U.S. Geological Survey for 1890–1891, p. 191–577.

McGinnies, W.J., 1960, Effect of mima-type microrelief on herbage production of five seeded grasses in western Colorado: Journal of Range Management, v. 13, no. 5, p. 231–234, doi:10.2307/3895047.

McGinnies, W.J., Osborn, L.W., and Berg, W.A., 1976, Plant-soil-microsite relationships on a salt-grass meadow: Journal of Range Management, v. 29, no. 5, p. 395–400, doi:10.2307/3897150.

McGuff, P.R., 1973, Association of archeological material with pimple mounds, *in* McGuff, P.R., and Cox, W.N., A survey of the archeological and historical resources of areas to be affected by the Clear Creek flood control project, Texas: Texas Archeological Survey Research Report no. 28: Austin, University of Texas, Appendix III, p. 62–72.

McGuff, P.R., and Cox, W.N., 1973, A survey of the archaeological and historical resources of areas to be affected by the Clear Creek Flood Control Project, Texas: Research Report no. 28: Austin, University of Texas, Texas Archaeology Survey, 72 p.

McKee, B., 1972, The Enigmatic Mounds, in Cascadia—The Geologic Evolution of the Pacific Northwest: New York, McGraw-Hill, p. 302–304.

McMillan, B.A., 2007, Plant assemblage structure on barrier island 'Pimple' dunes at the Virginia coast Reserve Long-Term Ecological Research site [unpublished Ph.D. thesis]: Norfolk, Virginia, Old Dominion University, 159 p.

McMillan, B.A., and Day, F.P., 2010, Micro-environment and plant assemblage structure on Virginia's barrier island "pimple" dunes: Northeastern Naturalist, v. 17, no. 3, p. 473–492, doi:10.1656/045.017.0308.

Mead, S.H., 1878, Curious mounds in California: Popular: The Scientific Monthly, v. 14, no. 15, p. 233.

Means, T.H., and Holmes, J.G., 1901, Soil survey around Fresno, California: U.S. Department of Agriculture Field Operation Bureau of Soils for 1900, 2nd Report, p. 333–384.

Meany, E.S., 1926, Diary of Wilkes in the northwest: Washington Historical Quarterly, 1925–1926, 99 p.

Melton, F.A., 1928, Unusual aerial photographs from southwestern Arkansas: Geological Society of America Bulletin, v. 39, p. 205.

Melton, F.A., 1929a, Natural mounds of southern Arkansas, northern Louisiana, and eastern Texas: Geological Society of America Bulletin, v. 40, p. 184–185.

Melton, F.A., 1929b, Natural mounds of northeastern Texas, southern Arkansas, and northern Louisiana: Oklahoma Academy of Sciences Proceedings, v. 9, p. 119–130.

Melton, F.A., 1934, Erosional soil hillocks: Geological Society of America Proceedings (for 1933), p. 98.

Melton, F.A., 1935, Vegetation and soil mounds: Geographical Review, v. 25, p. 430–433, doi:10.2307/209311.

Melton, F.A., 1954, Natural mounds of northeastern Texas, southern Arkansas, and northern Louisiana (revised): The Hopper: Oklahoma Geology Notes, v. 14, no. 7, p. 88–121.

Mielke, H.W., 1977, Mound building by pocket gophers (Geomyidae)—Their impact on soils and vegetation in North America: Journal of Biogeography, v. 4, p. 171–180, doi:10.2307/3038161.

Miller, T.W., 1994, Multi-dimensional analysis of Laramie Basin mima-like mounds, Wyoming [unpublished M.A. thesis]: Laramie, University of Wyoming, 113 p.

Moorhead, D.L., Fisher, F.M., and Whitford, W.G., 1988, Cover of spring annuals on nitrogen-rich kangaroo rat mounds in a Chihuahua Desert grassland: American Midland Naturalist, v. 120, p. 443–447, doi:10.2307/2426018.

Moorhead, D.L., Fisher, F.M., and Whitford, W.G., 1990, Factors affecting annual plant assemblages on banner-tailed kangaroo rat mounds: Journal of Arid Environments, v. 18, p. 165–173.

Morse, D.F., and Morse, P.A., 1983, Archaeology of the Central Mississippi Valley: New York: Academic Press, 345 p.

Mun, H.T., and Whitford, W.G., 1990, Factors affecting annual plants assemblages on banner-tailed kangaroo rat mounds: Journal of Arid Environments, v. 18, p. 165–173.

Murray, D.F., 1967, Gravel mounds at Rocky Flats, Colorado: The Mountain Geologist, v. 4, p. 99–107.

Murray, G.E., 1948, Geology of De Soto and Red River Parishes: Geological Bulletin no. 25: Baton Rouge, Louisiana, Department of Conservation, Louisiana Geological Survey, 312 p.

Myrum, R., and Ferry, W., 1999, Soil survey of upper Deschutes River area, Oregon, including parts of Deschutes, Jefferson, and Klamath counties: U.S. Department of Agriculture, Natural Resources Conservation Service, in cooperation with U.S. Forest Service, Bureau of Land Management, and Oregon Agricultural Experiment Station: Washington, D.C., U.S. Government Printing Office, 274 p.

Nelson, C.A., 1977, The origin and characteristics of soil mounds and patterned ground in north central Oregon [unpublished M.S. thesis]: Corvallis, Oregon State University, 45 p.

Nelson, J.W., Guernsey, J.E., Holmes, L.C., and Eckmann, E.C., 1919a, Reconnaissance Soil Survey of the Lower [*sic*, Upper] San Joaquin Valley, California: U.S. Department of Agriculture Bureau of Soils, Field Operations of the Bureau Soils for 1915, 17th Report, p. 2583–2733.

Nelson, J.W., Pendleton, R.L., Dunn, J.E., Strahorn, A.T., and Watson, E.B., 1919b, Soil Survey of the Riverside Area, California: U.S. Department of Agriculture Bureau of Soils, Field Operations of the Bureau of Soils for 1915, 17th Report, p. 2367–2450.

Nelson, J.W., Dean, W.C., and Eckmann, E.C., 1923, Reconnaissance Soil Survey of the Upper [*sic*, Lower] San Joaquin Valley, California: U.S. Department of Agriculture Bureau of Soils, Field Operations of Bureau Soils for 1917, 19th Report, p. 2535–2644.

Newcomb, R.C., 1940, Hypothesis for the periglacial "fissure polygon" origin of the Tenino mounds, Thurston County, Washington: Geological Society of the Oregon Country, v. 6, no. 21, p. 182.

Newcomb, R.C., 1952, On the origin of the Mima mounds, Thurston County region, Washington: The Journal of Geology, v. 60, no. 5, p. 461–472, doi:10.1086/625998.

Niiler, E., 1998, Mounds that confound: San Diego Union–Tribune [newspaper], Quest, Section E, 29 July.

Nikiforoff, C.C., 1928, The perpetually frozen subsoil of Siberia: Soil Science, v. 26, p. 61–82, doi:10.1097/00010694-192807000-00005.

Nikiforoff, C.C., 1937, The inversion of the great soil zones in western Washington: Geographical Review, v. 27, p. 200–213, doi:10.2307/210090.

Nikiforoff, C.C., 1941, Hardpan and microrelief in certain soil complexes of California: U.S. Department of Agriculture Technical Bulletin no. 745, p. 1–45.

Noble, J., 1951, The mysterious circles of Shasta: Westways, v. 43, no. 1, p. 18–19.

Noble, J., 1985, The mysterious circles of Shasta, in Neill, T.O., editor and compiler, Out of Time and Place: St. Paul, Minnesota, Llewellyn Publications, p. 55–59.

Noble, J.B., and Molenaar, D., 1965, Guidebook for Field Conference J, Pacific Northwest, Day 8: International Association of Quaternary Research (INQUA), 7th Congress, p. 59–68.

Norgren, J.A, Simonson, G.H., Thomas, B.R., Lindsay, M.G., Lovell, B.B., and Anderson, D.W., 1969, Deschutes drainage basin general soil map report with irrigable areas: Appendix I-5, in Oregon's Long-Range Requirements for Water: U.S. Department of Agriculture, Soil Conservation Service, in cooperation with the Oregon State Water Resources Board: Salem, Oregon, State Water Resources Board, 116 p.

O'Brien, M.J., 1996, Paradigms of the Past: The Story of Missouri Archaeology: Columbia, University of Missouri Press, 561 p.

O'Brien, M.J., Lyman, R.L., and Holland, T.D., 1989, Geoarchaeological evidence for prairie-mound formation in the Mississippi alluvial valley, southeastern Missouri: Quaternary Research, v. 31, p. 83–93, doi:10.1016/0033-5894(89)90087-2.

Olmsted, R.K., 1963, Silt mounds of Missoula flood surfaces: Geological Society of America Bulletin, v. 74, no. 1, p. 47–53, doi:10.1130/0016-7606(1963)74[47:SMOMFS]2.0.CO;2.

Orcutt, C.R., 1885, Aquatic plants of San Diego: Science, v. 5, p. 441, doi:10.1126/science.ns-5.121.441.

Othberg, K.L., 1991, Geology and geomorphology of the Boise Valley and adjoining areas, western Snake River Plain, Idaho [unpublished Ph.D. thesis]: Boise, University of Idaho, 124 p.

Owen, D.D., 1844a, A report of a geological reconnaissance of the Chippewa land district of Wisconsin [and northern part of Iowa]: Philadelphia, Lippincott-Grambo, 72 p.

Owen, D.D., 1844b, Report of a geological exploration of part of Iowa, Wisconsin and Illinois in the autumn of the year 1839: 28th Congress, 1st session, Senate Document 407, 191 p.

Owen, D.D., 1858, First Report of a Geological Reconnaissance of the Northern Counties of Arkansas (Made During Years 1857 and 1858): Littlerock, Arkansas, Johnson & Yerkes, State Printers, 141 p.

Owen, D.D., 1860, Second Report of a Geological Reconnaissance of the middle and southern Counties of Arkansas made during the years 1859 and 1860: Philadelphia, C. Sherman & Son, Printers, 433 p.

Paeth, R.C., 1967, Depositional origin of Mima mounds [at Upper and Lower Table Rocks, Oregon, and Mima Prairie, Washington] [unpublished M.S. thesis]: Corvallis, Oregon State University, 61 p.

Page, W.D., Swan, F.H., III, Hanson, K.L., Muller, D., and Blum, R.L., 1977, Prairie mounds (mima mounds, hogwallows) in the Central Valley, *in*

Huntington, G.L., Begg, E.L., Harden, J.W., and Marchand, D.E. (Singer, M.J., ed.) Soil Development, Geomorphology, and Cenozoic History of the northeastern San Joaquin Valley and Adjacent Areas, California: A Guidebook for the Joint Field Session of the American Society of Agronomy, Soil Science Society of America, and the Geological Society of America, p. 247–266.

Parsons, R.B., and Herriman, R.C., 1976, Geomorphic surfaces and soil development in the upper Rogue River Valley, Oregon: Soil Science Society of America Journal, v. 40, p. 933–938, doi:10.2136/sssaj1976 .03615995004000060034x.

Pearson, H., 1967, Mima mounds: National Parks and Conservation Magazine, v. 41, no. 233, p. 13.

Peterson, F.F., 1961, Solodized solonetz soils occurring in the uplands of the Palouse Loess [unpublished Ph.D. thesis]: Pullman, Washington State University, 280 p.

Péwé, T.L., 1948, Origin of the Mima mounds: The Scientific Monthly, v. 66, no. 4, p. 293–296.

Piper, C.V., 1905, The basalt mounds of the Columbia lava: Science, v. 21, p. 824–825, doi:10.1126/science.21.543.824.

Porsild, A.E., 1938, Earth mounds in unglaciated Arctic northwestern America: Geographical Review, v. 28, p. 46–58, doi:10.2307/210565.

Price, W.A., 1949, Pocket gophers as architects of mima (pimple) mounds of the western United States: The Texas Journal of Science, v. 1, no. 1, p. 1–17.

Price, W.A., 1950, Origin of pimple mounds, by E.L. Krinitzsky: American Journal of Science, v. 248, no. 5, p. 355–360, doi:10.2475/ajs.248.5.355.

Prokopovich, N.P., 1969, Pleistocene permafrost in California's Central Valley?: Geological Society of America Abstracts with Programs, Part 5, p. 66.

Purdue, A.H., 1905, Concerning natural mounds: Science, v. 21, p. 823–824, doi:10.1126/science.21.543.823.

Pyrch, J.B., 1973, The characteristics and genesis of stone stripes in North Central Oregon [unpublished M.S. thesis]: Portland, Oregon, Portland State University, 134 p.

Quinn, J.H., 1961, Prairie mounds of Arkansas: Arkansas Archeological Society Newsletter, v. 2, no. 6, p. 1–8.

Quinn, J.H., 1968, Prairie mounds: Encyclopedia of Geomorphology (Fairbridge, R.W., ed.): New York, Reinhold, p. 888–890.

Reagan, A.B., 1907, A probable origin of the small mounds of the lower Mississippi and Texas coast: Indiana Academy of Science Proceedings, for 1907, p. 99–100.

Reddy, M., 1971, The mysterious Mima mounds: Pacific Search, v. 6, no. 1, p. 9.

Reed, S.E., n.d., Testing a biologic hypothesis of origin of Mima mound-vernal pool topography on the Merced River chronosequence, Central Valley, California [Pending dissertation]: Berkeley, University of California.

Reed, S., and Amundson, R., 2007, Sediment, gophers, and time: A model for the origin and persistence of Mima mound-vernal pool topography in the Great Central Valley, in Schlising, R.A., and Alexander, D.G., eds., Vernal Pool Landscapes: Studies from the Herbarium, no. 14: Chico, California, Chico State University, p. 15–27.

Reed, S., and Amundson, R., 2012, this volume, Using LIDAR to model Mima mound evolution and regional energy balances in the Great Central Valley, California, in Horwath Burnham, J.L., and Johnson, D.L., eds., Mima Mounds: The Case for Polygenesis and Bioturbation: Geological Society of America Special Paper 490, doi:10.1130/2012.2490(01).

Reed, S.E., Vollmar, J., Heimsath, A.M., and Amundson, R., 2008, Using airborne-based LIDAR to test a biologic hypothesis of Mima mound formation in the Great Central Valley, California: Geological Society of America Abstracts with Programs, v. 40, no. 6, p. 209.

Reider, R.G., 1992, Mima-like mounds as salt-movement structures, Laramie Basin, Wyoming, USA: American Quaternary Association 12th biennial meeting: Program with Abstracts (Geological Association of Canada), p. 74.

Reider, R.G., Huss, J.M., and Miller, T.W., 1996, A groundwater vortex hypothesis for mima-like mounds, Laramie Basin, Wyoming: Geomorphology, v. 16, no. 4, p. 295–317, doi:10.1016/0169-555X(95)00142-R.

Reider, R.G., Huss, J.M., and Miller, T.W., 1999, Stratigraphy, soils, and age relationships of Mima-like mounds, Laramie Basin, Wyoming: Physical Geography, v. 20, no. 1, p. 83–96.

Retzer, J.L., 1945, Morphology and origin of some California mounds: Proceedings of the Soil Science Society of America, v. 10, p. 360–367, doi: 10.2136/sssaj1946.03615995001000C00062x.

Rice, T.D., and Griswold, L., 1904, Soil Survey of Acadia Parish, Louisiana: U.S. Department of Agriculture Field Operations of Bureau of Soils for 1903, 5th Report, p. 461–485.

Rich, J.L., 1934, Soil mottlings and mounds in northeastern Texas as seen from the air: Geographical Review, v. 24, p. 576–583, doi:10.2307/208847.

Ricks, D.K., Burras, L., Konen, M.E., and Bolender, A.J., 1997, Genesis and morphology of mima mounds and associated soils at Kalsow Prairie, Iowa: Agronomy Abstracts, American Society of Agronomy: Madison, Wisconsin, p. 255.

Riddell, J.L., 1839, Observations on the geology of the Trinity Country, Texas, made during an excursion there in April and May 1839: American Journal of Science and Arts, v. 37, p. 211–217.

Riefner, R.E., Jr., and Pryor, D., 1996, New locations and interpretation of vernal pools in southern California: Phytologia, v. 80, no. 4, p. 296–327.

Riefner, R.E., Jr., Boyd, S., and Shlemon, R.J., 2007, Notes on native vascular plants from Mima mound-vernal pool terrain and the importance of preserving coastal terraces in Orange County, California: Aliso, v. 24, p. 19–28.

Ritchie, A.M., 1953, The erosional origin of the Mima mounds of southwest Washington: The Journal of Geology, v. 61, no. 1, p. 41–50, doi: 10.1086/626035.

Robertson, J.B., 1867, Memorial and explorations of the Hon. J.B. Robertson, in relation to the agricultural, mineral, and manufacturing resources of the state: With the report of Joint Committee: New Orleans, J.O. Nixon, State Printer, 30 p.

Rockie, W.A., 1942, Pitting on Alaskan farm lands—A new erosion problem: Geographical Review, v. 32, no. 1, p. 128–134, doi:10.2307/210363.

Rogers, G.O., 1893, Drift mounds near Olympia, Washington: The American Geologist, v. 11, no. 6, p. 393–399.

Ross, B.A., Tester, J.R., and Breckenridge, W.J., 1968, Ecology of mima-type mounds in northwestern Minnesota: Ecology, v. 49, p. 172–177, doi:10.2307/1933579.

Saucier, R.T., 1974, Quaternary geology of the lower Mississippi Valley: Arkansas Archeological Survey Research Series, no. 6, p. 1–26.

Saucier, R.T., 1978, Sand dunes and related eolian features of the lower Mississippi River alluvial valley: Geoscience and Man, v. 19, p. 23–40.

Saucier, R.T., 1991, Formation of Mima mounds: A seismic hypothesis: Comment: Geology, v. 19, p. 284, doi:10.1130/0091-7613(1991)019<0284 :CAROFO>2.3.CO;2.

Saucier, R.T., 1994, Geomorphology and Quaternary geologic history of the lower Mississippi Valley: Volume 1: Vicksburg, Mississippi, Waterways Experimental Station, Corps of Engineers, 364 p.

Sasaki, S., 1979, Earth hummocks of northern Hokkaido: Quaternary Research, v. 18, no. 1, p. 31–37, doi:10.4116/jaqua.18.31.

Schaetzl, R.J., and Anderson, S., 2005, Soils—Genesis and Geomorphology: Cambridge, UK, Cambridge University Press, 817 p.

Scheffer, V.B., 1947, The mystery of the mima mounds: The Scientific Monthly, v. 65, no. 5, p. 283–294.

Scheffer, V.B., 1948, Mima mounds: A reply: The Journal of Geology, v. 56, no. 3, p. 231–234, doi:10.1086/625506.

Scheffer, V.B., 1954, Son exclusivos del oeste de Norteamerica los micromonticulos de tipa Mima?: Investigaciones Zoológicas Chilenas, v. 2, p. 89–94.

Scheffer, V.B., 1956a, Weather or gopher: Natural History, v. 65, no. 5, p. 278.

Scheffer, V.B., 1956b, 1st das mikrorelief der Mima-hugel auf das westliche Nordameriko beschrankt?: Saugetierkundliche Mitteilungen Sonderdruck, v. 4, no. 1, p. 17–21.

Scheffer, V.B., 1958, Do fossorial rodents originate mima-type microrelief?: American Midland Naturalist, v. 59, p. 505–510, doi:10.2307/2422495.

Scheffer, V.B., 1960, Spatial patterns in the world of nature: Pacific Discovery, v. 13, p. 18–20.

Scheffer, V.B., 1969, Mima mounds: Their mysterious origin: Pacific Search, v. 3, no. 5, p. 3.

Scheffer, V.B., 1981, Mima prairie's mystery mounds: Backpacker, v. 4–5, p. 96.

Scheffer, V.B., 1983, Tapestries in nature: Pacific Discovery, v. 36, no. 2, p. 16–24.

Scheffer, V.B., 1984, A case of prairie pimples: Pacific Discovery, v. 37, p. 4–8.

Scheffer, V.B., and Kruckeberg, A., 1966, The mima mounds: Bioscience, v. 16, p. 800–801.

Schmidt, E.W., 1935, A group of Minnesota lowland mounds: Minnesota History, v. 16, p. 307–312.

Schmidt, E.W., 1937, Lowland mounds in the Northfield area: Northfield, Minnesota, Mohn Printing, 34 p.

Schroder, G.D., and Geluso, K.N., 1975, Spatial distribution of *Dipodomys spectabilis* mounds: Journal of Mammalogy, v. 56, no. 2, p. 363–368, doi:10.2307/1379366.

Seifert, C.L., 2007, Investigation of the prairie mounds of the south-central United States [unpublished M.S. thesis]: Memphis, Tennessee, University of Memphis, 129 p.

Seifert, C.L., and Cox, R.T., 2007, Mid-continent prairie mounds are nebkhas (coppice dunes): Geological Society of America Abstracts with Programs, v. 39, no. 2, p. 87.

Seifert, C.L., Cox, R.T., Foreman, S.L., Foti, T.L., Waskiewicz, T.A., and McColgan, A.T., 2009, Relict nebkhas (pimple mounds) record prolonged late Holocene drought in the forested region of south-central United States: Quaternary Research, v. 71, p. 329–339, doi:10.1016/j.yqres.2009.01.006.

Seven, R., 2008, Mima mounds: Mystery hides in vast prairie: Seattle Times, 6 July: http://article.wn.com/view/2008/07/06/Mima_Mounds_Mystery _hides_in_vast_prairie_2/.

Shaw, C.F., 1928, Profile development and relationship of soils in California, *in* Deemer, R.B., ed., First International Congress of Soil Science, Washington, D.C., 1927: Proceedings and Papers, v. 4, Commissions 5–6, Miscellaneous Papers, p. 291–317.

Shaw, C.F., 1937, Letter, *in* Hertlein, L.G., and Grant, U.S., IV, 1944: The Geology and Paleontology of the marine Pliocene of San Diego, California, Part I, Geology: San Diego Society of Natural History Memoirs 2, p. 18 (note 66).

Shaw, E.W., 1957, Washington's mystery mounds: Ford Times, v. 49, no. 2, p. 14–15.

Shepard, E.M., 1905, The New Madrid Earthquake: The Journal of Geology, v. 13, p. 45–62, doi:10.1086/621205.

Shlemon, R.J., 1967, Quaternary geology of northern Sacramento County, California: Annual Field Trip Guidebook of the Geological Society of Sacramento, May, 60 p.

Shlemon, R.J., Begg, E.L., and Huntington, G.L., 1973, Fracture traces: Pacific Discovery, v. 26, p. 31–32.

Shlemon, R.J., Kuhn, G.G., Boka, B., and Riefner, R.E., Jr., 1997, Origin of a Mima-mound field, San Clemente State Park, Orange County, California: A test of the seismic hypothesis: Program with Abstracts, Association of Engineering Geologists 40th Annual Meeting, p. 148–149.

Sletten, R.S., Bliss, L.C., Schlichte, K., Frenzen, P., Marrett, D.J., Ping, C.-L., and Mitchem, D., 1994, Mt. St. Helens and mima mounds tour guide book: American Society of Agronomy, Crop Science Society of America, and Soil Science Society of America 86th Annual Meeting, Seattle, Washington, November 13, 55 p.

Slusher, D.F., 1967, "Pimple mounds" of Louisiana: Soil Survey Horizons, v. 8, no. 1, p. 3–5.

Smies, E.H., Tharp, W.E., Anderson, A.C., and Emerson, F.V., 1925, Soil survey of Sabine Parish, Louisiana: U.S. Department of Agriculture Bureau of Soils, Field Operations of Bureau of Soils for 1919, 21st Report, p. 1041–1098.

Smith, H.I., 1910, The archaeology of the Yakima Valley: American Museum of Natural History, v. 6, Part 1, 167 p.

Smith, H.M., and Marshall, R.M., 1938, Soil Survey of Bee County, Texas: U.S. Department of Agriculture Bureau of Chemistry and Soils, in cooperation with Texas Agricultural Experiment Station, Series 1932, no. 30, 34 p.

Smith, H.T.U., 1949, Physical effects of Pleistocene climatic changes in nonglaciated areas: Eolian phenomena, frost action, and stream terracing: Geological Society of America Bulletin, v. 60, p. 1485–1516, doi:10.1130/0016-7606(1949)60[1485:PEOPCC]2.0.CO;2.

Smith, L.H., Dwyer, C.H., and Schafer, G., 1945, Soil Survey of Kittitas County, Washington: U.S. Department of Agriculture Bureau of Plant Industry, Soils, and Agricultural Engineering in cooperation with the Washington Agricultural Experiment Station, Series 1937, no. 13, 69 p.

Sobecki, T.M., 1980, The distribution and genesis of calcic horizons in some soils of the Texas coast prairie [unpublished M.S thesis]: College Station, Texas A&M University, 168 p.

Sobecki, T.M., and Wilding, L.P., 1982, Calcic horizon distribution and soil classification in selected soils of the Texas Coast Prairie: Soil Science Society of America Journal, v. 46, p. 1222–1227, doi:10.2136/sssaj1982 .03615995004600060021x.

Sobecki, T.M., and Wilding, L.P., 1983, Formation of calcic and argillic horizons in selected soils of the Texas Coast Prairie: Soil Science Society of America Journal, v. 47, p. 707–715, doi:10.2136/sssaj1983 .03615995004700040021x.

Southard, A.R., and Williams, J.S., 1969, An explanation of stony earth movement on the Franklin ramp, Preston Quadrangle, Idaho: Geological Society of America Abstracts with Programs, v. 1, no. 5, p. 77.

Southard, A.R., and Williams, J.S., 1970, Patterned ground indicates unstable landscapes: Journal of Soil and Water Conservation, v. 25, no. 5, p. 194–196.

Spackman, L.K., 1982, Genesis and morphology of soils associated with formation of Laramie Basin (Mima-like) mounds [unpublished M.S. thesis]: Laramie, University of Wyoming, 81 p.

Spackman, L.K., and Munn, L.C., 1984, Genesis and morphology of soils associated with formation of Laramie Basin (mima-like) mounds in Wyoming: Soil Science Society of America Journal, v. 48, no. 6, p. 1384–1392, doi:10.2136/sssaj1984.03615995004800060038x.

Spillman, W.J., 1905, Natural mounds: Science, v. 21, no. 538, p. 632, doi: 10.1126/science.21.538.632-a.

Stallings, J.H., 1948, The Mima mounds: The Scientific Monthly, v. 66, no. 3, p. 269–270.

Stephenson, L.W., and Crider, A.F., 1916, Geology and Ground Waters of Northeastern Arkansas: U.S. Geological Survey Water-Supply Paper 399: Washington, D.C., Government Printing Office, 315 p.

Stockman, D.D., 1981, Soil Survey of Lincoln County, Washington: U.S. Department of Agriculture, Soil Conservation Service in cooperation with the Washington State University Agricultural Research Center: Washington, D.C., Government Printing Office, 94 p.

Storie, R.E., Cole, R.C., Owen, B.C., Koehler, L.F., Anderson, A.C., Leighty, W.J., and Retzer, J.L., 1944, Soil Survey of the Santa Cruz Area, California: U.S. Department of Agriculture Bureau of Plant Industry, Soils, and Agricultural Engineering in cooperation with the University of California Agricultural Experiment Station, Series 1935, no. 25, 90 p.

Storm, L., 2004, Prairie fires and earth mounds: The ethnoecology of Upper Chehalis prairies: Douglasia, v. 28, no. 3, p. 6–9.

Stout, M.L., 1965, Origin of mound topography, western United States—New theory: Abstracts for 1965: Geological Society of America Special Paper 87, p. 169.

Strahorn, A.T., Westover, H.L., Holmes, L.C., Eckmann, E.C., Nelson, J.W., and Van Duyne, C., 1912, Soil Survey of the Madera Area, California: U.S. Department of Agriculture Bureau of Soils, Field Operations of the Bureau of Soils for 1910, 12th Report, p. 1717–1753.

Strahorn, A.T., Holmes, L.C., Eckmann, E.C., Nelson, J.W., and Kolbe, L.A., 1914, Soil Survey of the Medford area, Oregon: U.S. Department of Agriculture Bureau of Soil, Field Operations of the Bureau of Soils for Year 1911, 13th Report, p. 2287–2356.

Stuart, J., 1837, A sketch of the Cherokee and Choctaw Indians: Little Rock, Woodruff and Pew, 42 p.

Surr, G., 1920, Hog wallows: California Cultivator, v. 55, p. 845.

Sweet, A.T., and McBeth, I.G., 1911, Soil Survey of the Klamath Reclamation Project, Oregon: U.S. Department of Agriculture Bureau of Soils, Field Operations of the Bureau of Soils for 1908, 10th Report, p. 1373–1413.

Tallyn, L.A.K., 1980, Scabland mounds of the Cheney quadrangle [unpublished M.S. thesis]: Cheney, Washington, Eastern Washington University, 94 p.

Tapler, J.F., Jr., 1996, Sodium analysis of Mima-like mounds, Laramie Basin, Wyoming [unpublished M.A. thesis]: Laramie, University of Wyoming, 61 p.

Taylor, A.E., and Cobb, W.B., 1921, Soil Survey of Hempstead County, Arkansas: U.S. Department of Agriculture Bureau of Soils, Field Operations of the Bureau of Soils for 1916, 18th Report, p. 1189–1237.

Taylor, S., 1843, Description of ancient remains, animal mounds, and embankments, principally in the counties of Grant, Iowa, and Richmond, in Wisconsin Territory: American Journal of Science, v. 44, p. 21–40.

Tester, J.R., and Breckenridge, W.J., 1964, behavior patterns of the Manitoba Toad, *Bufo Hemiophrys,* in northwestern Minnesota: Annales Academiae Scientiarum Fennicae, Series A, IV: Biologica, v. 71, no. 31, p. 421–431.

Thackray, G.D., 2008, Influences of lower level treeline migration on eastern Idaho mima-mound initiation: Geological Society of America Abstracts with Programs, v. 40, no. 6, p. 209.

Thoburn, J.B., 1937, The origin of the "Natural" mounds of Oklahoma and adjacent states: Chronicles of Oklahoma, v. 15, no. 3, p. 322–343.

Thomassy, R., 1860 [1978], Géologie pratique de la Louisiane: Arno Press, New York [reprint edition of the 1860 edition published by the author, Nouvelle-Orléans and by Lacroix at Boudry, Paris]: Librairie scientifique, Industrielle et Agricole, 263 p., 6 maps.

Thoroddsen, T., 1914, An account of the physical geography of Iceland, with special reference to the plant life, *in* Kolderup, R.L., and Warming, E., editors and compilers, The botany of Iceland, part 1, section 2: Copenhagen, J. Frimodt, p. 233–265.

Thornbury, W.D., 1965, Regional Geomorphology of the United States: New York, John Wiley and Sons, 609 p.

Thorp, J., 1949, Effects of certain animals that live in soil: The Scientific Monthly, v. 68, p. 180–191.

Tillman, B.W., Burn, R.R., Cobb, W.B., Lounsbury, C., and Strickland, G.G., 1919, Soil Survey of Jefferson County, Arkansas: U.S. Department of Agriculture Bureau of Soils, Field Operations of Bureau of Soils for 1915, 17th Report, p. 1163–1197.

Tillman, B.W., Hayes, F.A., and Hutton, F.Z., 1923, Soil Survey of Drew County, Arkansas: U.S. Department of Agriculture Bureau of Soils, Field Operations of Bureau of Soils for 1917, 19th Report, p. 1279–1322.

Topinka, L., 2009, Tom McCall Nature Preserve, Rowena Plateau, Columbia Gorge, Oregon: http://columbiariverimages.com/Regions/Places/tom_mccall_nature_preserve.html (accessed January 9, 2012).

Treasher, R., and Reynolds, R., 1937, Natural mounds of the Tenino (Washington) area: Field Guide: Geological Society of the Oregon Country Geology Newsletter, v. 3, no. 7, p. 72–75.

Tugel, A.J., 1993, Soil Survey of Sacramento County, California: U.S. Department of Agriculture–Soil Conservation Service in cooperation with the University of California Agricultural Experiment Station, 399 p.

Tullis, J.A., 1995, Characteristics and origin of earth mounds on the eastern Snake River plain, Idaho [unpublished M.S. thesis]: Pocatello, Idaho State University, 164 p.

Turner, A.J., 1982, Soil Survey of Willacy County, Texas: U.S. Department of Agriculture–Soil Conservation in cooperation with Texas Agricultural Experiment Station: Washington, D.C., U.S. Government Printing Office, 149 p.

Turner, H.W., 1896, Further contributions to the geology of the Sierra Nevada: U.S. Geological Survey 17th Annual Report, 1895–1896, part 1, p. 521–762.

Udden, J.A., 1906, The origin of the small sand mounds in the Gulf Coast country: Science, v. 23, p. 849–851, doi:10.1126/science.23.596.849.

Upham, W., 1904, Glacial and modified drift in and near Seattle, Tacoma, and Olympia: American Geologist, v. 34, p. 211–214.

Van Duyne, C., and Byers, W.C., 1915, Soil survey of Harrison County, Texas: U.S. Department of Agriculture Bureau of Soils, Field Operations of Bureau of Soils for 1912, 14th Report, p. 1055–1097.

Van Duyne, C., Agee, J.H., and Ashton, F.W., 1919, Soil Survey of Franklin County, Washington: U.S. Department of Agriculture, Bureau of Soils, Field Operations of the Bureau of Soils for 1914, 16th Report, p. 2531–2627.

Veatch, A.C., 1900, The Shreveport area: Section 3, Special Report no. 2, *in* Harris, G.D., and Veatch, A.C., A Preliminary Report on the Geology of Louisiana: Baton Rouge, Louisiana, p. 149–208.

Veatch, A.C., 1905, The question of the origin of the natural mounds of Louisiana: Science, v. 21, p. 310–311.

Veatch, A.C., 1906a, Formation of natural mounds, *in* Geology and Underground Water Resources of Northern Louisiana, with Notes on Adjoining Districts: Louisiana Geological Survey Bulletin, v. 4, p. 57–64.

Veatch, A.C., 1906b, Formation of natural mounds, *in* Geology and Underground Water Resources of Northern Louisiana and Southern Arkansas: U.S. Geological Survey Professional Paper 46: Washington, D.C., Government Printing Office, p. 55–59.

Veatch, A.C., 1906c, On the human origin of the small mounds of the lower Mississippi Valley and Texas: Science, v. 23, no. 575, p. 34–36, doi:10.1126/science.23.575.34-a.

Vitek, J.D., 1973, Patterned ground: A quantitative analysis of pattern: Proceedings of the Association of American Geographers, v. 5, p. 272–275.

Vitek, J.D., 1978, Morphology and pattern of earth mounds in south-central Colorado: Arctic and Alpine Research, v. 10, no. 4, p. 701–714, doi:10.2307/1550738.

Vodrazka, F.M., Stephens, F.H., Goddard, W.K., and Spotts, J.W., 1971, Soil Survey of Franklin County, Arkansas: U.S. Department of Agriculture–Soil Conservation Service–U.S. Forest Service in cooperation with Arkansas Agriculture Experimental Station: Washington, D.C., Government Printing Office, 94 p.

Voellinger, L.R., Gearhart, R.L., Frederick, C., and Moore, D.W., Jr., 1987, Archaeological and Historical Investigations at Cinco Ranch, Fort Bend and Harris Counties, Texas: Espey, Huston & Associates, Inc., document no. 870219: Austin, Texas, 101 p.

Vogel, G., 2004, An assessment of prairie mound origin theories at University of Arkansas Experimental Farms [Fayetteville, AR]: Unpublished, but online at http://projectpast.org/gvogel/uafpm/UAFPM.html.

Vogel, G., 2005, A view from the bottomlands: Physical and social landscapes and late Prehistoric mound centers in the Northern Caddo area [unpublished Ph.D. dissertation]: Fayetteville, University of Arkansas, 470 p.

von Lozinski, W., 1933, Palsenfelder und periglaziale bodenbildung: Neues Jahrbuch fur Mineralogie, Geologie und Paläontologie Beilage-bände, v. 71, part B, p. 18–47.

Vorhies, C.T., 1922, Life history of the kangaroo rat, *Dipodomys spectabilis spectabilis* Meriam: U.S. Department of Agriculture Bulletin no. 1091, 40 p.

Wallace, A.R., 1877a, Glacial drift in California: Nature, v. 15, p. 274–275, doi:10.1038/015274c0.

Wallace, A.R., 1877b, The "Hog Wallows" of California: Nature, v. 15, p. 431–432, doi:10.1038/015431c0.

Wallace, R.E., 1991, Ground-Squirrel Mounds and Related Patterned Ground along the San Andreas Fault in Central California: U.S. Geological Survey Open-File Report 91-0149, 25 p.

Warrick, S., 1984, Editorial [on Mima mounds]: Pacific Discovery, v. 37, p. 1–2.

Washburn, A.L., 1980, Geocryology: A Survey of Periglacial Processes and Environments: New York, John Wiley (Halsted Press Book), 406 p.

Washburn, A.L., 1988, Mima Mounds, an evaluation of proposed origins with special reference to the Puget Lowlands: Report of Investigations, State of Washington Department of Natural Resources, Division of Geology and Earth Resources Report no. 29: Olympia, Washington, 53 p.

Waters, A.C., and Flagler, C.W., 1929, Origin of small mounds on the Columbia River Plateau: American Journal of Science (5th Series), v. 18, no. 105, p. 209–224.

Watson, E.B., and Allen, R.T., 1912, Soil Survey of Morris County, Texas: U.S. Department of Agriculture, Bureau of Soils, Field Operations of the Bureau of Soils for Year 1909, 11th Report, p. 985–1004.

Watson, E.B., and Cosby, S.W., 1924, Soil Survey of the Big Valley [Lassen-Modoc Cos.], California: U.S. Department of Agriculture, Bureau of Soils, Field Operations of the Bureau of Soils for Year 1920, 22nd Report, p. 1005–1031.

Watson, E.B., and Party, 1919, Soil Survey of the Merced area, California: U.S. Department of Agriculture, Bureau of Soils, Field Operations of the Bureau of Soils for 1914, 16th Report, p. 2785–2850.

Watson, E.B., and Smith, A., 1921, Soil Survey of the Santa Maria Area, California: U.S. Department of Agriculture Bureau of Soils, Field Operations of the Bureau of Soils for 1916, 18th Report, p. 2531–2574.

Watson, E.B., Wank, M.E., and Smith, A., 1923, Soil Survey of the Shasta Valley, area, California: U.S. Department of Agriculture, Bureau of Soils, Field Operations of the Bureau of Soils for Year 1919, 22nd Report, p. 99–152.

WDFW (Washington Department of Fish and Wildlife), 1995, Washington State recovery plan for the Pygmy Rabbit. Wildlife Management Program: Olympia, Washington Department of Fish and Wildlife, 73 p.

Webb, F.S., 1957, Surface geology of the Eufaula-Texanna area, McIntosh and Pittsburg counties, Oklahoma [unpublished M.S. thesis]: Norman, University of Oklahoma, 60 p.

Webster, C.L., 1889a, Ancient mounds and earthworks in Floyd and Cerro Gordo counties, Iowa: Smithsonian Annual Reports for Years 1882–88, v. 42, pt. 1, p. 575–589.

Webster, C.L., 1889b, Indian graves in Floyd and Chickasaw counties, Iowa: Smithsonian Annual Reports for Years 1882–88, v. 42, part 1, p. 590–592.

Webster, C.L., 1889c, Ancient mounds in Johnson County, Iowa: Smithsonian Annual Reports for Years 1882–88, v. 42, part 1, p. 593–597.

Webster, C.L., 1889d, Ancient mounds in Iowa and Wisconsin: Smithsonian Annual Reports for Years 1882–88, v. 42, part 1, p. 598–601.

Webster, C.L., 1889e, Mounds of the western Prairies: Annual Report of the Board of Regents Smithsonian Institution, for year ending June 30, 1887, v. 42, Part 1, p. 603–604.

Webster, C.L., 1897, History of Floyd County, Iowa: Intelligencer Print (newspaper), Charles City, Iowa, Part 1, p. 14–27.

Wedding, K.C., 1985, Schmidt's Minnesota lowland mounds reconsidered: The Minnesota Archaeologist, v. 44, no. 1, p. 29–33.

Welch, R.N., 1942, Geology of Vernon Parish: Geological Bulletin no. 22: Baton Rouge, Louisiana, Department of Conservation and Louisiana Geological Survey, 90 p.

Wentworth, I.H., 1906, A few notes on 'Indian mounds' in Texas: Science, v. 23, p. 818–819, doi:10.1126/science.23.595.818-a.

Westman, B.J., 1946, Observations made of the unique Indian mounds near Bray, California, unpublished manuscript (4 p.) on file: Yreka, California, Siskiyou County Historical Society.

Wheat, J.B., 1953, The Addicks Dam site: An archaeological survey of the Addicks Dam basin, southeast Texas, River Basin Surveys Papers 4, no. 1: Bureau of Ethnology Bulletin, v. 154, p. 143–252.

White, C.A., 1870, Report on the Geological Survey of the State of Iowa… within the years 1866, 1867, 1868, and 1869: Volumes 1 and 2: Des Moines, Iowa, Mills and Company.

White, E.M., and Agnew, A.F., 1968, Contemporary formation of patterned ground by soils in South Dakota: Geological Society of America Bulletin, v. 79, no. 7, p. 941–944, doi:10.1130/0016-7606(1968)79[941:CFOPGB]2.0.CO;2.

White, K.L., and Wiegand, K.C., 1989, Geomorphic analysis of floodplain sand mounds, Navasota River, Texas: Bulletin of the Association of Engineering Geologists, v. 26, no. 2, p. 477–499.

Whitesides, V.S., Jr., 1957, Surface geology of the Stidham area, McIntosh County, Oklahoma [unpublished M.S. thesis]: Norman, University of Oklahoma, 109 p.

Whitney, D.J., 1921a, Water made hog wallows: California Cultivator, v. 56, p. 40.

Whitney, D.J., 1921b, Hog wallows again: California Cultivator, v. 56, p. 451.

Whitney, D.J., 1948, San Joaquin Valley hog wallows: The Scientific Monthly, v. 66, p. 356–357.

Whitney, J.D., 1865, Geology of California: Report of Progress and Synopsis of the Fieldwork, from 1860 to 1864, v. 1, 498 p. (cf. p. 367).

Whittaker, W.E., and Storey, G.R., 2005, Ground-penetrating radar survey of the possible 13AM446 Mound, Effigy Mounds National Monument, Allamakee County, Iowa, Contract Report 1234: Iowa City, The University of Iowa, Office of State Archaeologist, 11 p.

Wiegand, K.C., and White, K.L., 1989, Geomorphic analysis of floodplain sand mounds, Navasota River, Texas: Geological Society of America Abstracts with Programs, v. 21, no. 1, p. 44.

Wilder, H.J., and Shaw, C.F., 1908, Soil Survey of the Fayetteville area, Arkansas: U.S. Department of Agriculture Bureau of Soils, Field Operations of the Bureau of Soils for 1915, 8th Report, p. 1163–1197.

Wilkes, C., 1845, Narrative of the United States Exploring Expedition During the Year 1838–1842: Philadelphia, Lea & Blanchard, v. 4, 539 p.

Williams, G., ed., 1973, London Correspondence Inward from Sir George Simpson, 1841–42: The Hudson's Bay Record Society, London: Glasgow, Scotland, The University Press, 212 p.

Williams, H., 1990, The jiggle effect: The Seattle Times, 4 June, p. E1, E3.

Williams, J.S., 1958, Stone stripes at low altitudes, northern Utah: Geological Society of America Bulletin, v. 69, no. 12, part 2, p. 1749.

Williams, W.M., 1877, Hog wallows and prairie mounds: Nature, v. 16, p. 6–7, doi:10.1038/016006c0.

Willson, C.J.Z., 1999, Intracommunity differences in big sagebrush (*Artemisia tridentata* Nutt.) morphology and gas exchange [unpublished M.S. thesis]: Laramie, University of Wyoming, 61 p.

Wilson, M.D., 1978, Patterned ground of erosional origin southwestern Idaho: Third International Conference on Permafrost Programme: Edmonton, Alberta, p. 60.

Wilson, M.D., and Slupetzky, H., 1977, Origin of patterned ground near Boise, Idaho: Geological Society of America Abstracts with Programs, v. 9, no. 6, p. 775–776.

Winchell, N.H., 1911, The aborigines of Minnesota: Minnesota Historical Society, St. Paul: St. Paul, Minnesota, Pioneer, 761 p.

Winchell, N.H., and Upham, W., 1881, Geology of Minnesota, v. 1: Minneapolis, Minnesota, Johnson, Smith and Harrison, State Printers, 673 p.

Winward, A.H., and Youtie, B.A., 1978, Ecological inventory of the Lawrence Memorial Grassland Preserve: Proceedings of the Oregon Academy of Science, v. 14, p. 50–65.

Withrow, R., 2005, The enigmatic lowland mounds of Dakota and Rice Counties: Dakota County: Historia y Sociedad (Rio Piedras, San Juan, P.R.), v. 46, no. 2, p. 1–9.

Zedler, P.H., 1987, The ecology of southern California vernal pools: A community profile: U.S. Department of the Interior, Fish & Wildlife Service, Biological Report 85 (7.11), 136 p.

Zedler, P.H., and Ebert, T.A., 1979, A survey of vernal pools of Kearny Mesa, San Diego: Spring 1979 Progress Report: San Diego State University Foundation [Grant 06309 Caltrans].

Appendix F. Mima-mound–related masters' and doctoral theses, with a genetic and/or content précis of each[1]

MASTERS' THESES

Aldritt, J.E., 1986, Earth-mound characteristics and possible origins, Birch Creek Valley, East-Central Idaho [unpublished M.A. thesis]: Lawrence, University of Kansas, 114 p. (Advisor Curt Sorenson, Department of Geography.) Mounds are of polygenetic origin produced mainly by interactions between ground squirrels and the soil duripan, but where vegetation also likely plays a role.

Allgood, F.P., 1972, Genesis and morphology of mounded soils [unpublished M.S. thesis]: Stillwater, Oklahoma State University, 98 p. (Advisor Fenton Gray, Agronomy Department.) Mounds constitute an environmental equilibrium created by organisms that inhabit them to escape seasonal wetness.

Almaraz, R.A., 1986, The characteristics of patterned ground in the southern Cascade Mountains [unpublished M.S. thesis]: Ashland, Southern Oregon State College (now Southern Oregon University), 68 p. (Advisor John W. Mairs, Interdisciplinary Studies.) Mounds formed by polygenetic freeze-thaw constructional processes under a periglacial climate; on andesite-basalt bedrock in Jenny Creek–Iron Gate Reservoir areas near Oregon-California state line (Jackson County, Oregon, Siskiyou County, California).

Archeuleta, T.E., 1980, Analysis of Mima mounds in northeastern Arkansas [unpublished M.S. thesis]: New Orleans, Louisiana, University of New Orleans, 59 p. (Advisors W.W. Craig, Department of Earth Sciences, and Lou Pakiser, U.S. Geological Survey.) Batesville area mounds are constructional features formed subaqueously by a system of eddies and counter-rotational subeddies.

Arnold, J.J., 1960, Prairie mounds and their climatic implications [M.S. thesis]: Fayetteville, University of Arkansas, 36 p. (Advisor James A. Quinn, Geology Department.) Mounds attributed to paleoclimatic desert coppicing processes.

Bork, J.L., 1978, A survey of the vascular plants and vertebrates of Upper Table Rock [unpublished M.S. thesis]: Ashland, Southern Oregon State College (now Southern Oregon University), 91 p. (Advisor Frank A. Lang, Department of General Studies, Science, and Math.) Mound origins not treated, but notes that pocket gophers are common on mounds; thesis includes useful data on animals and vernal pools.

Butler, A.C., 1979, Mima mounds grasslands of the upper coastal prairie of Texas [unpublished M.S. thesis]: College Station, Texas A&M University, 94 p. (Advisor Fred Smeins, Department of Ecosystem Science and Management.) Ecological study of plant community-soil interrelationships on relict Mima-mounded Texas coastal prairies; mound origins not discussed.

Carty, D.J., 1980, Characteristics of pimple mounds associated with the Morey soil of SE Texas [unpublished M.S. thesis]: College Station, Texas A&M University, 211 p. (Advisor J.B. Dixon, Department of Soils and Crop Sciences.) Mounds are of uncertain origin, but likely caused by water and/or wind reworking, followed possibly by coppicing.

Collins, O.B., 1972, Climax vegetation and soils of the Blackland Prairie of Texas [unpublished M.S. thesis]: College Station, Texas A&M University, 102 p. (Advisor F.E. Smeins, Department of Ecosystem Science and Management.) Erosional origin of mounds supported.

Giles, L.J., 1970, The ecology of the mounds on Mima Prairie with special reference to Douglas-fir invasion [unpublished M.S. thesis]: Seattle, University of Washington, 99 p. (Advisor Arthur R. Krukeberg, Department of Botany.) Ecological study of Douglas-fir invasions, without treating mound origins.

Goebel, J.E., 1971, Genesis of microknoll topography in northeast Texas [unpublished M.S. thesis]: Commerce, East Texas State University (now Texas A&M University at Commerce), 207 p. (Advisor W. Farrin Hoover, Department of Earth Science.) Erosion and random drainage patterns left microknolls (mounds) as topographic highs.

Goodzari, N.K., 1978, Geomorphological and soil analysis of soil mounds in southwest Louisiana [unpublished M.S. thesis]: Baton Rouge, Louisiana State University, 70 p. (Advisor Richard H. Kesel, Department of Geography.) She supported the erosional origin of mounds.

Hill, J.P., 2001, Ecophysiological basis of phenotypic differentiation in *Artemisia tridentata* ssp. *wyomingensis* growing on and off Mima-like mounds in southeast Wyoming [unpublished M.S. thesis]: Winston-Salem, North Carolina, Wake Forest University, 73 p. (Advisor William "Bill" K. Smith, Department of Biology.) Mounds produced by burrowing animals.

Hirschey, S.J., and Sinclair, K.A., 1992, A hydrologic investigation of the Scatter Creek-Black River area, southern Thurston County, Washington State [unpublished M.A. thesis]: Olympia, Washington, The Evergreen State College, 126 p. (Advisors James M. Stroh, Paul Ray Butler, and Linton L. Waldrick, Department of Environmental Studies.) Little information on mounds, but thesis provides a hydrologic background for most mounded prairies in the area.

Horwath (now Horwarth Burnham), J.L., 2002, An assessment of Mima type mounds, their soils, and associated vegetation, Newton County, Missouri [unpublished M.S. thesis]: Urbana, University of Illinois, 265 p. (Advisor Donald L. Johnson, Department of Geography.) Mounds are polygenetic in origin, with burrowing animals dominant.

Huss, J.M., 1994, Stratigraphic and sodium analyses of Laramie Basin Mima-like mounds, Wyoming [unpublished M.A. thesis]: Laramie, University of Wyoming, 176 p. (Advisor Richard Reider, Department of Geography and Recreation.) Mounds form by groundwater vortices. Good bibliography and timeline of historic mound studies.

Irvine, L.L-A., 2005, A study of pimple mounds in southern Saskatchewan [unpublished M.A. thesis]: Regina, Saskatchewan (Canada), University of Regina, 188 p. (Advisors Janis Dale, Department of Geology, and David Sauchyn, Department of Geography.) Mima mounds are polygenetic, with burrowing animals dominant.

Johnson, C.B., 1982, Soil mounds and patterned ground of the Lawrence Memorial Grassland Preserve [unpublished M.S. thesis]: Corvallis, Oregon State University, 52 p. (Advisor Charles L. Rosenfeld, Department of Geography.) Mounds are polygenetic involving wind, rain, and frost action.

Klein, J.L., 1998, Vegetation patterning relative to Mima-like mounds and its implications for a tallgrass prairie restoration in southwestern Missouri [unpublished M.S. thesis]: Madison, University of Wisconsin, 167 p. (Advisor Evelyn Howell, Land Resources, Institute for Environmental Studies.) Focus is mound vegetation; mounds are polygenetic in origin, with depositional processes dominant, possibly involving burrowings.

Koons, F.C., 1926, Origin of the sand mounds of the pimpled plains of Louisiana and Texas [unpublished M.S. thesis]: Chicago, University of Chicago, 36 p. (Advisor J Harlan Bretz, Department of Geology and Paleontology.) Mounds are produced by pocket gopher burrowing.

---

[1]Includes entirely, largely, or partly mound-related studies.

Kowalski, C.D., 1978, Soil development on Pleistocene and Holocene geomorphic surfaces of the Laramie Basin, Wyoming [unpublished M.A. thesis]: Laramie, University of Wyoming, 131 p. (Advisor Richard Reider, Department of Geography.) Mounds are not treated, but this study sets the geomorphic stage for multiple subsequent Laramie Basin mound studies by this advisor's students.

Lea, P.D., 1984, Pleistocene glaciation at the southern margin of the Puget lobe, western Washington [unpublished M.S. thesis]: Seattle, University of Washington, 96 p. (Advisor Stephen C. Porter, Department of Geological Sciences.) Limited treatment of mounds. Nothing on origins, but thesis provides geological background for many mounded prairies in the area, including Mima Prairie.

McFaul, M., 1979, A geomorphic and pedological interpretation of the Mima-mounded prairies, South Puget Lowland, Washington State [unpublished M.A. thesis]: Laramie, University of Wyoming, 77 p. (Advisor Richard Reider, Department of Geography and Recreation.) Analysis of two dissected mounds at Rock Prairie (Colvin Ranch) suggests a periglacial origin, where a histosol developed on ice-wedge polygons; upon dissipation and thawing of polygons, mounds were formed.

Miller, T.W., 1994, Multi-dimensional analysis of Laramie Basin Mima-like mounds, Wyoming [unpublished M.A. thesis]: Laramie, University of Wyoming, 113 p. (Advisor Richard Reider, Department of Geography and Recreation.) Mounds likely formed by hydrostatic ground-water pressures; good bibliography and timeline of historic mound studies.

Nelson, C.A., 1977, The origin and characteristics of soil mounds and patterned ground in north central Oregon [unpublished M.S. thesis]: Corvallis, Oregon State University, 45 p. (Advisor Charles L. Rosenfeld, Department of Geography.) Mounds and stone rings both owed primarily to freeze-thaw processes in a cold Pleistocene environment, with polygenetic processes operating today.

Paeth, R.C., 1967, Depositional origin of Mima mounds [at Upper and Lower Table Rocks, Oregon, and Mima Prairie, Washington] [unpublished M.S. thesis]: Corvallis, Oregon State University, 61 p. (Advisor Ellis G. Knox, Department of Soils.) Argues for ice-rafted deposition of mounds at all three areas.

Pyrch, J.B., 1973, The characteristics and genesis of stone stripes in North Central Oregon [unpublished M.S. thesis]: Portland, Oregon, Portland State University, 134 p. (Advisor Larry W. Price, Geography Department.) Stone stripes cannot be explained by current processes, and likely were formed under periglacial processes, as evidenced by associated sorted circles, nets, polygons, and mounds.

Seifert, C.L., 2007, Investigation of the prairie mounds of the south-central United States [unpublished M.S. thesis]: Memphis, Tennessee, University of Memphis, 129 p. (Advisor Randall Cox, Department of Earth Sciences.) Grain size analyses of four mounds in different parts of Arkansas suggest eolian, nebkha-like coppiced origins.

Sobecki, T.M., 1980, The distribution and genesis of calcic horizons in some soils of the Texas coast prairie [unpublished M.S thesis]: College Station, Texas A&M University, 168 p. (Advisor Larry P. Wilding, Department of Soils and Crop Sciences.) Soil analyses indicate that calcic horizons occur only under micro highs (pimple mounds) and due to lateral moisture flow patterns (wicking), whereas argillic horizons occur only in wetter micro lows (intermounds).

Spackman, L.K., 1982, Genesis and morphology of soils associated with formation of Laramie Basin (Mima-like) mounds [unpublished M.S. thesis]: Laramie, University of Wyoming, 81 p. (Advisor Larry Munn, Department of Plant Sciences.) Mounds formed in a periglacial environment, and are formed by hydrostatic pressures of water entrapped between permafrost or bedrock and a downward moving freeze plane, which forced gravel-sand slurries to the surface as "mushroom" structures (mounds).

Tallyn, L.A.K., 1980, Scabland mounds of the Cheney quadrangle [unpublished M.S. thesis]: Cheney, Eastern Washington University, 94 p. (Advisors, Eugene Kiver, Dale Stradling, and Martin Mumma, Department of Geology.) Mound genesis uncertain, but probably due to periglacial processes.

Tapler, J.F., Jr., 1996, Sodium analysis of Mima-like mounds, Laramie Basin, Wyoming [unpublished M.A. thesis]: Laramie, University of Wyoming, 61 p. (Advisor Richard G. Reider, Department of Geography and Recreation.) Concludes that mound formation is due to ground-water vortices.

Tullis, J.A., 1995, Characteristics and origin of earth mounds on the eastern Snake River plain, Idaho [unpublished M.S. thesis]: Pocatello, Idaho State University, 164 p. (Advisor H.T. Ore, Department of Geology.) Mounds form by polygenetic processes, primarily by differential frost heaving during a cold climate period, later involving animal burrowing and disturbance, coppicing, and erosion.

Willson, C.J.Z., 1999, Intracommunity differences in big sagebrush (*Artemisia tridentata* Nutt.) morphology and gas exchange [unpublished M.S. thesis]: Laramie, University of Wyoming, 61 p. (Advisor William "Bill" K. Smith, Department of Botany.) Ecological study of plants on mounds that are typically disturbed, mounds assumed to be formed by burrowing animals.

DOCTORAL THESES

Aten, L.E., 1979, Indians of the upper Texas coast: Ethnohistoric and archeological framework [unpublished Ph.D. thesis]: Austin, University of Texas, 560 p. (Advisor Dee Ann Story, Department of Anthropology.) Two pimple mounds excavated, with archaeological implications discussed, but little on mound genesis.

Bernard, H.A., 1950, Quaternary geology of southeast Texas [unpublished Ph.D. dissertation]: Baton Rouge, Louisiana State University, 165 p. (Advisor Grover E. Murray, Department of Geology.) Main focus was on four classic (Fisk-named) coastal terraces of Louisiana and east Texas, but good discussion on mounds (p. 4, 102–109), which he believed were (are) caused by slope erosion.

Brotherson, J.D., 1969, Species composition, distribution and phytosociology of Kalsow Prairie, a mesic tall-grass prairie in Iowa [unpublished Ph.D. thesis]: Ames, Iowa State University. (Advisor Roger "Jake" Landers, Department of Botany.) He mapped 128 Mima mounds on this 65 ha prairie preserve.

Fosberg, M.A., 1963, Genesis of some soils associated with low and big sagebrush complexes in the Brown, Chestnut and Chernozem-prairie zones in southcentral and southwestern Idaho [unpublished Ph.D. dissertation]: Madison, University of Wisconsin, 308 p. (Advisor Francis D. Hole, Department of Soil Science.) Mounds are caused by Pleistocene periglacial processes.

Frederking, R.L., 1973, Spatial variation of the presence and form of earth mounds on a selected Alp surface: Sangre de Cristo Mountains, Colorado [unpublished Ph.D. dissertation]: Iowa City, University of Iowa, 201 p. (Advisor Neil E. Salisbury, Department of Geography.) Mounds are caused by periglacial processes.

Greene, G.S., 1975, Prehistoric utilization in the channeled scablands of Eastern Washington [unpublished Ph.D. dissertation]: Pullman, Washington State University, 149 p. (Advisor Richard D. Daugherty, Department of Anthropology.) Treats Mima mound occupations and burials within them on Channeled Scablands.

Henderson, J.A., 1974, Composition, distribution and succession of subalpine meadows in Mount Rainier National Park [unpublished Ph.D. dissertation]: Corvallis, Oregon State University, 153 p. (Advisors W.W. Chilcote and J.F. Franklin, Department of Botany and Plant Pathology.) Mounds at Sunrise Visitor's Center are briefly considered, and attributed to pocket gophers.

Holland, R.F., 1978, Biogeography and ecology of vernal pools in California [chapter-published Ph.D. dissertation]: Davis, University of California, 121 p. (Advisor Subodh Jain, Department of Ecology.) Although the main focus is on the distribution and ecology of vernal pools in California's Great Central Valley, many of the vernal pools mapped and discussed owe their origin in part to Mima mound topography. Hence, many maps in this thesis indirectly provide Mima mound distributions.

Holland, W.C., 1943, Physiography of Beauregard and Allen Parishes, Louisiana [unpublished Ph.D. dissertation]: Baton Rouge, Louisiana State University, 178 p. (Advisor Richard Russell, School of Geology.) Excellent section on pimple mounds and summary of theories; mounds are due to fluvial erosion.

McMillan, B.A., 2007, Plant assemblage structure on barrier island 'Pimple' dunes at the Virginia Coast Reserve Long-Term Ecological Research Site [unpublished Ph.D. thesis]: Norfolk, Virginia, Old Dominion University, 159 p. (Advisor Frank P. Day, Department of Ecological Sciences.) The small rounded dunes of the East Coast, called pimples, are habitation loci for plant and animal assemblages and activities. They are close genetic analogs to the mounds on North Padre Island, Texas. Both exemplify the hybrid mound concept as per Johnson and Horwath Burnham (this volume).

Othberg, K.L., 1991, Geology and geomorphology of the Boise Valley and adjoining areas, western Snake River Plain [unpublished Ph.D. dissertation]: Boise, University of Idaho, 124 p. (Advisor William B. Hall, Department of Geological Sciences.) Only several pages on mounds (p. 64–65, 73), but this study provides the geological foundation for the innumerable mounds that dot the region. Mound origins not treated.

Peterson, F.F., 1961, Solodized solonetz soils occurring on the uplands of the Palouse loess [unpublished Ph.D. dissertation]: Pullman, Washington State University, 280 p. (Advisor Henry W. Smith, Department of Agronomy.) Treats mound formation in Solodized-Solonetz soils, in Palouse loess and Channeled Scabland region; mounds are caused by burrowing animals, plant-fungal bioturbations, and erosion in both "normal" and Solodized soils.

Reed, S.E., n.d., Testing a biologic hypothesis of origin of Mima mound–vernal pool topography on the Merced River chronosequence, Central Valley, California [pending dissertation]: Berkeley, University of California. (Advisor Ron Amundson, Department of Environmental Science, Policy and Management.) Treats the pocket gopher origin of mounds and vernal pools in the densely mounded footslopes of the Sierra Nevada Mountains on the east side of California's Great Central Valley. A useful and novel LIDAR-quantitative assessment of this famously mounded area (see Reed and Amundson, this volume).

Vitek, J.D., 1973, The mounds of south-central Colorado—An investigation of geographic and geomorphic characteristics [unpublished Ph.D. dissertation]: Iowa City, University of Iowa, 175 p. (Advisor Neil E. Salisbury, Department of Geography.) Discusses natural mounds in the Sangre de Cristo Mountains and San Luis Basin, Colorado. Mounds are caused by periglacial freeze-thaw processes.

Vogel, G., 2005, A view from the bottomlands: Physical and social landscapes and late Prehistoric mound centers in the Northern Caddo area [unpublished Ph.D. dissertation]: Fayetteville, University of Arkansas, 470 p. (Advisor Marvin Kay, Department of Anthropology.) Focus is on artificial "Moundbuilder" mounds along Arkansas River. His 1930–1950s air photos show many natural mounds, often surrounding, and on same terraces as, artificial mounds. Leans toward a fluvial and or eolian erosion mode.